# *Rough Rice and Sea Island Cotton*



Of the many books that have been written about the Southern plantation economy none have heretofore directly addressed the vitally important coasting trade that enabled the movement of agricultural goods from field to market. Charles E. Pearson's refreshing study is about plantation owners, cotton and rice factors, marketing and finance, and the sailing captains and their vessels that traversed the low country tidal waterways in the antebellum period. This is an important study that provides a much-needed perspective on an aspect of Southern agriculture that has been ignored for far too long. *Rough Rice and Sea Island Cotton* is thus a highly original and thoroughly researched work that is also very readable. It will certainly be of interest to anyone wishing to know more about the plantation South.

—Buddy Sullivan, Coastal Georgia Historian

MERCER UNIVERSITY PRESS

*Endowed by*

TOM WATSON BROWN
*and*
THE WATSON-BROWN FOUNDATION, INC.

# *Rough Rice and Sea Island Cotton*

## The Georgia Coasting Trade, 1800–1861

Charles E. Pearson

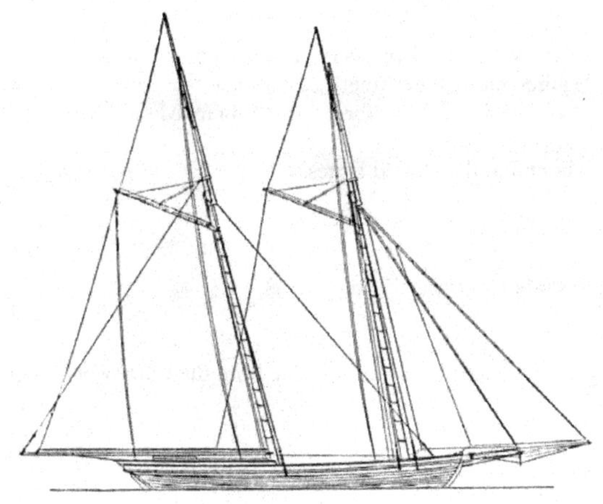

MERCER UNIVERSITY PRESS
*Macon, Georgia*

MUP/ P719


Published by Mercer University Press
1501 Mercer University Drive
Macon, Georgia 31207


29 28 27 26 25 5 4 3 2 1

Books published by Mercer University Press are printed on acid-free paper that meets the requirements of the American National Standard for Information Sciences—Permanence of Paper for Printed Library Materials.

Printed and bound in the United States.

This book is set in Adobe Caslon.

Cover/jacket design by Burt&Burt.

ISBN 978-0-88146-961-5
Cataloging-in-Publication Data is available from the Library of Congress

# Contents

# Preface

In the years before the Civil War, a small fleet of sailing vessels plied the waters of the Southeastern coast of the United States, transporting cargoes between the region's two principal ports, Charleston and Savannah, and the small port towns and plantations of the region. These ships, consisting almost entirely of single-masted sloops and two-masted schooners, were indispensable in the economy of the region where roads were few and overland travel was difficult. These small vessels were involved in the "coasting trade," a critical facet of American economy since the earliest years of settlement. This coasting fleet became particularly important to the state of Georgia in the late eighteenth century with the introduction of Sea Island cotton cultivation and the expansion of rice agriculture along the coast. These two crops became the most important commodities transported by the coasting vessels sailing into the state's principal port, Savannah. This regional coasting trade was disrupted by the American Civil War, and it never recovered to its prewar importance.

This study examines this local coasting trade at its peak between 1800 and 1861. The region of interest extends from about the Combahee River in lower South Carolina to the St. Johns River in Northeastern Florida. This region was commonly served by the coasting vessels sailing out of Savannah. During the principal shipping season, between October and March, "coasters" sailed into Savannah from plantations and small towns all along this stretch of coast to unload their Sea Island cotton and "rough" (unmilled) rice at the docks lining the city's riverfront.

Seafaring in America has attracted a huge amount of attention from both scholars and popular writers. In this vast literature, however, the "coasting trade," as it was known, has been neglected. The coasting ships were involved in mundane trading work, and their day-to-day activities were not as romantic or adventurous as those sailing the high seas. From the more scholarly point of view, the study of the American coasting trade, particularly at a local level, as is done in this work, is made difficult due to a lack of primary records relating to the trade. American laws regulating the coasting trade were intended to protect and encourage it with as little interference as possible. As a result, American vessels carrying American products within a single customs district or between districts in adjacent states rarely had to maintain or file any type of official documents. Thus, the daily activities of these ships and the cargoes they carried were almost never accounted for in official records. In order to study the local coasting trade in any detail, anywhere in the United States, one has to turn to other types of records. In this study, the most important source of information consists of shipping lists published in the newspapers of the region's port towns, especially Savannah and Charleston. These lists record the arrivals and departures of coasting ships and, typically, the shipmasters and the cargo carried. Savannah and Charleston newspapers published such shipping information for the entire period of interest, 1800 to 1861, and have been relied on extensively. This shipping information has proven especially useful when compiled over long periods so that patterns in trading activity can be identified and followed, both for individual ships or ship captains, as well as for the entire coasting fleet.

Relying on these newspaper shipping lists, as well as a range of other primary and secondary sources, including public documents, family histories, and personal accounts, this study opens a window on the little-examined regional coastal trade in Georgia. The focus is on the men, the ships, and the cargoes involved in the Georgia coasting trade. In doing so, it attempts to go beyond a simple dissertation of the economics of the trade and tries to provide details on the lives and activities of the men and ships involved. This detailed information is displayed against the overall economic patterns expressed in the Georgia trade over the first six decades of the nineteenth century. In addition, this study represents an effort to expand on the generally neglected topic of Southern maritime history.

# Acknowledgments

The impetus for this study arose out of a personal, family interest. My great-great-grandfather Charles Stevens was a coasting captain who owned and sailed several sloops and schooners in the Georgia coasting trade in the two decades before the Civil War. Growing up, I heard numerous stories about Charles Stevens and his life and exploits from older members of my family on St. Simons Island, so I wish to first thank my large, extended family for stimulating my interest in the coasting trade, particularly my grandparents, the late Reginald and Banford Taylor.

Numerous individuals and institutions provided assistance, encouragement, and support throughout my research. I thank the staff of several institutions that have helped answer questions and provided valuable source materials: the Georgia Historical Society; the National Archives, particularly the staff in the Maritime Records section, including Charles Johnson, Rebecca Livingston, and Rick Peuser; the Library at the Mariners' Museum, Newport News, Virginia; the Archives and Genealogy sections of the New Bedford Free Public Library; the Library at the New Bedford Whaling Museum; the Peabody Essex Library, Salem, Massachusetts; the Collections Research Center at the Mystic Seaport Museum, Mystic, Connecticut; the Southern Historical Collection, Wilson Library, University of North Carolina, Chapel Hill; the Special Collections & Archives at the W. E. B. Du Bois Library, University of Massachusetts, Amherst; the University of Georgia Libraries, Athens; the Georgia Archives, Atlanta; the University of Virginia Library, Charlottesville; the Bortz Library, Hampden-Sydney College, Hampden-Sydney, Virginia; the South Caroliniana Library, University of South Carolina, Columbia; the Bryan-Lang Historical Library, Woodbine, Georgia, especially John H. Christian; the Library of Congress, particularly Virginia Steele Wood; the South Carolina Historical Society, Charleston; the Glynn County Court House, Brunswick; and the Chatham County Court House, Savannah.

I also wish to thank several individuals for their comments and contributions. These include Judy Wood, Buddy Sullivan, Farris Cadle, Carole Melson, Sylvia Coates, Catherine McDonell, Marguerite Mathews, Robert Culver, Benny Bartley, and Terri Rogers.

I would be remiss if I did not thank my wife, Sharon, who has had to live with "coasting trade" comments and questions for several years. Throughout, she has provided support and encouragement as well as careful editorial comments. There are certainly others who provided help over the several years this study has been in progress I might have overlooked. I thank them, along with everyone else who contributed. Any errors are, of course, my own.

**Figure 1.1** Captain Charles Stevens (1816–1865) and his daughter Mary Henrietta (1855–1937). A native of Denmark, Charles Stevens came to Georgia in the 1830s, settled on St. Simons Island and sailed as a coasting captain until the Civil War. Photograph taken circa 1860.

Photograph in the possession of the author.

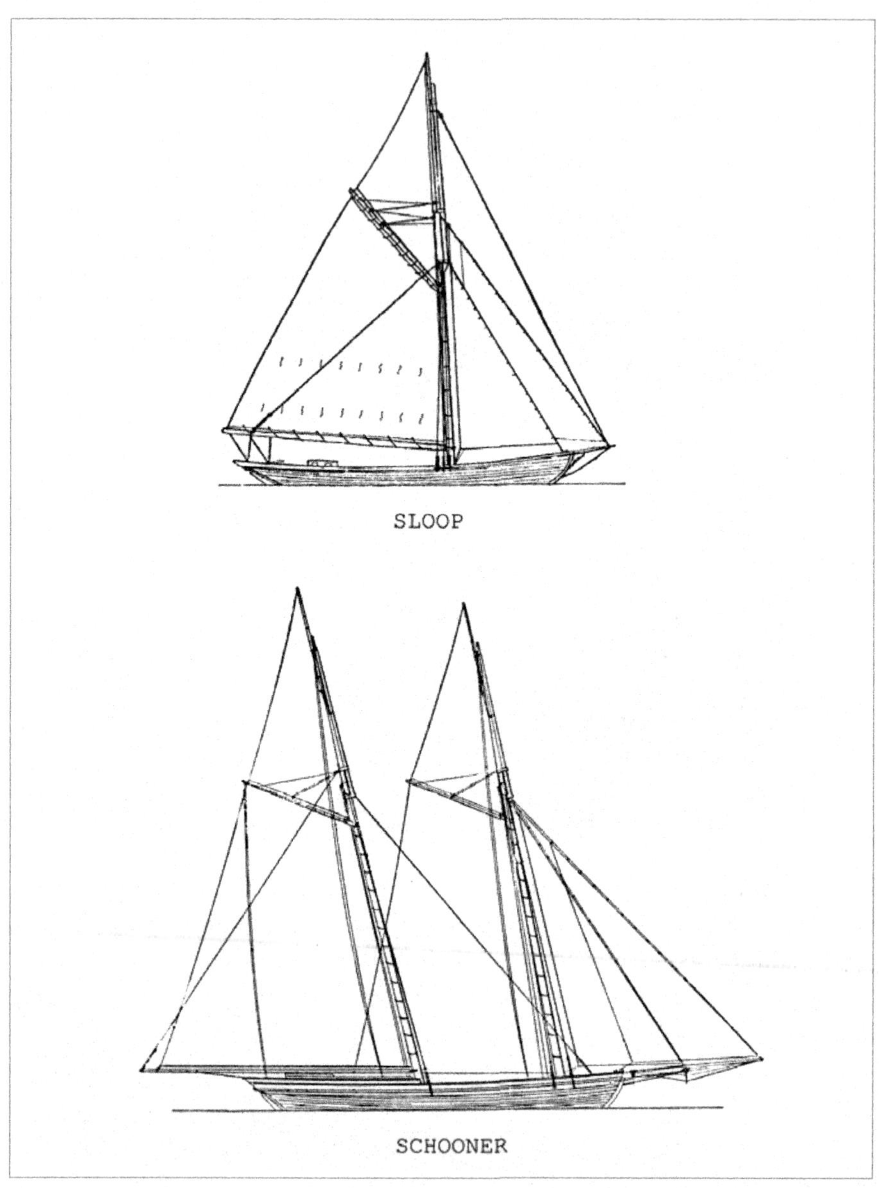


**Figure 1.2** A fore-and-aft rigged single-masted sloop and a two-masted schooner typical of those sailing in the Georgia coasting trade between 1800 and 1861. The sloop is similar in design to Charles Stevens' vessel *Splendid*.

Source: Pearson, Charles E. "Captain Charles Stevens and the Antebellum Georgia Coasting Trade," *The Georgia Historical Quarterly* 75, no. 3 (1991):493.

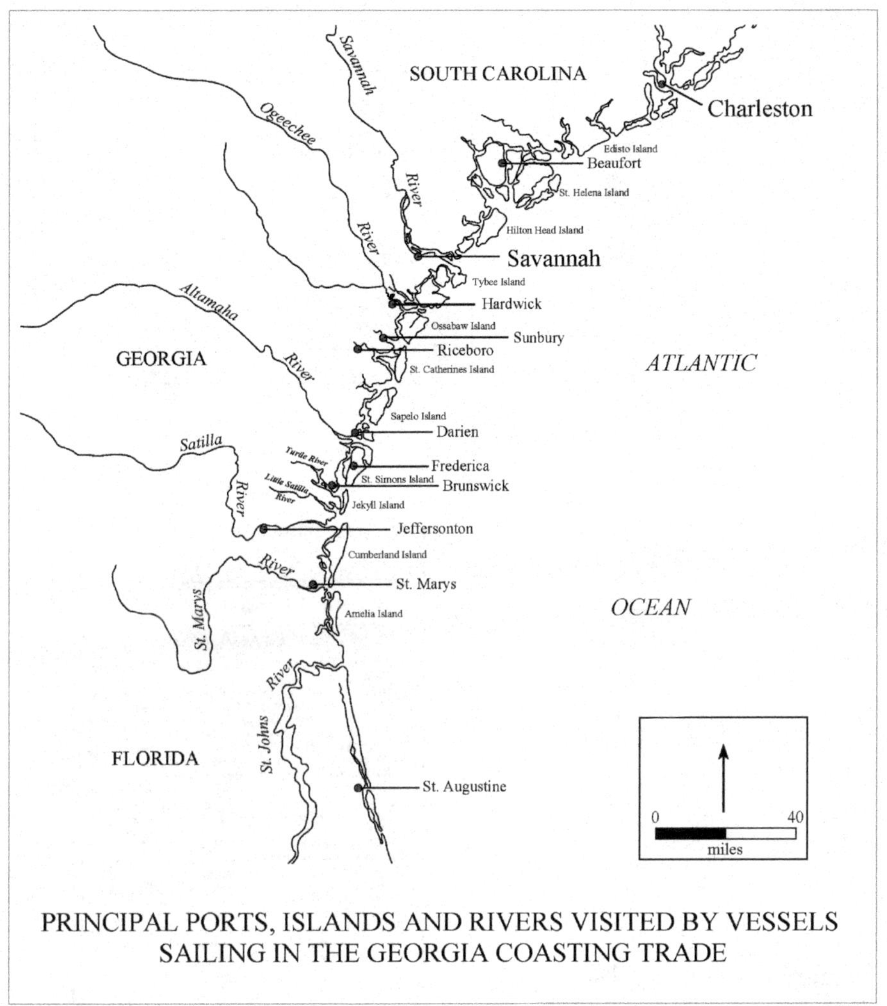


**Figure 1.3** The area traveled by vessels working in the Georgia coasting trade showing the principal ports, islands, and rivers visited by the sailing coasters.

Map drawn by the author.

Republican & Savannah Evening Ledger
April 20, 1809

PORT OF SAVANNAH.

ARRIVED,

| | |
|---|---|
| Ship Hope, Lovett, | Beverly |
| Sloop Fame, Swain, | Riceborough |
| —— Fish, Read, | Frederica |
| —— Prospect, Turner, | Riceborough |

CLEARED,

| | |
|---|---|
| Ship Pallas, Edwards, | Newport R. I. |
| Sch. Industry, Bragdon, | Philadelphia |
| ——Edmund, Champlen, | New-York |
| Sloop Factor, Bates, | Hardwick |
| —— Patsey, Brooks, | Amelia |

Daily Savannah Republican
January 22, 1835

MARINE LIST.

PORT OF SAVANNAH.

ARRIVED.

Br ship Kingston, Wills—sailed 13th Dec. alt to G Barnsley

Schr Idonius, Cannet, Rutledge's plantation. 2000 bushels Rough Rice to P De Villers.

Sloop Jackson Boles. Riceborough. 227 bales Sea Island Cotton Hides. Wood. &c to J F Kackler, R & W King. J Cumming G Anderson & Son, R Habersham. P De Villers. E Reed. A Porter and the Master

Sloop Science, Burr, Satilla, 161 bales Sea Island Cotton and Syrup to E Reed. R & W King G Anderson & Son R Habersham. A G Oemler.

Sloop President. Reed, Darien. 181 bales Cotton and 3000 bushels Rough Rice to B W Delamater & Co J Cumming.

Sloop William. Luce Darien. 284 bales Cotton to W Patterson.

Sloop Angel, Look, Darien 200 bales Cotton and 55 hides to Elias Bliss.

Sloop George, Lane, Darien. 230 bales Cotton to C Baldwin & Co

CLEARED.

Sloop Cashier, Paine Darien.

SAILED.

Ship Macedonia. Weeks, Mobile
Brig Sea Island, Sedrick, Boston.

DEPARTED.

Steam boat Augusta Norris, Augusta.

MEMORANDA.

Ship Floriand and brig Sadi. hence at New York 13th inst. being detained by ice.

Brig Angenora sailed from Baltimore for this port.

Ship Zephyr. up at Liverpool for this port

Brig Nelson. up at New York for this port

CHARLESTON Jan 20.—Arrived. ship Lafayette New York.

In the Offing, ship Martha, from Havana.

Cleared, brig Caledonia, Havre

**Figure 1.4** Shipping lists appearing in 1809 and 1835 Savannah newspapers.

Source: *Republican & Savannah Evening Ledger* April 20, 1809, and *Daily Savannah Republican* January 22, 1835, Film AN13.S45 S28, October 1, 1808, to April 21, 1810, and January 1835 to December 1835, Newspaper Microfilm Collections, University of Georgia Libraries, Athens.

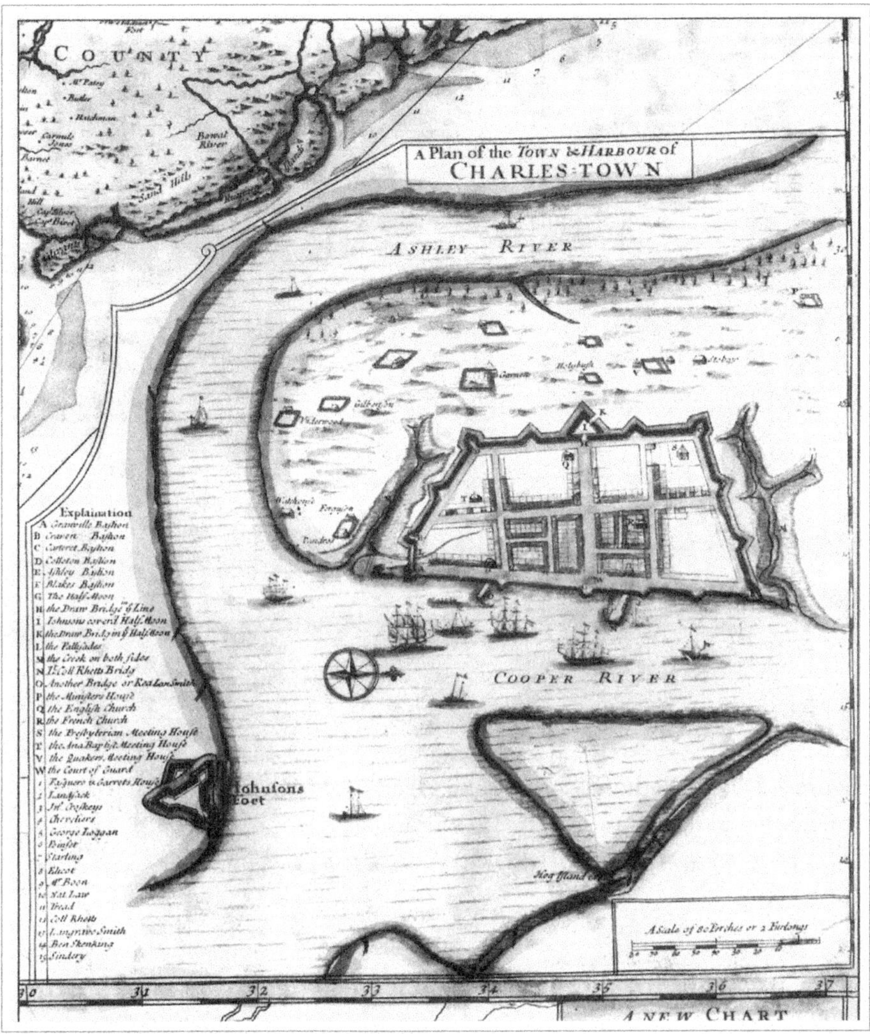


**Figure 2.1** Detail of a circa 1711 map of "Charles Town" by Edw. Crisp showing several single-masted sloops and two-masted schooner-like vessels.

Source: Crisp, Edw., pub. *A Compleat Description of the Province of Carolina in 3 Parts*. [map]. Scale not given. London: Edw. Crisp, 1711. Map G3870 1711.C6, Geography and Map Division, Library of Congress, Washington, DC.

**Figure 2.2** A 1736 drawing by Philip Georg Fredrich von Reck showing a two-masted periagua used on the Georgia coast.

Source: Hvidt, Kristian, ed. *Von Reck's Voyage-Georgia in 1736*. Savannah: Beehive Press, 1980.

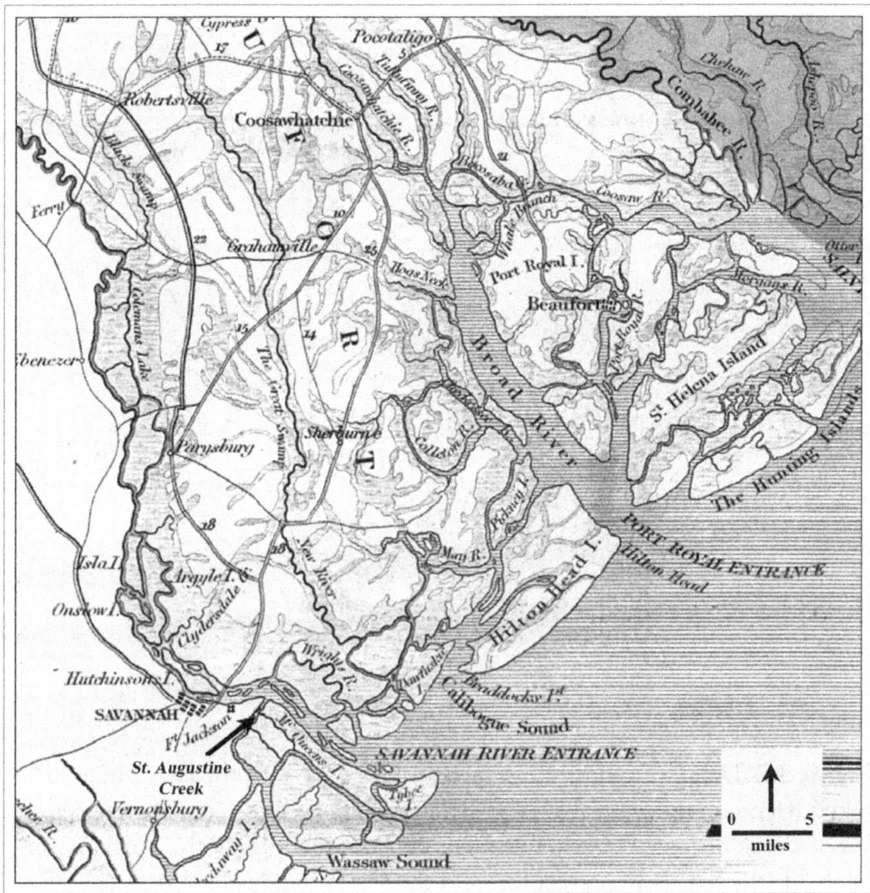


**Figure 3.1** Detail of an 1839 map showing the waterways on the lower South Carolina coast traveled by the coasters sailing out of Savannah. Note St. Augustine Creek just east of Savannah, the route typically used by coasters to enter the inland passage when sailing south from the Savannah River.

Source: Burr, David H. "Map of North and South Carolina Exhibiting the Post Offices, Post Roads, Canals, Rail Roads &c." [map]. Scale ca. 1:650,000. In *The American Atlas*, by David H. Burr. London: J. Arrowsmith, 1839. Map G3900 1839.B8 RR 273, Geography and Map Division, Library of Congress, Washington, DC.

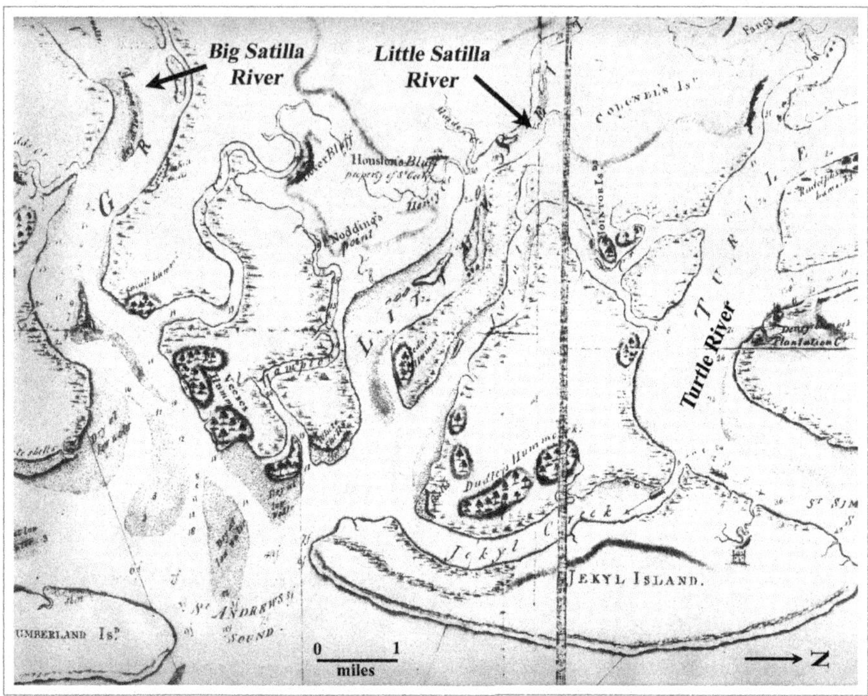


**Figure 3.2** Detail of a circa 1780 map showing the inland passage behind Jekyll Island on the central Georgia coast, where coasters could use either Jekyll Creek or Joiner Creek. Note soundings extending up the Turtle River (right) Little Satilla River (center) and the Great Satilla River (left), all of which were traveled by coasting vessels.

Source: Anonymous. *Map of the Coast of Georgia, Bordering on Camden and Glynn Counties, Showing Also the Course and Soundings of the Alatamaha, Turtle, Crooked, St. Mary's, Great Satilla, and Little Satilla Rivers.* [map]. Scale not given. 1780. Map G3922.C6 178-.M3, Geography and Map Division, Library of Congress, Washington, DC.

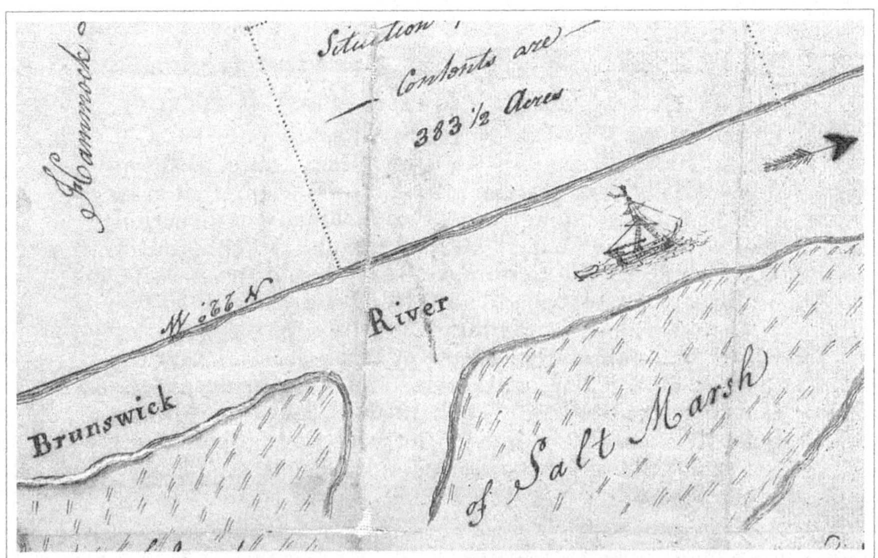


**Figure 3.3** Detail of a circa 1797 map of Brunswick and Brunswick Harbor depicting a single-masted sloop towing a small launch or rowboat.

Source: Purvis, George. *An Accurate Survey of the Town and Commons of Brunswick in the County of Glynn and the State of Georgia*. [map]. Scale not given. 1797. Historic Maps, Surveyor General, RG 3-3-65, Georgia Archives, Atlanta.

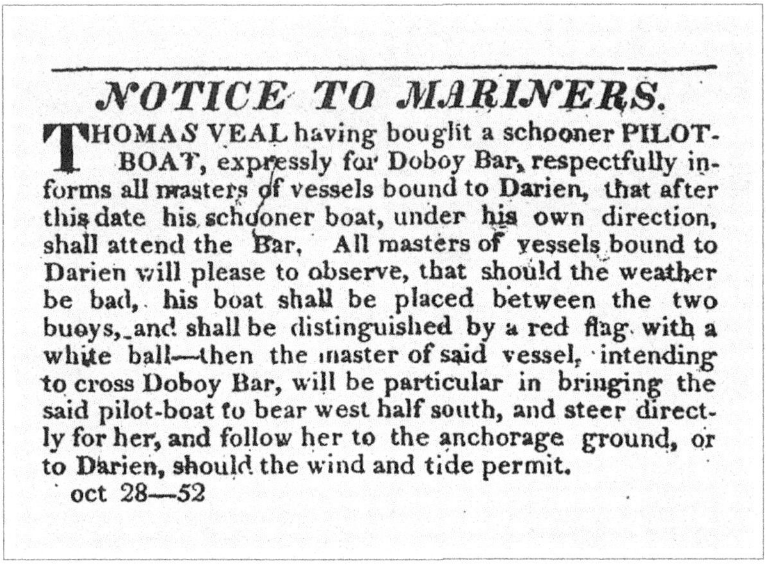

## NOTICE TO MARINERS.

THOMAS VEAL having bought a schooner PILOT-BOAT, expressly for Doboy Bar, respectfully informs all masters of vessels bound to Darien, that after this date his schooner boat, under his own direction, shall attend the Bar. All masters of vessels bound to Darien will please to observe, that should the weather be bad, his boat shall be placed between the two buoys, and shall be distinguished by a red flag with a white ball—then the master of said vessel, intending to cross Doboy Bar, will be particular in bringing the said pilot-boat to bear west half south, and steer directly for her, and follow her to the anchorage ground, or to Darien, should the wind and tide permit.

oct 28—52

**Figure 3.4** "Notice to Mariners" appearing in the 1820 *Darien Gazette* reporting that Thomas Veal was available to pilot vessels over Doboy Bar.

Source: *Darien Gazette* November 18, 1820. Film AN13.D22 D31, November 1818 to September 1828, Newspaper Microfilm Collections, University of Georgia Libraries, Athens.

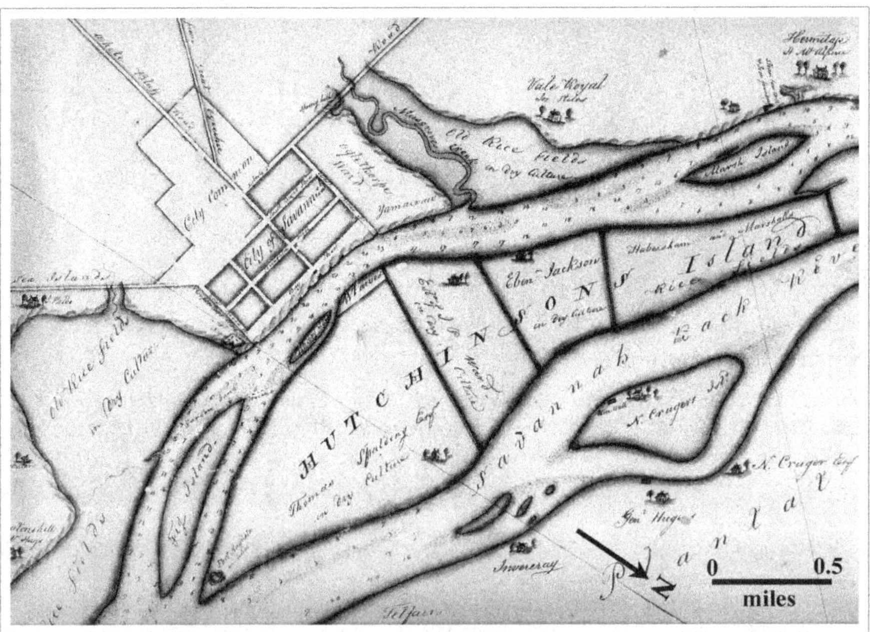


**Figure 3.5** Detail of 1825 *Chart of the Savannah River* by John McKinnon showing rice plantations on Hutchinson's Island and along the Savannah and Back ("Savannah Back River") rivers. Note the "rice mill" on "N. Cruger's Isl." in the Back River. By the late 1820s, this island was known as Pennyworth Island.

Source: McKinnon, John. *Chart of Savannah River, 1825.* [map]. Scale ca 1:33,350. Savannah: John McKinnon, 1825. Reproduced by the Georgia Historical Society, 1998.

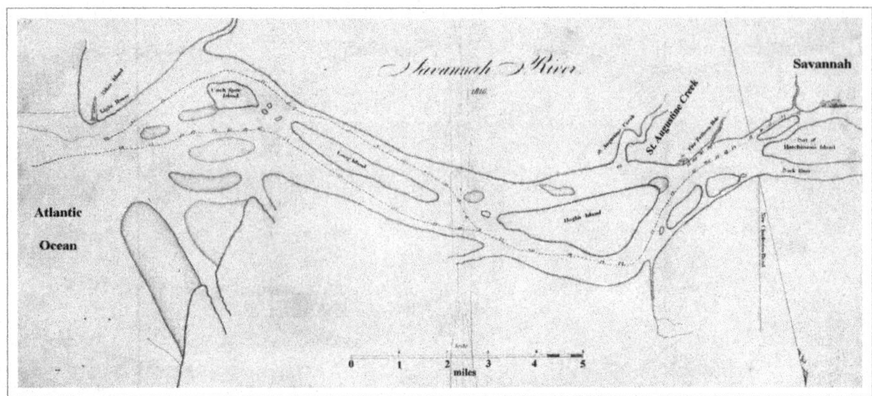


**Figure 3.6** Detail of an 1816 map showing the long and tortuous sailing route from the Savannah River entrance near Tybee Island to Savannah. North is to the bottom.

Source: Blunt, Edmund M. "Savannah River, 1816." [map]. Scale 1:62,500. In *The American Coast Pilot*, by Edmund M. Blunt. Newburyport, Mass.: E. Blunt, 1816. Map S3, Hargrett Rare Book & Manuscript Library, University of Georgia Libraries, Athens.

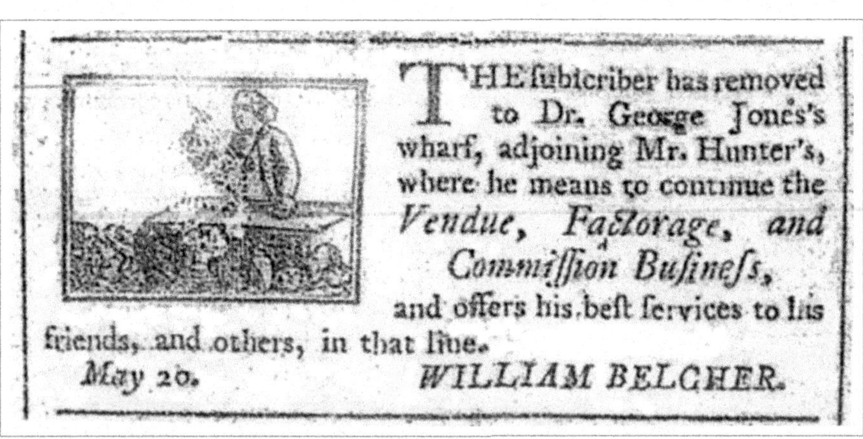

THE ſubſcriber has removed to Dr. George Jones's wharf, adjoining Mr. Hunter's, where he means to continue the *Vendue, Factorage, and Commiſſion Buſineſs,* and offers his beſt ſervices to his friends, and others, in that line.

*May* 20. *WILLIAM BELCHER.*

**Figure 3.7** Advertisement by William Belcher appearing in the Savannah newspaper *Georgia Gazette* January 2, 1800.

Source: Film N-US GA20, *Georgia Gazette* January 2, 1800, Microfilm Collection, University of Virginia Libraries, Charlottesville.

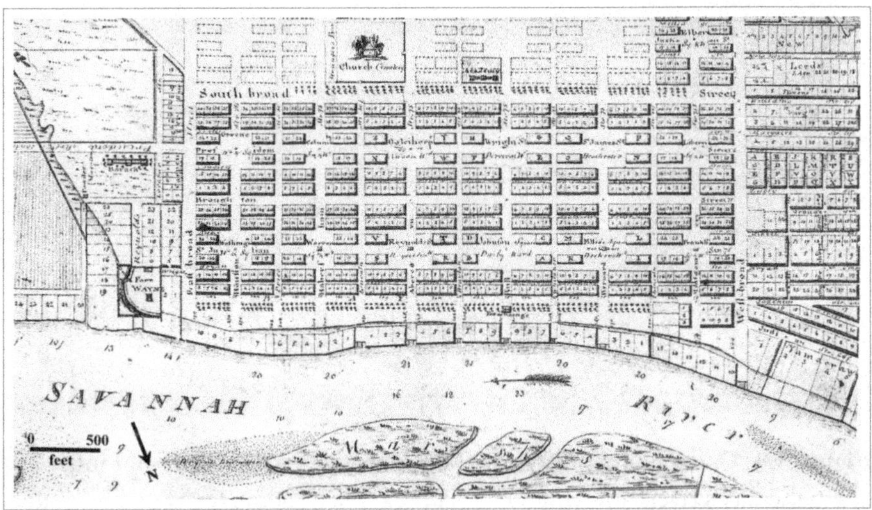


**Figure 3.8** Detail of an 1813 map of Savannah showing the numerous numbered wharf lots lining the riverfront. Note the Exchange building standing near the center of the waterfront.

Source: Anonymous. *Plan of the City of Savannah in Chatham County State of Georgia*. Savannah: 1813. Scale not given. Reproduction. GHS 1018 Charles Ellis and Elizabeth Stewart Waring papers, Vol. 02, Plate 14. Georgia Historical Society, Savannah.

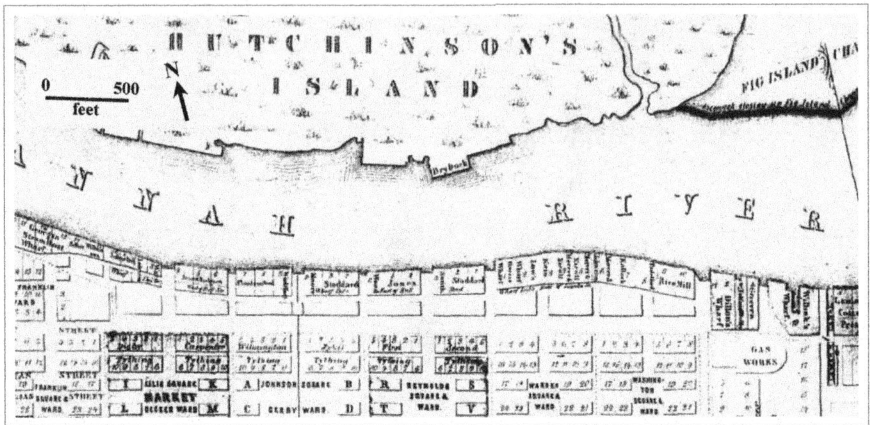


**Figure 3.9** Detail of an 1856 map of Savannah showing waterfront lots along the city's riverfront. The Georgia Steamboat Company wharf is located at the extreme left and the shipyard and wharf operated by Henry Willink, the largest shipbuilder in Savannah, is at the far right.

Source: John M. Cooper & Co., pub. *Map of the City of Savannah*. Savannah: 1856. Scale 1:6000. Reproduction. GHS 1018 Charles Ellis and Elizabeth Stewart Waring papers, Vol. 02, Plate 26. Georgia Historical Society, Savannah.

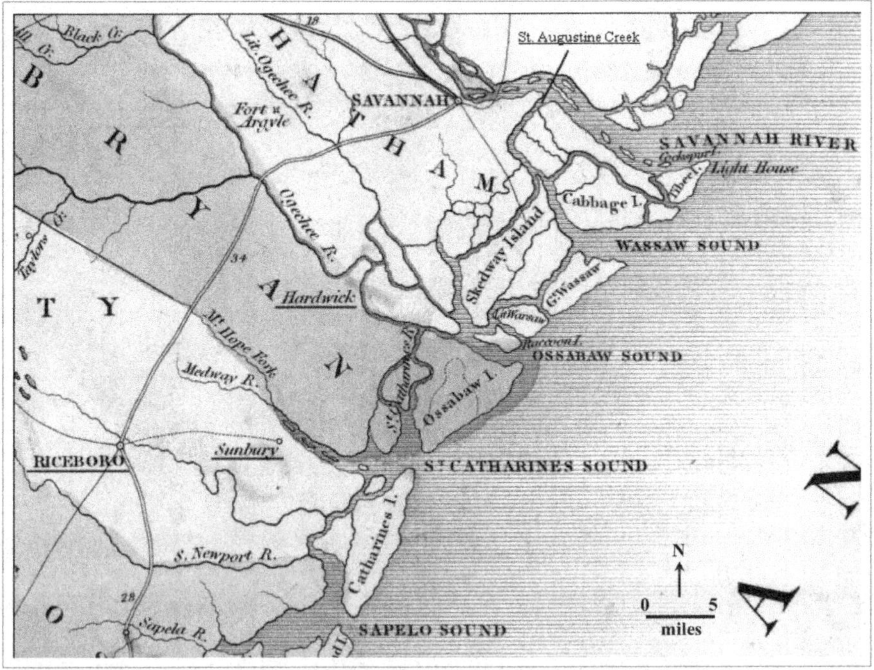


**Figure 3.10** Detail of an 1839 map showing the principal towns, islands, and rivers visited by the sailing coasters along the upper Georgia coast. St. Augustine Creek, the route commonly used by coasters to enter the inland passage when sailing south, is shown just east of Savannah. The towns of Hardwick on the Ogeechee River and Sunbury on the Medway River were established during the Colonial Period. Riceboro, located where the Savannah to Darien stage road crossed the North Newport River, became the area's principal port for coasters after about 1810.

Source: Burr, David H. "Map of Georgia & Alabama Exhibiting the Post Offices, Post Roads, Canals, Rail Roads &c." [map]. Scale ca. 1:650,000. In *The American Atlas*, by David H. Burr. London: J. Arrowsmith, 1839. Map G3920 1839.B8 RR 200, Geography and Map Division, Library of Congress, Washington, DC.

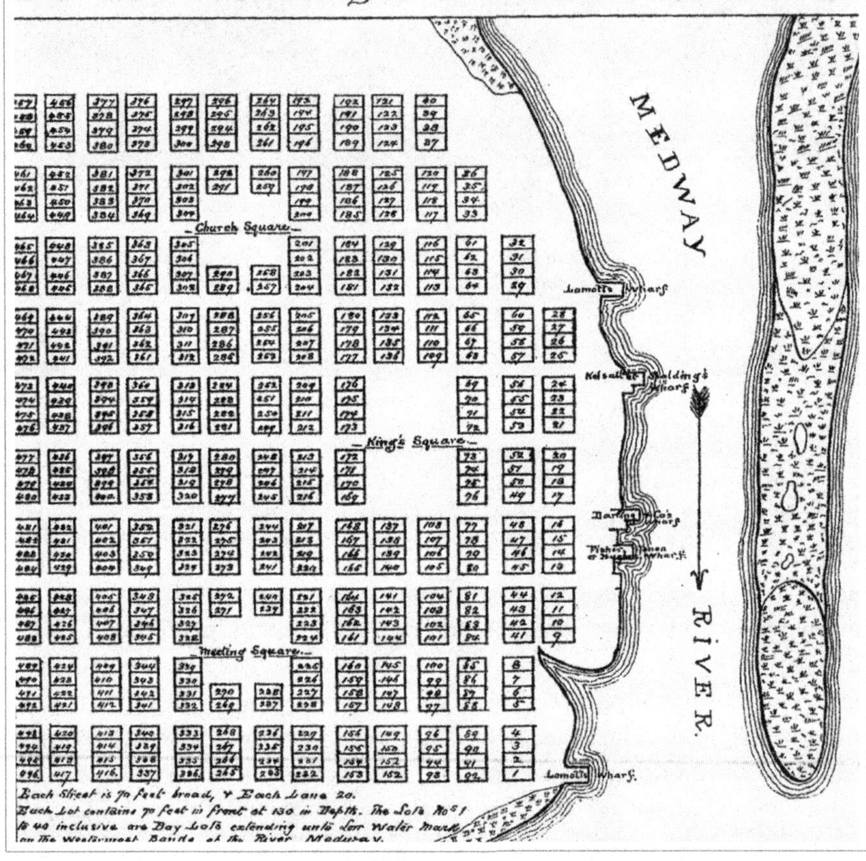


**Figure 3.11** "Plan of the Town of Sunbury" showing the layout of the town as originally planned. Although most of the lots shown were never occupied, the presence of five wharfs along the Medway River testifies to the town's importance as a trading center during the colonial period.

Source: Jones, Charles Colcock, Jr. *The Dead Towns of Georgia.* 1878. Reprint, Savannah: The Oglethorpe Press, 1997, 141.

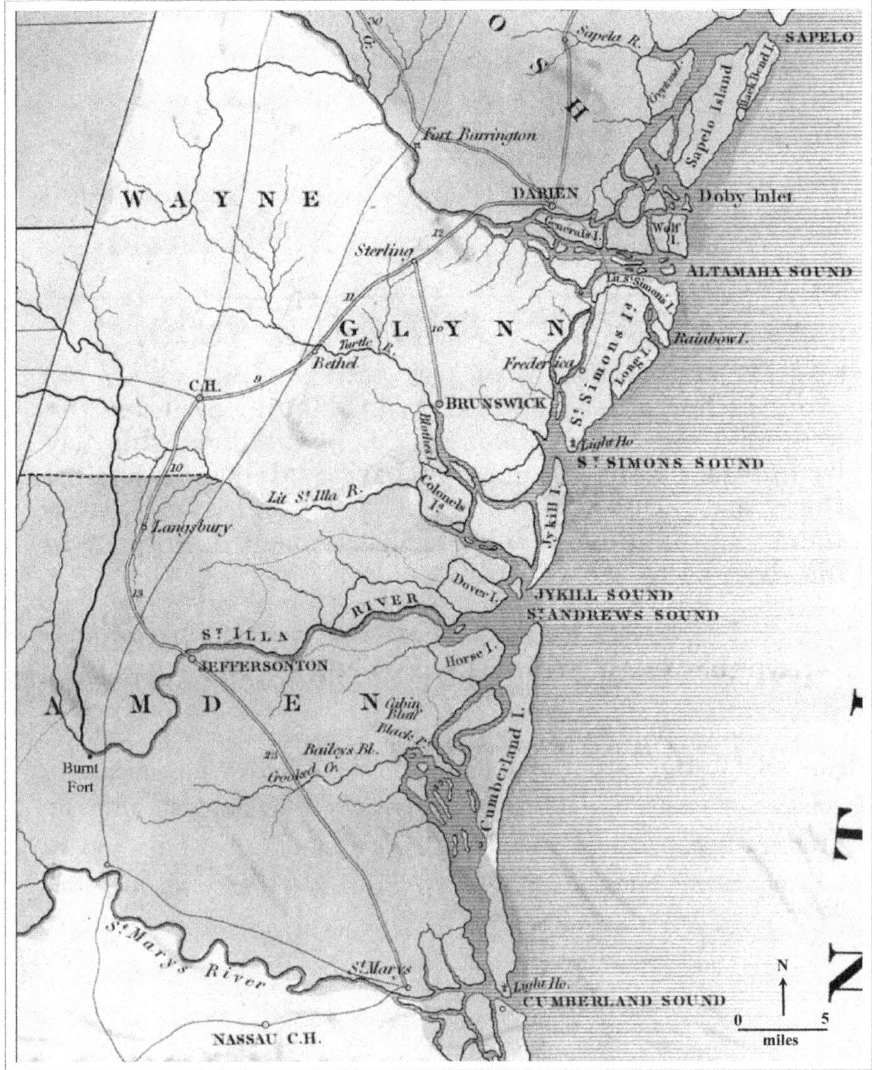


**Figure 3.12** Detail of an 1839 map showing the towns, islands, and rivers visited by the sailing coasters along the central and lower Georgia coast. The colonial town of Frederica on St. Simons Island existed as a minor port-of-call for coasters after 1800. The principal ports for coasters were Darien on the Altamaha River, Brunswick on the Turtle River, and St. Marys on the St. Mary's River.

Source: Burr, David H. "Map of Georgia & Alabama Exhibiting the Post Offices, Post Roads, Canals, Rail Roads & c." [map]. Scale ca. 1:650,000. In: David H. Burr. *The American Atlas*. London: J. Arrowsmith, 1839. Map G3920 1839.B8 RR 200, Geography and Map Division, Library of Congress, Washington, DC.

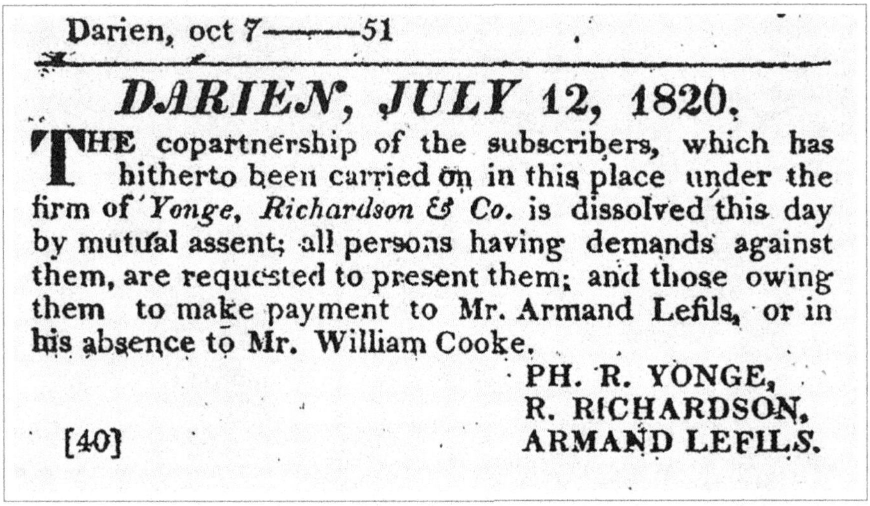

Darien, oct 7——51

DARIEN, JULY 12, 1820.

THE copartnership of the subscribers, which has hitherto been carried on in this place under the firm of Yonge, Richardson & Co. is dissolved this day by mutual assent; all persons having demands against them, are requested to present them; and those owing them to make payment to Mr. Armand Lefils, or in his absence to Mr. William Cooke.

PH. R. YONGE,
R. RICHARDSON,
ARMAND LEFILS.

[40]

**Figure 3.13** Advertisement from the 1820 *Darien Gazette* announcing the dissolution of Yonge, Richardson & Co., one of the few factorage firms in Darien.

Source: *Darien Gazette* November 18, 1820, Film AN13.D22 D31, Newspaper Microfilm Collections, University of Georgia Libraries, Athens.

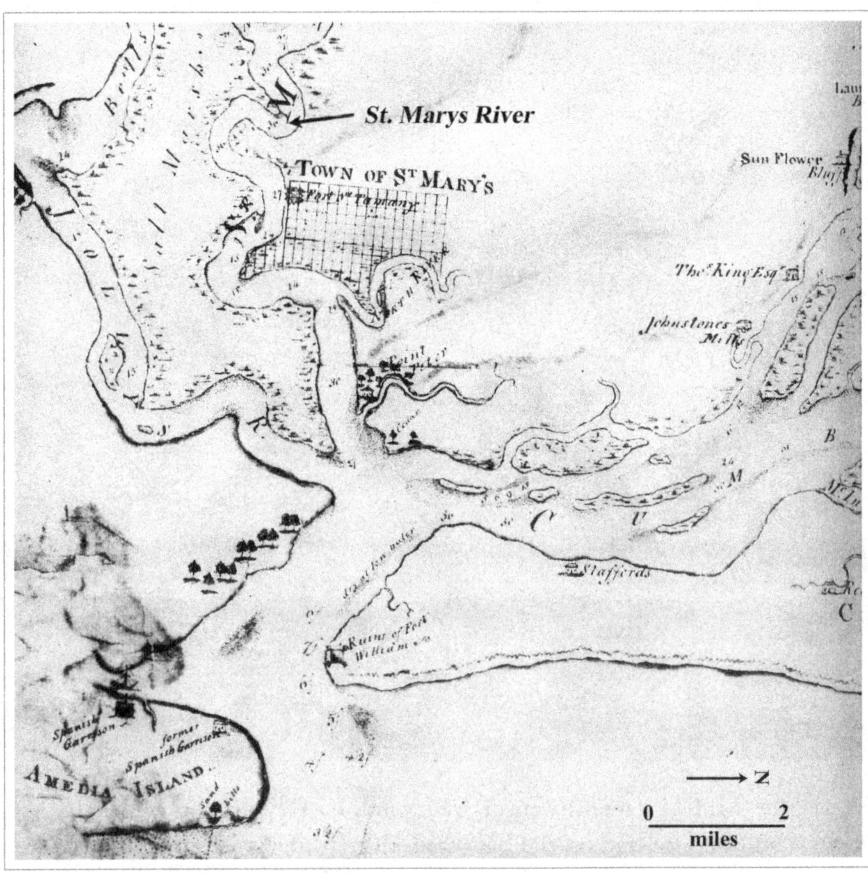


**Figure 3.14** Detail of a circa 1780 map showing the town of St. Marys and the entrance to the St. Marys River between Cumberland Island and Amelia Island, which was then a Spanish territory. Note the soundings provided along the principal waterways, including the portion of the inland passage behind Cumberland Island.

Source: Anonymous. *Map of the Coast of Georgia, Bordering on Camden and Glynn Counties, Showing Also the Course and Soundings of the Alatamaha, Turtle, Crooked, St. Mary's, Great Satilla, and Little Satilla Rivers*. [map]. Scale not given. 1780. Map G3922.C6 178- .M3, Geography and Map Division, Library of Congress, Washington, DC.

**Figure 4.1** Mid-nineteenth century photograph of the Robert Habersham & Sons building located on the Savannah riverfront. This company was one of the most prominent factors in Savannah and dealt extensively with the coastal rice and Sea Island cotton crops. In front of the building are bales ("bags") of Sea Island cotton and several barrels, probably "tierces" (one-third of a hogshead) containing "clean rice."

Source: R. Habersham and Co., Undated. Georgia Historical Society photograph collection. GHS 1361-PH-08-16-1561. Georgia Historical Society, Savannah.

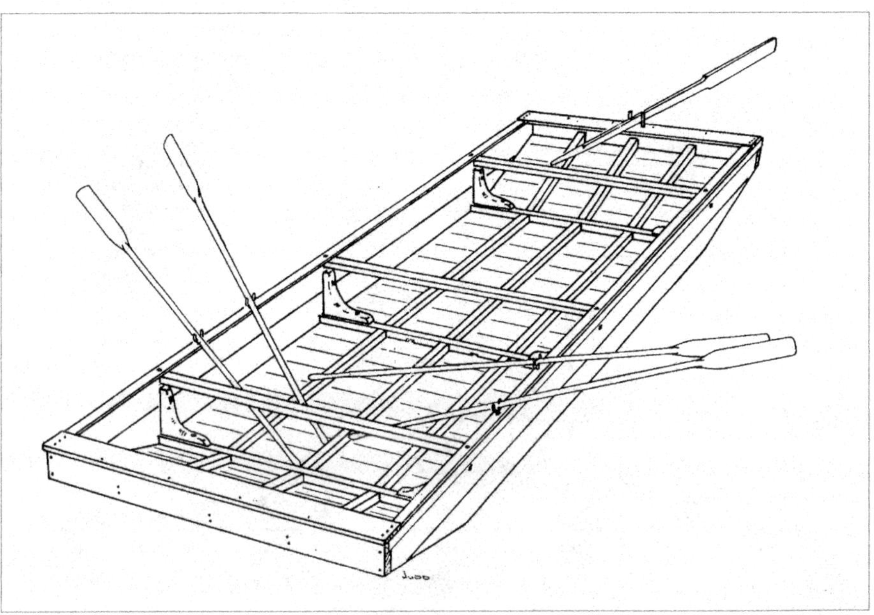


**Figure 4.2** Plantation flat used to transport rice and other crops and goods on coastal plantations. These flats were also used to haul crops short distances into Savannah from nearby plantations and mills.

Source: Fleetwood, William C., Jr. *Tidecraft: The Boats of South Carolina, Georgia and Northeastern Florida, 1550-1950.* Tybee Island, Ga.: WGB Marine Press, 1995:102. Illustration by William R. Judd.

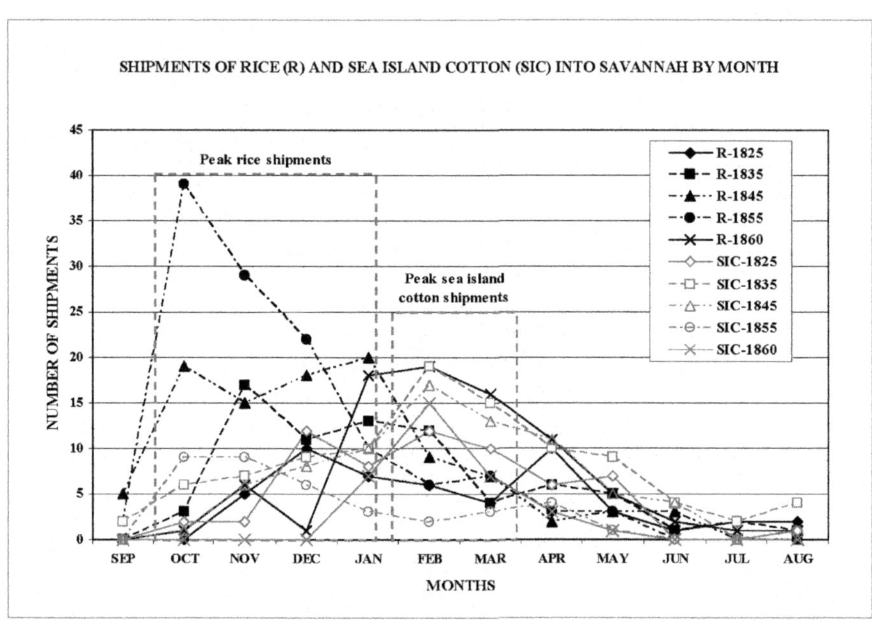


**Figure 4.3** Chart showing the number of shipments of rice (filled symbols) and Sea Island cotton (open symbols) carried into Savannah each month by coasting vessels for five different years. Data from the *Daily Georgian* 1825, 1835, 1845, and 1855 and the *Daily Morning News* and *Savannah Daily Republican* 1860.

Source: *Daily Georgian* 1825, 1835, 1845, and 1855, Film AN13.S45 G43; *Daily Morning News* 1860, Film AN13.S45 S26; and *Savannah Daily Republican* 1860, Film AN13.S45 S28, Newspaper Microfilm Collections, University of Georgia Libraries, Athens.

Charleston, November 29th 1854

James H. Couper Esqr

To Robert Mure Dr

Invoice of Sundries for Hopeton shipped per Schr Hopeton Luce master for Hopeton Plantation consigned to by order & for a/c & risk of James H. Couper Esqr these

| | |
|---|---|
| 800 Bushels corn @ 95¢ $760.00 Drayage $5.00 | 765.00 |
| 1 Bale Blankets 5 pcs @ $35. $175.00 4 wrappers @ $1 $4.00 | 179.00 |
| 2 Bales Cont'g 55 Blankets @ $35. | 120.31 |
| 52 " @ $25 | 81.25 |
| 18 Grey " @ $1 | 18.00 |
| 1 Double corn sheller $14 1 Blacksmith's Bellows $16 | 30.00 |
| 1 Grant's patent Fan $34 12 Kegs nails $^4/_6$ $^3/_8$ $^3/_{12}$ $^2/_{20}$ @ 5¼ $63 | 97.00 |
| 2 Sacks Cont'g 5 Bushels T I Salt @ 75¢ & bags | 4.50 |
| 1 Tin Can Cont'g 5 Galls Castor Oil $7.50 1 Box Cont'g 2 pt Bots Laudanum 1.50 | 9.00 |
| 2 pt Bots Paregoric 1.50 1 pt Bot. Ether 75 2 qts Bots Syrup Squills 2.50 | 4.75 |
| 3 oz Quinine 12.75 6 ½lb Bottles Mustard (Eng) 3.00 52lb Flax seed & bags 4.81 | 20.56 |
| Tin Can 1.25 Box 31 5 Cases Cont'g 240 pairs Plantation shoes @$1 $240 | 241.56 |
| 55 " " " @ 85 | 46.75 |
| Charges | $1617.68 |
| To Drayage to Schooner $1.50. Drayage on domestics from RR 1.50 | 3.00 |
| E E Charleston November 29th 1854 | $1620.68 |

Robt Mure

**Figure 4.4** Transcription of an 1854 invoice for cargo shipped aboard the coasting schooner *Hopeton* from Charleston to Hopeton Plantation on the lower Altamaha River.

Source: Estate of James Hamilton, Glynn County Wills and Appraisements Book F, 178.

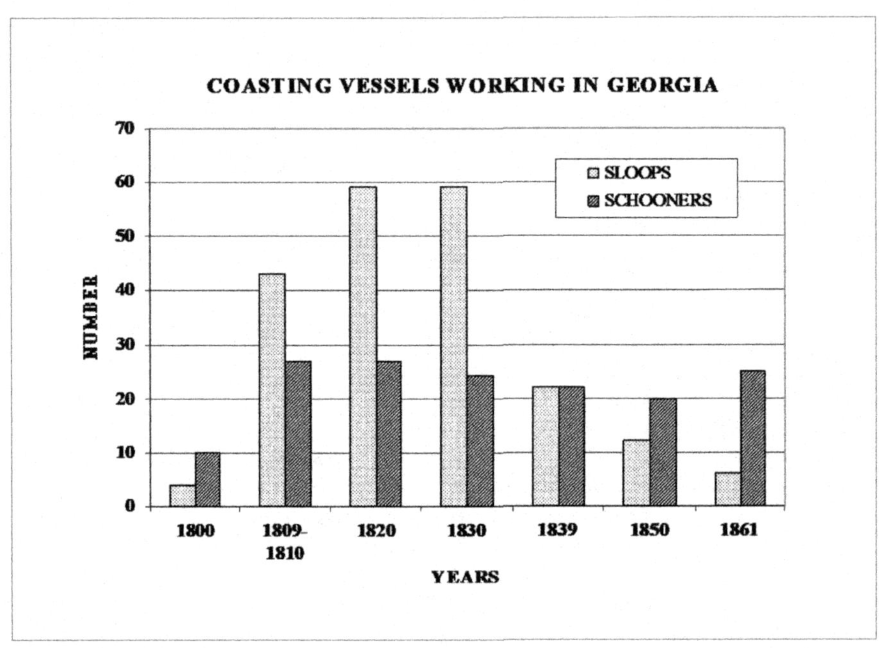


**Figure 5.1** The numbers of local coasting sloops and schooners sailing into Savannah at ten-year intervals between 1800 and 1861.

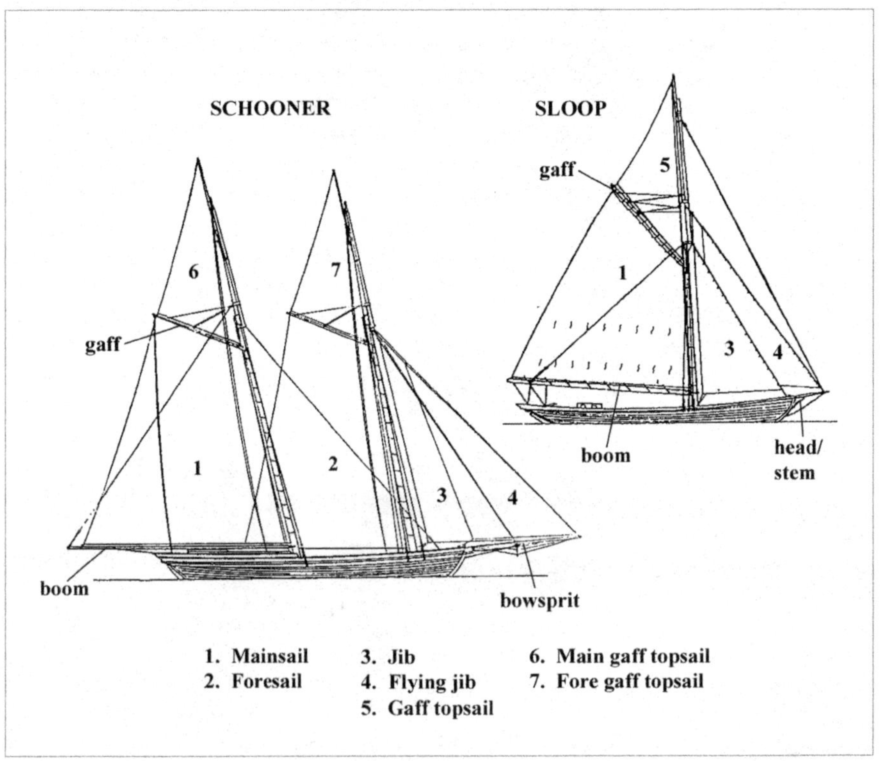


**Figure 5.2** Basic sail plans of a small coasting sloop and schooner. Labeling by the author.

Source: Pearson, Charles E. "Captain Charles Stevens and the Antebellum Georgia Coasting Trade," *The Georgia Historical Quarterly* 75, no. 3 (1991):493.

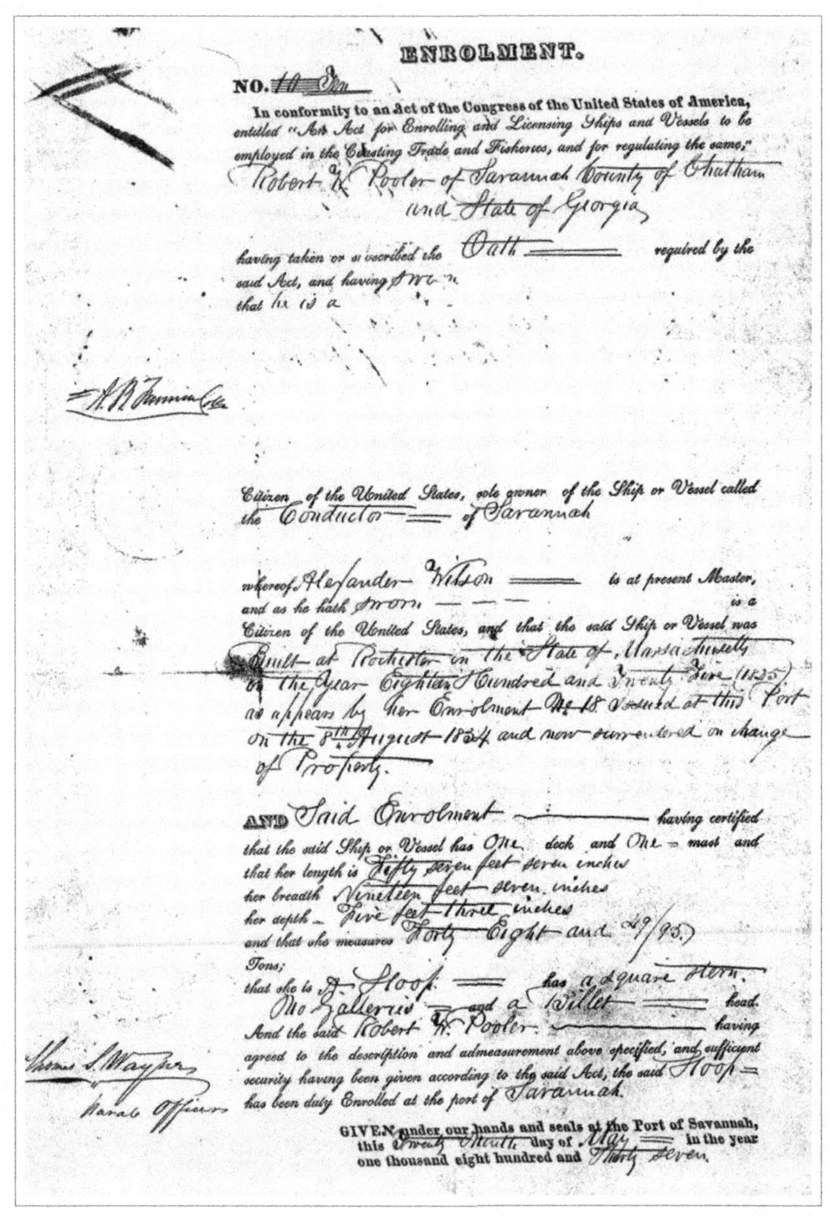

ENROLMENT.

NO. ~~10~~ Ten

In conformity to an Act of the Congress of the United States of America, entitled "An Act for Enrolling and Licensing Ships and Vessels to be employed in the Coasting Trade and Fisheries, and for regulating the same," Robert W. Pooler of Savannah County of Chatham and State of Georgia

having taken or subscribed the Oath required by the said Act, and having sworn that he is a

A. B. Fannin Collr

Citizen of the United States, sole owner of the Ship or Vessel called the Conductor of Savannah

whereof Alexander Wilson is at present Master, and as he hath sworn is a Citizen of the United States, and that the said Ship or Vessel was Built at Rochester in the State of Massachusetts in the Year Eighteen Hundred and Twenty five (1825) as appears by her Enrolment No 18 issued at this Port on the 8th August 1834 and now surrendered on change of Property.

AND Said Enrolment having certified that the said Ship or Vessel has One deck and One mast and that her length is Fifty seven feet seven inches her breadth Nineteen feet seven inches her depth Five feet three inches and that she measures Forty Eight and 29/95 Tons; that she is a Sloop has a square stern. No Galleries and a Billet head. And the said Robert W. Pooler having agreed to the description and admeasurement above specified, and sufficient security having been given according to the said Act, the said Sloop has been duly Enrolled at the port of Savannah.

Thomas S. Wayne
Naval Officer

GIVEN under our hands and seals at the Port of Savannah, this Twenty Ninth day of May in the year one thousand eight hundred and Thirty seven.

**Figure 5.3** Enrollment document for the coasting sloop *Conductor*, issued at Savannah on May 29, 1837.

Source: Enrollment No 10, Port of Savannah, sloop *Conductor*, May 29, 1837, Record Group 41, Bureau of Marine Inspection and Navigation, Records Relating to Vessel Documentation, National Archives and Records Administration [hereinafter cited NARA], Washington, DC.

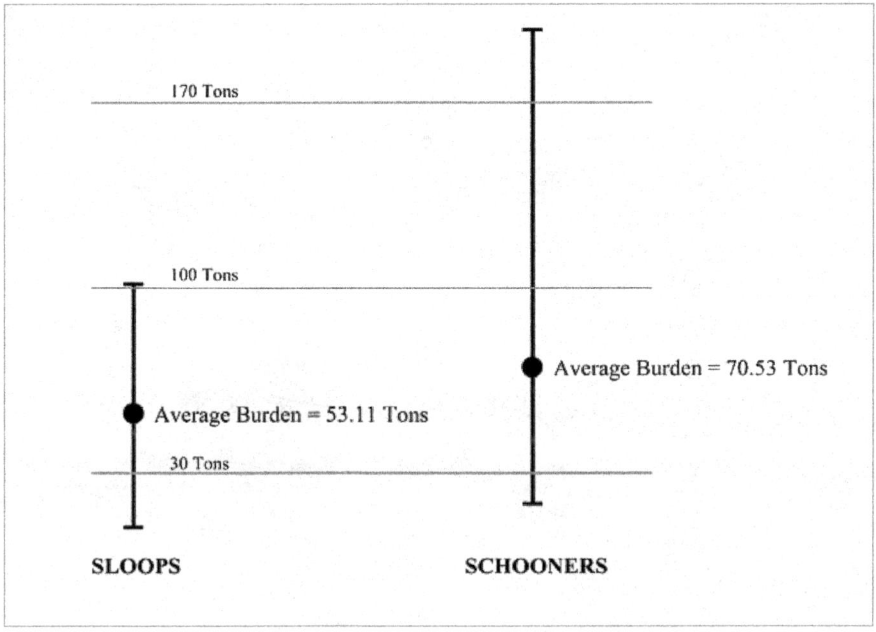


**Figure 5.4** Average and range of burdens of sloops and schooners sailing in the Georgia coasting trade between 1800 and 1861.

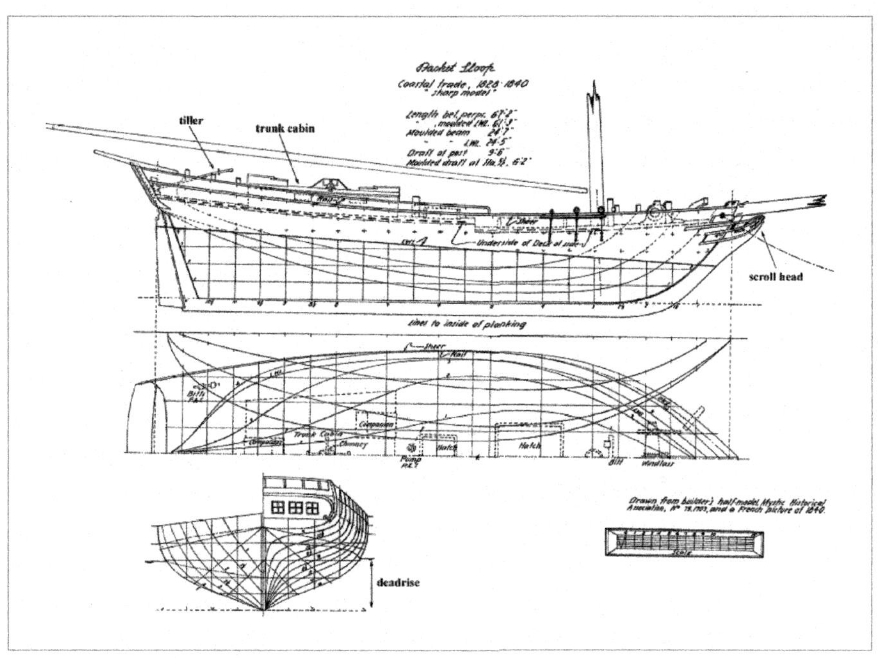


**Figure 5.5** Line drawings of a coastal packet sloop of the 1830–1840 period showing the position of the mast and principal deck features. Note the "scroll head," trunk cabin and the use of a tiller. This vessel had a deeper hull and greater deadrise than the typical sloop working in the Georgia coasting trade.

Source: Chapelle, Howard I. *The Search for Speed Under Sail, 1700–1855.* New York: W. W. Norton, 1967, 266.

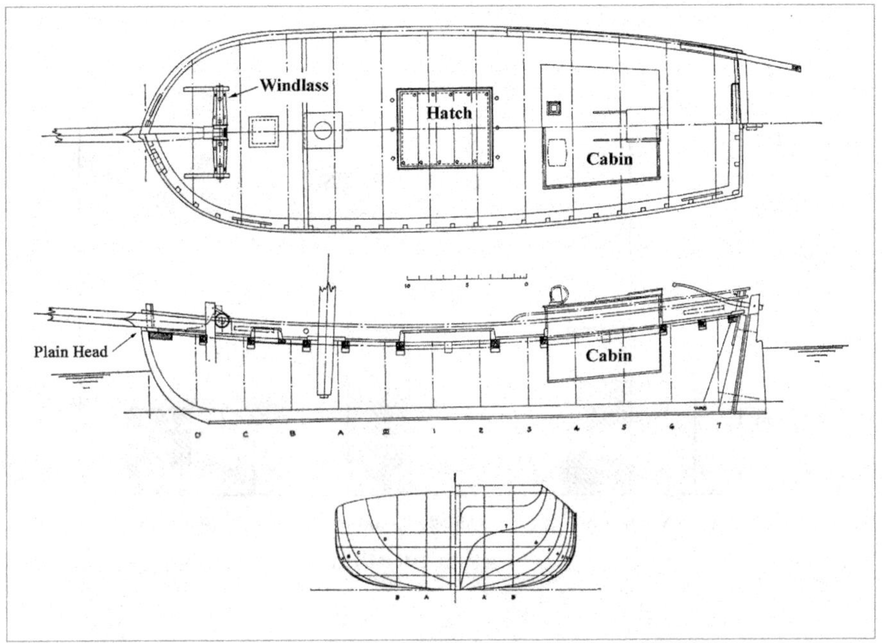


**Figure 5.6.** Projected deck plan, interior view, and cross-section of the 32 44/95-ton sloop *Mayflower*, built in 1823. Note the "plain head" and the flat-bottomed hull, characteristic features of the coasting sloops working in Georgia.

Source of drawing: Baker, William A. *Sloops & Shallops*. Columbia, S.C.: University of South Carolina Press, 1966, Figure 49.

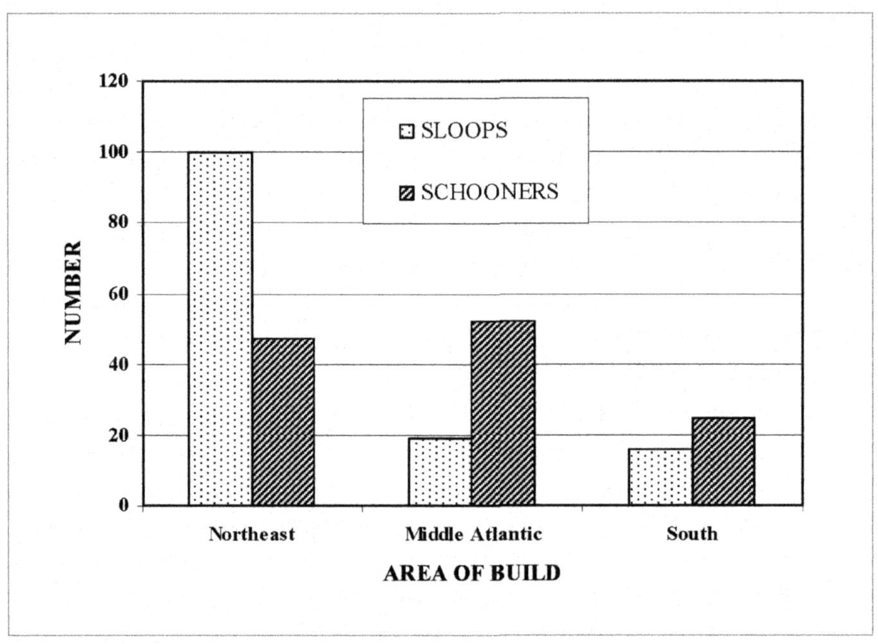


**Figure 5.7** Graph showing the area of build of vessels working in the Georgia coasting trade from 1800 to 1861.

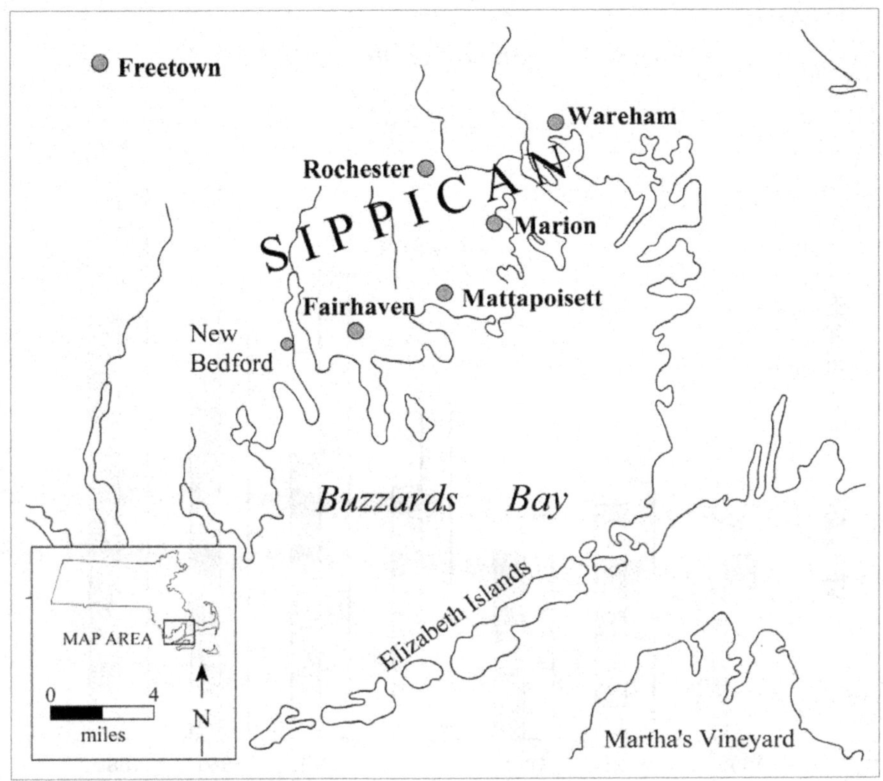


**Figure 5.8** The "Sippican" area of southern Massachusetts.

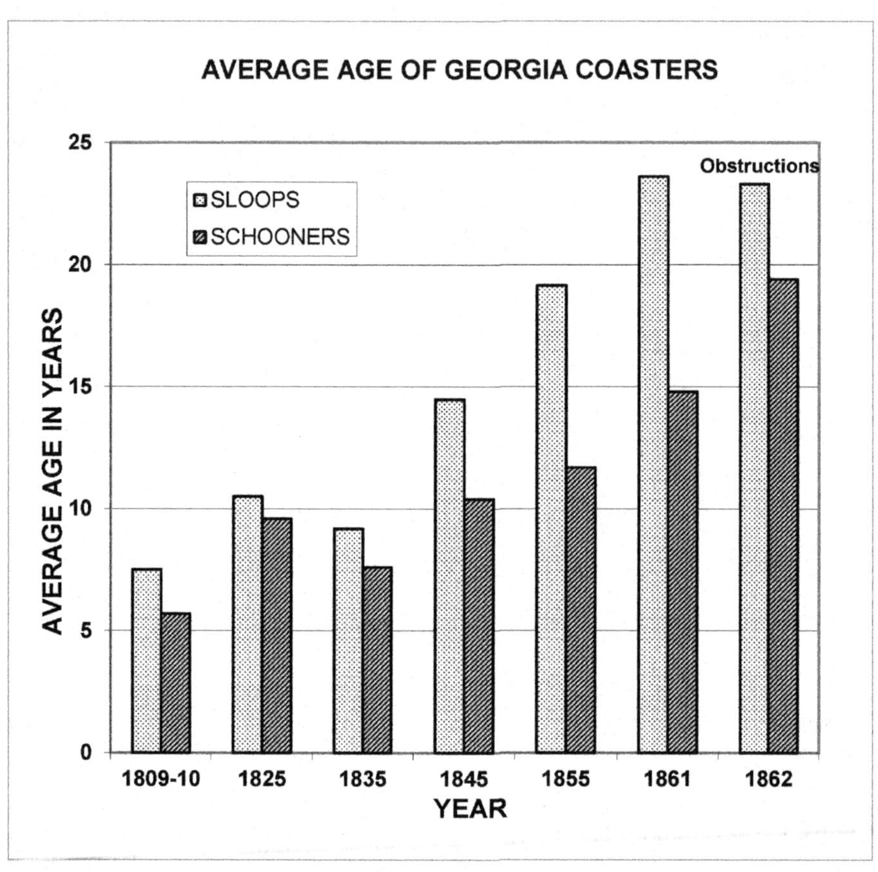


**Figure 5.9** The average age of Georgia coasters working in various years from 1809 to 1861. The average age of coasters sunk as obstructions in the Savannah River in 1862 is also shown.

**Figure 6.1** Partial page from accounts kept by Captain Charles Stevens recording transport of cargo aboard his schooner *Northern Belle* on July 10, 1860.

Source: Captain Charles Stevens Ledger Book Fragment. 1859–1860. Frewin, Stevens, and Taylor Family Papers. GHS 1909-27b. Georgia Historical Society, Savannah.

**Figure 6.2** Lewis Peter Wiggins (1819–1902), a native of Russia who came to Savannah in 1845 and served as a coasting captain until after the Civil War.

Source: Photograph from Culver, Robert Meldrim, Sr. "A Biography of Lewis Peter Wiggins, 1972," Ancestry.com, <https://www.ancestry.com/mediaui-viewer/collection/1030/tree/37999377/person/19218385492/media/ed80f6eb-5dfa-4d1e-9f38-42108bca2b27?_phsrc=yPe295&_phstart=successSource>, accessed December 15, 2022.

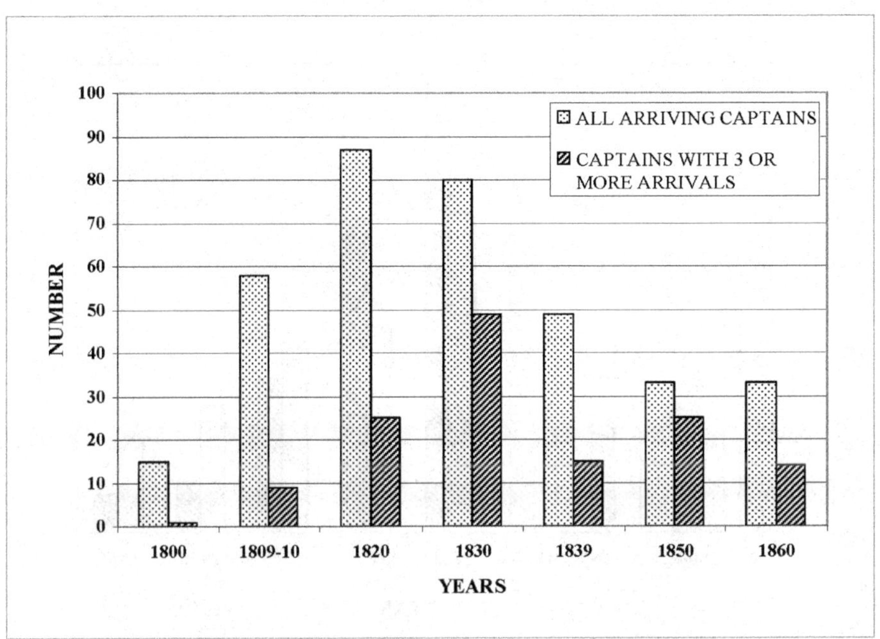


**Figure 6.3** The number of local coasting captains sailing into Savannah for various years between 1800 and 1860. Also shown are the numbers of captains making three or more arrivals each year.

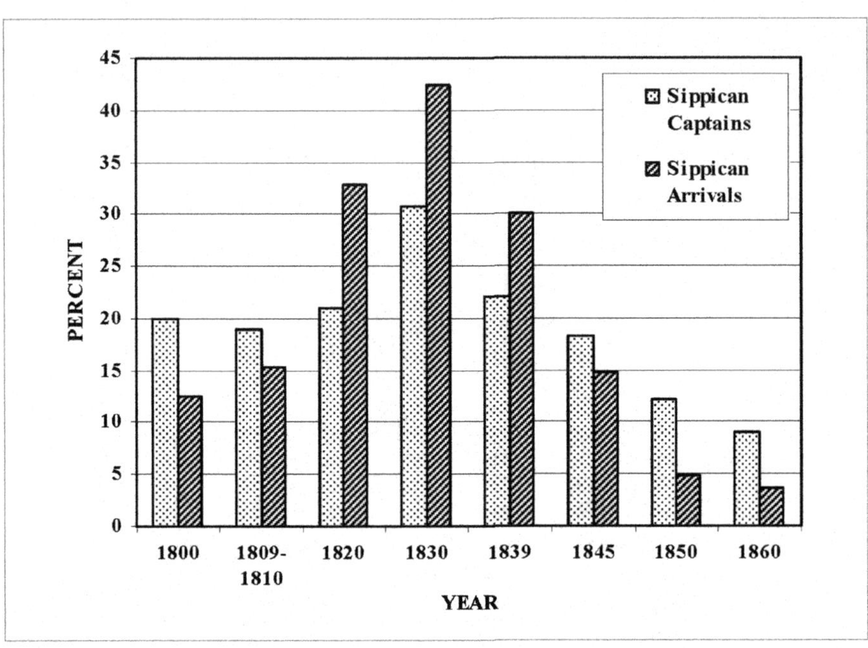


**Figure 6.4** Percentages of captains sailing coasters into Savannah from local ports identified as Sippican men and the proportion of arrivals they made into the city for various years.

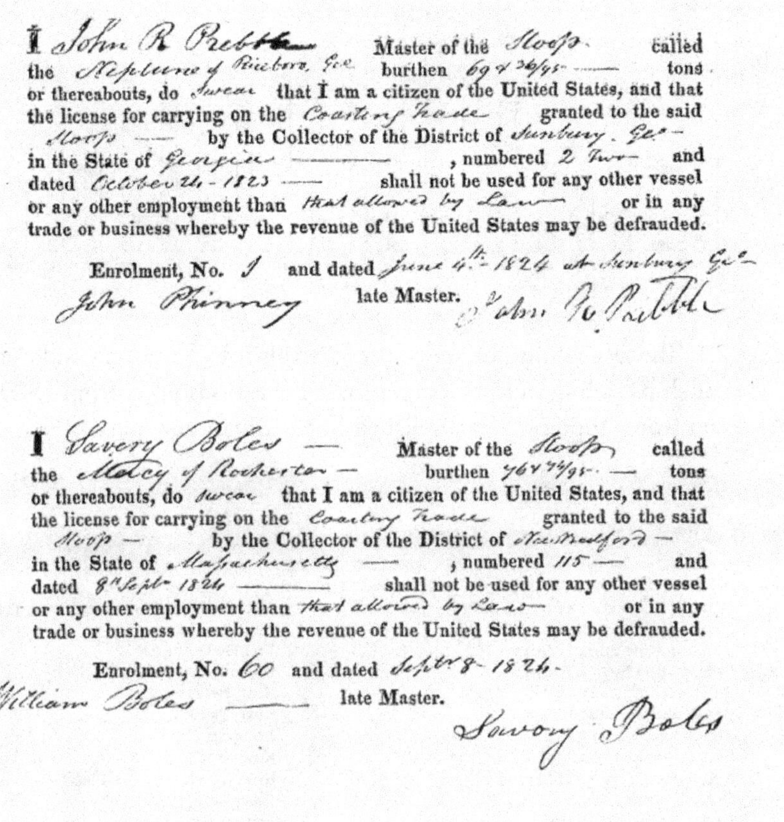

I John R Prebble Master of the Sloop called
the Neptune of Riceboro, Geo burthen 69 & 36/95 tons
or thereabouts, do swear that I am a citizen of the United States, and that
the license for carrying on the Coasting Trade granted to the said
Sloop by the Collector of the District of Sunbury, Geo
in the State of Georgia, numbered 2 Two and
dated October 24 - 1823 shall not be used for any other vessel
or any other employment than that allowed by Law or in any
trade or business whereby the revenue of the United States may be defrauded.

Enrolment, No. 1 and dated June 4th - 1824 at Sunbury Geo
John Phinney late Master. John R Prebble

I Savory Boles Master of the Sloop called
the Mercy of Rochester burthen 76 & 77/95 tons
or thereabouts, do swear that I am a citizen of the United States, and that
the license for carrying on the Coasting Trade granted to the said
Sloop by the Collector of the District of New Bedford
in the State of Massachusetts, numbered 115 and
dated 8th Septr 1824 shall not be used for any other vessel
or any other employment than that allowed by Law or in any
trade or business whereby the revenue of the United States may be defrauded.

Enrolment, No. 60 and dated Septr 8 - 1824.
William Boles late Master.
Savory Boles

**Figure 6.5** Oaths for coasting licenses issued to two prominent Sippican captains who sailed in the Georgia trade, John Prebble and Savory Boles.

Source: Port of New Bedford Custom House Papers, Licenses Oaths, Archives Room, New Bedford Free Public Library, New Bedford.

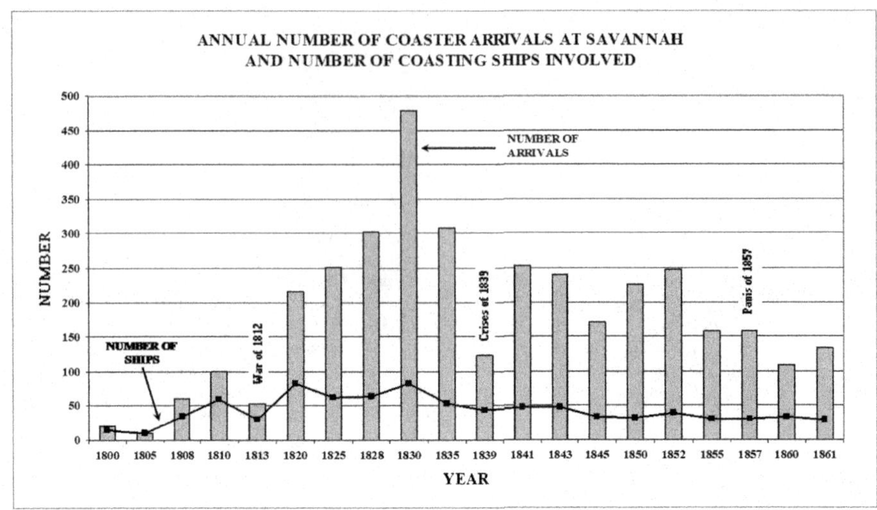


**Figure 7.1** The annual number of coaster arrivals into Savannah and the number of ships sailing in the Georgia trade for various years from 1800 to 1861. Data from shipping lists published in Savannah newspapers.

**MARINE LIST**

**Entered Inward**

| | |
|---|---|
| Slp. [Sloop] Patty, Landon [master] | Brunswick [port of origin] |
| Slp. Henry, Cherry | Charleston |
| Sch. [Schooner] Commerce, Hammond | Philadelphia |
| Sch., Sally, Brigham | Boston |
| Sch. Susan, Snow | New Bedford |
| Brig Huntress, Sammis | New York |
| Ship Mercury, Tate | London |
| Sch. Susan, Hammond | New Bedford |
| Sch. Laurel, Hitchcock | New York |
| Sch. Isaac, Donnald | Jamaica |
| Brig Mary, Hazard | Boston |
| Brig Minerva, Rhodes | ditto |
| Brig Arethuss, Smith | Charleston |
| Sch. Friendship, Vaughn | Jamaica |

**Cleared Out**

| | |
|---|---|
| Sch. Concord, Sands | New York |
| Sch. Deborah, Rhodes | Boston |
| Sch. Gosport, Jennings | Charleston |
| Sch. Hart, Smith | Tortola (one of the Virgin Islands) |
| Sch. Edward, Eldred | Cape Francois |
| Brig Sally, Bowman | Martinique |
| Sch. Sally, Brigham | Pasquotank |

**Figure 7.2** Transcribed "Marine List" from the *Georgia Gazette*, January 2, 1800.

Source: Film N-US GA20, *Georgia Gazette* January 2, 1800, Microfilm Collection, University of Virginia Libraries, Charlottesville.

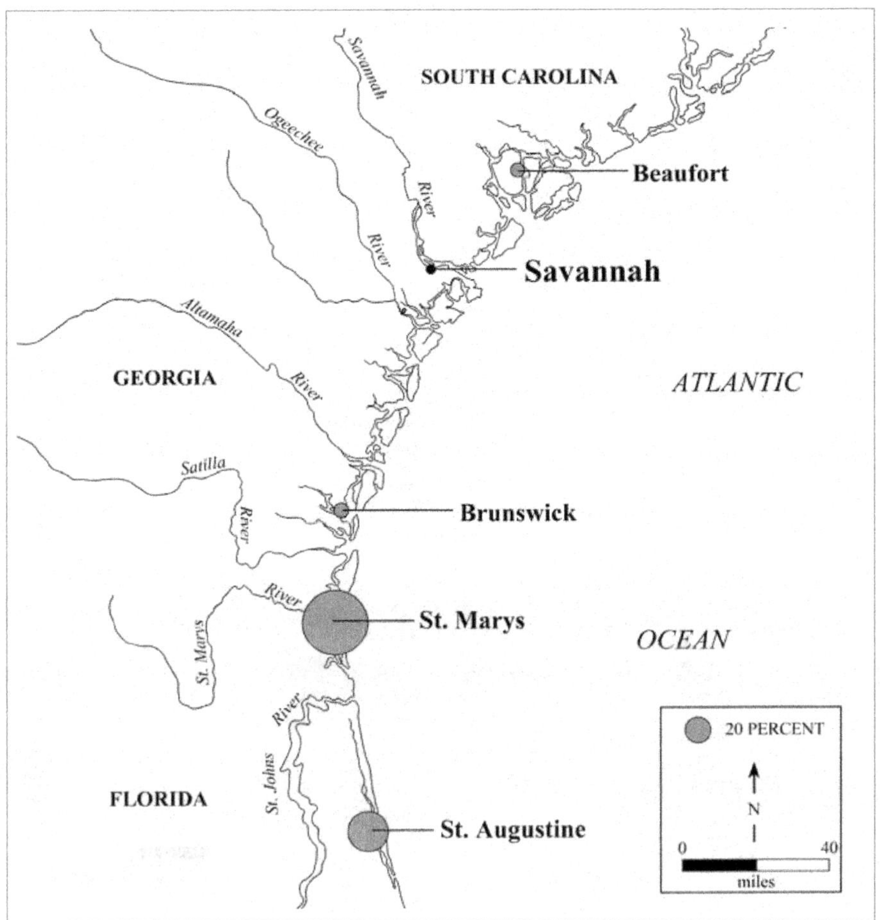


**Figure 7.3** Ports of origin for Georgia coasters sailing into Savannah in 1800 showing proportion of arrivals from each port. Data from the shipping lists published in the Savannah newspapers *Georgia Gazette* and *Columbian Museum & Savannah Advertiser* for the year 1800.

Source: Film N-US GA20, *Georgia Gazette* 1800, Microfilm Collection, University of Virginia Libraries, Charlottesville; and Film AN13.S45 S265, *Columbian Museum & Savannah Advertiser* 1800, Newspaper Microfilm Collections, University of Georgia Libraries, Athens.

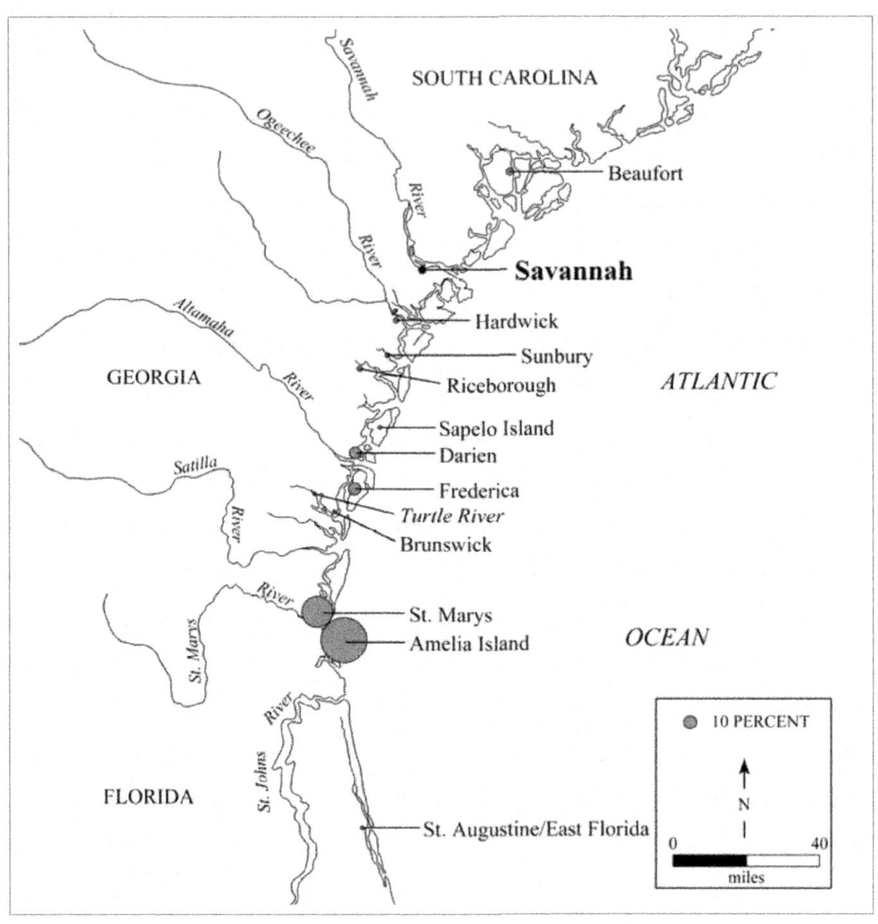


**Figure 7.4** Ports of origin for Georgia coasters sailing into Savannah in 1809–1810 showing proportion of arrivals from each port.

Source: Data from the *Republican & Savannah Evening Ledger* April 20, 1809–March 29, 1810, Film AN13.S45 S28, Newspaper Microfilm Collections, University of Georgia Libraries, Athens.

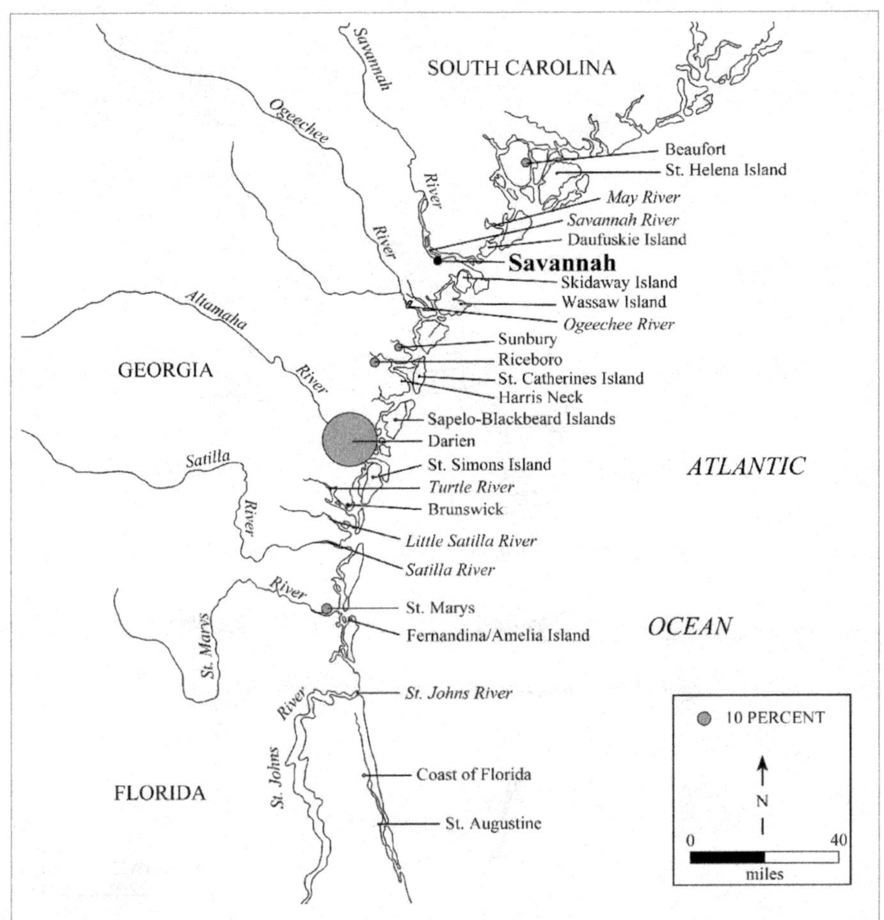


**Figure 7.5** Ports of origin for Georgia coasters sailing into Savannah in 1820 showing proportion of arrivals from each port.

Source: Data from the *Daily Georgian* January 1820–December 1820, Film AN13.S45 G43, Newspaper Microfilm Collections, University of Georgia Libraries, Athens.

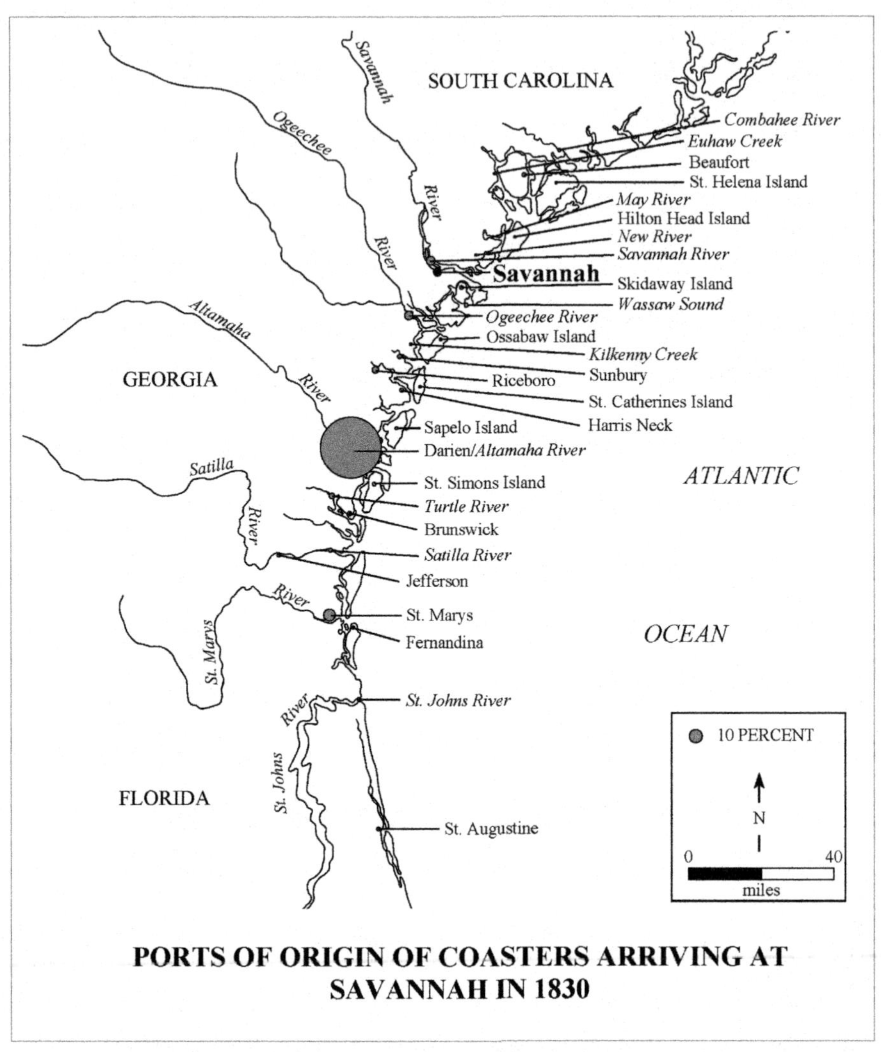


**Figure 7.6** Ports of origin for Georgia coasters sailing into Savannah in 1830 showing proportion of arrivals from each port.

Source: Data from the *Daily Savannah Republican* January 1830–December 1830, Film AN13. S45 S28, Newspaper Microfilm Collections, University of Georgia Libraries, Athens.

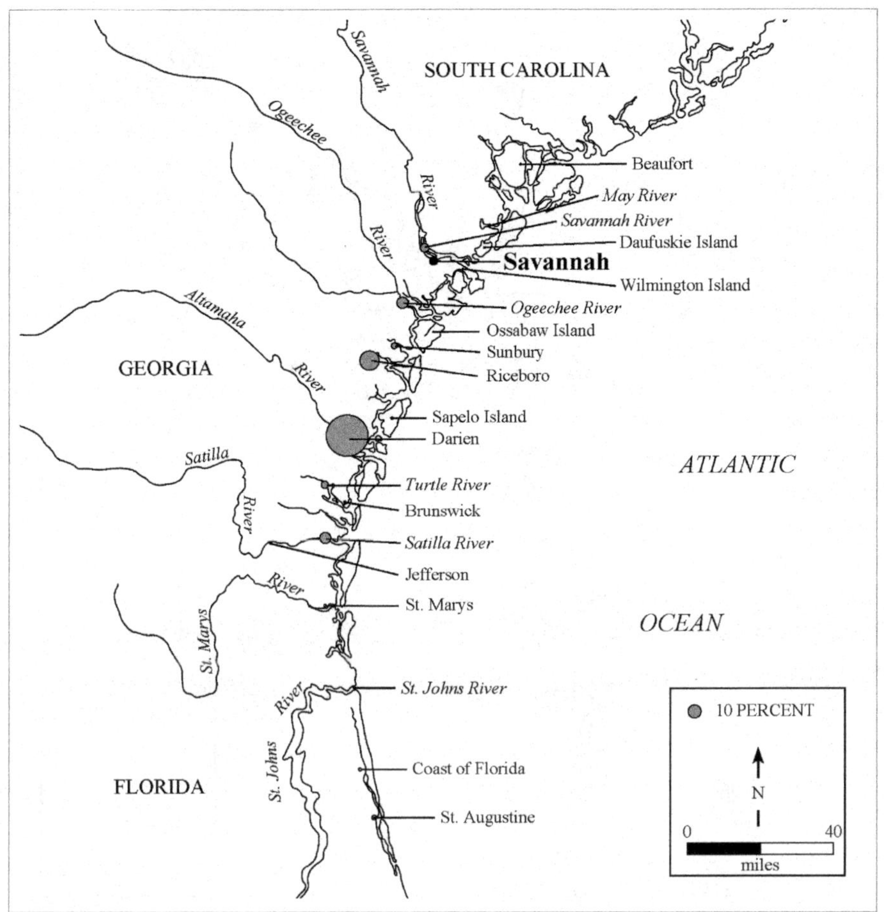


**Figure 8.1** Ports of origin for Georgia coasters sailing into Savannah in 1839 showing proportion of arrivals from each port.

Source: Data from the *Daily Georgian* January 1839–December 1839, Film AN13.S45 G43, Newspaper Microfilm Collections, University of Georgia Libraries, Athens.

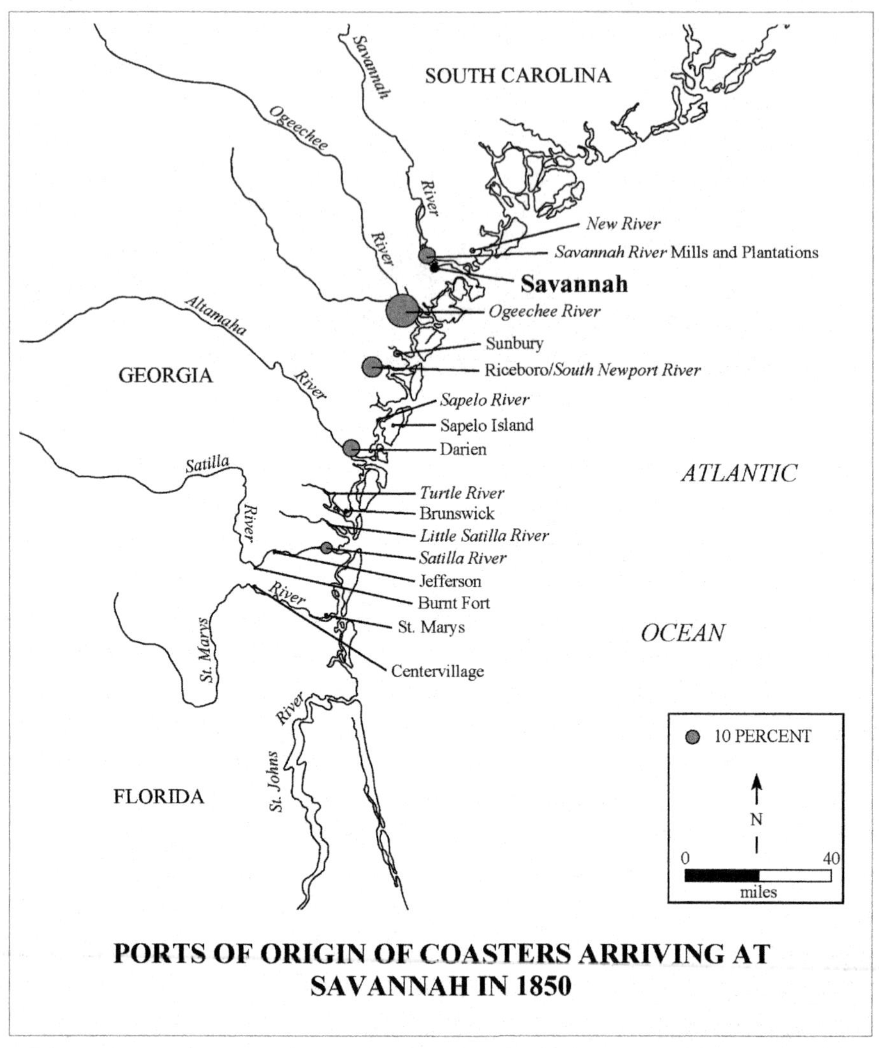


**Figure 8.2** Ports of origin for Georgia coasters sailing into Savannah in 1850 showing proportion of arrivals from each port.

Source: Data from the *Daily Georgian* January 1850–December 1850, Film AN13.S45 G43, Newspaper Microfilm Collections, University of Georgia Libraries, Athens.

**Figure 8.3** Ports of origin for Georgia coasters sailing into Savannah in 1861 showing proportion of arrivals from each port.

Source: Data from the *Daily Morning News* January 1861–December 1861, Film AN13.S45 S26, Newspaper Microfilm Collections, University of Georgia Libraries, Athens.

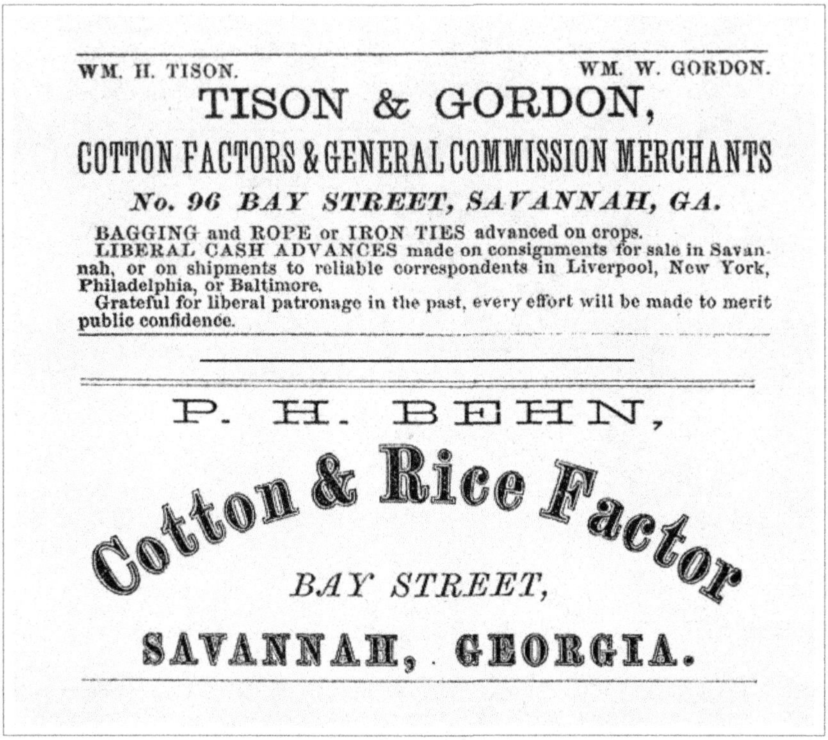

WM. H. TISON. WM. W. GORDON.

TISON & GORDON,

COTTON FACTORS & GENERAL COMMISSION MERCHANTS

*No. 96 BAY STREET, SAVANNAH, GA.*

BAGGING and ROPE or IRON TIES advanced on crops.

LIBERAL CASH ADVANCES made on consignments for sale in Savannah, or on shipments to reliable correspondents in Liverpool, New York, Philadelphia, or Baltimore.

Grateful for liberal patronage in the past, every effort will be made to merit public confidence.

P. H. BEHN,

Cotton & Rice Factor

*BAY STREET,*

SAVANNAH, GEORGIA.

**Figure 8.4** Advertisements for Tison & Gordon and P. H. Behn, two of the prominent Savannah factors dealing in coastal commodities in the 1850s and 1860s.

Source: Lee, F. D., and J. L. Agnew. *Historical Record of the City of Savannah*. Savannah: J. H. Estill, 1869, 8, 30.

# Chapter 1

# Introduction

On January 3, 1844, the Savannah newspaper the *Daily Georgian* reported the arrival of a forty-eight-foot, single-masted sloop named *Splendid*, laden with thirty-one bales of Sea Island cotton and twelve hundred bushels of "rough rice." The rice, consisting of threshed but unmilled grains, was dumped loose in the sloop's hold. The cotton, in its thick, sausage-shaped bags of burlap, was piled on deck. Relying on favorable winds and the incoming tide, the *Splendid*'s captain, twenty-eight-year-old Charles Stevens, had brought his vessel to the docks at the city of Savannah, seventeen miles from the sea. There, the eleven thousand pounds or so of valuable cotton and the fifty-four thousand pounds of the less lucrative rice were unloaded and placed into the hands of three of the largest merchant factors in the city: Andrew Low & Company, Elias Reed & Son, and J. W. Roberts.[1] Ultimately, the cotton would be reloaded onto ocean-going ships and carried to mills in the Northeast or across the Atlantic to England. The rice would first be milled at one of several "pounding mills" around Savannah, and then it might be shipped out or sold locally. Captain Stevens was bringing his ship and its cargo from the Satilla River, located on the lower Georgia coast eighty miles south of Savannah and site of several prosperous cotton and rice plantations.

Charles Stevens was one of many captains sailing in what was known as the "coasting trade." His arrival in Savannah with a cargo of cotton and rice exemplified an event that was repeated almost daily by many sailing vessels and steamboats throughout the first six decades of the nineteenth century (Figure 1.1). Stevens used his ship to carry a variety of products from the plantations, farms, and forests along the Georgia, South Carolina, and Florida coasts into the mercantile centers of Savannah and Charleston. Most of these products were agricultural, and rice and cotton, particularly the valuable long-

[1] The arrival of the *Splendid* is reported in the "Ship News" section of the Savannah newspaper *Daily Georgian*, January 3, 1844.

fiber variety known as Sea Island cotton, constituted the principal cargo of these coasting ships. On departing these major ports, Charles Stevens and the other coasting captains returned to small port towns and individual plantations, carrying an assortment of merchandise, ranging from calico to window sashes and from salt pork to Madeira wine. In an area and time where overland travel was difficult, the coasting vessels were indispensable in the movement of goods to and from market.

There is no comprehensive history of the coastwise trade in the United States, despite its importance in regional and national economies. Studies have examined limited aspects of the trade dealing with a specific port or region or by looking at particular people involved. These works commonly examine interregional trade, or trade between a region and a major port, such as New York, Philadelphia, or Boston; few deal with the truly local coasting trade.[2] A major reason for this neglect of the local trade, such as that existing between Savannah and the coasts of Georgia and adjacent South Carolina and Florida, is the almost total lack of official records documenting this commerce. Because of a peculiarity in the customs and navigation laws of the United States, American vessels carrying American products within a single customs district or between districts in adjacent states rarely had to file manifests or bills of lading with customs officials. As a result, the day-to-day activities of these ships almost never appear in official records. Consolidated information on the coastal cargoes arriving or leaving ports is available; but information on individual vessels is incomplete and difficult to find. What does exist must be gleaned from scattered sources. Fortunately, for some areas, including Savannah and Charleston, local shipping activity was documented in the often detailed newspaper accounts of the comings and goings of coasting vessels and also in the records, ledgers, letters, and journals of merchants and planters involved in receiving, shipping, or carrying goods by coasting vessel. Relying on these types of sources, particularly newspapers, this work

[2] Albion's, *Rise of New York Port*, although encompassing more than just the coastwise trade, serves as a model for a study of seaborne trade and its relationship to a single port city. Other single port studies include Maganzin, "New York City's Coastwise Trade"; Jensen, *Maritime Commerce*; Clowse, "Charleston Export Trade"; and Coker, *Charleston's Maritime Heritage*. Several studies have examined coastwise trade for specific regions, including Clark, "Coastwise and Caribbean Trade"; Middleton, *Tobacco Coast*; Pares, *Yankees and Creoles*. Hedges, *Browns of Providence Plantations*, provides a view of a family's involvement in maritime trade although much of it is related to overseas trade, not the coasting trade.

examines the coastal trade in Georgia, its history and economics, and the men, the ships, and the cargoes involved in it.

This study focuses on the coasting trade along the Southeastern Atlantic coast from 1800 to the beginning of the Civil War. More specifically, it deals with the activity of sailing vessels working in this trade (Figure 1.2). Steamboats entered the coasting trade in Georgia in the 1820s and by the 1840s, were carrying a considerable quantity of the coastal commerce. However, during these six decades, sailing ships remained important as carriers of cargo along this coast. The story of coastal steamboats is a complete study in itself, and attempting to include a comprehensive discussion of their activities would push the present work beyond manageable limits.

The term "coasting" refers to waterborne trade along a coast or major river and commonly refers to local trade, such as between Savannah and the nearby coasts of South Carolina and Georgia, or between New York City and adjacent states. However, the terms "coasting" and "coastwise" also refer to interregional coastal trade, such as the long-distance trade between Savannah and Northern ports such as New York or Boston. I use the term "coasting" to refer to truly local trade, occurring principally along the coast of Georgia and adjacent areas of Florida and South Carolina. Applying the term to this restricted geographical area has historical justification. During the nineteenth century, the term "coasting" commonly referred specifically to the trade between Savannah or Charleston and the neighboring coastal areas. The terms "Coasting Captain" or "Captain of Coasting Vessel" were used in official and unofficial records to describe the occupation of men involved in such local trade. Additionally, the terms "coasting" or "coaster" were regularly applied to both the ships and the men working in this trade. The term "coastwise" occasionally appears in early records, but it was more commonly applied to interregional trade than to the purely local trade considered here.

Restricting this study to the coast of Georgia, South Carolina, and Florida reflects reality in terms of the sphere of activity of the men and ships working in the trade. This region, extending from the Combahee River in South Carolina to the St. Johns River in Florida, plus inland areas reached by rivers, constituted the principal economic hinterland of Savannah (Figure 1.3). Local products from these areas were funneled mainly into or through Savannah after about 1820, and the city, in turn, served as this region's major mercantile and financial center. Although Savannah is the principal market center of concern, Charleston must be considered because, before the 1820s, that city received a considerable proportion of the plantation products from the Georgia coast. Coasting vessels carried produce from the Georgia coast into

Savannah before 1800, but not until the 1820s did the city develop the financial and marketing infrastructure required to serve the region's needs fully. By the 1820s, increasing numbers of coastal planters had begun to transfer their dealings from factors and commission agents in Charleston to those in Savannah.[3]

This coastal region was also a unique agricultural area during the first sixty years of the nineteenth century. The principal crops grown, rice and Sea Island cotton, formed an agricultural dyad distinct from the rest of the South. The unique character of this region's agriculture, including the methods and manner of shipping goods to market, sets it apart as a distinctive region.

Finally, this region had reality in early federal laws regulating coastal shipping. One of the first important laws of the new United States regulating commerce, passed in September 1789, legislated that coasting vessels carrying products of the United States could trade within or between the customs districts in a single state or the immediately adjacent states with a minimum of paperwork. If a ship captain stayed within these prescribed limits, he could enter and leave any port without filing manifests or obtaining permits to sail. Captains sailing beyond these limits often had to provide manifests to local customs officials and secure permits to sail. The same paperwork requirements were placed on vessels coming from foreign ports or laden with foreign goods. Captains who worked mainly along the Georgia coast could transport local cargoes anywhere along the Georgia, Florida, and South Carolina coasts with little in the way of required paperwork, making life easier for the coasting captains, but, because few manifests were produced, leaving us with essentially no official record of their cargoes. Subsequently, in 1819, the district within which a coasting vessel could trade without filing manifests was expanded to include the entire eastern seaboard (a "great district"), making obtaining information on the coastwise trade even more difficult.[4]

Early legislation establishing high tariffs on foreign-owned vessels working in the trade restricted coastwise cargoes to vessels of American ownership built and commanded by an American citizen. Thus, a study of the coastal trade during the first half of the nineteenth century is a study of American-built ships owned and crewed by Americans.

Goods and passengers certainly traveled between Georgia coastal communities before 1800, but a small number of ships were regularly involved before this date. It was not until the development of Sea Island cotton

[3] House, *Planter Management*, 45.

[4] Johnson, *History of Domestic and Foreign Commerce*, 329–30, 337.

agriculture in the last decade of the eighteenth century, combining with the already established cultivation of rice, that a mature coasting trade arose along the Southeastern coast. The volume of this trade grew through the first half of the nineteenth century, reaching its peak in the decade before the Civil War. In 1825, Savannah reportedly exported 137,695 "bags" of cotton and 7,321 "tierces," or barrels, of rice, most of which were carried into the city aboard small coasting sloops and schooners. By 1859–1860, Savannah exports had grown fivefold, with the city shipping 547,037 bales of Sea Island and upland cotton, 35,588 "casks" of rice, and more than eleven million feet of lumber.[5] By this time, much of the cotton, particularly the upland, short-staple variety, was reaching the city from the interior via an expanding railroad network, and steamboats were carrying a considerable amount of the waterborne commerce. However, sailing vessels were still active, particularly as carriers of "rough" or unmilled rice, a cargo ill-suited for most steamboats. In any given week during the rice-shipping season, September to March, a half dozen or more sailing vessels could be found unloading their cargoes at the docks lining the Savannah riverfront.

The Civil War disrupted the Georgia coasting trade and essentially marked the end of any significant participation by sailing vessels. In summer 1861, federal naval forces arrived off Tybee Island at the mouth of the Savannah River and by November had gained control of the coast, blockading all the major inlets along Georgia, South Carolina, and Northeastern Florida. The blockade made it unsafe for coasting vessels to operate, and the movement of goods by ship ended. On December 13, 1861, the *Savannah Daily Morning News* reported the arrival of the schooner *Fort George Packet* with a cargo of 2,030 bushels of rough rice. The schooner's captain, Andrew Hanson, had brought his tiny, forty-one-foot ship and its cargo from Butler Island, located at the mouth of the Altamaha River on the central Georgia coast. Subsequent issues of Savannah newspapers note the arrival of a few steamers from coastal locations, but the *Fort George Packet* seems to have been the last sailing vessel to carry coastal cargo into Savannah before the trade was entirely shut down.[6]

[5] "Commerce &c. of Savannah," *DeBow's Review*, June 1851, 10:685, and "Commerce of Savannah, 1860," *Debow's Review*, November 1860, 29:669–70.

[6] *Daily Morning News*, December 13, 1861. For information on the *Fort George Packet*, including the name of her captain, see "Enrollment No. 2, Port of St. Marys, schooner *Fort George Packet*, October 15, 1838," RG 41, Bureau of Marine Inspection and Navigation (BMIN), NARA (cited as BMIN). Hereafter, specific enrollments and

After the Civil War, trade along Georgia's coast slowly recovered as steamers of various types and sizes and a small number of sailing vessels became involved, but the coasting trade never reached the volume of the prewar years mainly because neither Sea Island cotton nor rice cultivation revived to their earlier positions. Cotton agriculture disappeared almost entirely from the coast. Rice fields, dikes, and ditches had been damaged during the war or had fallen into disrepair from neglect. Local planters were financially strapped and unable to raise the capital needed to repair their lands or to hire the labor required to produce crops. Also, rice cultivation had expanded to Louisiana, and by the 1870s that state's large rice crop was seriously depressing the Georgia and South Carolina markets. Timber and naval stores remained important products from the coastal regions and their production and shipment expanded significantly after the Civil War. Most of the lumber shipped from the Georgia coast after the war was carried by ships traveling directly to markets in the American Northeast or overseas, having little effect on the local coasting trade. Also, the now well-developed railroad network carried products that had once been shipped by water.

Although dealing specifically with the topic of the coasting trade, this study brings to light aspects of the maritime history and use of sailing ships in a particular area of the Southern United States. With a few exceptions, a reading of the literature on American maritime history would suggest that the South had little participation in this history. The great emphasis has been on the Northeast, which had a strong maritime tradition and, admittedly, a great influence on United States maritime history. Still, many writers of American maritime history have simply ignored the South. The present work represents an effort to expand our understanding of this aspect of Southern history.

*A note on sources*: This study of the coasting trade is a paper chase in the truest sense; it can be described as a "newspaper chase" because much of what can be learned about the local coasting trade in Georgia is derived from shipping accounts printed in contemporary newspapers. In the eighteenth and nineteenth centuries, newspapers in port cities commonly published

registers are cited by the document number, the port of issue, the name of the vessel, and the date of issue. If the enrollment information is derived from the "Master Abstracts of Enrollments" for a port rather than the individual enrollment document, this information is noted in the citation as "Master Abstracts." Vessel documents for some ports have been compiled and published, principally during the late 1930s and the 1940s, by the Works Progress Administration (later the Work Projects Administration). These compiled sources are referenced by title. A full discussion of these various vessel documents is provided in chapter 4.

information on the arrivals and departures of ships. Papers continued reporting shipping news into the twentieth century; some still do. This information was considered important to readers, particularly merchants and businessmen, permitting them to follow the movement of ships and cargoes in which they had an interest. These "shipping lists" were also important economic records for local governments that could be compiled for various purposes. In terms of the coasting trade, these lists take on even greater importance because neither state nor federal governments maintained records on the day-to-day activities of most coasters. As it turns out, the most complete inventory of the vessels and captains in the Georgia trade, and the most complete record of the cargoes they carried and the routes they sailed, come from the listings of vessel arrivals and departures published in port newspapers. For Georgia, the most useful of these are the Savannah papers. Various newspapers were published in the city over the entire time span of interest here (1800 to 1861), several of which included columns on ship arrivals and departures. Newspapers with vessel shipping information were published at other port towns, such as Darien and Brunswick, but these reports are incomplete. Charleston newspapers are available and provide information on coasters like that found in the Savannah papers. I examined Charleston papers for the very early period of interest because many of the coasters carrying Georgia produce landed there during the first two decades of the nineteenth century. However, with this study's emphasis on the coasting trade in Georgia, the Savannah newspapers provided the bulk of information on the vessels, men, and cargoes in the trade.[7]

The shipping columns in the newspapers ran under various names such as "Port of Savannah," "Ship News," or "Marine Intelligence," but all provided the same type of information: a listing of vessels arriving in port and those that had departed since the previous issue. For daily papers, this would typically be those ships that arrived or departed the day before the paper's publication. For papers published less frequently, the shipping lists encompassed several previous days of arrivals and departures. Before 1815, shipping lists in

[7] Microfilm and online records of twenty different named newspapers were examined, encompassing more than sixty years of publication. These included twelve newspapers from Savannah, one from Brunswick, one from Darien, four from Charleston, and two from Jacksonville, Florida. The names of newspapers often changed slightly from year to year or even in the same year. For example, the *Savannah Republican* was also titled *Daily Republican*, *Savannah Daily Republican*, and *Daily Savannah Republican*. When a newspaper is referenced, the name given for that specific issue or the name used most over the year is the one cited. See the list of newspapers in the bibliography.

Savannah newspapers provided information on the type of vessel, its name, the name of the master (captain), the port of origin for arriving vessels, and the port of destination of departing ones. In later years, some newspapers expanded on the information in their shipping columns, distinguishing among vessels that had "Arrived," "Departed," "Cleared," and "Went to Sea" and often provided additional important details, such as information on any deaths or accidents that occurred (Figure 1.4).

Until 1810, Savannah newspapers rarely identify the cargoes carried by coasting vessels. This information was incomplete and only intermittently provided until after 1820. After that, not only was the main cargo typically identified but also the amount on board (see Figure 1.4). At about this same time, the name of the recipient of the cargo, generally identified as the "consignee," was included. Thus, for the first two decades of the nineteenth century, we can roughly ascertain how many local coasting vessels sailed into and out of Savannah by looking at ports of origin and destination. Still, it is usually impossible to determine the types or quantities of commodities they carried. In Charleston, some newspapers identified cargoes carried into the city before 1820, but before about 1810, these papers did not regularly list vessels involved in the local coasting trade.

Vessels of all types, not just coasters, are included in newspaper shipping lists. The local coasters can be identified by their port of origin, their port of destination, and the type of vessel. With very few exceptions, the sailing ships involved in the Georgia trade were single-masted sloops and two-masted schooners. Other, larger sailing vessels, such as brigs or barks, might occasionally arrive in Savannah from one of the Georgia coastal ports, but when cargo information is provided, it typically reveals these vessels were not carrying typical coastal agricultural commodities such as rice, cotton, or sugar.

Although the newspaper shipping lists are vital to any study of the local coasting trade, they do contain inaccuracies and omissions. In both the Charleston and Savannah papers, information on vessel departures is incomplete, and it is apparent that keeping an accurate record of departures was not always of great concern. For example, a vessel may appear two or three times as arriving in Savannah, but there are no intervening listings of its departure. By the 1840s, many newspapers did not include local coasting vessels in their lists of departures. Information on arrivals was much more important because, at least by the 1820s, arrival information commonly included some identification of the vessel's cargo—important commercial data not recorded elsewhere. Even arrival information cannot be considered entirely accurate. For example, instances have been found in the Savannah newspapers where a vessel is listed as departing or "clearing" for a coastal port, but careful examination of preceding issues reveals no notice of the vessel's arrival. A comparison among Savannah newspapers published on the same day shows that one occasionally includes a vessel arrival that others omit. Savannah newspapers before 1805 typically list numerous sloops and schooners arriving from or departing for Charleston. However, shipping lists in Charleston newspapers only occasionally mention either the arrival or departure of these same vessels. These types of omissions of shipping information are more common before 1820 and for local vessels rather than those traveling to and from overseas ports, or ports farther north along the Atlantic seaboard.

The number of coasting vessels never included in local newspaper shipping lists is impossible to determine, but this number is believed to be small. Ships making a single trip into Savannah might be missed, but those fully engaged in the trade normally made several trips a year, and it is unlikely that all arrivals would be omitted. Local vessels carrying coastal produce and goods between plantations or small port towns without newspapers may never be identified. A few might appear in the Savannah newspapers even if they did not sail into or out of that city because Savannah papers often published a section titled "Memoranda," "From our Correspondent," or something similar (see Figure 1.4). This section provided shipping information on ports such as Darien, Brunswick, or St. Marys in Georgia, or on distant ports, such as New York or New Bedford, if the information was of interest to Savannah.

Despite these problems, port newspapers still provide our most complete and comprehensive view of the activities of the coasters. By following the information in shipping lists over a season, a year, or many years, it is possible to ascertain the changing patterns in the coasting trade in terms of vessels, ports, and cargoes, as well as track the activities of individual ships or captains over time.

Second to newspapers as a source of information is a category of records collectively known as "vessel documents" or "vessel papers." Although no official records were made of the daily activities of the Georgia coasters, the vessels themselves did require official documents issued by the United States government that provide basic descriptive and ownership information. These documents include certificates of enrollment, certificates of registry, and licenses. Vessels with burdens greater than twenty tons involved in foreign trade had to obtain registers while similar-sized ships engaged in the domestic coastal trades were required to obtain enrollments and licenses. Smaller vessels, between five and twenty tons in burden, only had to obtain licenses. Enrollments and registers are especially critical to this study. These documents record a vessel's place and year of build; dimensions, tonnage, and "rig" (i.e., sail and mast configurations); the names of its owner and master (captain); and the number, date, and place of issue of the previous vessel document.[8] Licenses are of particular value in providing information on coasters of less than twenty tons burden even though they contain less information than enrollments and registers, typically just the vessel name, tonnage, homeport, and master. Unfortunately, very few licenses survive for coasting vessels sailing in Georgia during the first half of the nineteenth century.

These documents have been used to gather information on many of the coasters named in the Savannah and Charleston newspapers, including information on the types, sizes, and places of build of vessels sailing in the Georgia trade between 1800 and 1861, as well as the names of shipowners and masters. No in-depth study of the local coasting trade anywhere in the United States is possible without the combined use of vessel documents and contemporary newspaper shipping lists.

A small number of journals, account books, and letters left by merchants, planters, and ships' captains have been used to complement the information contained in newspapers and official vessel documents. These few records are all that remain to tell the story of the once-thriving Georgia coasting trade. Despite the scarcity of historical documents, these small ships and the men who sailed them were indispensable in the region's economy, linking the rural planter with the Savannah factor and the small farmer or town resident with the city merchant. In addition to commerce, many aspects of life along the coast depended in some manner upon the coasting fleet. It is the intent of this study to bring to light the workings of the nineteenth-century coasting trade in Georgia and tell the story of those men and ships involved in it.

[8] Discussions and descriptions of vessel documents can be found in numerous publications. A clear and concise description of these records is provided in "Vessel Documents," RG 41, BMIN, hereinafter cited "Vessel Documents." A more comprehensive discussion and description of enrollments, registers, and other maritime records is found in Stein, *American Maritime Documents*.

# Chapter 2

## "We have built...a number of Coasting Vessels": The Establishment and Early Development of the Georgia Coasting Trade

On January 2, 1800, the "Marine List" in the Savannah *Georgia Gazette* reported the arrival of the "Sloop Patty" under Captain Landon from the port of Brunswick on the central Georgia coast. The fifty-three-foot *Patty*, one of the small number of vessels sailing in the local coasting trade out of Savannah at the start of the nineteenth century, transported plantation products into the city and carried a variety of goods and merchandise out of the city to the region's small towns and plantations. In the first three decades of the nineteenth century, the trade along the Georgia coast in which the *Patty* sailed would undergo dramatic growth, primarily related to the expansion of cultivation of two crops, rice and Sea Island cotton. The seeds for this thriving trade were planted seventy-five years earlier, on February 12, 1733, when forty English settlers and their families, led by General James Edward Oglethorpe, sailed up the Savannah River with the incoming tide. Oglethorpe selected a piece of level ground on the south side of the river about seventeen miles inland as the site for the first settlement of the new colony of Georgia. Here, the bank of the Savannah River formed a steep, forty-foot-high bluff, and Oglethorpe noted that ships drawing twelve feet of water could easily anchor close to the bank. General Oglethorpe named his settlement "Savannah" after the river on which it was located. It was destined to become the economic, political, and social center of the new colony.[1]

When selecting the site for his settlement, James Oglethorpe prioritized a location with a safe harbor that offered water access to key coastal ports, such as Charleston, and river access into the interior. Water access was critical to the survival of the new colony as the movement of people, merchandise,

[1] Coulter, *Georgia*, 24.

and produce was principally by boat. Although located on the banks of a major river, the new town of Savannah, despite Oglethorpe's claims, could not be considered the most convenient of ports. Ships had to travel seventeen miles from the sea to reach the town, and currents, sandbars, and mud banks often impeded the voyage. Despite these difficulties, Savannah became an important port and commercial center. Critical to this growth was the Savannah River, stretching more than three hundred miles inland. The river became the great highway for trade and travel to the "upcountry," and the tiny settlement of Savannah became the entrepôt for goods shipped from and into the interior regions of Georgia and South Carolina. The movement of goods to and from the inland community of Augusta, established on the upper Savannah River in 1735, and trade with interior Native American groups, such as the Creek and Cherokee, relied on the Savannah River. However, for most of the eighteenth century, this inland trade did little to stimulate Savannah's growth. It was not until the expansion of short-staple cotton agriculture into the interior shortly after 1810 that river trade significantly affected Savannah's economic standing. Increasing amounts of this cotton began to be shipped downriver to merchants in Savannah, who sold the crop to Northern and overseas buyers. The volume of cotton shipped from the inland through Savannah grew dramatically in the early decades of the nineteenth century, from a negligible amount before 1810, to 157,099 bales in 1820–1821, to 296,535 bales in 1844–1845, and even more in later years.[2]

The natural setting along the lower Savannah River contributed to Savannah's growth and provided ideal conditions for rice agriculture. Rice plantations were well established on the lower Savannah River by the 1750s, and this area became among the greatest rice-growing regions on the Southern coast. A final contribution to Savannah's economic well-being occurred in the late 1830s as the city became the hub of an ever-expanding network of railroads that fed even more upland cotton, as well as other products of the interior, into the city.

From the initial location near Yamacraw Bluff, settlement in Georgia soon expanded along the banks of the Savannah River and then south along the coast, areas dependent upon travel by water. This coast—fringed by vast expanses of salt marshes interlaced with tidal streams and rivers separated by sounds and inlets and dotted by numerous islands—is known collectively as the Sea Islands. Several large rivers flow out of the interior into this region—

[2] Cotton export figure from 1820–1821 is from Kayser & Co., *Commercial Directory*, 44; export figure for 1844–1845 from is from *DeBow's Review* 1 (March 1846): 277.

the Edisto, Combahee, Savannah, Ogeechee, Altamaha, Satilla, and St. Marys. Overland travel in this area was difficult, and a network of roads developed slowly, so new settlers in Georgia depended almost entirely on watercraft to import necessities and to export the few products of the early colony.

As the last of the original colonies to be established and settled, Georgia was late in developing a coasting trade. The coastwise commerce of the British colonies in America had begun in those areas where settlements were first established, in Virginia, in the area around Chesapeake Bay, and in New England. Here, coastal trading arose to interchange various products and collect commodities for export overseas. Goods imported into the principal ports were distributed to smaller coastal and river ports, towns, and plantations, mostly by various types of watercraft.[3]

The Dutch settlers of New Amsterdam in the seventeenth century seem to have been the first to establish an extensive coastwise trade in colonial America. At the time, Holland was one of the greatest trading centers in Europe, and the Dutch settlements in New Amsterdam were an expression and extension of that commercial system. New Amsterdam merchants initiated a brisk trade with the English colonies located both to the North and South. Small sailing vessels carried all manner of imported European goods out of New Amsterdam to the English settlements and returned with tobacco, grain, fish, and furs.[4] In 1664, the English seized New Amsterdam, and by the end of the seventeenth century, Dutch activity in coastwise commerce had virtually ceased.

Even before the Dutch were eliminated, the English colonies had established a well-developed system of coastal trading. Shortly after their settlement, the New England colonies were producing a surplus of grain for export, which, along with fish, was carried to markets in Virginia, Maryland, and the West Indies. In exchange, sailing vessels carried tobacco from Virginia and Maryland and sugar, dyewood, and molasses from the West Indies to the ports of New England, such as Boston, Salem, and Providence.

A truly local coasting trade also developed in New England, New York, and Pennsylvania, with many small vessels sailing between the numerous

[3] Johnson, *History of Domestic and Foreign Commerce*, 162–63.

[4] "Public Records of Colony of Connecticut," 7:583, cited in Johnson, *History of Domestic and Foreign Commerce*, 168.

towns established along the region's rivers and coasts.[5] New England's abundance of suitable timber made shipbuilding inexpensive, and local yards became involved in constructing small vessels for coastwise commerce.

Although ports developed along the coast north of Delaware, only a few were important in overseas trade. Small vessels carried merchandise meant for export from the small settlements into the principal ports. These same ships distributed the imports carried into the major ports and became involved in the interchange of locally produced goods and foodstuffs intended for local consumption. The combined advantages of cheap ships, well-located markets, and little outside competition permitted the traders of the Northern colonies to monopolize the intercolonial trade in British America.[6]

The lack of good harbors and, more importantly, the economic system of the region restricted the development of coastwise commerce in the Southern colonies. The commerce of the South was based heavily upon large-scale agriculture, and the two main crops of the early colonial period, tobacco and rice, were commonly sent directly to England, typically carried in ships dispatched to the colonies by English merchants. The greater part of this overseas trade was from individual plantations, but the system of collecting and distributing imports and exports by small vessels was undeveloped. Further, the demand for Southern products in the North was small, meaning coastwise commerce between the two regions was of lesser importance than the trade with Europe and the West Indies.[7] Few major port cities developed in the Southern colonies, so the local shipbuilding industry remained small, as did the number of traders, merchants, and shipowners. Further, much of the limited Southern coasting trade was conducted by Northern merchants using Northern-built ships. It was not until after the colonial period, when cotton agriculture became important, that a well-developed interregional coastwise commerce between the North and the South arose. Northern sailing ships carried Southern cotton to the North's many mills and carried Northern produce, imported items, and goods of local manufacture to the South.

The new colony of Georgia depended almost entirely upon receiving merchandise and supplies by ship. Charleston was the primary source of these goods during the colony's initial years, and it remained important well into

[5] "Documents relating to Colonial History of New York," 5:603, and "Public Records of Colony of Connecticut," 7:583, both cited in Johnson, *History of Domestic and Foreign Commerce*, 168.

[6] Johnson, *History of Domestic and Foreign Commerce*, 163.

[7] Ibid., 164.

the nineteenth century. Founded in 1670, "Charles Town" became the major settlement of the colony of South Carolina and quickly became the main port on the Southern Atlantic coast. South Carolina depended on England for most manufactured goods, and most of this trade passed through Charleston. Additionally, an active trade between Charleston and the West Indies soon developed.[8] Located on Charleston Harbor, the city had easy access to the Atlantic Ocean and to the inland waterway passing inside the coastal islands. This passage provided a protected route of travel for vessels from North Carolina to Florida (Figure 2.1).

For much of its early history, Savannah's maritime trade was closely linked to Charleston. Goods coming into the Georgia colony were typically funneled through Charleston, and a small fleet of vessels carrying merchandise and passengers soon sailed regularly between the two. This aspect of seagoing trade was principally a one-way trade involving the movement of manufactured items and foodstuffs from Charleston to Savannah. The Georgia colony had few exports, so little commerce moved from Savannah to Charleston. The development of a true "coasting trade" in Georgia awaited the expansion of population and the development of plantations that produced the commodities needed to support it.[9]

Early expansion of settlement along the Savannah River above the town of Savannah led to canoe and flatboat trade along the river, but not to a true coasting trade. In 1735, Oglethorpe directed that the town of Augusta be laid out on the south bank of the Savannah River opposite Fort Moore, a Carolina fortification. Located 150 miles up the Savannah, Augusta would serve as a defensive outpost; it also was in a good position to control the Indian trade. Soon, shipments of hides and furs were being carried downriver to Savannah.[10]

The early vessels used in the Savannah River trade were large canoes, commonly known as "periaguas" or "trade boats," measuring up to forty feet long and five to seven feet wide. The periagua was typically a large dugout canoe made from one or more logs, often with the sides built up with planks to increase the freeboard. Occasionally, the term "periaguas" was applied to plank-built vessels if they were long and narrow like a dugout.[11] River towns west of Savannah, such as Purysburg, established in 1731, Ebenezer in 1734,

[8] Coker, *Charleston's Maritime Heritage*, 37.
[9] Coulter, *Georgia*, 60.
[10] Rahn, *River Highway*, 4.
[11] Fleetwood, *Tidecraft*, 31; Rahn, *River Highway*, 6.

and Abercorn in 1733, received goods from downriver by trade boat and shipped the products of the interior down to Savannah. In 1748, one of the trade boats operating between Augusta and Charleston was described as "an Indian trading boat, forty-two feet long, and upwards of seven feet wide, with a cabin in her stern."[12] As many as six or eight men rowed these large vessels.

Settlers of Savannah were immediately preoccupied with trade, related both to the inland trade via the Savannah River and to overseas trade with the West Indies and Europe.[13] The local coasting trade during the early years of the colony was limited because there were few settlements and little local produce to carry. In June 1735, the Trustees of the Georgia colony established two fortified towns around the mouth of the Altamaha River. These were Darien, known for a short time as New Inverness, established near the site of an earlier fort known as Fort King George on the north bank of the Altamaha near its mouth, and Frederica, situated on the Frederica River, the portion of the inland waterway passing along the western side of St. Simons Island (see Figure 1.3). In January 1736, a group of Scottish Highlanders arrived at the mouth of the Altamaha to construct a fort and houses near the remains of old Fort King George.[14] Frederica, located about twenty miles south of Darien, had been established the previous year. The walled and moated town at Frederica soon became the main settlement on the coast south of Savannah. Oglethorpe also established small outposts at several other locations along the lower coast to defend the colony; Fort St. George at the mouth of the St. Johns River in present-day Florida, Fort William at the south end of Cumberland Island, Fort St. Andrews at the island's northern end, and another fortification on St. Simons Island known as Fort St. Simons situated on the southern end of the island.[15]

Darien and Frederica were the only outposts with significant numbers of settlers. The inhabitants of these towns were granted surrounding lands for growing crops, primarily for local consumption. Both towns depended on Savannah for most of their supplies, almost all of which were received by water from Britain, passing first through Charleston.

Both Frederica and Darien declined after the defeat of the Spanish at the Battle of Bloody Marsh on St. Simons in July 1742. The last British troops were withdrawn from Frederica in 1767, by which time the fortifications and

[12] *South Carolina Gazette,* February 1, 1748, in Rahn, *River Highway*, 6.

[13] M. L. Granger, *Savannah Harbor*, 6.

[14] Sullivan, *Early Days*, 17.

[15] Lovell, *Golden Isles*, 106–107.

many of the houses were in ruin. Frederica never regained its former position and remained a settlement of only a few families, but it retained its status as a principal boat landing on St. Simons into the second decade of the twentieth century.[16]

Georgia's original charter established a Board of Trustees of twenty-one persons to govern the new colony. The trusteeship operated for twenty-one years, after which the colony reverted to the crown. In 1752, Georgia became a royal colony governed by an appointed governor, a Royal Council of twelve men, and a Commons House of Assembly of nineteen men elected by the people.[17] Few records from the Trustee period shed light on the nature and extent of the coasting trade in the colony. These records do indicate this trade was minimal and mostly involved the movement of goods from Charleston to Savannah and from Savannah to the several towns, settlements, and forts scattered along the coast and the Savannah River. Little in the way of agricultural surpluses were produced, but a few commodities such as tar, pitch, and, especially, deerskins were carried to Savannah for export.

During these early years, the most common watercraft used to carry goods and people between Savannah and the coastal settlements was a variant of the piragua. To increase these vessels' size and carrying capacities, the log used to form the hull was often split and a plank inserted between two pieces to produce a wider hull. These small coasting vessels traveled along the inland passage behind the Sea Islands and rarely ventured into the open Atlantic. Larger sailing vessels were not commonly used because of the dangers of traveling the inland route's tortuous and often shallow waters.[18]

In 1735, Francis Moore, while waiting for passage from Savannah to St. Simons Island, described the boats used in coastal travel, noting, "These Periaguas are long flat-bottomed Boats, carrying from 20–35 tons. They have a kind of forecastle and a Cabbin; but the rest open, and no deck. They have 2 masts, which they can strike and Sails like Schooners. They row generally with 2 Oars only…."[19]

Philip Georg Fredrich von Reck, who visited Georgia in 1736, provided a drawing of one of the coastal periaguas he encountered en route to St. Simons Island. His drawing depicts a two-masted open vessel with no deck (Figure 2.2). The sails consist of a triangular mainsail and a jib set on the short

[16] Ibid., 108, 111; Vanstory, *Georgia's Land*, 110–11.

[17] Coulter, *Georgia*, 18–19, 82.

[18] Fleetwood, *Tidecraft*, 38, 41.

[19] Ibid., 38.

bowsprit. This use of simple, triangular sails with no gaffs at their heads is an adaptation of the "Bermuda" rig, which became popular on small sailing craft in the early eighteenth century.[20]

One of the first trading companies in Georgia was Robert Williams & Company in Savannah, owned in 1736 by Robert and James Williams. This firm imported merchandise and retailed it throughout the colony. By 1738, Williams & Company imported goods such as sugar, molasses, and salt directly from the West Indies and exported lumber. Another early mercantile firm in Savannah was Harris & Habersham, founded in 1744 by James Habersham and Francis Harris. They struggled to succeed, as did the few other merchants in Savannah, mainly because the colony had so little to export. Harris & Habersham were ultimately successful, partly by catering to the local market, importing only items that were in demand, and securing lucrative government contracts to supply the military outposts in Georgia.[21]

The coasting trade in Georgia grew with the expansion of settlement along the coast south of Savannah, the development of farms and plantations growing marketable produce, and the organized exploitation of other natural resources, especially timber. Agricultural development was restricted under the Trustees because of the prohibition of slavery and the system of land ownership. The original charter prohibited slavery because it was considered incompatible with the founding ideals of the colony. The land laws allowed land to be inherited only by the eldest male (*tale male*); otherwise, it reverted to the Trustees. In 1738, a number of Savannah citizens petitioned the Trustees to allow absolute ownership of land and to permit the purchase of slaves. In 1748, the prohibition of slavery was partially lifted, allowing an individual to hold ownership in four Black persons for each White indentured servant. By 1750, all prohibitions against slavery in Georgia were lifted, and the land rules were changed to allow colonists to buy and sell land as they pleased.[22]

By 1750, small amounts of silk and rice were being shipped to market through Savannah, plus products such as tar and pitch were produced and exported in small quantities. In 1751, an assembly of citizens recommended constructing a new wharf in Savannah, suggesting increased concern for commerce.[23]

[20] Ibid., 41.

[21] M. Granger, *Savannah River Plantations*, 175, Coulter, *Georgia*, 62–63; Lambert, *James Habersham*, 58–65.

[22] Coulter, *Georgia*, 66–68.

[23] Ibid., 76, 79.

In June 1752, the Trustees gave up their charter and Georgia became a royal colony. This change in government, plus the lifting of restrictions on slave and land ownership, stimulated the colony's growth. In 1752, about 350 Puritans from South Carolina moved to a large tract of land they acquired near the coast between the Ogeechee and Altamaha rivers in present-day Liberty County. Here, they developed rice and indigo plantations and established the town of Sunbury, which became an official "port of entry" in 1761 and a busy local port. A "port of entry" was the principal location within a region or customs district where officials were available to oversee the entry and exit of merchandise and people, to collect import duties, and to issue vessel documents. In 1762, fifty-six vessels were reported to have visited Sunbury.[24]

Despite growth under the Crown, Georgia depended almost entirely on Britain for manufactured goods. In 1766, Governor James Wright wrote that no manufacturing of any type was undertaken in the colony because residents found it cheaper to import these goods and spend their time at other tasks, principally "planting Rice, Corn, Pease, & a small quantity of wheat & rye, & in making pitch, Tar & Turpentine & in making Shingles & Staves, & sawing Lumber & Scantling & Boards of every kind; & in raising stocks of Cattle, Mules, Horses & Hogs."[25] Most of these commodities were for home and local consumption, but some "products of the country" were transported to Charleston or Savannah for export, revealing the development of a nascent coasting trade. This trade is further disclosed in Wright's letter when he notes, "We have built one Ship, one Snow, one Brigantine & five or six Schooners, & a number of Coasting Vessels since I have presided here."[26] These "Coasting Vessels" are not described, but they may have appeared similar to the sailing periagua depicted in von Reck's 1736 drawing.[27]

One of the early products of the Georgia colony was silk; however, its production achieved only modest success. In 1750, the Salzburgers, a group of German immigrants who settled in communities west of Savannah, produced over 1,000 pounds of silk, and by 1764 more than 15,000 pounds were produced in the colony. After 1764, silk production declined and ended entirely by the American Revolution. More successful as an early product was indigo. Indigo had been grown in South Carolina since about 1700 but did not become an important export until the 1740s when it was introduced into

[24] Ibid., 96–97.

[25] Coleman and Ready, *Colonial Records*, 221.

[26] Ibid.

[27] Fleetwood., *Tidecraft*, 49–53.

Georgia. Major Horton of Jekyll Island was growing indigo as early as 1746, and in 1753, 2,395 pounds were exported from Savannah. Indigo exports from Savannah rose to a high of 19,900 pounds in 1772, but afterward, the expansion and profitability of rice cultivation and disruptions in the indigo trade with England caused by the American Revolution resulted in the decline of Georgia's indigo production.[28]

Rice was an important export crop of Georgia during the eighteenth century, and the early growth and maturation of the local coasting trade were largely due to the expansion of rice cultivation along the coast, particularly along the lower reaches of major rivers. Rice had been grown in South Carolina since the seventeenth century, and its cultivation expanded into Georgia soon after the colony's founding. However, its widespread cultivation in Georgia awaited the change in land laws and the lifting of the ban on slavery in 1749–1750.

Rice agriculture was ideally suited to the coastal regions of the Carolinas and Georgia, in part because of the semitropical climate but more because of the interplay of large volumes of freshwater outflow with the peculiarly high tidal rise on the lower stretches of the larger rivers. These phenomena provided the periodic freshwater flooding required for rice cultivation on a large scale using the tide flow technique, which expanded into Georgia before the 1820s. During the first half of the nineteenth century, rice and Sea Island cotton were the most important cargo carried by the Georgia coasters.[29]

Before 1750, a number of rice plantations were established along the lower Savannah River, and by 1770, rice grown in the tidewater regions of the Ogeechee and Altamaha rivers became the principal money crop of the colony.[30] That year, 22,129 barrels of "clean" rice (rice that had been milled to remove the outer hull and then polished) and 7,064 bushels of "rough" rice (unmilled rice) were exported from Savannah.[31] However, much of the rice grown in Georgia was carried by sailing vessels directly to Charleston, the principal market in the region. In 1762, Governor Wright wrote that the rice, as well as the indigo and deer skin production of Georgia, "is obliged to be carried from thence to Charles Town in South Carolina, in small Craft, for

[28] Information on silk production from Coulter, *Georgia*, 59; 1746 Savannah indigo export figure from Coleman and Ready, *Colonial Records*, 55; J. F. Smith, *Slavery and Rice*, 22–23, 28; Sullivan, *Early Days*, 45.

[29] See J. F. Smith, *Slavery and Rice* and Sullivan, *All Under Bank*, 6–10, for descriptions of the tide flow technique.

[30] Sullivan, *From Beautiful Zion*, 34.

[31] *Daily Georgian*, July 12, 1828.

want of ships here,...& so appears by the Custom House Entries as the product of South Carolina and not of Georgia, which it really is."[32] Governor Wright's comments on the shipment of Georgia products to South Carolina and their omission from Georgia export records held true for the next hundred years.

During the last decade of the eighteenth century, cotton and rice became the coasters' most important cargo. Cotton cultivation began in Georgia during the earliest years of White settlement but was used mainly for home consumption. As late as 1766, only eight bags of cotton were exported from Georgia, and two years later, only three hundred pounds were shipped from the colony.[33] Beginning in the 1770s, demand for cotton was stimulated by the development of textile machinery in England and the expansion of a textile industry there. Cultivation of cotton in the coastal areas of South Carolina, Georgia, and Northeast Florida in commercial quantities began in the late 1780s with the introduction of the variety known as "Sea Island cotton." Sea Island cotton was highly desirable because its fiber is very fine and measures one to one-and-a-half inches long, compared to a length of about a half-inch for short-staple or upland cotton.

Sea Island cotton cultivation expanded rapidly along the Georgia and South Carolina coasts, where conditions were amenable for its growth and where indigo production was in decline. Almost two dozen planters were growing this cotton in Georgia by 1789, and soon it was being raised in South Carolina.[34] The first exports of Sea Island cotton from Charleston were recorded in 1789 when thirty bags, representing nine thousand pounds, were shipped. Much of this cotton was probably grown in Georgia. By 1800, this number had soared almost nine-hundred-fold to 25,157 bags, or more than eight million pounds of cotton.[35] A lack of statistics for Georgia exports makes it impossible to determine when the first Sea Island cotton was shipped from Savannah, however, it was reported that ten thousand pounds were exported to England in 1791.[36]

Sea Island cotton brought a premium price on the market, from one-and-a-half to three times that of the short-staple, upland cotton. High profits from the sale of Sea Island cotton contributed to the rapid expansion of its

[32] Coleman and Ready, *Colonial Records*, 183.
[33] Gray, *History of Agriculture*, 1:184.
[34] Rosengarten, *Tombee*, 50.
[35] Gray, *History of Agriculture*, 2:680.
[36] *Columbian Museum & Savannah Advertiser*, October 15, 1799.

cultivation. Unlike the output of upland cotton, which experienced a phenomenal increase through the first half of the nineteenth century, Sea Island cotton reached its limits of production by about 1805. The crop was geographically limited to suitable lands on the Sea Islands and adjacent mainland, and it was impossible to expand production without the exorbitant costs of making less-than-desirable coastal soils suitable for its cultivation. Sea Island cotton became the almost exclusive type grown on the coastal islands and adjacent mainland and, as a consequence, was a major commodity carried by the local coasting vessels.

The Georgia trade grew in conjunction with the expansion of rice and cotton agriculture along the coast and with the increasing economic importance of Savannah. Information on the number of vessels sailing along the Georgia coast before the American Revolution is scarce. However, Georgia was not a seafaring colony; in 1766, the colony's residents reportedly owned only thirty-six vessels with a combined tonnage of about two thousand tons.[37] This represents an average of only about fifty-five tons of burden per vessel, and most were likely much smaller than this.

Through the eighteenth century, Georgia remained dependent on imports for almost all manufactured items, as well as a considerable variety of foodstuffs. Newspapers of the period rarely published information on the activities of vessels carrying local products in the coasting trade, but they often gave information on vessels arriving from more distant ports. For example, on February 16, 1786, the Savannah newspaper *Georgia Gazette* contained the following notice:

> Imported in the sloop Spencer, Francis Holmes, master, from Baltimore in Maryland and for sale by the subscribers, Coniac Brandy, bar iron, Russia feather beds, ship bread, superfine flour, Teneriffe wines, Irish linens. Partick Crookshank and Co.[38]

A week later, the newspaper reported the arrival of the schooner *Relief* with a cargo of "superfine and common flour, buck wheat meal, ship and corn bread, barrelled pork and beef, butter and hog's lard, vinegar in barrels, saltpetered hams, bar iron, narrow axes, tin and earthen ware, riding chairs and sulkey, saddles and bridles."[39]

[37] Coulter, *Georgia*, 102.
[38] Kilbourne, *Savannah Georgia Newspaper Clippings*, 10.
[39] Ibid., 11.

The American Revolution seriously disrupted the incipient local coasting trade along the Southeastern coast, but trade revived quickly after the war. By the end of the eighteenth century, rice was abundant and reasonably priced, as was cotton, and the output of the coastal area from South Carolina to Florida was considerable. In 1796, the value of exports out of Savannah was $501,383, rising to $2,155,982 by 1800.[40] This dramatic increase during the final years of the eighteenth century is related to the expansion of Sea Island cotton cultivation along the coast and the high prices it brought. The main markets for these coastal crops were overseas; Southern Europe depended on the region's rice, England on its cotton, and the West Indies on its lumber. For a time, the region's productivity and the demands of these markets made Charleston the greatest exporting point on the American continent. Goods were gathered into Charleston, mostly aboard small, single-masted sloops.[41]

The activities of coasters in Georgia before 1800 are not well documented, but after the American Revolution, their numbers increased to carry the rapidly expanding Sea Island cotton and rice crops to market in Charleston and, to a lesser extent, Savannah. As with any segment of commerce, one measure of growth and maturation is the implementation of rules and regulations. Since the early Colonial period, regulations and fees had governed the treatment of coastal products imported into Charleston and Savannah. On March 31, 1796, the *Charleston City Gazette* reported that some of the coasting captains themselves had agreed upon specific freight rates for classes of cargo in an effort to standardize costs and stabilize their trade. The paper noted:

> The Captains of the Coasting Vessels to the Southward, from South Edisto to May River, this day assembled, have agreed on and do constitute the following prices for freight of the under-mentioned articles, to be observed by each and every one of them, viz.

| | £ | s. | d. |
|---|---|---|---|
| Rice, per hundred | | | 9 |
| Pitch, Tar, Turpentine, Beef, Pork, Flour, per bbl. | | 2 | 0 |
| Iron, per hundred | | 1 | 0 |
| Boards, Staves, per thousand | | 13 | 8 |
| Grain, Salt, per bushel | | | 6 |

[40] Kayser & Co., *Commercial Directory*, 45.
[41] Phillips, *History of Transportation*, 26, 45.

| | | |
|---|---|---|
| Tobacco, sugar, Rum, per hogshead | 10 | 0 |
| Wine or Brandy in pipes | 10 | 0 |
| Bricks, per thousand | 15 | 6 |
| Cotton, per hundred | 2 | 4 |

By the close of the eighteenth century, the local coasting trade in Georgia took the form it would maintain for the next sixty years. A fleet of small, one-masted sloops and two-masted schooners were transporting Sea Island cotton and rice from coastal plantations into Charleston and Savannah. On their outbound voyages, these craft carried various types of merchandise and foodstuffs coastal residents and merchants needed. Most of these local coasters sailed into and out of Charleston; relatively few sailed into Savannah. It would be some years before that city developed the commercial infrastructure to attract and support this local trade. However, even in the late eighteenth century, a few factors and merchants in Savannah were handling merchandise and products in the coasting trade. One of these was David Robinson, who advertised in the *Georgia Gazette* in June 1792 that he was "commencing the factorage and commission business" and would "receive tobacco, rice, indigo, corn, [and] lumber" at the wharf and store he leased from Levi Sheftall.[42] At the time, Savannah merchants advertised all manner of wares for sale, some of which were shipped aboard coasters to smaller ports and individual plantations along the coast.

By the late eighteenth century, Savannah and Charleston newspapers were printing the shipping lists critical to our understanding of this nascent trade. By 1799, the *Columbian Museum & Savannah Advertiser* regularly printed a "Marine Register" column, naming the ships arriving and departing from Savannah. In addition to large numbers of arriving and departing ships and brigs in January that year, the Marine Register listed eighteen arrivals of sloops and schooners into the port and twenty-six departures. These are the types of ships that sailed in the local coasting trade. However, most sloops and schooners arriving in January 1799 were not engaged in the local trade; they were sailing from Northern ports, such as Boston, Philadelphia, or New York, or from islands in the Caribbean, such as Jamaica, Barbados, and the Bahamas. The few vessels arriving from local ports consisted of two from the town of St. Marys on the lower Georgia coast, two from St. Augustine in what was then Spanish Florida, and two from Charleston. Departures were similar, with most vessels sailing to the West Indies or to Northern ports.

[42] Kilbourne, *Savannah Georgia Newspaper Clippings*, 402.

Other than five departures for Charleston, the only local port of destination named in the January 1799 newspaper was a single departure for the town of St. Marys.

So few vessels sailed between Savannah and local ports and plantations because so little of the coastal production was being shipped into the city. The cultivation of Sea Island cotton was established along the entire coast, but in 1800, the quantities produced were still small. Rice was grown along the South Carolina coast and the upper Georgia coast by 1800. At this time, as during later years, small "plantation" boats, including craft such as sailing sloops, rowed canoes, and poled flats, were transporting some locally grown produce into Savannah from nearby plantations on the lower Savannah River. These vessels are rarely mentioned in newspaper shipping lists, and we know little about the types or quantities of cargoes they carried. Despite some local shipments into Savannah, at the beginning of the nineteenth century, most crops produced along the Georgia coast were shipped to Charleston, not Savannah. This was soon to change, and over the next thirty years, the local coasting trade at Savannah grew tremendously as the city began to receive increasing quantities of the expanding coastal crops of rice and cotton.

## Chapter 3

# "For Any Port to the Southward": The Routes of the Coasters and the Ports in the Georgia Trade

On January 2, 1810, the Savannah newspaper *Republican & Savannah Evening Ledger* advertised the sale of the sloop *Little Poll* and stated that if the vessel was not sold within a day or so, she would "take freight for any port to the Southward." By this time, the paper's readers knew what ports these would be because the geographical area served by the coasters working out of Savannah and Charleston was well established.[1] This area, extending from lower North Carolina to about St. Augustine, Florida, encompassed what became a distinctive agricultural region during the last decade of the eighteenth century—that area where rice and Sea Island cotton were grown. For those coasters working out of Savannah, the northern limit of the area typically served was at the Combahee River north of St. Helena Island in lower South Carolina (see Figure 1.3). The Charleston coasters concentrated their activities north of the Savannah River. Still, they did sail the entire region, from southern North Carolina to northeastern Florida, because of Charleston's early dominance as the region's mercantile and financial center and the development of business relationships between the city's merchants and bankers with planters all along the South's Atlantic coast. By the 1820s, Savannah began to capture an increasing market share south of Beaufort, South Carolina. As a result, the number of coasters sailing from Charleston to locations along the Georgia and Florida coasts decreased after 1820 but did not

[1] "Working out of" Savannah or Charleston is an important concept because many coasters were neither owned nor operated by residents of either city. A large number of the coasters sailing in the local trade out of these two ports were owned and manned by men who came south during the winter to haul plantation products and then returned north during the summer. The participation of these northern men and their ships in the Georgia trade is discussed in later chapters.

disappear entirely because some of this area's planters retained business and personal relationships with Charleston merchants up to the Civil War.

## The Geographic Area of the Georgia Coasting Trade

The area served by the Georgia coasters extended from about Charleston to St. Augustine, Florida. This is known as the Sea Island region and is characterized by numerous low-lying coastal islands separated from the mainland by vast expanses of saline and brackish marsh intertwined with tidal rivers and creeks. This is an area where a fortuitous set of natural circumstances and a unique geography joined to make it suitable for cultivating two important Southern crops: rice and Sea Island cotton. These natural circumstances included a semitropical climate, abundant rainfall, a high tidal range, contiguous areas of well-drained sandy soils and low swamplands, and several large freshwater rivers whose adjacent swamps could be converted to rice cultivation. The Sea Island region includes a narrow strip of mainland that is environmentally similar to the offshore islands. This mainland strip is crossed by several large rivers that flow out of the interior. The main rivers are the Combahee, Savannah, Ogeechee, Altamaha, Satilla, St. Marys, and St. Johns. This entire region is influenced by daily tidal changes of five to seven feet, among the greatest on the Atlantic coast.

The land in the Sea Island region tends to be low and flat, with areas of slightly elevated sandy soils separated by low, poorly drained lands and swamps. These soils are not particularly fertile but are suitable for cultivating Sea Island cotton. Similarly, the low and poorly drained swamplands are not suitable for agriculture, but when drained, they were ideal for rice cultivation. By the end of the eighteenth century, rice plantations were established in former coastal swamplands while Sea Island cotton plantations flourished on the islands and the higher land along a narrow fringe of the mainland.

The region's natural conditions made travel by water a necessity, and towns in the area were commonly established to take advantage of water access. The region's plantations also depended on water to ship their crops or to receive necessary goods; many had their own boat landings that were visited by the coasting ships. The importance of plantation boat landings is often shown in advertisements of property for sale. In 1804, an advertisement for the sale of the plantation named Mariana near the Crooked River in Camden County touted "to any part of which any vessel can come, drawing ten feet of water…."[2] Those plantations without landings depended on docking facilities of nearby communities. Among these were the old colonial towns of Frederica on St. Simons Island and Sunbury on the Medway River in Liberty County. Both retained their positions as landings for the receipt and shipment of goods long after they ceased to exist as population centers.

## The Sailing Routes of the Coasters

The thousands of miles of tidal waterways in the region and the lower reaches of the larger freshwater rivers provided travel routes for the coasters. Those waterways that formed what is known as the "inland passage" were the most heavily utilized. This passage was not a single channel but consisted of many tidal rivers connected by sounds that provide a protected passage between the Sea Islands and the mainland. The inland waterway was important as a trade route from the earliest European occupation and had been used by the region's prehistoric inhabitants. With their fore-and-aft sails, the shallow draft sloops and schooners working in the coasting trade were maneuverable and well adapted to sailing the sinuous and often shallow waters of the inland waterways.

The captains sailing the inland passage had to have some knowledge of the local geography, if only to select the appropriate route. At various points along the passage, a captain could be faced with myriad choices in tidal rivers and creeks—all of which might resemble one another, but only one of which was correct. Many waterways led into the salt marshes, where they ended; others became too shallow or narrow to travel. Plantation landings were often located along small creeks, and the captain had to know where these entered a main channel and on what level of the tide they could be safely sailed. Navigation was easier for captains after portions of the inland passage were marked (by the beginning of the nineteenth century), but they still had to know the water depths and current flow at different tidal stages along the route. This knowledge was gained only with experience. Captains who did not know the inland passage could rely on qualified pilots until they learned the sailing routes. In 1818, on a voyage to Savannah, the sloop *Harmony* paid two dollars for "pilotage down from Charleston" when it sailed along the inland passage. The *Harmony* was a Northern vessel whose captain, Paul Wing, was unfamiliar with the inside route. A year earlier, Captain Wing had paid a pilot $13.56 when he entered the Savannah River for the first time. The

[2] *Columbian Museum & Savannah Advertiser*, April 14, 1804.

*Harmony* sailed into the Savannah River several more times in the next year but never paid another pilot's fee, suggesting Paul Wing had familiarized himself with the route up the river during his first entrance.[3] The local coasting captains rarely relied on the services of pilots once they became familiar with a particular waterway.

By the beginning of the nineteenth century, captains sailing in the Georgia trade knew the segments of the inland route well. In lower South Carolina, these included the Coosaw, Morgan, and Beaufort rivers behind St. Helena Island; Mackay Creek and Calibogue Sound behind Hilton Head Island, and the Cooper River and the lower reaches of the New River inside Daufuskie Island (Figure 3.1). When sailing south from Savannah, coasters typically left the Savannah River and entered the inland passage at St. Augustine Creek, just east of Savannah and well inside the mouth of the Savannah River. St. Augustine Creek led to the Wilmington River, which could be followed southward behind Whitmarsh Island to the Skidaway River behind Skidaway Island, to the mouths of the Little Ogeechee River and the Ogeechee River, which empty into Ossabaw Sound. An alternate route involved bypassing the Skidaway River, continuing along the Wilmington River into Wassaw Sound, and then passing in front of Skidaway Island along one of the small waterways through the "Romerly Marshes" into Ossabaw Sound. Continuing southward to the Altamaha River, the inland passage followed the Bear River behind Ossabaw Island to St. Catherines Sound and then the North and South Newport rivers behind St. Catherines Island, the Front River or the Mud River, Teakettle Creek and Doboy Sound behind Sapelo Island, and then the Little Mud River between Doboy Sound and Altamaha Sound at the mouth of the Altamaha River. South of the Altamaha, the inland passage followed Buttermilk Sound to the Mackay and Frederica rivers inside of St. Simons Island, to the Brunswick River, before reaching Jekyll or Joiner creeks behind Jekyll Island (Figure 3.2), across St. Andrews Sound, and then along the Cumberland River and Cumberland Sound behind Cumberland Island to the entrance of the St. Marys River. From the St. Marys River, the inland passage extended south into Florida along the Amelia River behind Amelia Island.

In addition to sailing these tidal segments of the inland passage, coasters journeyed up the lower reaches of the region's navigable freshwater rivers to reach port towns and plantation landings. To reach the docks at Savannah, ships had to sail well up the Savannah River, about seventeen miles from the

[3] For accounts of the sloop *Harmony*, 1817–1818, see manuscript titled "Paul Wing Accounts, 1805–1824."

Atlantic. From Savannah, coasters traveled even farther up the river and its northern branch near the city, the Back River, to reach the many rice plantations and several rice mills located along both rivers. The approximate head of navigation on the Savannah River for sailing coasters was Okatee Bluff, a river landing on the South Carolina side, about twenty miles above the docks at Savannah. Coasters also sailed well inland on other major rivers, such as the Big Satilla and St. Marys, both with accessible river landings more than twenty miles inland.

The greatest advantage of the inland passage was that it provided a route protected from the weather dangers existing on the open sea. However, when traveling the inland course, vessels faced a sinuous route that could be difficult to sail depending upon the winds and tide. Travel along the inland passage was often tediously slow, so, when possible, coasters sailed in the Atlantic Ocean just offshore and within sight of land, or "outside" as it was generally called. This was particularly true when they had to travel longer distances and when weather conditions allowed it. Rarely do contemporary accounts specifically state if a coaster sailed outside or along the inland passage on a particular voyage. However, before 1840, some Savannah newspapers printed the sailing times of arriving coasters, and the very short times often given indicate that ships were sailing offshore in the Atlantic. In 1820, the sailing times of five vessels traveling from Amelia Island, Florida, to Savannah ranged from two to twelve days. The straight-line distance between the two ports is about 115 miles, and the two sloops that sailed this span in just two days could only have traveled in the Atlantic Ocean on the outside route. That same year, there were ninety-one coaster arrivals at Savannah from Darien. The times involved in sailing between the two ports ranged from nine days to just eight hours. The schooner *Mario*, which made the approximately seventy-five-mile trip in eight hours, would have been cruising at about eight knots, or around nine miles per hour. This seems an unbelievable speed for a sailing coaster to maintain over such a great distance, but if true, the schooner had to have been sailing outside with a fair wind and ideal sea conditions.[4]

Although protected, ships sailing the waterways of the inland passage and the sounds and rivers of the Sea Island region faced difficulties. Channels were often sinuous and contained sandbars, mudflats, shell banks, and

[4] *Daily Georgian*, 1820. The sailing times given here are from my compilation of the daily "Marine News" entries published in the *Daily Georgian* for 1820. In all following yearly compilations of daily shipping news entries, only the year or years are cited instead of a single issue.

narrows. Given the great tidal range in this region, many streams that could be navigated at high tide were entirely dry or so low as to be difficult or impossible to travel during low tide. Sailing vessels also had to contend with the constantly changing direction and strength of the tidal flow, which continues in one direction for about six hours before reversing. During much of any six hours, a coaster could take advantage of the tide and could sail or even drift with it, a seemingly common technique for coasters traveling on the North Newport River to and from the village of Riceboro in Liberty County, as it must have been on other waterways.[5] Coasters commonly had to sail against the tide. In strong tides and unfavorable winds, ships would lie at anchor waiting for a change in the tide or a more favorable breeze. Coasters often carried or towed small boats to pull a vessel against an adverse current (Figure 3.3). James Silva wrote about a trip he made up the St. Marys River aboard the sloop *Independence* in the 1840s. He noted, "Where the trees fringe the river and shut off the breeze progress was slow and had to be assisted by a process of towing contributed by two oarsmen in a row boat." The *Independence* took seven days to travel the thirty-five miles from the town of St. Marys to the interior landing at Trader's Hill and back, a voyage that Silva described as "a rather monotonous drift over a tea-colored river."[6] After 1820, steamboats were available to tow sailing vessels up the Savannah River, and they were also used at other ports. In 1828, an advertisement in the *Daily Georgian* stated, "A Steamboat can be obtained any time this week, for the purposes of towing vessels in and out," and during summer 1838, the Georgia Steamboat Company kept the steamer *Chatham* at Savannah specifically "for towing vessels down and up the river."[7] Large ships certainly took advantage of these tow boats, but how often the sailing coasters used them is unreported.

## Navigation Aids

Captains sailing in the trade had to know local conditions, where sandbars were, and the water depths at different tides, as well as simply being able to distinguish the correct route. This knowledge was best gained through experience, but captains were also aided by charts, increasingly more available and reliable through the first half of the nineteenth century, and by navigation aids, such as buoys, channel markers, and lighthouses. All were in place at

[5] The North Newport River information is from the 1818 account of Ebenezer Kellogg reported in Martin, "New Englander's Impression," 255.

[6] Silva, *Early Reminiscences*, 20.

[7] *Daily Georgian*, May 23, 1828, and August 25, 1838.

critical locations along the Southeastern coast by the late eighteenth century, and notices were periodically published in newspapers describing the placement or movement of these aides and the routes to sail when using them.

By 1800, lighthouses had been constructed at the entrances to Charleston and Savannah, and over the next several decades, more would be established at smaller ports. The first light marking the entrance to Charleston was built on Morris Island in 1673 and used burning pitch and oakum for the light. This original structure was replaced by other lighthouses in subsequent years. A brick lighthouse constructed on the north end of Hunting Island at the entrance to St. Helena Sound marked the route to Beaufort, South Carolina.[8]

The first lighthouse at the Savannah River entrance was erected on the north end of Tybee Island in 1736, shortly after the founding of the city of Savannah. Over the years, this lighthouse was replaced or rebuilt several times. Other lights and markers were placed along the Savannah River to mark the channel up to the town. One of these lights helped river traffic get past what was known as "The Wrecks," a location below Savannah where vessels were scuttled during the American Revolution as a defense measure.[9]

A sixty-foot-tall lighthouse and a keeper's cottage were completed on Sapelo Island by 1820. The Sapelo Island lighthouse aided vessels crossing Doboy Bar on their way to the growing port of Darien and points on the lower Altamaha River. A severe hurricane in September 1824 damaged the Sapelo Island lighthouse, but it was soon repaired and put back into service.[10]

In 1810, a seventy-five-foot-tall lighthouse and a keeper's cottage were constructed on the south end of St. Simons Island to mark the entrance to St. Simons Sound and aided navigation to the port of Brunswick and the Turtle River.[11]

In 1838, a sixty-foot-tall brick lighthouse was constructed on the north end of Little Cumberland Island, marking the entrance to St. Andrews Sound. This light provided navigation aid to those vessels traveling to the Big Satilla and the Little Satilla rivers. In 1820, a lighthouse was established at the southern end of Cumberland Island, marking the entrance into the St. Marys River and the town of St. Marys. This lighthouse was dismantled in

[8] South Carolina Information Highway, "South Carolina Lighthouses."

[9] Rowlett, "Lighthouses of the United States: Georgia."

[10] Sullivan, *Early Days*, 125–28; *Darien Gazette*, April 19, 1819.

[11] Rowlett, "Lighthouses of the United States: Georgia"; Vanstory, *Georgia's Land*, 161–62.

1838 and moved south to Fernandina on Amelia Island in Florida. By 1830, a lighthouse was constructed at the entrance to the St. Johns River in Florida, and markers and beacons were erected to guide ship traffic along the river. In 1857, a lightship was stationed off Dames Point at a shoal that impeded ship traffic into the St. Johns.[12]

Lighthouses were constructed only at entrances leading into the principal ports. Range markers and buoys were placed at entrances and along rivers as additional aids to commercial vessel traffic. In 1807, the *Georgia Republican* published a request for bids by the commissioner of pilotage to construct six buoys in the Savannah River to aid navigation. The bid notice described the required buoys, noting that they were to be built of the "best pitch pine" and must remain watertight for at least twelve months. To help prospective bidders, one complete example and one half-finished example of the desired buoy were available for inspection at the wharf of Robert and John Bolton on the Savannah riverfront. Similar buoys were in place at other locations all along the Southeastern coast.[13]

## Pilots and Pilot Boats

By the start of the nineteenth century, most of the larger port towns frequented by the Georgia coasters had pilots to bring ships into harbor entrances and to safe anchorage. Pilots had been working at the principal ports of Charleston and Savannah since the earliest years of settlement.[14] Pilots, individually or as small groups, owned the vessels they used in their work. Generally, these pilot boats were two-masted schooners with well-modeled and relatively deep hulls, built for speed with little regard for cargo capacity. Although pilot boats were not built specifically as cargo carriers, they occasionally carried coastal cargo under their pilot owners. Several of these boats were sold and used as coasters after their tenure as pilot craft.

During the first half of the nineteenth century, the Savannah pilots were governed by the Board of Commissioners of Pilotage. This board set rules and regulations, established fees, and issued and revoked licenses. Despite local port regulations, reports of fraud and illegal or unacceptable activity by pilots did occur. In February 1800, Thomas Pitt, secretary for the Savannah Board of Commissioners of Pilotage, published the names of the official pilots

[12] Rowlett, "Lighthouses of the United States: Georgia"; Lighthousefriends, "Amelia Island"; *Daily Savannah Republican*, January 12, 1835.

[13] *Georgia Republican*, November 7, 1807.

[14] Fleetwood, *Tidecraft*, 43, 56–60.

for the city because "persons were acting as pilots illegally." This official list named six pilots for Savannah: James Johnson, James Scranton, Isham Clay, Ephraim Coleman, Thomas Nottage, and John Jamieson.[15]

On January 26, 1825, the *Savannah Republican* published "A Correct List of Pilots, authorized to act for this port." The list contained the names of twenty-six men serving as pilots, two of whom are identified as "colored." The increase from six to twenty-six pilots in Savannah between 1800 and 1825 reflects the great growth in vessel traffic at the town during the intervening twenty-five years. Several men named as pilots in 1825 also worked as coasting captains at one time or another. These included Ebenezer Bolles, Elijah Broughton, Silas Briggs, Noah B. Sisson, James Pitcher, and Wright White. At least two of these men, Ebenezer Bolles and Silas Briggs, were natives of Rochester, Massachusetts, and regularly sailed coasters in the Georgia trade before taking up residence in Savannah.[16]

As early as 1803, Domingo Tedoir advertised his services as a pilot to the town of Darien, and by 1820, two pilot boats were available to bring vessels across Doboy Bar and into the Darien River. One of the Darien pilots in 1820 was Thomas Veal, who sailed his "schooner PILOT-BOAT" (Figure 3.4). In 1821, possibly while still working as a pilot, Captain Veal was sailing the sloop *Mary Ann* in the coasting trade, transporting cotton from Darien to Savannah.[17]

## Ports in the Georgia Coasting Trade

The coasters sailing in the Georgia trade during the first six decades of the nineteenth century visited the principal ports of Savannah and Charleston and a number of local ports along the Georgia, South Carolina, and Florida coasts. In addition, they called at many individual plantation landings. In 1800, Savannah, the only true "city" in Georgia, was the region's social, commercial, and population hub. At that time, however, Charleston was far more important in relation to the Southern coasting trade and the activities of many captains working in Georgia. At the start of the nineteenth century, many of

[15] *Columbian Museum & Savannah Advertiser*, February 11, 1800.

[16] Kayser & Co., *Commercial Directory*, 46. The Bolles and Briggs families were mariners from the area of Rochester, Massachusetts, whose names appear commonly as captains and owners of vessels engaged in the coasting trade along the southeastern Atlantic coast. See chapter 5 for a discussion of these Rochester men.

[17] *Columbian Museum & Savannah Advertiser*, March 22, 1803; *Darien Gazette*, November 4, 1820; *Daily Georgian*, 1821.

the Georgia planters in the coastal region dealt with Charleston merchants. They shipped their crops to Charleston and depended heavily on Charleston merchants and bankers for their supplies and financial aid. Savannah had neither the marketing nor banking facilities to serve the needs of these coastal planters. By the 1820s, however, these facilities were developing in Savannah, and increasing numbers of Georgia planters (and some in lower South Carolina) transferred their accounts and business to merchants in that city and began to ship their crops there.[18]

The few other towns and villages on the coast were of little commercial significance compared to Charleston or Savannah although they did accommodate local business activities. The towns and locations visited by the coasters are recorded in the newspaper listings of vessel arrivals and departures. In both the Savannah and Charleston papers, the place of origin for an arriving coaster was often a port town, such as St. Marys, Brunswick, or Beaufort. However, papers commonly gave a more general origin for a voyage. These were typically the coastal rivers, such as "Ogeechee," "Cumbahee" (now most commonly spelled Combahee), or "Great Satilla," but often the paper printed the name of an island, such as "Sapelo," "Ossabaw," or "Fuskey" (Daufuskie). In these instances, a vessel was sailing from one or more plantation landings on these islands or along these rivers. Occasionally a specific plantation or landing is named as the place of origin, most commonly for vessels arriving in Savannah from rice mills and plantations along the lower Savannah River. For example, in 1852, the sloops *Two Marys* and *B. S. Newcomb* arrived in Savannah several times from "Pennysworth Mill" or "Pennysworth Island," as well as from "Taylor's Plant.," "Berrien Plant.," and "Habersham Plant." On these voyages, the sloops carried rough rice or "tierces" (casks) of clean rice.[19] "Pennysworth Mill" refers to the steam-powered rice mill on Pennyworth Island in the Back River just upriver of Savannah. Before about 1826, the island was known as "Cruger's Island," after its owner, Nicholas Cruger (Figure 3.5). The steam mill at Pennyworth Island began operation in the 1830s, replacing the earlier tidal mill, and became one of the largest on the Savannah River. Berrien Plantation refers to the plantations of John MacPherson Berrien located about fifteen miles above Savannah. "Habersham Plantation"

[18] House, *Planter Management*, 45. House argues that Savannah did not become the region's principal financial and economic center until after 1830; however, the significant increase in coaster traffic into Savannah from local ports by 1820 suggests that the city was becoming vital to the regional economy before 1830.

[19] *Daily Georgian*, 1852.

presumably refers to Robert Habersham's lands at Deptford and Causton's Bluff just downriver of Savannah or his Upper Rice Mill at Yamacraw, just above the city.[20]

*Charleston*

Charleston, or "Charles Town," as it was originally known, was established as the capital of the South Carolina colony in 1670. The founding of Charles Town was tied to maritime commerce, and the city became the principal port on the Southern Atlantic coast. The colony depended on overseas trade with England for most manufactured goods, and an active overseas trade with the West Indies also developed, involving the import of sugar and rum.[21] The city was located on the inland waterway, which became the most important avenue of trade on the Southern frontier, serving as the route for Indian traders, for trade with Spanish Florida, and for moving people and goods into lower South Carolina and, later, into new lands in Georgia and Florida.[22]

Charleston's rise to prominence as a maritime center began in the second decade of the eighteenth century. It continued for the next one hundred years, a period associated with the expansion of indigo and rice cultivation, the colony's major exports through the mid-eighteenth century. England needed indigo for its developing textile industry, and rice was a desirable food product in England and on the Continent. The development of Sea Island cotton agriculture along the South Carolina, Georgia, and Florida coasts in the late eighteenth century and the great expansion of cotton cultivation in the interior in the first decade of the nineteenth century created a commercial boom in Charleston.[23]

In 1807, President Thomas Jefferson's embargo on overseas exports crippled Charleston's commerce, and some believe it marked the start of the decline of the city as a port. In 1810, Charleston was the sixth largest city in the United States, with a population of 24,700; however, by 1820, the population had increased by only a hundred people, reflecting a significant slowing in growth. By the end of the War of 1812, Charleston's importance as a port was in decline. The city was still the nation's leading exporter of cotton in

[20] Leech and Wood, *Archival Research*, 168 (Pennyworth Island and Pennyworth Mill); M. Granger, *Savannah River Plantations*, 16–17, 108–10 (Berrien and Habersham plantations).

[21] Coker, *Charleston's Maritime Heritage*, x, 37.

[22] Rowland et al., *History of Beaufort County*, 5.

[23] Coker, *Charleston's Maritime Heritage*, 43; G. C. Rogers, *Charleston in the Age*, 10–11; Rowland et al., *History of Beaufort County*, 162.

1815, but by 1822, New Orleans had taken the lead in cotton exports. By 1840, Charleston was third, after New Orleans and Mobile. The nationwide economic crisis known as the Panic of 1819, related to the end of the cotton boom, damaged Charleston's economy and forced many of the city's merchants out of business. Additionally, Charleston Harbor had a relatively shallow entrance, only sixteen feet, preventing the larger ships then being used in overseas trade from calling at the city. The city also had no easy connection to the interior, where the cultivation of cotton was expanding at a rapid rate.[24]

Although Charleston would never regain its pre-1820 importance as a port, the city remained closely tied to maritime activities. Numerous merchants in the city handled coastwise and overseas trade, and banks and insurance companies supported these commercial activities. In 1820, Charleston had six banks, three insurance companies, and eight ship chandlers. There were reported to be 146 "commercial houses" in the city, including 75 individuals or firms identified as "factors" or "commission merchants."[25] Many of these merchants were intimately involved in the local coasting trade through their dealings with coastal planters, the handling of coastal produce, and ship ownership. The expansion of cotton agriculture along the coast and in the interior in the early nineteenth century was particularly important to the city's maritime commerce. By the 1820s, regularly scheduled packets were sailing between Charleston and Northern ports with cotton as their principal cargo. In 1822–1824, the coastwise shipment of cotton from Charleston averaged about 16 percent of the city's total exports; by 1859, this number had risen to 45 percent. By 1850, about 84 percent of Charleston imports and 64 percent of exports were to or from Northern ports like Philadelphia, New York, or Boston. Some ships loaded with cotton went directly to European ports from Charleston, but many went to New York, where the cotton was loaded on ships bound for Europe.[26]

Savannah and the new colony of Georgia depended heavily on goods shipped from Charleston. Charleston retained close trade connections with Savannah, even though the cities became commercial rivals in the eighteenth century. Newspapers reveal the intensity of traffic by small sailing vessels between Charleston and Savannah. During 1800, for example, two Savannah

[24] Coker, *Charleston's Maritime Heritage*, 35, 154; 171, 174; Taylor, *Transportation Revolution*, 7–9.

[25] Kayser & Co., *Commercial Directory*, 4–6.

[26] Taylor, *Transportation Revolution*, 107, 197; Coker, *Charleston's Maritime Heritage*, 176.

newspapers, the *Georgia Gazette* and the *Columbian Museum & Advertiser*, listed seventy-two arrivals in Savannah by sloops and schooners from locations between Charleston and Florida. Fifty-two, fully 72 percent of these arriving vessels, sailed from Charleston.

Before the War of 1812, coasters sailing out of Charleston dominated the local trade along the Southern coast all the way to Florida. However, as early as 1800, some of this trade was being transferred to Savannah. Although the number of coasting vessels sailing between Charleston and points south of the Savannah River declined before 1820, vessel traffic between Charleston and Savannah remained high. These ships were mainly involved in the transport of passengers, merchandise, and foodstuffs from Charleston to Savannah. By 1820, steamboats were in use in South Carolina, and in 1821, steamers began regular service between Charleston and Savannah.[27] With the entry of steamboats into the Charleston-Savannah trade, the numbers of small sailing ships traveling between the two ports declined.

### *Beaufort and Lower South Carolina*

The town of Beaufort is located on the waters of Port Royal Sound in lower South Carolina. Port Royal Sound is considered one of the best harbors on the lower Atlantic coast. Beaufort was established there in 1711 to capitalize on this great natural advantage. Beaufort did not develop into the great regional port envisioned, but it did become the most important port in the local coasting trade on the lower South Carolina coast.

Settlement in the Beaufort region began to expand in the 1730s with the spread of rice cultivation into the region. This early rice agriculture involved inland swamp cultivation that was typically practiced in South Carolina then. The more productive tide flow rice cultivation was not introduced until the mid-eighteenth century. Ideal conditions for tidal rice cultivation were found in the lower Beaufort district and along the Savannah and Back rivers, making this region one of the country's most productive rice-growing areas. With the establishment of rice agriculture, large numbers of enslaved Black workers began to be moved into the Beaufort area.[28]

Savannah was only fifty miles from Beaufort by the inland water route, so a close maritime connection developed between the two towns. Additionally, many South Carolina rice planters from the Beaufort region moved to Georgia after the prohibition on slavery was lifted in 1749, establishing

[27] Rowland et al., *History of Beaufort County*, 262.

[28] Ibid., 112–13, 119.

plantations along the Savannah River and farther south along other rivers, such as the Ogeechee and Altamaha.

Several mercantile firms in Beaufort developed close ties with merchants in Savannah, and by the 1760s the firm of Kelsal and Darling had branches in both cities. As Savannah grew as a mercantile center, merchants and planters in lower South Carolina increasingly traded there. By the end of the eighteenth century, boat captains regularly sailed between Beaufort and Savannah, and an increasing number of planters in lower South Carolina dealt with Georgia merchants.[29]

In 1805, the *Charleston Courier* noted that Beaufort had 656 White inhabitants, 944 Black inhabitants, 186 students, 120 dwellings, 13 stores, and 9 workshops. Beaufort would remain small and secluded, even though many rice and Sea Island cotton plantations were established in this region. The cotton and rice grown here were transported to Savannah and Charleston by small sloops and schooners. In 1820, the Savannah's *Daily Georgian* listed fourteen arrivals of coasters from "Beaufort." In addition, the *Daily Georgian* recorded arrivals from "St. Helena" (St. Helena Island) and "Fuscay" (Daufuskie Island), two of the Sea Islands in Beaufort County.[30] In 1821, steamers traveling between Charleston and Savannah began to call at Beaufort. The steamboats quickly captured much of the passenger traffic formerly carried by the sailing coasters, but coasters continued to transport plantation crops out of the region.

The coasters sailing out of Savannah called at other locations along the South Carolina coast in the Beaufort region. In 1827–1828, for example, the *Daily Georgian* reported twenty-nine coaster arrivals from locations in lower South Carolina, such as the Combahee River, the May River, Euhaw Creek, and Daufuskie Island. Savannah newspapers also reported numerous coaster arrivals from rice mills and plantations along the Savannah and Back rivers, some of which were located on the South Carolina side.

Okatee Bluff, located twenty miles above the city of Savannah, was at the normal head of navigation on the Savannah River for the sailing coasters and served as a landing for shipping produce from the area's plantations and for receiving the merchandise. Savannah newspapers occasionally mention the arrival of coasters from Okatee Bluff with regional produce.

[29] Ibid., 181.

[30] 1805 census data from Rowland et al., *History of Beaufort County*, 263; shipping data from the *Daily Georgian*, 1820.

*Savannah*

Savannah is the port of principal interest in this study. Established as the capital of the new colony of Georgia in 1733, it remained the main population, financial, and governmental center into the nineteenth century. In 1741, a survey of the colony reported that "besides ware-houses and huts" in Savannah, there were "at least one hundred and thirty houses," and the town was "excellently situated for trade, the navigation of the river being very secure, and ships of three hundred tons can lie within six yards of the town, and the worm does not eat them."[31] The firm of Harris and Habersham, established in 1744, is reported to have constructed the first wharf for seagoing vessels in Savannah. Others quickly followed.[32]

The legalization of slavery in 1749 stimulated the growth and expansion of agriculture, particularly of rice, and led to increases in exports. By 1770, 73 ships and 119 sloops and schooners with a total tonnage of 10,514 tons were loaded with exports at Savannah. In 1769, thirty-two merchants were advertising in the *Georgia Gazette*, more than a dozen warehouses stood along the riverfront, and several wharves existed to handle the city's increasing maritime traffic.[33]

The expansion of rice cultivation contributed significantly to the growth in exports from Savannah. In 1755, Savannah exported 2,299 barrels of rice; by 1778, it was exporting 17,783 barrels. The shift from upland, swamp rice agriculture to tide flow agriculture contributed to the increase in rice production. By the end of the eighteenth century, most of the tidal marshes lining the lower Savannah River had been converted to rice agriculture.[34]

Cotton was grown in Georgia in small quantities by the mid-eighteenth century and was first exported from Savannah in 1764 when eight bales were shipped to England. It was not until late in the eighteenth century, with the introduction of Sea Island cotton cultivation, that cotton became an important export from the city.[35]

Export figures for Savannah are intermittent or often unreliable for most of the eighteenth century and for the early years of the nineteenth. Even those that are available cannot be considered a true measure of the production of the Georgia colony and, later, the state. As has been stated previously, much

[31] "Collections of the Georgia Historical Society," 60.

[32] Coleman, *History of Georgia*, 51.

[33] "Collections of the Georgia Historical Society," 74; Coleman, *History of Georgia*, 51.

[34] Reese, *Colonial Georgia*, 130.

[35] M. Granger, *Savannah River Plantations*, xiv.

of the rice and Sea Island cotton, as well as minor commodities such as syrup, molasses, sugar, and hides, were shipped from Georgia locations directly to merchants in Charleston, becoming incorporated into export figures for that city, with a general presumption that they reflect the output of South Carolina. How much of Georgia's production bypassed Savannah and was shipped to Charleston during these early years is unknown, but it was a considerable proportion of the colony's entire output.

By 1773, the value of goods shipped out of Savannah had risen to $193,395; a twenty-two-fold increase over the $8,897 worth of goods shipped in 1750 near the end of Trustee rule.[36] These goods were carried to Charleston and to ports in the Northeast, Europe, and West Indies. The West Indies, in particular, became a market for Georgia tar, pitch, lumber, pork, and cattle. By the 1770s, there was a considerable trade between the two places. In 1773–1774, eighty-eight West Indian vessels reportedly sailed into Savannah, a particularly important trade as it brought much needed gold and silver currency to Georgia merchants.[37]

A British fleet arrived off the mouth of the Savannah River in January 1776, abruptly ending Savannah's maritime trade. In late 1778, British forces occupied the city and most of Georgia, and they remained in control of Savannah until July 1782, near the end of the American Revolution. The city's maritime trade revived quickly after the war. In 1786, the value of all Savannah exports was $321,377, less than it had been in 1773. In 1791, when President George Washington visited Georgia, he reported that agricultural production had been restored, surpassing the prewar levels. He also commented on the large number of vessels in Savannah Harbor. Two principal reasons for the revival of maritime trade in Savannah were the continued expansion of rice cultivation and the introduction of Sea Island cotton to the region. Although these two crops were the most valuable shipped through Savannah, other commodities such as tobacco, lumber, and indigo were exported.[38]

Few good harbors existed along the Southeastern coast; the entrances to major rivers and the rivers themselves were often too shallow to permit the passage of large vessels. The main entrance to the Savannah River, located between Hilton Head and Tybee islands, had a depth of almost twenty feet at low water and was considered one of the best along the Southern coast.

[36] Kayser & Co., *Commercial Directory*, 45.

[37] Coulter, *Georgia*, 101–102.

[38] Mid-Atlantic Technology, *Archaeological Data Recovery*, 9; Harden, *History of Savannah*, 471.

Despite the advantages of a deep entrance, the town of Savannah was located about seventeen miles from the sea, and the route vessels had to travel up the river was described as "very narrow and very tortuous."[39] A short distance below the city, the Savannah River divided into the Back River, which flowed on the South Carolina or north side of Hutchinson and Fig islands, and the Front River (or the Savannah River proper), which flowed along the south side, fronting the city (Figure 3.6).[40]

Once inside the sea bar, vessels faced several obstructions in the form of sand and mud bars on the route to Savannah. Among these obstacles was "The Wrecks," where the water was only about eight feet deep. "The Wrecks" consisted of mud- and sand-bars that developed around the hulls of several vessels scuttled during the Revolutionary War to block the channel from British attack. Vessels drawing less than about fourteen feet of water could generally travel the complete distance to Savannah at high tide, but ships with greater drafts often had to stop at intermediate points on the river where they would discharge or receive their cargo aboard lighters.[41]

In addition to lights, buoys, and markers, other efforts were made to improve navigation of the Savannah River. Before about 1815, local governments undertook these attempts. Lotteries were authorized as one technique to raise money for improvements along the Savannah. These efforts produced meager results, so beginning in 1799, the state chartered several navigation companies to undertake the work. Most of this activity was conducted along inland portions of rivers such as the Savannah, the Ogeechee, and the Oconee. Ultimately, most of these improvements were temporary. Through the Army Corps of Engineers, the federal government initiated improvement work on the lower Savannah River in 1826 when Congress appropriated $50,000. This effort included dredging and building dams to close the connections between the Back River and Front River to increase river flow to scour and deepen the channel in the area of "The Wrecks." This work continued over the next decade with little success. It was not until 1855 that Congress again authorized funding to clear the obstructions at The Wrecks. The improvement work began in 1856 and was completed in 1860. However, these improvements were

[39] Hardee, *Reminiscences*, 122.

[40] Haunton, "Savannah in the 1850s," 81–82.

[41] Barber and Gann, *Savannah District*, 42; Haunton, "Savannah in the 1850s," 82–83.

undone during the Civil War when the river was again obstructed with vessels scuttled by Confederate authorities.[42]

At the end of the eighteenth century, Savannah was still a small city, not much more than a town, but it was the political, social, and economic center of Georgia. In 1798, the *Georgia Gazette* noted the town contained 618 dwellings housing 6,227 inhabitants. Over half of this number, 3,216, were enslaved African Americans; 238 were "free persons of color."[43] The city depended heavily on sea trade, and wharves and docks lined the bustling waterfront along the river. A small but active group of merchants and mercantile houses specialized in the two important crops, rice and Sea Island cotton. These companies typically advertised themselves in newspapers as "Factors" or "Commission Merchants." In the *Georgia Gazette* in 1800, William Belcher advertised that he was continuing "the Vendue, Factorage, and Commission Business" at Dr. George Jones's wharf; Ebenezer Stark noted he was going into the factorage and commission business on Alexander Watts wharf, and merchants Joseph Machin and John Macintosh advertised to readers that they had "formed a co-partnership for the Factorage and Commission Business" to be known as Machin & Macintosh (Figure 3.7). The firm of Meins & Mackay advertised that they had "New Rice—for sale in whole and half barrels" while Johnston, Robertson & Co. stated that "Cash will be given for Best Quality Sea Island cotton." John J. Gray advertised his business as a "Vendue and Commission Store," noting that he had been appointed "Vendue Master" in the city and would "buy and sell any kinds of Country produce, or transact any business in the commission line." The term "vendue" referred to public sale or auction, and the city had apparently authorized John Gray to conduct public property sales. City newspapers of the period include numerous announcements of public sales, and occasionally listed among the articles of property sold were coasting sloops or schooners.[44]

In addition to the occasional visit by a coasting vessel from a local port, on any given day in 1800, ships of all sizes from Northeastern ports such as New York or Boston or overseas harbors such as London; Liverpool; or Kingston, Jamaica, might be unloading merchandise at Savannah. These imported

[42] Barber and Gann, *Savannah District*, 24–27, 42, 44–45; M. L. Granger, *Savannah Harbor*, 32.

[43] This population figure was published in the *Georgia Gazette*, February 6, 1800, but Haunton reports that Savannah's population was 5,166 in 1800, less than two years later ("Savannah in the 1850s," 13).

[44] Advertisements from the *Georgia Gazette*, January 2, 1800, October 2, 1800, September 18, 1800, and February 26, 1801.

goods consisted of the great range of manufactured items and foodstuffs not produced locally, ranging from clothing to ceramics and from nails to cognac. A variety of Savannah firms received these imported goods, and many were also involved in the purchase and export of local products. Newspaper advertisements identify these companies and provide insights into the nature of their business and the range of imported commodities available to Savannah residents. In the first month of 1800, David Sandige, Hamilton & Monroe, Dickson & Johnstone, John Miller, Alexander Watt & Co., B and F. Metcalfe, and others all advertised that they had merchandise for sale received by ships directly from London, Liverpool, and Glasgow. The mercantile firm of McLeod & Miller took delivery of a variety of merchandise from Charleston aboard the schooner *Cotton Planter*, and Taylor, Miller & Co. was receiving brandy, ship chandlery, and linseed oil from New York, and sugar and salt from the island of St. Bartholomew. This firm also advertised that it wanted to purchase Sea Island cotton.[45]

The mercantile system in Savannah began to expand after 1800, and a banking infrastructure slowly developed. In January 1808, a notice appeared in the *Public Intelligencer* reporting the formation and incorporation of the "Planter's Bank," the first bank in the city.[46] That year, thirty-one different sloops and schooners made sixty arrivals into Savannah from local ports between Beaufort and St. Augustine, most carrying rice and Sea Island cotton. The number of local coaster arrivals into Savannah had tripled since 1800, in part a reflection of the growing number of factors and commission merchants in Savannah who handled coastal products. A map of Savannah dated 1813 shows the extensive waterfront development at this early date. Immediately in front of the town were thirty-four numbered wharf lots. There were additional lots east of the town and west, along the area known as "Yamacraw" (Figure 3.8). Standing in the center of the Savannah riverfront was the "Exchange," a five-story commercial building that housed the offices of city government. Although the Exchange was the most prominent building on the riverfront, it was not considered an inspiring structure, variously described as a "tasteless and illy constructed manufactory," as "uncouth," and as a "barn looking edifice with a clumsy nondescript sort of watch tower rising from the middle."[47]

[45] *Georgia Gazette*, January 2, 9, and 23, 1800.

[46] *Public Intelligencer*, January 22, 1808.

[47] Shipping information from the *Republican & Savannah Evening Ledger*, 1808. Haunton describes the Exchange building in "Savannah in the 1850s" (24).

Although Savannah's commercial facilities were expanding in the first decade of the nineteenth century, they lagged far behind those in Charleston. In 1811, a Scotsman, J. B. Dunlop, visited Charleston and Savannah. He wrote of Savannah, "With regard to trade Savannah is entirely in her infancy. There are few or no speculative merchants in it and those who term themselves such are merely Agents employed by the people of Charleston or the merchants of Liverpool & Glasgow to buy up the Cottons which the State annually produces."[48]

The War of 1812 disrupted Savannah's overseas maritime commerce, and British naval forces seriously hampered the local coasting trade. However, Savannah's economy recovered quickly after 1815, and rice and Sea Island cotton soon regained their former importance as exports. By 1820, productive rice plantations were established along the lower reaches of most of the region's major rivers. The rice fields along the lower Savannah River were so extensive that they began to encroach on the city of Savannah, leading to a fear of disease from water-laden fields. This concern led to an ordinance stating that no rice fields using tide flow agriculture could be within two miles of the city limits.[49]

In 1820, the city's population was 7,523 (3,866 Whites, 3,075 enslaved, 582 free Blacks), representing a 21 percent increase in the total population in the twenty-two years since 1798. Most of this increase was in the White (a 39 percent increase) and free Black populations (an increase of 140 percent). The enslaved population of the city decreased slightly during these two decades. This changing makeup in the city's racial population is one reflection of increasing urbanization, including the expansion of the merchant and other business classes. In 1819–1820, Savannah exported 11,895 bales of Sea Island cotton, 134,528 bales of upland cotton, and 14,918 tierces of rice.[50]

In January 1820 "occurred the largest fire which ever ravaged the city." The fire destroyed 463 buildings with losses estimated at four million dollars. Many of the commercial buildings along the Savannah waterfront were damaged or destroyed. The 1820 blaze was just one of several fires in Savannah that seriously damaged residential and commercial establishments.[51]

[48] Mohl, "Scotsman Visits Georgia," 261.

[49] Mid-Atlantic Technology, *Archaeological Data Recovery*, 9.

[50] "The City of Savannah," *DeBow's Review*, July 1853, 15:107 (population); Kayser & Co., *Commercial Directory*, 12, 44 (export figures).

[51] White, *Statistics*, 170–71; Coulter, "Great Savannah Fire," 1–2.

Despite its growth, Savannah remained one-third the size of its principal economic rival, Charleston, which had a population of 24,780 in 1820. During the crop year of 1819–1820, Charleston exported 21,396 bales of Sea Island cotton, almost twice as many as Savannah. However, Charleston exported only 124,604 bales of upland cotton that year, less than Savannah did. This reflects Savannah's receipt of so much short-staple cotton from the interior, mainly along the Savannah River.[52]

The fire of 1820 and a serious yellow fever epidemic that year slowed Savannah's population growth, but its commercial position continued to expand. By 1823, Savannah had three banks, and seventy-seven individuals and firms operated in the city as "commercial houses." These commercial houses included nine identified as "Factors and Commission Merchants," fifty-one identified as "Commission Merchants," one identified as a "Factor," eight as "Merchants," and four as "Auctioneers & Merchants."[53]

The number of local coasters sailing into Savannah increased slowly during the first decade of the nineteenth century. It was after the War of 1812 and, particularly, during the 1820s that local coaster traffic at the city grew dramatically. In 1820, the *Daily Georgian* reported 216 arrivals of sloops and schooners from local ports. This number increased to 302 in 1827–1828. Coastal plantations were producing greater quantities of rice and Sea Island cotton; increasing amounts of upland cotton were being shipped downriver to coastal ports, especially Darien, where it was transferred to small ships for transport to Savannah or elsewhere. In addition, as dock facilities expanded and warehouses, counting rooms, commission merchants, and factors increased, more and more coastal planters began to ship their produce to Savannah instead of Charleston. By 1825, Savannah exported 137,695 bags of cotton (representing 49,570,200 pounds) and 7,231 tierces of rice.[54]

In 1830, the population of Savannah was 7,776, only slightly more than in 1820. Although the population of the city showed little increase, the activity of the sailing coasters reached its peak about this time. The 1830 *Daily Savannah Republican* reported 479 arrivals of sailing coasters from local ports, more than double the number of arrivals in 1820 and, as it turned out, the largest number of arrivals to occur in any year between 1800 and 1861. The

[52] Kayser & Co., *Commercial Directory*, 12, 44 (export figures). As noted elsewhere, some considerable but unknown amount of the Sea Island cotton exported from Charleston came from Georgia plantations whose owners dealt with Charleston factors and shipped their crops to that city.

[53] Kayser & Co., *Commercial Directory*, 39.

[54] White, *Statistics*, 159.

shipping lists in the *Daily Savannah Republican* reveal that the fifty-nine sloops and twenty-four schooners sailing in the local trade in 1830 carried at least 9,614,565 pounds of rice, 2,456,745 pounds of Sea Island cotton, and 13,705,815 pounds of upland cotton into Savannah. Occasionally, the newspaper did not name or quantify the cargo aboard a vessel, so these amounts are certainly understated. Even so, they demonstrate the tremendous quantity of local produce moved by the coasters. The number of sailing vessels working in the Georgia trade decreased after 1830, as they were supplanted by steamboats that transported increasing quantities of coastal cargoes into Savannah.[55]

The three decades from 1830 to 1861 were truly a "Golden Age" for Savannah and the coastal region. During these years, the economic, labor, and services infrastructure that had been developing in Savannah since 1800 reached maturity. These decades saw unprecedented productions of rice and cotton that brought great wealth to many coastal planters and city merchants. In 1838, the population of Savannah was 12,758, and the following year, the city exported 199,176 bags of cotton (71,703,360 pounds) and 21,332 tierces of rice. By 1848, the amounts of cotton and rice shipped from the city had both increased by close to 50 percent to 104,590,190 pounds of cotton and 30,136 tierces of rice. Savannah, by this time, was the greatest exporter of lumber on the South Atlantic coast, with 16,449,558 feet of lumber shipped from the city in 1848.[56]

Savannah had five banks by 1849, and a new customhouse, to be completed in 1852, was under construction at the corner of Bull and Bay streets. In 1849, the city boasted several rice mills, sawmills, and cotton presses to handle the huge amounts of rice, cotton, and lumber moving through the port. Among these facilities were the Upper Steam Rice Mill of Blake & Habersham, the Savannah Steam Rice Mill, the Savannah Steam Sawmill, the Oglethorpe Steam Sawmill, the New Eagle Steam Sawmill, the Vale Royal Steam Saw and Planing Mill, and the Steamboat Company of Georgia cotton press. The city also had two iron foundries, several insurance companies, packet lines, and steamboat companies. These steamboat companies included the Daily Unites States Mail Steam Packet Line, traveling between Savannah

[55] "The City of Savannah," *DeBow's Review*, July 1853, 15:107. The 1830 US Census gives the population as 7,303, and a city census puts it at 7,173, suggesting that the city's population declined slightly between 1820 and 1830 (see Haunton, "Savannah in the 1850s," Table 1 and 4n13). Shipping information on the vessels and cargoes is derived from the "Marine List" published in the *Daily Savannah Republican* for 1830.

[56] Eisterhold, "Savannah: Lumber Center," 526; White, *Statistics*, 159.

and Charleston; the Semi-weekly United States Mail Steam-packet Line, traveling between Savannah and Palatka, Florida; the Semi-weekly Steam-packet Line; the Steamboat Company of Georgia; and the Iron Steamboat Company of Georgia, all of which plied the Savannah River between Savannah and Augusta. The Iron Steamboat Company of Georgia maintained a boatyard and engine shop in Savannah, employing one hundred people. In addition to these organized companies, five or six steamers worked independently.[57]

By the 1850s, Savannah's riverfront was fully developed. A map from 1856 shows more than forty lots along the Savannah River in front of the city and several across the river on Hutchinson's Island (Figure 3.9).[58] These wharf lots were owned and occupied by cotton presses, foundries, shipbuilders, and rice mills, as well as by the city's major mercantile firms.

Savannah experienced great growth during the decade from 1850 to 1860. Five new banks were organized, and insurance companies increased from fourteen to fifty-nine. By 1860, there were about three dozen "industrial" firms in the city with a yearly production of $1,907,367. These included carriage and wagon manufacturers; sash, door, and blind manufacturers; flour and grist mills; cooperage firms; steam-powered cotton presses; and shipyards. During the decade, the city's population grew from 15,312 to 22,295. Exports of rice rose from 30,748 casks in 1853–1854 to 35,588 in 1859–1860. Shipments of Sea Island cotton out of Savannah rose from 11,642 bales in 1850–1851 to 24,941 in 1859–1860, and exports of upland cotton increased from 305,792 bales to 522,096 in the same period.[59]

The economic growth in Savannah during the 1850s was not expressed in the number of coasters sailing in the local trade. By this time, steamers were carrying a considerable proportion of the coastal cargo once hauled only aboard sailing vessels, and railroads were now transporting most of the cotton from the interior, cotton that had once been shipped down rivers to coastal ports and loaded on coasters. In 1850, twelve sloops and twenty schooners made 225 voyages into Savannah with local products. Ten years later, in 1860, eight sloops and twenty-five schooners sailed in the trade out of Savannah,

[57] Population figures from *Daily Georgian*, September 5, 1838; White, *Statistics*, 156–59.

[58] Cooper, "Map of the City of Savannah."

[59] Haunton, "Savannah in the 1850s," 121–22, 130, and Table 1 (population); "Exports of Rice and Lumber from Savannah," *DeBow's Review*, October 1855, 19:461; "Commerce of Savannah, 1860," *DeBow's Review*, November 1860, 29:669–70; Donnell, *Chronological and Statistical*, 385, 492 (upland and Sea Island cotton export data).

just one more than in 1850, but these ships together made only 108 trips into the city, about half the number made ten years earlier. Newspaper shipping lists indicate the sailing coasters carried 12,317,850 pounds of rough rice, 720,360 pounds of Sea Island cotton, and 33,120 pounds of upland cotton into Savannah in 1860. Relying on 1860 commodity prices, this represented approximately $217,000 of rice and $342,212 of cotton. The value of rice carried was almost twice that transported by the coasters into the city in 1830, but the value of the cotton was less than one-fifth the value of the over sixteen million pounds carried by coasters into Savannah in 1830. These figures establish that steamboats had taken much of the coastal cotton trade from the sailing coasters, but very few steamers carried rough rice, leaving that cargo to the deep-hulled sailing vessels.[60]

### *Coasting Ports in the Vicinity of Savannah*

In addition to the city of Savannah, the sailing coasters called at many of the plantations and rice mills located on both banks of the lower Savannah and Back rivers, as well as those on the river islands, including Hutchinson, Argyle, Onslow, and Isla. A map titled *Chart of the Savannah River*, produced by John McKinnon in 1825, shows the locations of the major plantations and rice mills along the lower river. These plantations included Deptford Hill, Brewton Hill, Hermitage, Whitehall, Coleraine, Drakies, and Mulberry Grove on the Georgia side of the river and the Screven, Huger, Rutlege, Ancrum, and Taylor properties on the South Carolina side. On the river islands were plantations owned by Thomas Spalding, Thomas Young, Thomas Gibbons, John Potter, John P. Williamson, and Hugh Rose.[61]

Rarely are these plantations and mills named in newspaper shipping lists before the 1820s. In the case of the mills, most did not exist before that time. In regard to the river plantations, the newspapers apparently made no effort to record coasters coming from these nearby locations during these early years. In 1820, the *Daily Georgian* did not list a single coaster arriving from a Savannah River plantation, although many must have. By 1825, the *Savannah*

[60] Shipping data for 1850 are from the “Ship News” column of the *Daily Georgian* for that year and similar data for 1860 come from the “Shipping Column” of the *Daily Morning News*. The 1850 and 1860 newspapers provide quantified information on cargoes carried for most, but not all, arrivals, such that the quantities of rice and cotton given here represent minimal amounts. The estimates of value are based on average rice, Sea Island cotton, and upland cotton prices for 1850 and 1860 given in Gray, *History of Agriculture*, 2:1027–1031.

[61] McKinnon, “Chart of Savannah River, 1825.”

*Republican* reported coasters arriving from three different rice mills on the Savannah River. These were Ancrum's Mill, located at Laurel Hill Plantation on the South Carolina side of the river; Paragon Mill, located at Drakies Plantation in Georgia; and Clifton's Mill, at Clifton Plantation, also in Georgia.[62] After 1825, newspapers regularly named Savannah River plantations and rice mills in newspaper shipping lists as ports of origin for coasters.

Occasionally, other locations in the vicinity of Savannah appear as ports of origin or destinations of coasters. Skidaway Island, Wilmington Island, and the Isle of Wight, all located east of the city, appear infrequently in shipping lists. By the early nineteenth century, roads were built to all these islands, after which planters commonly shipped their crops overland to the city.

### *Hardwick and the Ogeechee River*

The Ogeechee River, the first large river south of Savannah, forms the boundary between present-day Chatham and Bryan counties. Along its lower fifteen miles or so, the Ogeechee is a sluggish stream influenced by the tides, and beginning in the mid-eighteenth century, this lower stretch of the river, known as Bryan Neck or Ogeechee Neck, became a center of rice cultivation. Not far above its mouth, the Ogeechee forms a large, hairpin bend known as Seven Mile Bend. In 1754, the leading planters of Ogeechee Neck petitioned Georgia authorities to establish a town at Seven Mile Bend to serve as a port for the region. The port was to be known as Georgetown. Shortly thereafter, surveyor Henry Yonge laid out lots for the new town, renamed Hardwicke by Royal Governor John Reynolds (Figure 3.10). Governor Reynolds was so impressed with the location of Hardwicke that he proposed that the colony's capital be moved there from Savannah.[63]

Hardwicke never became the thriving port envisioned by Reynolds. It grew into only a small village that served as a minor commercial center and landing for the lower Ogeechee region. In 1793, it was named the "port of entry" for the new customs Collection District of Hardwick, a designation based on the hope that the town would develop into a true commercial center, influenced by the expansion of rice cultivation along the lower Ogeechee. The state legislature designated Hardwick as the location for a public warehouse and as a point for the inspection and shipment of tobacco. None of these public buildings were ever erected, tobacco never became an important local crop, and in 1797, the county seat was moved from Hardwick to the

[62] M. Granger, *Savannah River Plantations*, 156–57, 208.
[63] Sullivan, *From Beautiful Zion*, vii, 38–39.

community of "Cross-Roads" located upriver near the bridge crossing the Ogeechee.[64]

Although few people lived at Hardwick, it had some sort of landing facilities because a few coasters sailed from there into Savannah throughout the first decade of the nineteenth century. By 1813, coasters begin to arrive in Savannah from "the Great Ogeechee," which might refer to Hardwick but more likely refers to landings at rice plantations along the lower Ogeechee River. The mention of Hardwick as a port of origin or destination for local coasters ends around the War of 1812. The Ogeechee River, however, became one of the most commonly named origins of coasters arriving after 1812. Coasters also occasionally arrived in Savannah from the "Little Ogeechee," a smaller stream just north of the Ogeechee River proper. This distinction between this smaller river and the Ogeechee River itself was not always maintained, and many arrivals were identified as coming only from "Ogeechee," which might reference either river. In 1830, the *Daily Georgian* reported that six different sloops made thirty-two arrivals in Savannah from the Ogeechee River. Each of these vessels carried rough rice as cargo on every voyage. A few vessels also carried clean rice and Sea Island cotton. Rough rice was the primary crop carried from the Ogeechee River by 1830 and reflects the full development of tide flow rice agriculture along the lower river.

The sailings of the coasters show that the Ogeechee River continued as a center of rice production up to the Civil War. In 1860, the *Daily Georgian* reported coasters made twenty-nine arrivals into Savannah from the Ogeechee River, and every vessel carried rough rice on every voyage. In that year, the *Daily Georgian* listed more coaster arrivals from the Ogeechee River than from any other local port south of Charleston.

### *Sunbury*

The town of Sunbury is located on a prominent bluff on the south side of the Medway River in present-day Liberty County, Georgia. It was established on land deeded by Captain Mark Carr in 1758 to a board of trustees composed of prominent citizens and planters (Figure 3.11). Many of the early residents of Sunbury were from the nearby settlement of Midway that had been established by a group of Puritan Congregationalists in 1752. These individuals established homes in Sunbury because of its location along the

[64] "Georgia Customs Districts," 159–64, RG 36, Bureau of Customs, NARA; Jones, *Dead Towns*, 227–28.

navigable Medway River, a location considered more "salubrious" and disease-free than interior areas.[65]

Sunbury prospered as a regional commercial center and as a seaport. The Medway River empties into St. Catherines Sound and was considered ideal for navigation. In 1773, Governor James Wright noted that "fifteen feet of water may be carried up to the town distant twelve miles from the sea."[66] Wharves were soon built at Sunbury, and the town became an official "port of entry" in 1761. Soon afterward, it was reported to have had eighty dwellings and several stores. The town was visited by vessels from ports around the Atlantic rim, but its principal trade was with the West Indies and the Northern colonies.[67]

During the American Revolution, the region around Sunbury known as St. John's Parish was a center of patriotic activity, so it was incorporated into the renamed "Liberty County" after independence. In 1778, the British captured and occupied the coast of Georgia. By 1783, they had abandoned the Sunbury area. The region slowly revived as plantations were reestablished, but Sunbury never returned to its former importance as a port and declined in the last decade of the eighteenth century. Once shipped through Sunbury, regional crops were sent to Riceboro (or "Riceborough"), about eighteen miles up the nearby North Newport River, or transported overland to Darien by the new Riceboro bridge. In 1797, the county seat was shifted from Sunbury to Riceboro, and by 1817, Sunbury contained fewer than twenty households. The once-numerous wharves along the riverfront had all but disappeared. In the 1840s, the town was described as "a deserted village, inhabited by not more than six or eight families."[68]

Soon after Sunbury's founding, an active coasting trade developed with Savannah and other regional ports, such as Beaufort and Charleston. This trade decreased as the town declined, but coasters continued to sail between Savannah and Sunbury after 1800, principally transporting the increasing amounts of Sea Island cotton grown in Liberty County. In the first three decades of the nineteenth century, ten or so different vessels were making up to fifteen voyages each year between the two towns. After 1830, trade between Sunbury and Savannah began to decline, and by the 1850s, it was common

[65] Jones, *Dead Towns*, 149–54.

[66] Ibid., 157.

[67] Coulter, *Georgia*, 96–97; Jones, *Dead Towns*, 145, 154–58, 173; Vanstory, *Georgia's Land*, 29–31; White, *Statistics*, 372.

[68] Jones, *Dead Towns*; Martin, "New Englander's Impression," 255; Vanstory, *Georgia's Land*, 36; White, *Statistics*, 372.

for just one, or possibly two, coasters, together making only four or five voyages a year, to handle all trade between the two ports.

*Riceboro*

The town of Riceboro (commonly spelled "Riceborough" in early records) is in Liberty County on the south side of the North Newport River, about eighteen miles from the river's entrance at St. Catherines Sound. The town was established at the "North New Port Bridge," where the Darien-to-Savannah road crossed the river (see Figure 3.10). In 1797, Matthew McAllister donated property at "the Bridge" to construct a courthouse and jail because of his desire to "promote the growth of the town of Riceborough and benefit the inhabitants thereof." That year, the county seat was moved to Riceboro from nearby Sunbury. Riceboro was more centrally located near the bridge where the stage road provided transportation to the North and South, and the North Newport River provided water access to Savannah and other ports.[69]

At the head of navigation for coasters, Riceboro was of minor importance as a port until after 1810. It then developed into an active regional shipping center and was visited frequently by vessels sailing in the local trade. Ebenezer Kellogg, a native of Massachusetts, visited Riceboro in January 1818 and commented on the town's shipping. He wrote that the North Newport River was "often not more than twenty or thirty yards wide, [and] sloops enter and drift up with the tide. The channel is too crooked and narrow to sail up." Several coasting sloops were at Riceboro when Kellogg visited, and he noted these ships "come yearly from New England, and spend the winter in carrying produce from different parts of the coast to Savannah...."[70]

In 1837, Adiel Sherwood reported that the Riceboro settlement had about thirty dwelling houses, and stores occupied by "30 whites and several blacks." He also noted, "Sloops come up to the town, 20 miles from the ocean by the river." In the late 1840s, Riceboro was described as "the principal shipping port of the county" with "three or four stores" and a population that did "not exceed 25 whites and as many blacks."[71]

Although Riceboro became important for the local coasters, travel to the town was inconvenient for sailing vessels. Access was by the North Newport River, a sinuous tidal stream that narrows dramatically below the Riceboro

[69] Jones, *Dead Towns*, 216.
[70] Martin, "New Englander's Impression," 255.
[71] Sherwood, *Gazetteer*, 218; White, *Statistics*, 372.

landing. The straight-line distance from the entrance of the North Newport River in St. Catherines Sound to Riceboro is about eighteen miles, but the water distance was much greater because of the numerous bends and twists in the river. As Ebenezer Kellogg observed in 1818, coasters traveling to Riceboro often relied on the tide to carry them up or down the river.

One of the earliest mentions of Riceboro as a port of call for coasters out of Savannah appears in November 1808, when the sloop *Polly & Betsey* departed the city, bound for "Riceborough." After that date, the town became more frequently mentioned as a port of call for coasters, and advertisements for vessels sailing to Riceboro began to appear regularly in Savannah newspapers. After the War of 1812, Riceboro typically ranked from second to fourth as a local port in terms of the volume of coaster traffic with Savannah. Generally, only Darien, St. Marys, and the "Ogeechee River" had more trade. A few coasting ships, from two to six or seven, made as few as ten to as many twenty-five voyages from Riceboro into Savannah each year through the 1850s. Liberty County became a center for Sea Island cotton cultivation, and much of it was shipped through Riceboro. In 1825, the *Daily Georgian* listed twenty-four arrivals of coasters into Savannah from Riceboro, second only to Darien as a port of origin for coasters that year.[72]

In addition to Sea Island cotton, the coasters sailing from Riceboro often carried rough and milled rice, corn, sugar, syrup, hides, wood, peas, and "groundnuts," or peanuts. Except for cotton, these commodities formed minor cargoes in the years before about 1850. Liberty County lacked the large freshwater rivers that provided an environment for tide flow rice cultivation, and it lagged behind other Georgia counties in annual rice production. However, in the late 1840s, coasters began carrying increasing amounts of rough rice from the town. In 1848, rough rice was carried on eight of the fifteen trips from Riceboro, increasing to twenty-four of the thirty-two voyages in 1850. In 1859, coasters transported 25,650 bushels of rough rice and 873 bales of Sea Island cotton from Riceboro to Savannah. This rough rice represented about 692,500 pounds of clean rice, or somewhat less than 30 percent of the entire production reported for Liberty County.[73]

[72] The 1808 account of the *Polly & Betsey* is from the *Republican & Savannah Evening Ledger*, November 3, 1808.

[73] Riceboro cargo information from the *Daily Georgian* 1820, 1827–1828, 1841–1854, and the *Daily Morning News*, 1858–1859. Stewart, *What Nature Suffers*, 116–17, 164, among others, points out the importance of Sea Island cotton agriculture in Liberty County to the neglect of rice. A figure of twenty-seven pounds of clean rice per bushel of

*St. Catherines and Sapelo Islands*

The two principal Sea Islands lying off the coast of Bryan, Liberty, and McIntosh counties are St. Catherines and Sapelo islands. Both are occasionally mentioned as ports of origin of coasters sailing into Savannah. St. Catherines Island had been the location of the Santa Catalina Mission, one of the most important Spanish missions established along the Georgia coast in the sixteenth century, in the region known as Guale. In the 1760s, the island became the property of Button Gwinnett, one of Georgia's signers of the Declaration of Independence. By the 1820s, St. Catherines was owned by George W. Waldburg and his brother Jacob. Their island lands were known for the quality of the Sea Island cotton grown there.[74]

By 1808, Savannah newspapers report coasters sailing to and from one or both islands.[75] Only one or two coaster arrivals from St. Catherines Island were reported for any given year, and in some years, none were listed. This doesn't mean crops were not produced on the island in those years; rather, it is likely the island's production, almost exclusively Sea Island cotton, was taken to a nearby landing, such as Sunbury, and shipped to Savannah from there, or the crop was carried to a factor in Charleston.

Like St. Catherines and Ossabaw islands, Sapelo Island had been retained by the Creek Indians as hunting grounds through an agreement made with James Oglethorpe when Georgia was founded. In 1747, these three islands were awarded to Mary Musgrove, a part Creek Indian who had served as Oglethorpe's interpreter, and her husband, Thomas Bosomworth. In 1760, Sapelo was sold at public auction to Grey Elliott, a member of the King's Council in the Colony of Georgia. Two years later, Patrick Mackay, an Indian trader, purchased Sapelo and adjacent Blackbeard Island. Mackay seems to have been the first to undertake extensive agriculture on the island. In 1790, Sapelo was acquired by several exiled members of the French nobility who fled France to escape the dangers of the French Revolution. These five men operated as "The Sapelo Company." In addition to Sapelo, they acquired Jekyll Island, which is located farther down the Georgia coast. These Frenchmen attempted agriculture and cattle raising on the two islands, but in the end, their venture on Sapelo was unsuccessful. In 1802, Thomas Spalding acquired

rough rice is used here (see House, *Planter Management*, 65, for the development of this figure).

[74] J. F. Smith, *Slavery and Rice*, 223; Vanstory, *Georgia's Land*, 21–26; D. H. Thomas, *St. Catherines*, 11–16.

[75] *Republican & Savannah Evening Ledger*, 1808.

four thousand acres on the southern end of Sapelo Island and all of Blackbeard Island, and in 1843, he purchased most of the remaining seven thousand acres of Sapelo.[76]

Thomas Spalding was born in 1774 at Frederica on St. Simons Island, where his father operated a successful mercantile firm in partnership with Donald Mackay. Thomas became one of the most noted agriculturalists in the South, and he was active in state and national politics, serving in the United States House of Representatives in the 1805–1806 term. He was a prolific writer on agrarian topics, noted for his pioneering efforts in the cultivation of sugarcane along the Georgia coast, for innovations in the farming of Sea Island cotton, and for his promotion and perfection of the use of "tabby" (a mixture of whole and burned oyster shell, sand, and water) as a building material. On Sapelo, Spalding established a prosperous plantation, growing mostly Sea Island cotton but also rice and sugarcane, which he processed into sugar at a mill of his own design and construction.[77]

Like most coastal planters, Thomas Spalding shipped his crops to market by water. Sapelo is named as a port of origin for coasters in Savannah newspapers more often than most of the Sea Islands, a reflection of Spalding's very productive plantation. Occasionally, as many as four coasters sailed into Savannah from the island in a given year. But, as is the case with the other large islands, in some years, no arrivals at all are reported. Often, the cargoes carried from Sapelo Island reflected the varied agricultural interests of Thomas Spalding. In 1820, the sloop *Union*, one of four coasters sailing from Sapelo that year, carried Sea Island cotton and "orange juice," the latter no doubt derived from groves planted by Spalding. Ten years later, the sloop *Argo* made two voyages from Sapelo. On one trip, the *Argo* carried 106 bales of Sea Island cotton, the principal money crop of the island, but on the other voyage, the sloop transported twenty-five hogsheads of sugar, the commodity that Spalding so strongly promoted for coastal planters.[78]

### *Harris Neck*

Harris Neck, occasionally listed as a port of call for coasters, is a narrow peninsula of land extending into the salt marshes in upper McIntosh County, southeast of the town of Riceboro. By the 1820s, Harris Neck was the site of

[76] Sullivan, *Early Days*, 80–81, 89.

[77] Coulter, *Thomas Spalding*, 70; Sullivan, *Early Days*, 96, 98, 102.

[78] Shipping information is from the 1820 *Daily Georgian* and the 1830 *Daily Savannah Republican.*

several cotton plantations. Savannah newspapers indicate coasters made just a few arrivals from there in any given year, and in some years, particularly after about 1850, no vessels are listed as sailing from Harris Neck. In 1830, The *Daily Savannah Republican* reported that three sloops, the *Angelica*, *Ann*, and *Two Friends*, made an unusually high ten voyages into Savannah from Harris Neck, carrying 241 bales of Sea Island cotton and twenty-one bales of "upland cotton." No rice or other cargoes are listed aboard the coasters sailing from Harris Neck in 1830, but in other years, vessels did transport other commodities, including wood (probably cord wood), hides, and, after 1847, naval stores, such as the 150 barrels of turpentine and eight barrels of spirits of turpentine carried by the sloop *Liberty* in 1848.[79]

The sailing coasters also traveled up the South Newport River to plantation landings west of Harris Neck, where rice and Sea Island cotton were grown. A few ships carried cargo to and from the tiny community of South Newport located at the head of navigation on the South Newport River. Coasters also visited cotton and rice plantations along the Sapelo River, which empties into Sapelo Sound about seven miles south of the South Newport River. These included Belleville Plantation, once the property of Richard Leake, one of the first planters to successfully grow Sea Island cotton.[80]

### *Darien*

Darien, located in McIntosh County on the north bank of the Darien River (a branch of the Altamaha River) about twelve miles from the Atlantic Ocean, was one of the first settlements established by James Oglethorpe. A group of Scottish Highlanders, brought to Georgia by Oglethorpe, began the settlement in January 1736 on the site of old Fort King George, originally established on the Altamaha to protect the South Carolina colony from the Spanish to the south.[81]

Vessels could sail directly to Darien from the entrance at Doboy Sound and along the Darien River, but the route was considered difficult because of shoals and the crookedness of the river. It was reported that "[t]he sinuosities of the present river channel are so numerous, that no wind will either bring a vessel up, or take a vessel down…."[82] Despite these obstacles to navigation,

[79] Sullivan, *Early Days*, 236–50. Information on the cargo of the *Liberty* is from the *Daily Georgian*, October 2, 1848.

[80] Sullivan, *Early Days*, 97, 216–17, 227–30, 252–56.

[81] Ibid., 6–22.

[82] *Daily Georgian*, August 21, 1827.

Darien's position on a branch of the Altamaha River, the largest river in Georgia, gave it a great advantage as a port (Figure 3.12). The Altamaha extends over 135 miles inland, and its two major branches, the Oconee River and the Ocmulgee River, arise in the foothills of the Appalachian Mountains in the northern part of the state. Each runs more than 250 miles before joining to form the Altamaha. The swamps and lowlands fringing the Altamaha supported great cypress stands, and the river provided access to the huge expanse of pine found a short distance inland.

Despite these natural advantages, Darien remained a small settlement until the development of Sea Island cotton agriculture and the establishment of rice plantations along the vast estuary of the lower Altamaha River after the American Revolution. The Altamaha became particularly important to Darien's economy with the spread of cotton agriculture into interior Georgia after 1800. By the early nineteenth century, the Altamaha had become, with the Savannah River, a main artery of commerce between interior Georgia and the coast. Flatboats, known as "Oconee boxes," and pole boats began carrying large amounts of upland cotton down the river from as far inland as Milledgeville, located on the Oconee River about 225 miles from Darien.[83] At Darien, the cotton was loaded aboard sloops and schooners and, later, on steamboats for shipment to Savannah or Charleston.

The lower twenty or so miles of the Altamaha River provided ideal conditions for tide flow rice agriculture. By the third decade of the nineteenth century, the area around Darien was the location of some of the most prosperous rice plantations in the state. Darien also became an important timber and lumber shipping port, drawing on the vast stands of cypress growing along the river and pine from the inland areas, which were floated down the Altamaha to sawmills in the town. The early Darien sawmills were powered by hand or water, but in 1817 a steam-powered mill was erected at the Upper Bluff, just above the town. This was the Darien Steam Saw Mill with a steam engine manufactured by Bolton & Watts of England and described as "perhaps the best in the United States."[84]

Darien was on the verge of an economic boom at the start of the nineteenth century. The town was to grow in importance as the regional commercial outlet for local produce and a source of supplies for local planters and residents. In 1800, however, the *Columbian Museum & Savannah Advertiser* does not list a single vessel sailing between Savannah and Darien. The town

[83] Morrison, "Raftsmen," 4, 9.

[84] *Daily Georgian*, May 27, 1822.

certainly depended on sailing vessels for supplies, but these seem to have come from Charleston or Northern ports. In 1803, Domingo Tedoir advertised his services as a pilot to Darien in the *Columbian Museum & Savannah Advertiser*, reflecting the town's growing importance as a seaport. Tedoir stated he possessed a "perfect knowledge of the Inlet of Doboy & River Altamaha." Tedoir's advertisement went on to praise the advantages of the Altamaha region noting, that although it was

> a river hitherto little known, but which from its good navigation, and central situation, running through an immense extent of country abounding on each side with the finest Rice, Cotton, Tobacco and Provision Lands, already considerably inhabited, and with some of the best Mill sites, and Pine timber in Georgia, promises shortly to become of the utmost importance.[85]

After 1800, Savannah newspapers report increasing numbers of coasting vessels sailing between there and Darien. Before 1810, more coasters were sailing from Savannah to Darien than from Darien to Savannah. This suggests that Savannah served as a source for merchandise, food, and supplies but that Charleston was the destination for much of the plantation produce shipped out of Darien. In 1806, Darien was named a "port of delivery" for the Customs District of Brunswick, a sign of its increasing importance as a seaport. The "port of delivery" designation indicated a location of lesser importance and offering fewer services than a "port of entry" within a customs district. It was not until about 1810 that information on the identity of coaster cargoes began to appear in Savannah newspapers. From April 1809 to April 1810, the *Republican & Savannah Evening Ledger* listed five different coasters that made nine trips from Darien to Savannah. One of these ships, the sloop *Willing Maid*, sailed from Darien to Savannah four times, and the schooner *Experiment* made two voyages. These vessels seem to have been regular Savannah-Darien traders. Over this year, however, only one of these nine listings included information on the cargo carried. This was for the sloop *Liberty*, whose arrival in Savannah was reported on March 20, 1810, with a cargo of "rice and cotton."

Lumber became an important export from Darien, and by 1819, the town had two steam-powered sawmills. Most of this sawn lumber was sent

[85] *Columbian Museum & Savannah Advertiser*, March 22, 1803.

directly to distant ports in the Northeast or the Caribbean; little was carried into Savannah or Charleston.[86]

Darien grew steadily during the first decade of the nineteenth century, and an increasing number of merchants established themselves in the town, providing supplies to area residents and planters. However, few local establishments had the resources to market plantation crops as commission merchants. Planters along the coast typically dealt directly with factors and commission merchants in Charleston, and later Savannah, for the sale of their crops and for many supplies or with agents of these city merchants who might be established in the smaller communities. One of the few early firms in Darien acting as a commission merchant for area planters was Vivion, Dunham & Co., formed in 1809. In January 1810, the *Republican & Savannah Evening Ledger* advertised that:

> The subscribers having entered into co-partnership under the firm of VIVION, DUNHAM & Co., with a view to transect business in the FACTORAGE and COMMISSION LINE, will receive and forward produce to any port in the United States on reasonable terms. Having large and convenient Warehouses, they will receive on Storage all kinds of Goods and Produce at reduced rates.

By 1810, docks and warehouses lined the waterfront at Darien. A fire in 1812 damaged buildings along the river, but the waterfront was soon rebuilt. In 1818, "40 houses were erected during 3 months," and the Bank of Darien was established with a capital of $150,000. The following year, in recognition of the town's growing importance, the McIntosh County seat was moved from Sapelo Bridge to Darien.[87]

Also in 1818, a newspaper, the *Darien Gazette*, was established. The *Gazette* published a "Marine News" column containing information on arriving and departing vessels, including coasters, and named the cargoes carried, the consignee receiving the cargo, and the number of days sailing from the port of origin. The *Darien Gazette* is one of the few newspapers published in the smaller ports visited by the Georgia coasters. As such, it provides information on the activities of the coasters relative to these small ports, information that is not always included in the Savannah and Charleston papers. A typical

[86] Sullivan, *Early Days*, 1, 146–47.

[87] *Darien Gazette*, August 31, 1819; Kayser & Co., *Commercial Directory*, 49; Sullivan, *Early Days*, 141.

"Marine News" column is the one that appeared in the *Darien Gazette* on November 4, 1820, reporting the arrival of four vessels during the previous week:

| Vessel | Master | Origin | Days | Cargo | Consignee |
|---|---|---|---|---|---|
| Schooner Lucretia | Courter | Charleston | 2 | assorted cargo | C.L. Champayne, W.B. Holzendorf, J. Lester, others |
| Sloop Linnet | Bowman | New York | 8 | assorted cargo and passengers | Hall, Cooke & Co. |
| Brig Rose-in-Bloom | Auchinlich | New York | 5 | assorted cargo and passengers | (none listed) |
| Schooner Post Boy | Jones | Washington, NC | 11 | in ballast | to master |

Two ships named in this "Marine News" listing, the schooner *Lucretia* and the sloop *Linnet*, occasionally carried plantation produce in the coasting trade but not on these voyages into Darien. Few coasters carried plantation produce into Darien or the other small ports; those goods were destined almost exclusively for Savannah or Charleston. The cargoes carried by coasters into the smaller ports generally consisted of "assorted cargo" as mentioned in the 1820 listing, which would have included a wide range of merchandise and foodstuffs. In addition to local coasters, larger merchant ships from ports like New York, Philadelphia, and Liverpool called at Darien. Many of these larger ships were loaded with lumber from the Darien sawmills.

The *Darien Gazette* also listed the arrivals of "boats" and "flats" that had floated down the Altamaha River carrying upland cotton. On March 29, 1819, the paper noted the arrival of one boat and two flats from the upcountry. These were a "Boat" named *True Blooded Yankee* from Fort Hawkins with 310 bales of cotton and two "Flats" from Twiggs County with 446 bales of cotton between them. In late 1818, the steamboat *Altamaha* left Darien and steamed up the Altamaha and the Oconee, reaching Milledgeville in January 1819. This was the first steamboat to work on the Altamaha and its tributary rivers, but soon other steamers were involved in carrying passengers and towing flats and barges of cotton downriver to Darien.[88]

[88] Rogers and Saunders, *Swamp Water*, 225–26; Sullivan, *Early Days*, 153.

The *Darien Gazette* occasionally printed advertisements related to coasting vessels. On December 9, 1820, the newspaper advertised "For Freight or Charter": "The Superior and fast sailing sloop CHAUNCEY, 88 tons burthen, stows a large cargo—for terms apply to captain Bulkley on board at street's wharf or Benj. Williams."

Most advertisements for departing coasting vessels noted they were sailing for Savannah or Charleston, but a few departed for Northern ports, such as New York or Newport, Rhode Island.

In 1823, there were eleven "commercial houses" in Darien, six of which were identified as commission merchants, and many of these individuals and firms were advertised in the *Darien Gazette* (Figure 3.13).[89] In 1821, a notice appeared in the *Darien Gazette* informing readers of the formation of a factorage and commission business called "Yonge, Atkinson & Co.," established by three Darien residents, Philip R. Yonge, George Atkinson, and R.J. Nichols. In 1824, Rufus R. Merrill advertised that he intended to enter the "mercantile business on commission" and had "leased the wharf and stores belonging to Thos Spalding" where he would provide "Wharfage and Storage on the most reasonable terms."[90]

Some Darien businesses acted as forwarding agents for area planters and as intermediaries, often receiving and holding goods and supplies from factors and merchants in Savannah and Charleston for those planters that could not be visited directly by sailing vessel or steamer. These businesses also maintained their own stocks for sale to residents, including planters. In 1819, Charles F. Sibbold advertised that he had "15000 weight of Prime Bacon" for sale that had arrived aboard the sloop *Mary Ann* from Charleston, and, in 1821, Charles Dewitt had dry goods, groceries, crockery, etc., for sale that had been received by the schooner *Cheves*. In the 1850s, the Darien firm of Mitchell & Collins provided a wide range of merchandise to area planters.[91]

Darien continued to prosper and grow during the 1820s and early 1830s. One expression of this growth is seen in the ninety-one departures of sailing coasters from Darien to Savannah in 1820, as reported in the newspaper the *Daily Georgian*. More coasters sailed into Savannah from Darien this year than from any other port; Riceboro and St. Marys were distant seconds with sixteen sailings each. The ninety-one voyages from Darien to Savannah were

[89] Kayser & Co., *Commercial Directory*, 50.

[90] *Darien Gazette*, February 17, 1821, and February 5, 1824.

[91] *Darien Gazette*, March 29, 1819, and January 6, 1821; Glynn County, Georgia, estate of James Hamilton, Wills and Appraisements, Book F.

ten times greater than the number reported just ten years earlier and reflect the expanding production of rice and Sea Island cotton along the lower Altamaha River and, even more importantly, the huge increase in the amount of upland cotton floated down river from the interior. This growth in maritime trade with Savannah was an expression of the increasing importance of that city as a regional market for coastal and inland planters. In 1826–1827, 47,065 bags of cotton were reportedly shipped from Darien, mostly upland cotton carried down the Altamaha River. By 1830, Georgia was the world leader in the export of cotton, and one-third of that total, approximately 50,000 bales, was shipped through Darien. Not all cotton shipped from Darien went to Savannah or Charleston; some was carried by ships directly to Northeastern ports or overseas. As early as 1828, it is reported that Darien exported 912 bales of cotton directly to England and 61 bales to New York. These direct exports from the town increased over the years. In 1833, Darien shipped 3,076 balcs to Liverpool, 2,698 to New York, and 40 to Providence, Rhode Island. In addition, Darien had become the most active lumber center in the state.[92]

In 1837, the population of Darien reportedly reached five hundred, but the period of prosperity and growth the town had experienced for thirty years was interrupted by several events. The nationwide financial crises, the Panic of 1837 and Crisis of 1839, depressed cotton prices, negatively affecting merchants and planters alike. In 1842, the Bank of Darien failed, and finally, newly constructed railroads between interior Georgia and Savannah began to divert cotton shipments previously carried down the Altamaha River. Even with the economic slowdown, cotton exports out of Darien remained high for a while. The *Brunswick Advocate* on February 22, 1838, reported that a total of 56,100 "bags" of cotton had been exported from Darien between June 1837 and February 1838. Most of this was upland cotton received by way of the Altamaha. Sailing coasters and steamboats transported most of this cotton, 33,496 bags, to nearby Savannah. The rest was carried to Charleston (9,694 bags), New York (8,060 bags), Providence, Rhode Island (500 bags), and 3,350 bags were shipped directly to Liverpool aboard some of the larger ships that called at Darien. Direct cotton exports out of Darien to Northern and overseas ports remained relatively high over the next several years: 10,537 bales were shipped to New York in 1840 and 13,656 bales to New York and

[92] Sherwood, *Gazetteer*, 161 (1826–1827 exports); Donnell, *Chronological and Statistical*, 129, 179 (1828 and 1833 export figures); Wayne, "Darien Waterfront"; Sullivan, *Early Days*, 154, 157.

Providence in 1843. After that, direct exports of cotton out of Darien fell dramatically. Only 1,411 bales were shipped in 1844, and just nine were shipped to New York in 1848. At this time, the overseas shipment of cotton from Darien seems to have ended entirely although it was still shipped to Savannah and Charleston aboard coasters and steamers. By the mid-1840s, Darien's economic boom was over, but rice, timber, and naval stores remained economically important up to the Civil War.[93]

For twenty-five years, from just after the War of 1812 to about 1840, more local coasters sailed into Savannah each year from Darien than from any other port. After 1840, Darien generally ranked just third or fourth in terms of the annual volume of coaster traffic with Savannah. The great rice-producing areas, such as the Ogeechee River, the Back River, and the small town of Riceboro, became the most often mentioned ports in the local trade. Some of this drop in coaster traffic from Darien relates to the overall economic decline of the town. Upland cotton had been Darien's greatest export to Savannah, but by the late 1830s, railroads began to transport cotton from the interior into Savannah, and smaller and smaller amounts were floated or towed downriver to Darien. Further, by this time, much of the cotton that did reach Darien was carried into Savannah by steamboats, not by coasters.

The pattern of coaster traffic out of Darien in the early 1840s continued to the Civil War. Through the mid-1850s, coasters typically made fewer than twenty or twenty-five voyages from Darien into Savannah, and after 1855, this number was usually less than ten. Rice was the principal cargo carried out of Darien after 1840, most coming from the large rice plantations located in the Altamaha River estuary near the town. These included plantations such as Broadfield, Hofwyl, New Hope, Elizafield, Hopeton, Altama, Wright's Island, Butlers Island, Rhett's Island, Champney's Island, Broughton Island, Cambers, Generals Island, Potosi Island, and Cathead.[94] When coasters carried rice from these plantations, their origin was often identified as "Darien" or "Altamaha River" in the Savannah shipping lists; rarely were individual plantations named. Cotton tended to be the second most important item carried by coasters out of Darien after 1840, to which were added small amounts of other commodities, including lumber, naval stores, hides, and peanuts.

In the years just before the Civil War, Darien retained its somewhat diminished position as a regional mercantile center and modest seaport.

[93] Sherwood, *Gazetteer*, 161 (1837 population); Donnell, *Chronological and Statistical*, 252, 300, 347 (1840s cotton export figures); Sullivan, *Early Days*, 160, 161, 171.

[94] Sullivan, "Darien Becomes," 5–7.

Steamboats out of Savannah visited the town regularly, and a few coasters continued to sail between it and Savannah. In 1860, on the eve of the Civil War, the *Daily Morning News* reported just four coasters sailing from Darien into Savannah, together making only seven voyages. These were the schooners *Charles W. Bentley*, *Challenge*, *Elias Reed*, and *J. Truman*, and rough rice was their only cargo (except for four hundred bushels of corn the *Elias Reed* carried on one voyage).

*Frederica and St. Simons Island*

Frederica, located on the bank of the Frederica River on the western side of St. Simons Island, was the frontier town and military outpost established by James Oglethorpe in 1735 to guard the southern flank of the new colony from the Spanish to the south. Within a few years, the town had a population of almost a thousand, including the "Regiment of Foot for the Defense of His Majesty's Plantations in America," who garrisoned the fort. After the defeat of Spanish invaders on St. Simons at the Battle of Bloody Marsh in 1742, there was a long period of peace and little need for the garrison at Fort Frederica. Most of the troops were removed; many settlers left Frederica, and the town went into decline. A fire in 1758 destroyed many houses in the town, and others fell into ruin or disrepair. In 1760, Donald Mackay and James Spalding headquartered their trading business at Frederica. Their firm, Mackay and Spalding, was involved in the Indian trade, and goods were shipped from Frederica to their trading posts scattered across the Southeast.[95]

Frederica and St. Simons were largely abandoned during the American Revolution. Many coastal planters, including St. Simons resident James Spalding, remained loyal to the British crown and fled to Florida. A number of these loyalists returned to Georgia after the war, including Spalding, who became one of the first to cultivate Sea Island cotton. A small number of plantations were established on St. Simons, but by the late 1700s, Frederica contained only a handful of residents although it remained the seat of Glynn County until 1797, when county government was shifted to Brunswick on the mainland.[96]

At the start of the nineteenth century, the population of Frederica consisted of only a few families. However, its position on the inland waterway was advantageous. It continued to serve as a principal boat landing for St. Simons and is often named as a point of origin or destination for Georgia

[95] Vanstory, *Georgia's Land*, 106–107.

[96] Ibid., 111–12.

coasters (see Figure 3.12). In 1810, for example, the *Republican & Savannah Evening Ledger* listed ten different arrivals of coasting vessels from Frederica. These were the sloops *Anubis* (master DuBignon), *Fish* (master Reed), *Mink* (master Parsons), *Raccoon* (master Massey), and *Sela* (master Shaddock) and the schooners *Jupiter* (master Harris) and *York* (master Dart). Although the newspaper does not name the cargoes carried by these ships, most of them undoubtedly carried cotton from the plantations on St. Simons. After 1810, "Frederica" rarely appears in newspapers as a port-of-call for local coasters and is replaced by "St. Simons Island." The reference to St. Simons Island could mean Frederica, but it also could refer to various landings at the several plantations on the island.

The 1820 census for Glynn County lists only four "heads of families" at Frederica. These are George Abbott, James Frewin (Fruin), John Cole, and Luke Blount. Abbott, Cole, and Blount were all farmers; James Frewin was the only one identified as "engaged in commerce." This occupation referred to Frewin's activities as a coasting captain with his schooner *Maria*, but he also seems to have operated a small store at Frederica in the 1820s.[97]

The population of Frederica remained small through the nineteenth century, with only a few families residing there. However, in addition to James Frewin, several men who served as captains or owners of coasting ships resided at Frederica or on St. Simons Island in the first six decades of the nineteenth century. These included McGregor Baisden, Charles Porquet, James A. D. Lawrence, Fenn and Michael Peck, Charles Stevens, and, possibly, Anthony Shaddock.[98]

After 1830 Savannah newspapers only occasionally list coasters sailing from St. Simons. Rarely were more than four or five arrivals reported from the island in any given year, and in some years, no sailing vessels carried cargoes from the island into Savannah. By the 1830s, St. Simons was regularly served by steamboats out of Savannah, and these, not the sailing coasters, carried much of the cotton crop from the island.

[97] Unless otherwise noted, United States Census data are derived from the original, unpublished manuscript versions presented online at Ancestry.com and are referenced by type of census (Population or Slave Schedule), county, and year; US Census Bureau, "US Census of Population, Glynn County, 1820"; Vanstory, *Georgia's Land*, 113. A biography of James Frewin is found in Pearson, "Captain James Frewin."

[98] US Census Bureau, "US Censuses of Population, Glynn County, 1820–1860"; Pearson, "Georgia Coasting Trade"; Pearson, *Charles Stevens*.

### *Brunswick and the Turtle River*

Brunswick is located on the mainland of the central Georgia coast opposite St. Simons Island. Located on the Turtle River, Brunswick was founded in 1771 on the property of Mark Carr. The town was considered well situated as a port; the Turtle River provided a deep channel to the town and convenient access to the Atlantic through St. Simons Sound between St. Simons and Jekyll islands. Additionally, the Turtle River opened into the principal waterways of the inland passage and provided access to inland areas (see Figure 3.12). The town of Brunswick was surveyed; streets were laid out soon after its formation, but it grew slowly. The Revolutionary War, which resulted in widespread abandonment of the coast, impeded the town's growth. After the war, Brunswick became part of the newly formed Glynn County, which extended along the coast from the Altamaha River to the Little Satilla River. The 1790 census shows that the entire population of Glynn County consisted of 413 Whites and 215 enslaved persons.[99]

In 1789, Brunswick was made a port of entry of the Brunswick District, one of four customs districts on the Georgia coast. The Brunswick District covered the central Georgia coast and included the Altamaha, Frederica, and Turtle rivers south of Sapelo Island to the southern end of Jekyll Island. Brunswick was the port of entry, and the small community of Frederica was named a port of delivery. The transfer of the county seat from Frederica to Brunswick in 1797 did little to stimulate the town's growth.[100]

Between 1800 and 1810, Brunswick and Glynn County began to experience some growth, partly stimulated by the expansion of cotton and rice cultivation in the region. An increasing number of vessels, carrying away the products of local planters and bringing in goods needed by the slowly expanding population, began to call at the town. By 1810, the population of Glynn County was 3,417, of whom 2,845 were enslaved. The fact that slaves represented more than 80 percent of the county's population reflects the increase in cotton and rice agriculture with their heavy dependence on enslaved labor. The decade between 1810 and 1820 saw little in the way of growth in the town or county. The total population of Glynn County in 1820 is reported to have been 3,418, an increase of only one individual in ten years. During the War of 1812, the British Navy blockaded parts of the Georgia coast,

[99] Vanstory, *Georgia's Land*, 118–19; US Census Bureau, "US Census of Population, Glynn County, 1790."

[100] "Georgia Customs Districts," 159–64, RG 36, Bureau of Customs, NARA; Vanstory, *Georgia's Land*, 119.

disrupting coastal shipping. The British also went ashore and occupied some of the Sea Islands for short periods and took away many enslaved persons. These activities dislocated the lives of coastal residents and impeded growth all along the Georgia coast. Although Brunswick was a port of entry, only three vessels were issued enrollments in the town in 1815. These were the schooners *Kitty Ann* and *Experiment* and the sloop *Lively*.[101]

By 1830, the population of Glynn County had grown to 4,567 persons. Of this number, 3,968, fully 87 percent of the entire county population, were enslaved Black people. Brunswick, the principal town in the county, was tiny in the 1820s, consisting of only "four or five dwelling-houses, and ten or twelve white inhabitants" in 1827. Brunswick saw growth and prosperity in the 1830s as tide flow rice agriculture and productive rice plantations developed along the Altamaha River at the northern border of Glynn County. At the same time, Sea Island cotton cultivation was well established on the uplands of the county. Brunswick became increasingly important as a port for the shipment of these products. Brunswick was incorporated in 1836, new mercantile businesses were established, and a newspaper, the *Brunswick Advocate*, began publication.[102]

Vessels were sailing regularly into and out of Brunswick before the 1830s. Most were local coasting ships carrying goods back and forth between the town and Savannah or Charleston. Often, these coasters visited plantations and landings located along the Turtle River west of Brunswick. In October 1837, Samuel A. Hooker, who operated a mercantile store in the community of Bethel, received by the sloop *Argo*, a regular trader to the town, a full supply of "Staple and Fancy Dry Goods, Negro cloths, Negro shoes, Groceries etc., etc." Bethel was located on the north bank of the Turtle River just west of Brunswick and served the Sea Island cotton planters in that area (see Figure 3.12). Other coasters commonly calling at Brunswick in the 1830s included the sloops *America* and *Bolivar* and schooners *Columbia*, *Nile*, *Betsey Marla*, and *Tiger*.[103]

[101] US Census Bureau, "US Censuses of Population, Glynn County, 1810 and 1820"; "Master Abstracts of Enrollments," Districts of Brunswick and St. Marys, 1815, RG 41, BMIN, NARA.

[102] 1827 description of Brunswick from US Congress, *American State Papers*, 6, *Naval Affairs* 3:277; US Census Bureau, "US Census of Population, Glynn County, 1830"; Vanstory, *Georgia's Land*, 120–21.

[103] *Brunswick Advocate*, October 17, 1837 (Samuel Hooker's advertisement); other shipping information is from the *Brunswick Advocate*, 1837–1839.

In Brunswick, as in other small towns along the Southeastern coast, there were individuals and firms acting as commission merchants and forwarding agents for local planters, and there were several merchants' stores providing supplies to planters and other residents. In the late 1830s, Robert Walsh & Co. promoted themselves in the *Brunswick Advocate* as "Commission Merchants, General Agents and Auctioneers." The firm of Rice, Parker & Co. advertised:

> TO PLANTERS, Planters in this vicinity wishing supplies for their Plantations, can be furnished with Merchandise in every variety, on as reasonable terms as can be provided in Savannah or Charleston at the store of
> RICE, PARKER & CO.

On November 16, 1837, the *Brunswick Advocate* reported that the sloop *Thomas Butler King* had arrived from Newport, Rhode Island, and was "intended as a pilot boat for this port." The sloop was built at Newport, Rhode Island, in 1837, and her namesake was one of Georgia's most prominent planters and politicians. Thomas Butler King owned a cotton plantation on St. Simons Island and rice lands along the Satilla River. He was an avid promoter of the town of Brunswick, a member of the state legislature, and a United States congressman recognized as an authority on naval affairs.[104]

It is unknown if Thomas B. King himself had any financial or business interest in the sloop, but the assignment of the vessel as a pilot boat at Brunswick in 1837 likely resulted from King's influence as a legislator and his strong interest in improving waterborne commerce at the town. As with other pilot boats, the *Thomas Butler King* occasionally sailed as a coaster, carrying crops and merchandise into and out of Brunswick.[105]

In 1840, the population of Glynn County had risen to 5,302 persons, including 4,644 enslaved individuals. The financial recession known as the Crisis of 1839 affected local businesses and planters, and the town of Brunswick declined. The Bank of Brunswick and the newspaper, the *Brunswick Advocate*, both closed in 1839, and canal and railroad projects pushed by Thomas King ended. With the depression of the early 1840s, the population of Brunswick declined to four hundred people.[106]

[104] *Ship Registers and Enrollments of Newport*, 614 (pilot boat *Thomas Butler King*); Steele, *T. Butler King*.

[105] *Brunswick Advocate*, 1837–1839.

[106] Vanstory, *Georgia's Land*, 121.

The 1850s were a period of prosperity for the Georgia coast in general, including Brunswick. The Bank of Brunswick reopened, another newspaper was begun, and new businesses were started and buildings constructed. This brief revival ended in fall 1861, when the United States Navy initiated the blockade of the Georgia coast and most residents of the Sea Islands evacuated to the interior. On March 10, 1862, Brunswick was occupied by federal forces.[107]

Savannah newspapers often record coasting ships arriving from or departing for the "Turtle River," referring to the river at whose mouth Brunswick is located. These vessels were calling at plantations located along the river, and in 1837, it was reported that "sloops go up this stream into Wayne County, and take down the produce to Savannah."[108] The Turtle River is short, extending less than forty miles inland (see Figure 3.12). The river did not carry the large flow of freshwater required for tidal irrigation, and little rice was grown along its banks. The river was the location of several large cotton plantations, including Anguilla Plantation, one of the earliest properties to cultivate Sea Island cotton. One of the earliest mentions of the Turtle River relative to the coasting trade appeared in the December 8, 1808, issue of the *Republican & Savannah Evening Ledger* with a report of the arrival of the schooner *Relief* in Savannah from the Turtle River. Newspapers provide no information on cargoes at this early date, but the *Relief* was sailing after Sea Island cotton had become established along the Turtle River, and this cotton probably comprised the schooner's principal cargo. Before the late 1820s, the Turtle River was rarely named as a port of call for Georgia coasters. Five sloops and schooners arrived in Savannah from there in 1827–1828, eight in 1839, six in 1845, and eight in 1852. Intervening years had similar numbers of arrivals. The vessels sailing from the Turtle River most often carried Sea Island cotton, such as the 595 bags brought into Savannah by the North Carolina-built schooner *Joseph* on three trips in 1845. "Upland" cotton is occasionally mentioned as a cargo, probably collected from farms or plantations in interior Glynn County or even farther inland. Turtle River cargoes also included hides, sugar and beeswax, and, after about 1850, various naval stores, such as turpentine and rosin. On a very few occasions, rough rice was

[107] Ibid., 122; Heard, "St. Simons Island," 251.

[108] Sherwood, *Gazetteer*, 245.

mentioned as a cargo of the coasters leaving the Turtle River, but it may have been loaded at other locations.[109]

*The Big and Little Satilla Rivers*

The Satilla River was a relatively common point of origin for coasters sailing into Savannah, particularly after 1830. Often referred to as the "St. Illa" or "Great Satilla River," this river extends through Camden County and far inland. It was navigable for small coasting vessels as far as the community of Burnt Fort, about forty-eight miles above its mouth, and for "boats," probably referring to flatboats or bateaux, for another sixty miles or so.[110] The lower reaches of the Satilla were ideal for tide flow rice agriculture and a number of rice plantations were established there by the 1820s. Just north of the Satilla River is the Little Satilla River, which forms the boundary between Glynn and Camden counties. Extending inland less than fifty miles, the Little Satilla was less suitable for rice cultivation than were the larger coastal rivers. Even so, rice plantations were established along the Little Satilla's lower portions and are occasionally mentioned as a port of call for coasters.

The town of Jeffersonton, or "Jefferson," was established in 1801 on a bluff on the south side of the Satilla River in Camden County, about twenty-five miles from the river's mouth (see Figure 3-13). Named in honor of Thomas Jefferson, the town had been expressly authorized by the Georgia Legislature a year earlier as the seat of Camden County.[111] Jeffersonton was situated near large rice plantations, and the shift of the seat of government there from the county's principal port of St. Marys was due mainly to the influence of prominent rice planters along the Satilla. Jeffersonton never grew to any great size; in 1849, it was reported to have "a court house, jail, three stores, &c. It is considered unhealthy, being surrounded by rice plantations." Despite this lack of growth, the county seat was kept at Jeffersonton until 1872 when it was shifted back to St. Marys.[112]

Jeffersonton was located at the head of navigation for most sailing vessels on the Satilla River, and coasters called there to load cargoes collected from the surrounding country. Despite this navigation advantage, Jeffersonton is only occasionally named as a port of origin or destination for Georgia coasters.

[109] Information on coaster sailings to and from the Turtle River comes from the *Daily Georgian*, 1827–1828, 1839, 1845, and 1852.

[110] Sherwood, *Gazetteer*, 224.

[111] Reddick, *Camden's Challenge*, 6.

[112] Vocelle, *History of Camden*, 50; *Statistics*, 140 (1849 report).

The few coasters arriving in Savannah from Jeffersonton normally carried cotton or hides and, after about 1850, naval stores.

The community of Burnt Fort was established on the Satilla River at the site of a burned fort occupied by South Carolinians before 1725. Burnt Fort was as far upriver as small coasters could travel and was near the upper end of tidal influence on the Satilla.[113] The community became a sawmilling and lumber center, and the timber goods shipped from there were commonly carried to distant ports. These timber products occasionally were transported into Savannah and Charleston or smaller coastal communities, such as Brunswick.

### *St. Marys and the St. Marys River*

The town of St. Marys is located on Buttermilk Bluff on the north side of the St. Marys River near its mouth. Settlement around what would become St. Marys was stimulated after 1783 when the British ceded East Florida back to Spain following the American Revolution. Many British nationals living in Florida, including Royalists uncertain about their future under Spain, had fled Georgia and South Carolina during the Revolutionary War and moved to the area around the mouth of the St. Marys River in 1783–1784; they intended to board ships for other British overseas possessions. Some would-be emigrants remained and settled along the lower St. Marys River. The town was established in 1787, but it was not until 1792 that the community was given the name "St. Marys."[114]

St. Marys is located on the southern border of Georgia and, before 1821, was also the national boundary of the United States and Spanish Florida. Although this position made the town an important entrepôt for trade between the two countries, the region along the St. Marys River was isolated. For many years, it was an area of lawlessness, well known for smuggling. The Spanish attempted to maintain law and order on their side of the river, but the numerous "vagrants, thugs, and sadists" occupying the area between the St. Marys and St. Johns rivers made those efforts difficult. In 1790, forty-four families, consisting of two hundred Whites and seventy-one Blacks, resided in the St. Marys area, and many of those 271 people reportedly engaged in illegal activities.[115]

[113] Sherwood, *Gazetteer*, 136.

[114] Bullard, *Robert Stafford*, 6; Reddick, *Camden's Challenge*, 6; Vocelle, *History of Camden*, 34–35.

[115] Bullard, *Robert Stafford*, 12.

The town of St. Marys grew into the main population and commercial center of Camden County. It became the most important port on the lower Georgia coast, partly because of the town's position at the nation's border and partly because of its natural harbor that allowed vessels of "heavy burthen" to sail up to the city wharves (Figure 3.14). In addition, the St. Marys River gave access to the interior country. This area became important as a source of cotton, timber resources, and hides from the large herds of cattle raised in what was known as the "wiregrass" region. The river was navigable for small sailing vessels for thirty miles while "boats" could travel upriver for twice that distance.[116]

By the first decade of the nineteenth century, Sea Island cotton was grown on the mainland around St. Marys and on nearby Cumberland Island and contributed to the town's importance as a shipping center. In 1800, the *Columbian Museum & Savannah Advertiser* listed ten arrivals of sloops and schooners at Savannah from St. Marys, second only to the fifty-two sailings from Charleston. This same year, the newspaper recorded twenty-six departures for St. Marys, again second only to departures for Charleston. The trade between Savannah and St. Marys in 1800 was dominated by a single vessel, the schooner *Alective*, which made twenty-two trips between the two ports under captains Rudolph (probably Robert Rudolph) and John Chevalier.

The St. Marys River was not well suited for rice cultivation, so the planters living along the river and near the town primarily grew cotton. As early as 1810, "excellent cotton" was being grown on Cumberland Island, just east of the town and the largest of the Georgia Sea Islands. St. Marys's position as a seaport was enhanced with the deterioration of Spanish control of Amelia Island and the port town of Fernandina, located just over the border. Residents of St. Marys became involved in smuggling of contraband, including enslaved Black people, from Florida into the United States. In 1821, Spanish East Florida was ceded to the United States, and St. Marys lost its commercial advantage as a border town. However, it did retain some importance as a local port. On May 7, 1822, the newly acquired portion of Northeast Florida was included within the St. Marys Collection District and the town was named a port of entry, making it the only port of entry between Darien and the Nassau River in Florida.[117] By the late 1830s, St. Marys had a population of four hundred Whites and two hundred Blacks, and $50,000 worth of goods were being brought into the town annually from the surrounding countryside.

[116] Sherwood, *Gazetteer*, 226.

[117] Vocelle, *History of Camden*, 50; Bullard, *Robert Stafford*, 64–66, 147.

These goods included "hides, tallow, wax, and furs" in addition to cotton, rice, and other agricultural products.[118]

Before 1830, St. Marys was among the most important ports in the regional coasting trade. In many years, more coasters sailed from St. Marys into Savannah than from any other port except Charleston. During the 1830s, however, the number of coasters sailing between St. Marys and Savannah dropped dramatically. This decline began in the 1820s, influenced by the acquisition of Florida by the United States in 1821, eliminating St. Marys's enhanced commercial advantage as a border community. Additionally, in the 1820s, steamers rather than sailing vessels began to carry increasing quantities of the goods, especially Sea Island cotton, from St. Marys. Because the St. Marys River was unsuitable for rice cultivation, the town had little rice to ship aboard the sailing coasters.[119] In 1835, the *Daily Savannah Republican* listed eighteen arrivals of coasters from St. Marys, fewer than sailed from either Darien, Riceboro, or the Ogeechee River. In 1845, Savannah papers list just one coaster making a single voyage from St. Marys into Savannah. This was the sloop *Splendid*, commanded by Captain Charles Stevens and carrying a cargo of fifty bales of Sea Island cotton.[120] Subsequent years show the same pattern, with sailing coasters rarely making more than two or three departures from St. Marys for Savannah per year. In some years, such as in 1855, 1860, and 1861, not a single coaster is reported to have sailed from St. Marys into Savannah, a reflection of the minimal importance of this trade by midcentury and an expression of St. Marys's great decline as a port.

Centerville, also known as Centervillage, was one of several inland landings coasters visited along the St. Marys River. This settlement was established in the early nineteenth century about forty miles above the river's mouth, at the intersection of roads crossing the river to Florida. Centerville was a short distance from the river, and Camp Pickney, the nearby landing on the St. Marys River, served as the shipping point for the town. Centerville was the site of sawmills and became the commercial center for surrounding farms and plantations. James Silva, writing about the 1840s and 1850s, noted that the merchants of the town dealt with farmers from Georgia and nearby Florida who "brought their cotton, cowhides, tallow, beeswax, coonskins and other marketable commodities and bartered them for agricultural implements, three legged pots and other things. This trade kept a fleet of small

[118] Bullard, *Robert Stafford*, 116; Sherwood, *Gazetteer*, 226.
[119] Gray, *History of Agriculture*, 2:680.
[120] *Daily Georgian*, March 18, 1845.

crafts going to and returning from Savannah, loaded both ways. I remember the schooner Elias Reed and the American Coin...."[121]

Centerville is mentioned as a port of call for coasters after the 1840s, but like Jeffersonton on the Satilla River, it typically appears in Savannah newspaper shipping lists a few times each year or not at all. In the early 1850s, one vessel, the seventy-foot sloop *Catherine Chard*, under Captain Lewis Wiggins, seems to have conducted much of the coasting trade with Centerville. In 1852, the *Catherine Chard* made at least four trips into Savannah from the town carrying Sea Island and upland cotton, hides, rosin, and spirits of turpentine.[122]

Other boat landings on the St. Marys River were Traders Hill and Coleraine. Traders Hill, located upriver of Centerville, was for many years the principal community in western Camden County, and it became the county seat of Charlton County when it was formed in 1854. Coleraine, located on the St. Marys River about sixty-five miles above its mouth, was a small community established on a site originally occupied by a Spanish post. Like Centerville, Coleraine was a commercial center where local farmers sent their goods to have them shipped to market by water. Neither Colerain nor Traders Hill was as important as Centerville in the coasting trade during the first half of the nineteenth century.[123]

### *Florida Coasting Ports*

Several localities along the Northeast Florida coast were involved in the coasting trade out of Charleston and Savannah. This area was where Sea Island cotton, a small amount of rice, and tropical crops such as oranges and limes were grown. The early Florida ports mentioned in Charleston and Savannah newspaper shipping lists included St. Augustine, Amelia Island (often listed only as "Amelia"), and the St. Johns River. Later, Fernandina, the port town on Amelia Island, and Jacksonville and Palatka on the St. Johns River are occasionally mentioned. Sometimes, newspapers list just "East Florida," "coast of Florida," or "Cape Florida" as origins or destinations for coaster voyages.

St. Augustine, the old Spanish capital of Florida, was only a minor port during the early nineteenth century. The harbor at St. Augustine was blocked by a shallow, shifting bar that restricted entry by the larger vessels coming into

[121] Silva, *Early Reminiscences*, 19–20.
[122] *Daily Georgian*, 1852.
[123] Vocelle, *History of Camden*, 43–44.

favor in the late eighteenth century. Smaller ships, such as the sloops and schooners sailing in the coasting trade, did call at St. Augustine, but by 1810, Amelia Island, located immediately south of the Georgia-Florida border, was much more important to the Georgia coasters. In 1800, the *Columbian Museum & Savannah Advertiser* lists St. Augustine eleven times as a port of origin or destination for coasters sailing into or out of Savannah. By 1809–1810, only two coaster arrivals from St. Augustine were reported in the *Republican & Savannah Evening Ledger*, as opposed to thirty-five from "Amelia." In subsequent years, St. Augustine is typically named no more than a few times, if at all, as a port of call for the Georgia coasters.

Amelia Island and its principal harbor, Fernandina, were particularly important trading centers during the later years of the eighteenth century, up to about 1815. Named by General James Oglethorpe in honor of Princess Amelia, a daughter of King George II, Amelia Island saw a turbulent history during this period. Florida was a Spanish possession until 1768, when it was ceded to Great Britain. Many South Carolina and Georgia planters who remained loyal to the British crown during the Revolutionary War moved to Florida, the Bahamas, or British possessions in the West Indies. In 1783, Florida was restored to Spain by the Treaty of Paris, after which many British planters from East Florida began to leave. Some moved north into the United States while others migrated to the Bahamas or other overseas British holdings.[124]

Eager to enrich Florida through trade, in 1802 the Spanish turned the Amelia Island port of Fernandina (sometimes known as Louisa before 1811) into a free port, open to ships of all nations. It quickly became a thriving trading center involved in both legal trade and smuggling. After the United States made the importation of slaves illegal in 1808, smuggling slaves from Florida into the United States became an important activity in East Florida and on Amelia Island. Some individuals, American as well as Spanish, engaged in transporting slaves, liquor, and foreign luxury goods across the St. Marys River into Georgia or by sailing vessel from Amelia Island to ports along the Atlantic seaboard. The extent of this trade along the South Atlantic coast is expressed in the many ships sailing between Amelia Island and the ports of Charleston and Savannah. In 1800, the *Savannah Museum & Columbian Advertiser* recorded almost fifty sailings of sloops and schooners between Savannah and St. Marys or St. Augustine, Florida. Many of these ships were transporting foreign goods obtainable legally or illegally at these two ports.

[124] Hoffman, *Florida's Frontiers*, 236.

Cotton exports out of Fernandina amounted to 77,000 pounds in 1805, 66,000 pounds in 1806, 60,900 pounds in 1807, and increasing amounts in following years. After 1807, some of this cotton was cultivated in East Florida, but most was grown in South Carolina and Georgia and carried to Amelia Island aboard coasting sloops and schooners. President Jefferson's Embargo Act of 1807, which prohibited direct American trade with British and French overseas ports, emboldened planters, merchants, and shippers to evade the embargo law by transporting even more cotton to Spanish Florida for overseas sale to these embargoed countries. On January 16, 1810, the *Charleston City Gazette* printed a report from a vessel arriving from Amelia Island stating there were more than 150 British and American vessels at the island doing business and that bales of cotton lined the beach ready to be shipped. When the embargo was lifted in 1810 and British and French ports were once again open to direct receipt of American cotton, the cotton exports from Amelia Island declined by 90 percent.[125]

American marines occupied the port of Fernandina in 1818, partly to try to stop the importation of illegal goods. However, the United States claimed the marines were holding East Florida "in trust" for the Spanish government. The American troops were withdrawn in 1819, primarily because of a yellow fever epidemic, but it was obvious Spain could not retain control of Florida. Under the Treaty of 1819, Spain agreed to cede East Florida and West Florida to the United States for five million dollars, but the transfer did not officially occur until summer 1821. After Florida became part of the United States, Amelia Island declined rapidly as a port of call for the coasters sailing out of Charleston and Savannah. Fernandina remained a port of entry for only a short time after American control was established. In 1822, it was annexed into the St. Marys Collection District.[126]

Although trade with Amelia Island largely ended after the United States acquired Florida, the local coasters sailing out of Savannah and Charleston did continue to visit other Florida ports. The Florida location most commonly mentioned after 1821 is the St. Johns River, which may refer to plantations or small landings along the river or to larger communities, such as Jacksonville.

[125] Ibid., 253–55, 267.
[126] Bullard, *Robert Stafford*, 65–66.

## The Regulation of the Coasting Trade

Among the earliest laws passed by the United States government were those regulating coastwise commerce. The Constitution specifically forbade individual states from levying tonnage duties on coastwise commerce without the consent of Congress, leaving the federal government with the power to regulate trade between the individual states and with foreign lands. The first measure to regulate this trade was passed by Congress on July 31, 1789, and signed by President Washington.[127] This bill established the United States Customs Service and was designed to raise revenue for the fledgling government and to protect the economic interests of American shipowners and builders. It stated that any vessel constructed in the "United States and owned by citizens thereof, or vessel not built in the United States but belonging to the citizens thereof on May 29, 1789, and continuing in their possession, was required upon entry at any port of the country to pay a tonnage duty of 6 cents per ton."[128] Vessels constructed after the passage of the law that were owned wholly or in part by foreign citizens were required to pay thirty cents per ton, and all other vessels paid fifty cents per ton. American-built and American-owned vessels engaged in the coasting trade or fisheries were required to pay tonnage duties no more than once a year while other vessels carrying American produce in the trade had to pay the duty on each entry. Although this act did not legally forbid foreign-owned and foreign-built vessels from working in the coasting trade, such onerous fees and requirements essentially eliminated them from involvement in it. This early law left the coasting trade almost exclusively to American vessels, and subsequent legislation continued this policy.[129]

In July 1789, Congress passed an act that established a series of districts and ports for collecting the required import and tonnage duties. Under this law, Massachusetts had the most districts, with twenty, while Georgia had four and South Carolina, three. Ports of Entry, where duties would be collected, were established within each customs district. These ports were selected based on their shipping activity. Over the years, the numbers and boundaries of customs districts, and the designated ports of entry changed as states entered the Union and as the patterns of shipping and commerce changed. A customs collector, appointed by the president, and a naval officer were assigned to each district. Larger ports also employed other positions,

[127] United States of America, *Public Statutes*, 1:27.

[128] Johnson, *History of Domestic and Foreign Commerce*, 327.

[129] Ibid.

such as surveyors, inspectors, weighers, etc. In 1842, in addition to James Hunter, the collector, customs officials at Savannah consisted of a deputy collector, two appraisers, a naval officer, a surveyor, a weigher and gauger, a storekeeper, and ten inspectors. W. T. Baker, the keeper of the lighthouse at Tybee Island, and William Carig, the captain of the "floating light" at the entrance to the Savannah River, were also considered customs employees.[130]

In September 1789, Congress passed the second important law to regulate coastwise navigation and commerce. Titled an "Act for Registering and Clearing Vessels, Regulating the Coastwise Trade, and for Other Purposes," this law established the rules for registering commercial vessels and for dealing with unregistered vessels working in the coastwise trade. The act established the category of "documented vessels," which were those American commercial vessels that were issued documents known as registers, enrollments and licenses to legally operate in maritime trades. Vessels of over five tons burden were required to obtain these documents to engage in foreign or coasting trades or in fisheries. The act mandated that all vessels of greater than twenty tons burden built and owned in the United States and those foreign-built vessels wholly owned by United States citizens had to obtain a "register" to engage in foreign trade or the whale fisheries and an "enrolment" to engage in coastwise trade or other fisheries. These documents were obtained from the collector of customs of the district in which the owner or one of the owners of the vessel resided. The masters or owners of enrolled vessels had to obtain a yearly license to trade between different customs districts, and, to secure the license, the yearly tonnage duty had to be paid and a bond of a thousand dollars be given that the vessel would not be involved in illicit trade. Vessels of less than twenty tons and greater than five tons had to obtain a yearly "license" and pay a bond of two hundred dollars to engage in trade between customs districts.[131]

Licensed and enrolled coasting vessels carrying products of the United States could trade within or between the customs districts in a single state or in immediately adjacent states with a minimum of paperwork. If a captain stayed within these prescribed limits, he could enter and leave any port without filing manifests or obtaining permits to sail. Unfortunately, this provision of the act has meant that official records of the activities of the local coasting

[130] Stein, *American Maritime Documents*, 59; *Daily Georgian*, September 13, 1842.

[131] See Stein, *American Maritime Documents*, for a comprehensive discussion of vessel documentation.

trade, as it occurred within the customs districts of a single state, or adjacent states, are almost nonexistent.

The register and enrollment documents issued to vessels are important historical records. They are almost identical documents and provide information on the ownership, place, and year of construction and dimensions of the vessel. Licenses typically contain only the name, tonnage and ownership of a vessel, but they do normally state whether a vessel was engaged in the "coasting" trade or in the "fisheries," information that is typically not included in enrollments.[132]

Laws regulating the coastwise trade were modified over time. In 1790, the law respecting tonnage duties for American-owned ships was altered. Under the new law, enrolled United States vessels carrying American goods within the districts of a state, or an adjacent state, paid no duties.[133] In 1792 and 1793, new laws were passed that more specifically regulated coastwise commerce. The 1793 law stipulated that new enrollments and licenses had to be obtained anytime a vessel was sold, altered, or transferred to some other activity and that any enrolled vessel sailing to a foreign port without exchanging its enrollment for a register would be liable to seizure and forfeiture. Additionally, the 1793 act specified that the bond required of all vessels licensed in the coasting trade would range from a hundred to a thousand dollars, depending on the size of the vessel, and the master would have to take an oath that he was a citizen of the United States. As with the earlier laws, coasting vessels carrying American goods between districts within a state or an adjoining state were not required to submit any paperwork to customs officials, secure any type of permit to depart, or make any report of arrival. However, the master was required to have a manifest of the cargo available for exhibit should it be demanded by authorities.[134]

Customhouses also issued "master carpenter's certificates," documents prepared when the construction of a new vessel was complete. The certificate, which included the date and place of construction and the vessel's name and basic dimensions, was signed by the builder or carpenter. This document was required before any enrollment or register could be issued.[135]

[132] Ibid., 86, 88.

[133] Johnson, *History of Domestic and Foreign Commerce*, 330; United States of America, *Public Statutes*, 1:135.

[134] Johnson, *History of Domestic and Foreign Commerce*, 331; Stein, *American Maritime Documents*, 74.

[135] Stein, *American Maritime Documents*, 110.

The provisions of the 1793 laws concerning enrollment and licensing of vessels in the coastwise trade remained in effect, with only minor modifications, throughout the nineteenth century. The various navigation laws were important in many respects, not the least of which was that they incorporated one of the few ways the American government could raise revenue. Importantly, they were intended to protect and encourage American commerce by making American-produced goods easier and cheaper to transport aboard American vessels under the command of American masters.[136]

As a result of the 1789 legislation, four customs (or "collection") districts were established in Georgia. These were Savannah, Sunbury, Brunswick, and St. Marys. These towns were selected based upon their importance as shipping ports at the time. Customs officers were assigned to each district, and these men were required to issue vessel documents (enrollments, licenses, registers, master carpenter's certificates, etc.) and to collect required customs duties. The collection district of Savannah extended from the Savannah River south to Ossabaw Island. It encompassed the Savannah River, the Ogeechee ("Great Ogeechee"), and Little Ogeechee rivers, plus the islands of Tybee, Little Tybee, "Warsaw" (Wassaw), and Ossabaw. Savannah was named the port of entry for the district. The collection district of Sunbury was the area from south of the Ogeechee River to Sapelo Island (in present-day Bryan, Liberty, and McIntosh counties) and encompassed the Medway, North Newport, South Newport, and Sapelo rivers, and the inlets into St. Catherines and Sapelo sounds. Sunbury, located on the Medway River, was named the port of entry.

The Brunswick district covered the central Georgia coast, encompassing the area south of Sapelo Island to the southern end of Jekyll Island and included the Altamaha, Frederica, and Turtle rivers. Brunswick was the port of entry, and Frederica on St. Simons Island was named a "port of delivery." In 1806, the town of Darien, near the mouth of the Altamaha River, was named a port of delivery.

The collection district of St. Marys covered the southern Georgia coast and included the Big ("Great") Satilla, Little Satilla, Crooked, and St. Marys rivers and the inlets into St. Andrews and Amelia sounds. The town of St. Marys on the St. Marys River was named the port of entry.

These four districts were established based on the shipping patterns in Georgia in 1789. These patterns changed over time, as did the boundaries of collection districts. For example, in 1790, the boundary of the district of St.

[136] Johnson, *History of Domestic and Foreign Commerce*, 332.

Marys was shifted northward to include the territory from the southern end of Jekyll Island, and, in 1822, the southern boundary was shifted to encompass the northeastern portion of Florida to the Nassau River, reflecting the recent acquisition of East Florida by the United States. In 1793, the new collection district of Hardwick was established with the tiny community of Hardwick on the Ogeechee River as the port of entry. In 1844, this district and the Sunbury district were abolished and annexed to the Savannah district. In 1818, Darien replaced Brunswick as the port of entry in the Brunswick district, and the district was abolished and annexed to the St. Marys district in 1844. The shift from Brunswick to Darien in 1818 reflects the boom in shipping that occurred at the latter town after the War of 1812 due to the tremendous amount of upland cotton being shipped down the Altamaha River from the interior.[137]

[137] "Georgia Customs Districts," 159–64, RG 36, Bureau of Customs, NARA.

# Chapter 4

## "The Products of the Country": The Cargoes in the Georgia Coasting Trade

The thirty-one bales of Sea Island cotton and twelve hundred bushels of rough rice carried into Savannah by Captain Charles Stevens aboard his sloop *Splendid* in January 1844 represented the most lucrative crops produced in tidewater Georgia in the first half of the nineteenth century. The sailing coasters carried Sea Island cotton and rough rice more than any other cargo (Figure 4.1). The cultivation of Sea Island cotton began after the American Revolution, but rice had been an important crop since the colony's early years. With indigo, rice represented the first of the staple crops of the settlers of coastal Georgia and South Carolina. These products were supplemented by naval stores, timber, and food crops (corn, potatoes, and oranges), as well as livestock, mostly cattle and hogs. Rice remained an important tidewater crop for over a century, but in Georgia, the cultivation of indigo was in decline by the late eighteenth century, about the time Sea Island cotton gained favor among coastal planters. Upland, or short-staple cotton, grown primarily inland, became the dominant agricultural product of the American South and was an important cargo of the coasting vessels after about 1810. This cotton was not grown in large amounts on the Georgia coast, but it was shipped down the state's rivers to several small coastal ports, such as Darien, in large quantities. The coasters carried it into Savannah and Charleston from these port towns.

Most coastal products inbound into Savannah were destined for export to Northeastern ports or overseas. Fairly small amounts were intended for local consumption in and around Savannah. These "products of the country" constituted the coasters' main cargo from small coastal ports and plantations into Savannah and Charleston and, overall, represented a primary source of income for most of the coasting captains and shipowners. They did, however, form only part of the cargo involved in the coastal trading system. The goods carried by local vessels from Savannah and Charleston differed entirely from those carried into these ports. Outbound cargoes consisted of dry goods, foodstuffs, building materials, hardware, and other merchandise and consumer items not available or produced locally. After about 1820, Savannah newspapers printed fairly complete information on inbound vessels in their shipping columns, which included information on the types and quantities of cargoes carried (at least the principal ones), the ports of origin of vessels, and the merchants or persons for whom the shipments were intended. These newspaper entries provide valuable information on the cargoes inbound into Savannah, and they represent a primary source of information in this study. Information on the cargoes outbound from Savannah and Charleston to coastal towns and plantations was rarely printed in newspapers, so our understanding of the goods moving in the coasting trade as derived from newspaper entries is incomplete. Some information on the makeup of this outbound trade can be gleaned from other sources, such as the few newspapers published in small port towns and from documents such as plantation or merchant journals and account books that list goods shipped and received by coasting vessels.

Much has been written about the cultivation of the coastal crops, particularly rice and Sea Island cotton, and their relationships to the system of slavery and the exclusive planter class. Little has been written about these tidewater products from the point of view of the principal concern of this work, the coasting trade. Of particular interest here is how these goods were prepared, packaged, and handled as cargo for the vessels carrying them to market, the seasonality of this market, and the economics of their transport. These aspects of coastal commodity production and marketing are rarely addressed in contemporary documents. However, sufficient details exist in available sources to enable an informed view of how the planters prepared their crops for shipment, how they were handled as cargo by the coasting captains, and what were the basic financial aspects involved in their transport.

## Indigo

With rice, indigo represented the principal cash crop of Georgia farmers during the colonial period. The semitropical climate and light, sandy soils of the coastal lands were well suited to cultivating the indigo plant (*Indigofera tinctorial*). Indigo was grown in South Carolina before 1700 but was largely supplanted by rice in the early 1700s. In the 1740s, a series of events led to the revival of indigo cultivation and contributed to its spread in the Sea Island region of South Carolina and the recently established colony of Georgia.[1] These events included a decline in rice prices, partly related to the disruption of the rice trade from an almost continuous naval war between the English in South Carolina and the Spanish in Florida. This warfare also shut off British supplies of indigo from the traditional French and Spanish sources. As a result, British textile manufacturers turned to the South Carolina and Georgia colonies for indigo even though they considered it inferior to that produced in East Florida and the French West Indies. To stimulate and maintain the plant's cultivation in the colonies, Parliament in 1748 instituted the Imperial Indigo Bounty, which authorized the payment of subsidies to growers. These subsidies remained in effect until the American Revolution.[2]

Only modest amounts of indigo were exported from Charleston before 1754, but in 1755, 177,000 pounds of the dye were shipped from that city, plus 23,000 pounds shipped from the ports of Georgetown and Beaufort. Subsequently, indigo cultivation in coastal South Carolina continued to expand. However, by the early 1790s, Sea Island cotton began to replace indigo as a cash crop. Indigo cultivation lingered in South Carolina until the end of the eighteenth century, with Charleston exporting 6,892 pounds in 1800; apparently, the last year indigo appears as an export from the city.[3]

After its introduction into Georgia in about 1740, indigo cultivation spread rapidly through the coastal section of the colony. Indigo production in Georgia never equaled that of South Carolina, but records show 2,395 pounds of indigo were exported from Savannah in 1753. This was a trifling amount compared to the 2,996 barrels of rice, representing about 689,080 pounds, shipped from the city the same year. Indigo exports from Savannah rose to 9,633 pounds by 1763 and to 19,900 pounds by 1772. Subsequently, indigo production in Georgia declined because of the expansion and profitability of rice cultivation and the introduction of Sea Island cotton. As in South Carolina, the American Revolution cut off the primary market for indigo and ended the lucrative bounties paid to Georgia planters. By 1791, Britain relied almost exclusively on indigo from East India.[4]

[1] Bonner, *Georgia Agriculture*, 18.

[2] Rowland et al., *History of Beaufort County*, 161–62.

[3] Ibid., 161–62, 280.

[4] Kayser & Co., *Commercial Directory*, 45; J. F. Smith, *Slavery and Rice*, 22–23, 28; Sullivan, *Early Days*, 45; "Agricultural Memoir," *DeBow's Review*, July 1855, 19:113.

Indigo grows best in rich, light soils containing little clay—exactly the type of soil commonly found in the Sea Island region.[5] Planting normally began in early April, and the plant was first harvested when about three feet tall and in full bloom, usually by early July. Commonly, plants were cut more than once, often as many as three times through September. Producing dye from the indigo plant was a simple procedure requiring the construction of facilities to ferment and process the plant. Building these facilities involved a modest capital investment by the planter that was much less than the expense required for rice cultivation. Indigo processing was normally done in two or three water-tight vats constructed of wood, each about twelve feet square and four to five feet deep. One vat, known as the steeping vat, was often elevated so that it could drain into the other vat, commonly known as the "beater" or "battery." Bundles of indigo plants were placed into the water-filled steeping vat, where they were allowed to steep and ferment, drawing the coloring out of the plants, which settled toward the bottom of the vat. The steeping process normally took twelve to fifteen hours.

A drain near the bottom of the steeping vat was opened, and the heavy, dye-infused water was drained into the beater vat. Here, the liquid was churned or agitated with paddles attached to a shaft extending across the vat. The churning added oxygen to the liquid, causing the indigo coloring to settle out as a precipitate. Lime (or a similar substance) was generally added to aid in precipitation. As the coloring settled to the bottom, the water was carefully drawn off through a series of holes extending down the side of the vat. Once the water drained, the indigo sediment in the bottom of the tank was scooped out, strained, and then hung in cloth bags to drain further. Once drained, the indigo paste was pressed, cut into small blocks, and placed in a drying house where it was slowly dried. When the pieces were fully dry, they were boxed for shipment to market, usually in November or December.

Processing indigo was a disagreeable and dangerous task. The fermentation of the plants attracted flies and other insects, and the smell was extremely disagreeable. The drainings from the vats were toxic and often killed fish and animals. Enslaved Black workers performed all the physical work involved in indigo cultivation and processing in Georgia, and the indigo must have adversely affected them, too.

[5] Bonner, *Georgia Agriculture*, 19–20; "Agricultural Memoir," *DeBow's Review*, July 1855, 19:113; "The Cultivation and Preparation of Indigo," *DeBow's Review*, August 1855, 19:243; Sullivan, *Early Days*, 45.

As late as 1792, Savannah merchant David Robinson advertised that he was "commencing the factorage and commission business" and would "receive tobacco, rice, indigo, corn, and lumber."[6] However, by this time, indigo was disappearing as a commercial crop in Georgia. No listing of indigo as an export from Savannah in 1800 or later years has been found, and indigo doesn't appear as cargo of the coasting vessels arriving into Savannah during these years. While it is unlikely that indigo was being produced commercially to any measurable extent in Georgia after 1800, it was being grown there as late as 1825, possibly only for home use, as indicated by an advertisement published that year in the *Savannah Republican* for "INDIGO SEED" being sold by Samuel Richardson "on the White Bluff Road."[7] As late as 1825, indigo was treated as a unique commodity in the commercial regulations of Savannah. The regulations for "Wharfage on Landing and Shipping Country Produce" that year contain an entry for indigo, noting that "For each package of indigo," the charge was four cents. While Savannah regulations do not specify the type of container used to package indigo, the Charleston regulations for 1820 list wharfage rates for indigo "per cask, barrel, box, case, or other package." These regulations for indigo were probably long outdated and had not been removed from printed policies. No specific description of how indigo was packaged as cargo for coasting vessels sailing after 1800 has been found.[8]

## Rice

Rice was grown in Georgia from the earliest settlement until the late nineteenth century. It was an important staple for coastal planters for over a century, from about 1750 to the Civil War, and it became one of the two most important commodities of the Georgia coasting vessels. The subtropical climate of coastal Georgia and South Carolina, with mild winters, hot summers, excessive humidity, and high precipitation, was ideal for rice cultivation.[9] Before the American Revolution, rice cultivation was largely confined to

[6] *Georgia Gazette*, June 28, 1792.

[7] *Savannah Republican*, February 11, 1825.

[8] Information for Savannah in 1825 comes from the *Savannah Republican*, January 4, 1825, and the information on Charleston in 1820 is from Kayser & Co., *Commercial Directory*, 42. Sullivan, *Early Days*, 45, notes that indigo produced on the Georgia coast was packed in barrels.

[9] There is considerable literature dealing with rice cultivation along the southeastern coast. Principal sources for this discussion include James M. Clifton, *Life and Labor*; Hoffman and Hoffman, *North by South*; House, *Planter Management*; J. F. Smith, *Slavery and Rice*; Stewart, *What Nature Suffers*; and Sullivan, *Early Days* and *All Under Bank*.

freshwater swamps above tidal influence that were cleared, drained, and diked for planting. Swamps or creeks lying at slightly higher elevations were dammed to produce lakes or ponds from which water could be drawn to irrigate the cultivated land. This swamp technique for growing rice began to be supplanted by the tide flow method in the middle of the eighteenth century. This technique relied on the interplay of large volumes of freshwater outflow with the peculiarly high tidal rise on the lower stretches of the region's major rivers. Essentially, high tides produced the periodic freshwater flooding required for rice cultivation. The effects of tides along the South Carolina and Georgia coast extend as much as ten to twenty miles inland on the major rivers in the area, including the Combahee, Savannah, Ogeechee, Altamaha, and Satilla. Within this zone, the strong flow of an incoming tide turns back the river's current, forcing it to overflow its banks and flood the adjacent swamplands with salt-free river water. Rice planters took advantage of this fortuitous circumstance, and harnessing this natural freshwater overflow formed the mechanical basis of the coastal rice industry. Ultimately, much of the low river-valley land within the tidal zones of these two states was turned to rice cultivation.

The exact date of the introduction of rice cultivation into South Carolina is unknown and obscured by legends and unverified accounts. There is evidence that rice was grown there as early as 1677. Numerous accounts mention its cultivation in the 1680s and 1690s, and it was being exported from Charleston by 1695.[10] The early rice was of inferior quality, and it was not until the introduction of superior seed from Madagascar in the 1690s that rice culture developed as a profitable staple crop. The economic possibilities of rice cultivation were soon realized, and its rapid expansion in South Carolina changed life in the colony. Enslaved Black laborers worked the rice fields, and as rice-growing spread, the number of enslaved people in the colony increased dramatically. By 1708, it is reported there were as many enslaved people in South Carolina as Whites, and for several decades thereafter, the slave population grew faster than the White population. By the 1720s, rice cultivation had spread to most of the suitable areas along the central and northern South Carolina coast and north into North Carolina. Georgia's ban on slavery limited rice cultivation there. When this ban was lifted in 1749, many South Carolina planters began to purchase land along the mouths of the rivers below Savannah for rice cultivation. During the British occupation of Florida (1763–1783), many South Carolina and Georgia rice planters settled there. But, after

[10] Coker, *Charleston's Maritime Heritage*, 43.

a brief period of success, these planters abandoned rice planting in Florida. Within a half-century of its introduction, Carolina rice cultivation reached its geographic limits, from the Cape Fear River in North Carolina to the St. Marys River at the lower border of Georgia. The boundaries of rice cultivation along the Southeastern coast remained virtually unchanged for the next century.

In Georgia, the early swamp technique of rice cultivation did produce small quantities of rice for export. Rice plantations had been established along the lower Savannah River before 1750, and rice was shipped from the colony as early as 1741. Rice became a meaningful commercial export from Savannah in 1753, when 2,996 barrels were shipped. In 1755, 2,299 barrels of "clean" (milled) rice and 237 bushels of "rough" (unmilled) rice were exported from the city. How much rice these figures represent is debatable due to conflicting statements as to how many pounds were in a barrel. Amounts range from 230 to 600 pounds. While the actual number of pounds of rice exported from Savannah is impossible to determine, the amount was insignificant relative to the 92,210 barrels exported from Charleston in 1755. Fifteen years later, in 1770, Savannah exports had risen to 22,129 barrels of clean rice and 7,064 bushels of rough rice. Rice was now grown in the tidewater regions of the Savannah, Ogeechee, and Altamaha rivers and irrigated by the tide flow technique.[11]

While rice had become the principal money crop of the Georgia colony by 1770, other exports were also important. In 1770, Savannah also exported 22,336 pounds of indigo, 284,840 pounds of deerskins, 44,539 pounds of tanned leather, 13,447 pounds of tobacco, and 18,405 pounds of sago powder. Smaller quantities of other commodities were shipped out of Savannah and other Georgia ports. These included naval stores (pitch, tar, and turpentine), timber, lumber, shingles, staves, lime, ground nuts (peanuts), myrtle wax, turkeys, horses, oxen, chickens, beaver skins, raccoon skins, otter skins, cow horns, leather, and sturgeon.[12]

The rice exports listed for Savannah in the eighteenth century are not a true measure of the actual production of the crop in Georgia. Many Georgia planters shipped their rice directly to Charleston for sale, bypassing Savannah, because they were from South Carolina and had established ties with

[11] Fraser, *Savannah in the Old South*, 40; Savannah export figures are from M. Granger, *Savannah River Plantations*, 62; Gray, *History of Agriculture*, 2:1022; Kayser & Co., *Commercial Directory*, 45.

[12] *Daily Georgian*, July 24, 1828; Coulter, *Georgia*, 102.

merchants in Charleston. In addition, Savannah lagged far behind Charleston in terms of both the physical and financial facilities needed to process and market the rice crop.[13]

The techniques of growing rice by tidal irrigation were fully developed in the 1820s, and by the 1830s, most rice in Georgia and the Carolinas was being grown on tide flow lands; its cultivation on inland swamps had been largely abandoned. With the adoption of tidal irrigation, plus improvements in machinery for harvesting and processing, rice cultivation began to reach its full potential on Georgia coastal rivers, and the production and export of rice dramatically expanded.[14] In 1848, there were about fifty rice plantations on the Savannah River (half of which were in South Carolina), nineteen on the Ogeechee River, and thirty-five on the Altamaha and Satilla rivers.[15] Table 4.1 provides information on the agricultural production of the principal coastal counties involved in the Georgia coasting trade in 1840, 1850, and 1860. Rice production increased between 1840 and 1850 in these counties, reflecting the final perfection of the tide flow technique.

Successful rice cultivation relied on a combination of technical skills, including understanding tidal hydrology, building dikes and canals, and expertly placing wooden culverts, known as "trunks," to manage field flooding and drainage. It also required a considerable financial outlay to purchase sufficient land to grow an income-producing crop. The labor needed to clear land, dig ditches, and build and maintain dikes and trunks required a large force of enslaved workers. As a result, successful tide flow rice cultivation was limited to a small number of wealthy planters.[16]

Preparing the land for tidal irrigation was laborious, time-consuming, and expensive, and South Carolina and Georgia planters determined it could be accomplished only with enslaved laborers. Initially, a planter selected a tract low enough to be flooded by river water at high tide but high enough to be drained at low tide. Then, enslaved laborers cut trees and cleared and removed underbrush before expending a great deal of effort digging drainage

[13] See House, *Planter Management*, for a discussion of the differences between the commercial and financial infrastructures of Savannah and Charleston although, as noted elsewhere, evidence derived from the volume of local coaster traffic suggests that the maturation of the mercantile system in Savannah occurred in the 1820s, about a decade earlier than proposed by House.

[14] House, *Planter Management*, 20–25; Rowland et al., *History of Beaufort County*, 321.

[15] Bancroft, *Census of the City of Savannah*, 38.

[16] Clifton, *Life and Labor*, xii–xiii.

ditches and building earth dikes and banks around fields. Openings, or trunks, were placed in dikes to permit flooding and draining of fields. These trunks were rectangular wooden culverts from twenty to thirty feet long with flap doors at either end. To flood a field, the outer trunk door was opened and the inner door let down. At high tide, water pressure would push open the inner door and flood the field. At low tide, the inner door would be forced closed by the water trying to flow out of the field, thus damming the water in the field. The opposite procedure was employed to drain fields.

Each field was kept as level as possible, and dikes and ditches were constructed in fields to ensure the proper water flow. Smaller dikes, known as "cross dams," were erected to enclose fields an acre or so in size. These cross dams were fitted with trunks to regulate water flow. Large ditches lined the perimeter of major fields, and smaller ditches drained into the main ditch. Deep canals were often dug from the river to supply water to backfields. These canals also served as thoroughfares for the "flats" or "lighters" used to move crops, construction material, and men around the fields (Figure 4.2).

In Georgia, preparing rice fields for planting usually began soon after Christmas. Fields were allowed to drain and dry out, then plowed and broken up with harrows, leveled, and then trenched with hoes or plows. Some planters left their fields plowed and dry before planting, and others flooded their fields just before planting to kill weeds. Rice planting was done between March and May, depending on the weather. Seeds were sown and covered in the previously dug trenches. Following planting, the fields were periodically flooded to kill weeds and to encourage the growth of the rice plants. Between floodings, enslaved work crews hoed the fields to remove weeds.

A final (usually the third or fourth) flood of the rice fields in August continued several weeks until the rice ripened. The water was then drained, and the harvest began. The rice was cut with scythes and left in the fields for several days to dry. It was then tied into small sheaves and carried out of the fields to the threshing yard or barn. The bulk of the rice shipped from South Carolina and Georgia plantations was "rough" rice, referring to grain that had been only threshed, winnowed, and cleaned of chaff and straw. The Georgia rice was typically threshed using flail sticks on a prepared hard earth surface or a raised wooden threshing floor. After threshing, the grain was winnowed to remove the chaff either by dumping the grain from an elevated room, allowing the wind to blow the chaff away, or by using large fans driven by animals, tide, or steam to blow away the chaff. Threshing floors and storage barns were built near rivers or streams so the grain could be easily loaded into vessels for transport to Savannah or Charleston. Threshing and winnowing were normally completed in September or October, when the rough rice was loaded into baskets or sacks, carried to the awaiting vessel, and, typically, dumped loose into the hold. Rough rice was measured and sold by the bushel, representing about forty-five pounds of rice.[17]

Rice "pounding mills" were constructed where the rough rice was processed into clean rice. Typically, one forty-five-pound bushel of rough rice produced about twenty-seven pounds of clean rice after milling.[18] The early mills were driven by tidal flow and used machinery that raised and dropped large timbers (known as pestles) into mortars filled with rice. Later, many of the tidal mills were converted to steam power, and in the 1830s, superior technology using revolving stone cylinders emerged.[19]

One of the earliest rice-pounding mills in Georgia was Drakies Mill, known to have been operating by 1809 at Drakies Plantation on the Savannah River, about ten miles upriver of Savannah. By 1825, several rice mills were located along the Savannah River near the city.[20]

These early, tide-driven mills were not extremely efficient. As late as 1828, a Savannah newspaper lamented that these mills were not sufficient for Georgia planters, forcing them to ship their rice to Charleston mills.[21] In 1828, Alexander Telfair, Robert Habersham, and Amos Scudder built what may have been Savannah's first steam-powered pounding mill, the Savannah Steam Rice Mill.[22] Over time, other steam-powered mills were erected near Savannah. One of the largest was the Pennyworth or Hamilton Mill on Pennyworth Island. The machinery at the Pennyworth Mill, manufactured by the West Point Foundry of New York for five thousand dollars, consisted of a thirty-horsepower steam engine powered by four boilers.[23]

Many rice mills were constructed on plantations to handle only that plantation's crop. However, some mills were commercial ventures, pounding rice for planters for a "toll," or a percentage of the clean rice produced. In 1839, James Hamilton published a notice that his Pennyworth steam mill was prepared to "receive Rice to beat on toll at the same rates as the city mills"—

[17] J. F. Smith, *Slavery and Rice*, 55; House, *Planter Management*, 62–66.

[18] House, *Planter Management*, 65.

[19] Sullivan, *Early Days*, 179.

[20] M. Granger, *Savannah River Plantations*, 155.

[21] *Daily Georgian*, August 12, 1828.

[22] "Savannah Steam Rice Mill," Mary Lane Morrison Papers, Georgia Historical Society.

[23] Leech and Wood, *Archival Research*, 166, 168–70; Rowland et al., *History of Beaufort County*, 321–22; *Daily Georgian*, December 4, 1839.

and he would deliver the milled rice to Savannah at no charge. Millers typically charged 10 percent of the sale value of the rice for milling and storage.[24]

The production of clean rice also produced "small" or broken rice and rice flour, both of which could be sold but were usually unprofitable. Therefore, small rice and rice flour were often consumed on the plantation, primarily by the enslaved laborers. Only rarely do accounts appear of coasting vessels coming into Savannah carrying rice flour, and these typically sailed from the toll mills on the Savannah River.

When clean rice was shipped, it was loaded into casks, or "tierces." A tierce was one-third of a hogshead, and it generally held about 600 pounds of grain, but there was some variation in this amount. Records for the estate of Glynn County planter James Troup show he shipped 282 tierces of rice from his Altamaha River plantation on March 31, 1851, each holding an average of 588 pounds of rice. A year earlier, the 1,513 tierces shipped from nearby Hopeton Plantation each held an average of 708 pounds of rice.[25] These weight variations were related to several factors, including slight differences in the size of the handmade barrels, how carefully the barrels were filled, and the weight of the grain itself, a reflection of how fully it had been dried for shipment.

The weight of the wooden casks themselves ranged from about seventy-five to eighty pounds and added considerably to the weight of a cargo of clean rice, so the ship's captain had to account for the additional weight as he loaded his vessel. These casks were made locally, as indicated in an advertisement by Robert Habersham & Sons appearing in the *Savannah Daily Republican* of July 19, 1860. The advertisement noted that the firm "Wanted—Contracts for one to five hundred thousand Pine Staves suitable for Rice Casks." Mills that pounded rice needed tierces to hold their processed grain, and some manufactured their own. In 1824, the Upper Steam Saw and Rice Mill at Darien advertised that it sought to purchase "25,000 rice barrel staves and 10,000 hoop poles suitable for rice barrels." Various records suggest that about twenty-one or twenty-two bushels of rough rice (approximately 945 to 990 pounds) were required to produce one tierce of clean rice.[26]

While planters could make a greater net profit on clean rice, shipping their crop in rough form precluded the need for a mill and eliminated the

[24] *Daily Georgian*, December 4, 1839; Rowland et al., *History of Beaufort County*, 322.

[25] Glynn County, Georgia, estate of James Troup, Wills and Appraisements, Book E, 251; ibid., estate of James Hamilton, 211–13; ibid, Book F, 179, 183.

[26] *Darien Gazette*, June 22, 1824.

difficulties and expenses associated with processing the grain and maintaining the mill equipment. It was much easier to ship the rice in the rough form and leave these concerns with the factor, and most of the rice shipped into Savannah and Charleston was rough rice.

Because of the high costs required to develop a profitable rice plantation, there were fewer than six hundred rice planters in the South in 1850; 427 were in South Carolina. Georgia, with five major rice-producing rivers, had 125 planters in 1850, eight of whom produced over one million pounds in 1849. The number of rice planters in these two states decreased over the 1850s. By 1860, only ninety-six remained in Georgia. However, the production of the largest plantations significantly increased over the decade of the 1850s, and by 1860, the lower Savannah River was the center of rice production in the United States. That year, South Carolina planter Louis Manigault noted forty-two plantations along the river, with a combined 17,758 acres in production.[27]

Newspaper records of coasters arriving in Savannah indicate that shipments of new rice into the city sometimes began as early as mid-September but were typically somewhat later (Figure 4.3). For example, in 1858, the October 6 issue of the *Savannah Morning News* reported "New Rice —The first load of the new crop this season was received at the lower rice mill yesterday." This rice came from the plantation of Dr. J. P. Screven on the lower Savannah River. In some years, the first shipments of rice aboard sailing coasters were not reported until mid or late October, possibly reflecting late harvests. However, in these years, plantation flats may have carried some rice into the city on earlier dates. In 1825, the earliest shipment of rice into the city aboard a coaster was not reported in the *Daily Georgian* until November 11, an unusually late date, when the sloop *Bolivar* arrived with a cargo of "rice, etc." from Darien. However, at the time, Savannah newspapers were not consistently publishing information on the cargoes of arriving coasters, so an earlier arrival of rice into the city may not have been listed. Occasionally, unusual circumstances delayed the transport of rice into Savannah, such as in 1854 when a serious yellow fever epidemic struck Savannah; a quarantine essentially ended all local coaster traffic into the city that summer and fall. That year, the first reported shipment of rice into the city was not until November 2, when the schooner *Cotton Plant* arrived with thirty-one hundred bushels of rough rice from the Ogeechee River after the danger from yellow fever had passed.

[27] Clifton, *Life and Labor*, xiii–xvi.

Additionally, a hurricane in September 1854 seriously damaged the Georgia rice crop, significantly reducing the amount available for shipment that year.[28]

Typically, most rice crops were shipped between October and December or January, continuing at a lesser pace through February or March. It was not uncommon for shipments of small amounts of rice to continue into Savannah until as late as May or June (see Figure 4.3). Occasionally, rice shipments are reported in July and August, but these represent rice held back from the market for one reason or another. By the 1840s, steamboats carried a considerable amount of the cotton crop produced along the Georgia coast; however, because they lacked the deep hulls found on sailing vessels, steamers rarely carried rough rice.

A few accounts describe the actual loading of coasting vessels with rice. One is by Alice Ravenel Huger Smith, who wrote about life on her family's South Carolina rice plantation in the 1850s. She describes loading a schooner with rough rice that was being sent to the "city" (probably Charleston), where it would be pounded into clean rice. Smith notes:

> It was always a great pleasure to watch the loading of one of these schooners by long lines of men and women with baskets on the head, boarding the schooner by one plank and going ashore by the other. Close to the open hatchway stood the ten-bushel tub with the captain or mate on one side and the overseer or key-keeper on the other. As each darkey passed the tub she would tilt the basket and pour the golden rice into the tub. When the measure was full, it would be "struck" by passing over the top a smooth board and then tilted into the hold. With each tilt the striker would cry "one" or "two" or "three" or "four" or "tally," and the captain and the overseer would enter it accordingly. On some plantations the schooner could go up to the mill itself, and would be loaded through a long chute.[29]

Smith's description is probably typical of the process involved in loading rough rice on most plantations. However, how the holds of coasting vessels may have been fitted to accommodate rough rice is unreported. It seems likely that the holds were lined with boards and, possibly, subdivided to keep the

[28] Numerous reports of the yellow fever epidemic appeared in the *Daily Georgian* for 1854; Lee and Agnew, *Historical Record*, 134; the arrival of the *Cotton Plant* was reported in the *Daily Georgian*, November 2, 1854.

[29] A. R. H. Smith, *Carolina Rice Plantation*, 62–63.

cargo from shifting or from spilling into the bilges. A well-constructed and sealed cargo hold would make it easier to remove the rice once in port at Savannah or Charleston. Employees or enslaved laborers of the merchant receiving the rice typically unloaded arriving vessels. In the 1850s, William Gordon, a clerk with the Savannah merchant firm Tison & Mackay, noted that enslaved women measured the rice delivered to the rice mill where he worked. Vessels sometimes carried rough rice from more than one plantation on a single voyage, so there must have been divisions erected in a ship's hold to keep the crops separate in the event there were later questions about quality, spoilage, etc. Savannah River rice planter Charles Manigault wrote to his son Louis on December 13, 1855, noting that a vessel named *Catherine* would arrive at the family's Gowrie plantation from Charleston for a load of rice. Manigault stated that the *Catherine* could carry five thousand bushels, but he had promised her only two thousand because the vessel was already engaged to pick up three thousand bushels for John Rutledge, another planter on the lower Savannah River.[30]

Because the rice was carried loose in the hold, a record of the amount dumped into a vessel was critical. However, there were some discrepancies in amounts shipped. On January 7, 1845, Roswell King, Jr., recorded shipping 1,944 bushels of rough rice and sixteen barrels of syrup from his plantation, South Hampton, in Liberty County, to Savannah aboard the sloop *Eagle*. On January 13, the *Daily Georgian* reported the arrival of the *Eagle* with the sixteen barrels of syrup, but only 1,800 bushels of rough rice. Presumably, the difference in the amount of rice shipped and the amount received would be worked out between the planter and his factor in Savannah. The newspaper reported the *Eagle* was arriving from the small port town of Riceboro, suggesting the newspaper was using the name because it was the nearest port to King's plantation landing, where the rice would almost certainly have been loaded.[31] Charles Manigault, writing in 1852 on the costs entailed in growing and milling rice, recorded some of the problems he saw in shipping the grain. Manigault wrote:

> As soon as Rough Rice leaves the Plantation [it] is in the hands of those who Care Nothing about it—and in all its subsequent movements it encounters loss—theft—or injury. From the Planters Barn to the Vessel, & in unloading at the Toll Mill there will surely occur

[30] Clifton, *Life and Labor*, 201; Haunton, "Savannah in the 1850s," 98 (William Gordon account).

[31] Roswell King's shipment is reported in Sullivan, *All Under Bank*, 42.

> some little damage. And on her voyage from the Plantation to the Toll Mill some little is sure to get Dirty, or perhaps Wet, at the bottom of the Vessel. And while lieing [*sic*] at the Mill Wharf, perhaps several days, waiting her turn to discharge, some pilfering or waste may occur.[32]

He also wrote that there were instances where, on unloading at a mill, coaster captains kept for their private use any rice above the amount measured into the vessel at a plantation.[33]

How much rice any individual coasting vessel could carry was rarely specifically stated although a few newspaper advertisements and notices address this point. Before 1800, it was common to describe the carrying capacity of a coasting vessel in terms of the number of "barrels" it could carry, referring specifically to barrels of rice.[34] After 1800, it became more common to list only the measured burden of the vessel in tons or, infrequently, its carrying capacity in bushels of rough rice. As late as 1820, however, the carrying capacity of coasters was sometimes given in number of barrels. This was the case of the schooner *Choctaw* offered for auction in Savannah on November 25, 1820. The *Daily Georgian* advertised that the *Choctaw* was two years old, constructed of live oak and red cedar, and had a "burthern [of] about 300 barrels." More common by this time were listings of the number of bushels of rice a vessel could carry. A November 9, 1841, advertisement in the *Daily Georgian* stated that the sloop *Visitor* was available to sail and could "take 3000 bushels of rice." The *Visitor*, built in Greenwich, New Jersey, in 1833, was sixty-three feet, nine inches long, and had a burden of 58 72/95 tons. The three thousand bushels represent about 135,000 pounds of rice, meaning the *Visitor* could carry about twenty-three hundred pounds, or fifty-one bushels, per ton of burden.[35]

By the late 1820s, Savannah newspapers were beginning to provide consistent information on the amount of rice carried into the city by individual coasters. Every vessel was not filled on every trip, but the maximum amount of rough rice that could be carried by a particular coaster can be estimated by following its arrivals over several years. For example, the sixty-two-foot sloop *Science* was active in the coasting trade under several owners and masters from

[32] Clifton, *Life and Labor*, 109.

[33] Ibid., 110.

[34] Fleetwood, *Tidecraft*, 51.

[35] Dimensions of the *Visitor* are from Enrollment No. 13, Port of Savannah, sloop *Visitor*, August 23, 1837.

the mid-1830s to 1860. An examination of her arrivals into Savannah in the 1840s shows that the sloop commonly carried more than three thousand bushels of rough rice per voyage, and the most she ever brought into the city was thirty-five hundred bushels, which she did several times. This pattern continued until the mid-1850s, when the *Science* began to arrive with as much as four thousand bushels of rice. On one voyage in December 1858, she carried forty-three hundred bushels.[36] The maximum capacity of the *Science* appears to have been about thirty-five hundred bushels of rough rice until the mid-1850s, when she began to carry more rice. It appears as if the hold of the vessel was altered sometime in the 1850s to accommodate larger cargoes. The forty-three hundred bushels of rice carried by the *Science* in 1858 represents 193,500 pounds of grain. With a burden of seventy tons, this means the sloop could carry about sixty-one bushels (2,764 pounds) of rough rice per ton of burden. An examination of several coasting vessels operating out of Savannah between the 1830s and 1850s shows the average maximum amount of rough rice carried was about fifty-five bushels, or 2,502 pounds, per ton of burden. However, coasters came into Savannah with a maximum cargo of rough rice only once or twice a season. The rest of the time vessels carried lesser amounts of rough rice, in addition to other cargoes.

Loading a vessel with rice at a plantation landing seems to have been done quickly. Richard Arnold, a planter who owned two large plantations along the Ogeechee River in Bryan County, noted on March 23, 1847, that his slaves "put on board Capt Thompson's Sloop Science 2200 Bushels Rough Rice" and they still had time in the day to hoe "one task Sugar Cane." Later, on December 30, Arnold's journal notes that his slaves loaded two thousand bushels of his own rice on the *Science* at Cherry Hill Plantation, in addition to six hundred bushels on account with a Mr. Dorsey. The following day, the *Science* was moved to Arnold's other Ogeechee River plantation, White Hall, where an additional 410 bushels of rough rice were put aboard. The *Science* had arrived at Cherry Hill Plantation on December 29 with a cargo of eighty empty barrels that Arnold planned to use for holding syrup made from his sugarcane crop. In March 1853, Louis Manigault wrote that 4,830 bushels of rice were loaded on the schooner *John W. Anderson* in one day at the Manigault's plantation on Argyle Island on the lower Savannah River. This was an exceptional amount because Manigault wrote that the schooner's captain,

[36] *Daily Georgian*, 1842–1854; *Daily Morning News*, 1855, 1858–1861; *Savannah Daily Republican*, 1856–1857.

William Watson, stated that no more than four thousand bushels had ever been loaded on the schooner in one day before.[37]

These accounts demonstrate the quickness with which vessels were loaded. The entries for the sloop *Science* show that rice for two individual shippers was loaded on a single vessel. Transporting cargo from more than one shipper was not uncommon because coaster captains attempted to take on as much cargo as possible before returning to Savannah or Charleston. It also meant that the captains had to keep track of the various shipments to ensure that the rice was correctly accounted for when delivered at Savannah. Presumably, written accounts or logs of some type would have been necessary, but these records are almost nonexistent.

While rough rice was dumped loose into the holds of vessels, clean rice, which was shipped in large barrels, or tierces, was probably carried on deck as well as in the hold. Captains would have stored the heavy, six- to seven-hundred–pound tierces on deck whenever possible, simply because of the labor involved in lowering and hoisting them into and out of the hold. The tierces were easy enough to roll onto the vessel's deck, but a boom and pulleys had to be rigged to load them inside the hull.

The cost of shipping rice by sailing coaster was based on the bushel or tierce. Specific information on the freight rate charged for carrying rice is relatively rare before the 1840s. In 1841, the sloop *Visitor* advertised that it would carry rough rice into Savannah for 3.5 cents per bushel, but a more typical rate was five or six cents a bushel, a cost that continued through 1860. In 1823, the rate for transporting clean rice from Darien to Savannah was 62.5 cents per tierce, although in 1834 the sloop *Conductor* charged only twenty-five cents. The *Conductor* was carrying the rice on the short haul from Savannah River mills into the city, a probable explanation for the low rate. After 1850, the freight rate on clean rice was commonly one dollar per tierce, at least for Georgia planters shipping their rice to Savannah. Rates may have been slightly higher on longer deliveries, such as the $1.125 per tierce paid for clean rice shipped from Hopeton Plantation on the Altamaha River to the Charleston factor Robert Mure & Company in 1850.[38]

[37] Hoffman and Hoffman, *North by South*, 76–77, 131 (Richard Arnold); Clifton, *Life and Labor*, 140 (Louis Manigault).

[38] The sloop *Visitor* advertisement is from the *Daily Georgian*, November 9, 1841; other freight rates are taken from Kayser & Co., *Commercial Directory*; House, *Planter Management*; Pearson, "Georgia Coasting Trade"; Unidentified Cash Book, Georgia Historical Society; Glynn County, Georgia, estate of James Hamilton, Wills and Appraisements, Book E, 211–12.

## Sea Island Cotton

Cotton cultivation began in Georgia during the earliest years of White settlement and was used mainly for home consumption; only very small amounts were exported. Thomas Jones, storekeeper in Georgia for the Trustees, reported in 1741 that he shipped to England a "Bag of Cotton" from Doctor Patrick Graham's Mulberry Grove Plantation.[39] As late as 1766, only eight bags of cotton were exported from Georgia, and two years later three hundred pounds were reportedly shipped from the colony.[40] The expansion of the textile industry in England, beginning in the 1770s, stimulated a demand for cotton. Commercial cotton production in the coastal areas of Georgia, South Carolina, and Florida began with the development of the perennial, long-staple variety known as Sea Island cotton (*Gossypium barbadense* L.). The fiber of Sea Island cotton is very fine and silky and measures one to one and a half inches long, compared to a length of about half inch for short-staple, or upland cotton.

Conflicting claims and accounts make it difficult to know who first cultivated Sea Island ("black seed") cotton and in what year. Thomas Spalding, prominent planter from Sapelo Island, writing in 1828, stated that his father, James Spalding, was among the first to successfully grow Sea Island cotton, planting it on St. Simons Island in 1787. Spalding said the cottonseed had been sent the previous year to his father and several other coastal planters by friends in the Bahamas, where it had arrived from the island of Anguilla in the Leeward Islands. He also noted that the cotton did not produce well the first year but "ratooned or grew from its root the following year." The seeds from this second year's growth were successfully planted, and continued seed selection from the most vigorous plants resulted in cotton more fully adapted to the local conditions than the original from the Bahamas.[41]

Others claim that Sea Island cotton was first grown, at least commercially, elsewhere. Francis Levett stated he grew Sea Island cotton at his plantation near Harris Neck in McIntosh County in 1787. Levett acquired the land in McIntosh County in the 1760s, moved to East Florida in 1769, and then fled to the Bahamas during the Revolutionary War. He returned to Georgia at the war's conclusion, probably with the knowledge of long-staple cotton and with cottonseeds obtained in the Bahamas or from a seed collector

[39] M. Granger, *Savannah River Plantations*, 61.

[40] Gray, *History of Agriculture*, 1:184.

[41] Thomas Spalding's account was originally published in *Athenian*, June 17, 1828 (Athens, Georgia), found in Sullivan, *Early Days*, 116.

in South America. In 1799, an unidentified Chatham County correspondent to the Savannah newspaper *Columbian Museum & Savannah Advertiser* stated that several area planters had turned to cultivating Sea Island cotton in the 1780s because of the unprofitability of indigo. He wrote that, in 1791, ten thousand pounds of the cotton were shipped to England by "Messrs. Johnston and Robertson on account of Francis Levett, Esq., which established the character of Georgia Sea Island cotton; being the first shipment of any consequence." Other claims state that a John Earle of Skidaway Island planted some variety of black-seed cotton as early as 1767 for domestic use and had begun experimenting with the cultivation of Sea Island cotton in 1787. Richard Leake was growing Sea Island cotton on his Jekyll Island plantation by 1788, and within a year or so, several planters on the Georgia coastal islands were successfully producing the crop. In 1790, Nicholas Turnbull is reported to have successfully cultivated forty acres of Sea Island cotton on his plantation on Whitmarsh Island near Savannah.[42]

S. G. Stephens argues that a variety of long-staple cotton derived from the West Indies was being grown along the South Carolina and Georgia coast for local use well before the mid-eighteenth century. He states this cotton resulted from an outcross of *G. barbadense* with a native West Indian wild form of *G. hirsutum* that produces long, fine fibers. When this cotton was transplanted to the South Atlantic coast, there was a natural selection for variants that would produce seeds before the first frost. In the late eighteenth century, coastal planters' careful seed selection ultimately resulted in the commercially successful Sea Island cotton.[43]

Despite the conflicting claims, it appears that the commercial cultivation of Sea Island cotton began on the Georgia coast in 1786 or 1787 using seeds initially obtained from the West Indies. It also appears that the earliest growers of Sea Island cotton in Georgia included several who had close ties with planters in East Florida, where they may have learned about the cotton and its techniques of cultivation. Finally, careful seed selection led to a plant well adapted to the conditions found in the Sea Island region.

Sea Island cotton grows best in light sandy soil, salt air, and moderate temperatures, conditions found along the tidewater region from South

[42] Francis Levett's claim is reported in Gray, *History of Agriculture*, 2:676; Sullivan, *Early Days*, 230–31; and the *Columbian Museum & Savannah Advertiser*, October 15, 1799. John Earl's and Richard Leake's claims are from Gray, *History of Agriculture*, 2:674–677. Information on Nicholas Turnbull is from Turnbull, "Beginning of Cotton Cultivation," 39–45; and M. Granger, *Savannah River Plantations*, 33.

[43] Stephens, "Origin of Sea Island," 391–99.

Carolina to Northeastern Florida. Its cultivation quickly expanded from the coastal islands to a narrow strip of the adjacent mainland extending from Georgetown, South Carolina, south to the St. Johns River in Florida. Reportedly, the first mainland plantation to successfully grow Sea Island cotton was located along the Turtle River in Glynn County, just west of the port town of Brunswick. Robert Hazlehurst grew cotton here and named his plantation "Anguilla," after the long-staple, black-seed cotton that was to become a mainstay for the coastal economy. Seed selection remained an important aspect of cotton cultivation, and by the 1820s, at least six seed types were identified. The Sea Island cotton from the mainland was considered lower quality than that grown on the coastal islands. Similarly, the cotton from the islands on the lower Georgia coast was generally considered poorer quality than that coming from the islands north of St. Simons.[44]

Sea Island cotton cultivation expanded rapidly along the Georgia and South Carolina coasts, where conditions were most amenable for its growth and where indigo production had declined. By 1789, twenty planters were growing the cotton in Georgia, and in 1790, William Elliott of Hilton Head Island raised the first successful crop in South Carolina.[45]

Joseph Bancroft, in an 1848 account of the commerce of Savannah, states that the first shipment of Sea Island cotton from that city occurred in 1788. This cotton was grown by Alexander Bisset of St. Simons Island and shipped from Savannah by Thomas Miller.[46] In 1789, the first exports of Sea Island cotton from Charleston, thirty bags, were recorded. Some, if not all, of this cotton shipped from Charleston came from plantations on the Georgia Sea Islands, where its cultivation was already established. Sea Island cotton exports from Charleston soared to 25,157 bags, or over eight million pounds of cotton by 1800. Unlike the output of upland cotton that experienced a phenomenal increase through the first half of the nineteenth century, Sea Island cotton reached its geographic limits of cultivation around 1805. The Sea Island crop of 1830 was about the same as that of 1805.[47]

A great advantage of Sea Island cotton, in addition to its long fiber, is that it has a smooth seed that can be easily separated from the boll by hand or with simple roller gins. This task is much more difficult with short staple, or upland, cotton whose seed is difficult to separate from the fiber.

[44] Bullard, *Robert Stafford*, 147; Rosengarten, *Tombee*, 51.

[45] Rosengarten, *Tombee*, 50, 54.

[46] Bancroft, *Census of the City of Savannah*, 37.

[47] Gray, *History of Agriculture*, 2:680.

Because of its long, fine fiber, Sea Island cotton brought a premium price on the market, as much as three times that of upland cotton, and profits from its cultivation were high. Prices for Sea Island cotton in Charleston in the first years of the nineteenth century were extraordinarily high, ranging from forty-four to fifty-two cents per pound. These prices decreased to about thirteen cents per pound in 1813 and then rose to an extraordinary high of sixty-three cents in 1818. Prices fell to between fifteen and twenty-five cents per pound in the 1820s, where they remained until rising again in the late 1840s, reaching sixty cents per pound in 1860.[48]

Sea Island cotton became the almost exclusive type grown on the coastal islands and adjacent mainland and, as a consequence, was a major commodity carried by the Georgia coasting vessels. Some short-staple, or upland, cotton was grown on the coast, and soon it became the dominant variety grown in the interior of the entire South. One stimulus for upland cotton cultivation was Eli Whitney's invention of a new type of gin, which was more efficient at separating the seeds from the tightly adhering fibers. Whitney's invention was quickly pirated, and by 1795 local mechanics across the South were producing the new gins, with an immediate effect on the cultivation of upland cotton. In 1794, the United States produced 16,719 bales, or about eight million pounds of cotton, much of it Sea Island cotton. By 1800, national production had risen to almost thirty-six million pounds, mostly short-staple cotton grown in the uplands of Georgia and South Carolina. After this, the production of short-staple cotton increased dramatically as its cultivation expanded west. By 1820, cotton was the most valuable commodity shipped out of the South, and a considerable quantity passed through the port of Savannah. In 1817, Ebenezer Kellogg, a resident of Massachusetts traveling through Georgia, noted while in Savannah: "At this place I hear nothing but talk of the price of cotton. The Georgians are madly devoted to cotton. Last year there were exported from Savannah a hundred and six thousand bales of upland, and ten thousand of Sea Island."[49]

Before 1800, upland cotton brought high prices, only slightly lower than that of Sea Island cotton. After 1800, the price of upland cotton dropped to about twenty cents per pound and remained there or lower for the next several decades.

Upland cotton became an important cargo of the Georgia coasters because it was brought down the state's rivers in large quantities to towns and

[48] Rosengarten, *Tombee*, 49; Gray, *History of Agriculture*, 2:697, 1030.

[49] Martin, "New Englander's Impression," 252.

landings near the coast, where it was taken aboard sailing vessels for the trip to market in Savannah or Charleston. Most of these shipments originated at Darien on the Altamaha River; smaller quantities were shipped from other river towns, such as Jeffersonton on the Satilla River.

Sea Island cotton, normally planted in early spring, was hoed frequently after sprouting. The cotton harvest season corresponded closely to that of rice, beginning in September or October although the long picking season for Sea Island cotton could extend into January. While Sea Island cotton brought a much higher price than upland cotton, it bore less fruit, producing about half as much per plant as the upland variety. Rosengarten notes that fifty pounds a day was considered a good picking while others state that a good hand could pick an average of seventy-five to a hundred pounds per day. Regardless, it was about half of what could be picked in fields of upland cotton. After picking, the cotton was dried in the sun; sorted to remove trash, leaves, and stained fiber; and then "whipped." Whipping involved thrashing the cotton with a stick or feeding it into a machine, known as a whipper, to remove additional trash from the fiber.[50]

Coastal planters faced numerous risks in their efforts to produce a successful crop of Sea Island cotton. These included market risks, unpredictable weather and other natural phenomena, and damage that could occur during shipping or storage. Late frosts could injure or kill young cotton plants, and the sandy soils of the region lost moisture quickly. Even short periods without rain could seriously damage the plants. Droughts in 1802 and 1816 resulted in very poor crops; in the latter year, only about half of the crop was harvested. At the opposite extreme, heavy or extended periods of rain could easily destroy a crop. Rain and wind from hurricanes in the late summer and fall were particularly damaging. The hurricane of September 1804 devastated much of the coastal cotton crop, which had already been damaged by a long spell of dry weather. Storms in 1813 and 1824 also seriously damaged Sea Island cotton crops.

Various pests attacked Sea Island cotton, and there are accounts of "the caterpillar" and its damage to the crop. The cotton caterpillar first appeared along the Georgia coast in 1793, and outbreaks periodically occurred afterward. A very serious caterpillar infestation in 1826 destroyed much of Georgia's coastal cotton crop. In October 1838, the *Brunswick Advocate* reported that the caterpillar "had appeared in Camden County, in the Buffalo Swamp and along the Satilla." The appearance of this pest came after a serious

[50] Rosengarten, *Tombee*, 72.

drought, and it was feared the crop along the entire coast would be reduced by as much as half.

The sailing coasters were a principal means of conveying news and information to residents along the coast, including information on crop conditions and yields. On September 4, 1828, the *Daily Georgian* reported, "We learn by Capt. Chevalier from St. Marys, that the Caterpiller [*sic*] has made its appearance on the Florida side of the St. Marys River and he was told by a respectable planter that he should lose at least one-third of his crop by them." "Capt. Chevalier" was John Chevalier, a resident of St. Marys who sailed in the coasting trade for more than thirty years.[51]

After the cotton was cleaned, roller gins extracted cotton fibers from seeds. The ginning process used a frame-mounted pair of wooden rollers, turning in opposite directions via handle or foot pedal. As cotton was fed between the rollers, fibers were pulled from seeds, which were too large to pass through, effectively separating the two. The smooth seed of Sea Island cotton made it easier to gin than upland cotton. Most coastal plantations could easily afford the simple roller gins required although some growers paid others to gin their crop. In 1800, Joseph Clay erected several gins on his Vale Royal Plantation, just upriver of Savannah, to process his own cotton as well as that of other planters. Clay charged "two pence" per pound to gin Sea Island cotton.[52]

After ginning, the cotton was packed into bags or "round bales," using iron pestles or by the packer stomping the cotton with his feet. Screw presses were not used to pack Sea Island cotton because the pressure they produced could damage the long, delicate fibers. Edward Thomas described packing Sea Island cotton at his family's plantation, located on the South Newport River in McIntosh County, Georgia. He noted that young boys "would take the newly ginned cotton to the strong men with the iron pestles, who stood in a strong bag of stout bagging—no presses those days—until the contents were hard and fast...."[53]

There was no established standard weight for these bags or bales, and they ranged considerably in mass. Lewis C. Gray wrote that Sea Island cotton was packed in "round bales" weighing from 300 to 400 pounds; the bales produced by Thomas Chaplin at his Tombee Plantation on St. Helena Island,

[51] Keber, *Seas of Gold*, 204–205, 213, 234, 245–46; Rosengarten, *Tombee*, 74–75; *Brunswick Advocate*, October 4, 1838.

[52] M. Granger, *Savannah River Plantations*, 461.

[53] Rosengarten, *Tombee*, 73; E. J. Thomas, *Memoirs*, 12.

South Carolina, in the 1840s and 1850s ranged in weight from 209 to 333 pounds and averaged 309 pounds. In January 1852, twenty-five bags of Sea Island cotton shipped to Charleston aboard the schooner *Hopeton* from Hamilton Plantation on St. Simons Island ranged from 359 to 394 pounds and averaged 370 pounds. Two years later, thirty bags of Sea Island cotton from Hamilton Plantation ranged in weight from 316 to 350 pounds and averaged 334 pounds. The best that can be said is that bales or bags of Sea Island cotton weighed from as little as 200 pounds to as much as 400 pounds, with an average weight around 350 pounds.[54]

Upland cotton, unlike Sea Island cotton, could be mechanically pressed into compact bales without serious damage to the fiber. Machinery for pressing upland cotton into square bales began to be used in the early nineteenth century. Even so, there was no standard weight for upland cotton bales before 1860. In 1824, bales of upland cotton imported into Liverpool averaged 266 pounds; by 1832, the average was 319 pounds. In 1835, the average weight for bales of upland cotton in the South Atlantic states ranged from 300 to 325 pounds while those from the Gulf Coast were somewhat heavier, averaging 400 to 450 pounds.[55]

James A. Silva, a native of St. Marys, Georgia, describes the Sea Island cotton bags used along the Georgia coast in his recollections of coasting traders on the St. Marys River in the 1840s. Silva noted that schooners:

> passed regularly by St. Marys with decks filled with cotton bales and some even lashed out on the bowsprit. All of this was Sea Island cotton packed in large manila bags, making round piles about six feet in length and two and a half feet in diameter, weighing about three hundred pounds. At each end of the bale were two lugs or ears for handling.[56]

Hand packing remained popular for Sea Island cotton, and coastal planters preferred to produce "old-fashioned round bales."[57] After the 1830s, newspapers commonly used the term "bale" when referring to shipments of Sea Island cotton. However, "bale" was used simply for convenience because plantation accounts reveal that planters continued to ship their Sea Island cotton in "bags," as described by James Silva.

[54] Gray, *History of Agriculture*, 2:736; Rosengarten, *Tombee*, 73; Glynn County, Georgia, estate of James Hamilton, Wills and Appraisements, Book F, 19, 180, 182.

[55] Gray, *History of Agriculture*, 2:705.

[56] Silva, *Early Reminiscences*, 20.

[57] Rosengarten, *Tombee*, 70.

The bagging used to contain the cotton was typically made from hemp, as was the cord used for binding. This bagging represented a consistent expense for cotton planters, and it was typically purchased in Savannah or Charleston and received by sailing coaster. Often, the bags were stenciled with a distinguishing letter or mark to designate the shipper. In the late 1830s, prominent Georgia resident Charles C. Jones purchased bagging from his Savannah factor, R & W. King, in October or November, just before the cotton on his Liberty County properties was ready to pack. This bagging cost twenty-two to twenty-five cents per yard, and in May 1840, Jones paid Captain John Grovenstein fifty cents for transporting two pieces of this bagging from Savannah aboard his sloop *Macon*.[58]

Cotton shipments, like those of rice, occurred primarily in the fall and winter. The earliest cotton shipments typically began to arrive in Savannah in October; however, the laborious and time-consuming process of cleaning, ginning, and packing the crop meant that the bulk of the Sea Island cotton was shipped between January and April, peaking a month or so after the shipments of rough rice (see Figure 4.3). The shipping columns in the *Daily Georgian* note the first arrival of cotton aboard a coaster in 1825 was on October 27; in 1835, it was October 20, and in 1845, it was October 22, when the sloop *B. S. Newcomb* arrived from the Turtle River in Glynn County with six bales of Sea Island cotton. The October cotton shipments presumably reflect planters' efforts to take advantage of the higher prices often brought for the early crop. Like those of rice, cotton shipments continued at a lesser pace well into the summer.

Records of how the cotton was carried on the coasting vessels are almost nonexistent. Like tierces of rice, cotton was likely carried on the main deck whenever possible, simply because of the inconvenience of loading and unloading the heavy and bulky bags into and out of the ship's hold. James Silva's account noted above indicates that some cotton was carried on the decks of vessels, sometimes to the point of overflowing. As with other cargo, at individual landings, the plantation hands would load the cotton aboard while in Savannah, a factor's employees or bondsmen would do the unloading.

How much cotton a particular coasting vessel could carry is difficult to determine because the amount transported varied so much on any given trip. For example, the forty-eight-foot-long sloop *Splendid* sailed into Savannah many times during the 1840s with Sea Island cotton. The sloop carried cotton

[58] "Maybank Plantation Accounts," October 1836–May 1840, Charles C. Jones Plantation Books, Howard-Tilton Library, Tulane University.

on just over half of her trips, and the number of bales or bags ranged from a single bale to a high of 135 bales on a voyage in February 1849. Assuming a bag/bale weighed about 350 pounds, the largest quantity of 135 bags would have weighed about 23.6 tons or approximately 0.61 tons of cotton per ton of the sloop's burden (38 40/95 tons). Some vessels carried very large amounts of cotton. In 1852, the seventy-foot sloop *Catherine Chard* brought large quantities of cotton into Savannah several times. The 337 bales she carried into the city in March made up her largest cargo for the year. This shipment constituted approximately 59 tons of cotton, or about 0.91 tons of cotton per ton of measured burden (64 64/95 tons).[59]

An examination of several coasting vessels operating between the 1830s and 1850s indicates the average maximum amount of cotton carried was about 4.1 bales, or 0.72 tons (1,435 pounds) per ton of burden. As noted previously, cotton is thought to have been commonly carried on the main deck of the vessels. It was probably only when very large quantities of cotton were aboard that it became necessary to load some into the hold. Vessels rarely came into port with what appears to be anything near a maximum cargo of cotton.

Throughout the six decades of interest, the rate charged by the Georgia coasters for carrying cotton was generally about one dollar per bale or bag. In 1823, the freight on transporting a bale of cotton from Darien to Savannah was reported to be one dollar. This was the same rate paid by the estate of James Hamilton to ship twenty-five bags of Sea Island cotton from St. Simons to the firm of Robert Mure & Co. in Charleston in January 1852. A few records show that the freight on cotton fluctuated slightly. For example, in March 1852, the estate of James Troup paid seventy-five cents per bale to ship thirty bales of cotton from near Darien to Savannah aboard the sloop *Liberty*. In 1841, an advertisement in the *Daily Georgian* stated that thc sloop *Visitor* would transport Sea Island cotton for fifty cents a bale. This is the same advertisement reporting the sloop would carry rough rice at the reduced rate of 3.5 cents per pound. It may be that the *Visitor*'s owner and captain, William P. Eaton of Savannah, was undercutting the normal freight rates in an effort to attract business.[60]

[59] Pearson, "Georgia Coasting Trade," 494; *Daily Georgian*, 1852.

[60] Kayser & Co., *Commercial Directory*, 50 (1823 freight rates); Glynn County, Georgia, estate of James Hamilton, Wills and Appraisements, Book F, 19; Glynn County, Georgia, estate of James Troup, Wills and Appraisements, Book F, 38; *Daily Georgian*, November 9, 1841 (ad for sloop *Visitor*).

In addition to cotton itself, cottonseeds represented a minor product carried by the coasters. Planters would save the seed from well-producing Sea Island cotton for future plantings or for sale; factors also commonly supplied cottonseed to their clients. In 1838, James Gould of St. Simons Island advertised that he had "From five to six hundred bushels of cotton seed, of the growth of 1836" for sale at fifty cents per bushel. Gould's advertisement noted the seed would be delivered "at Frederica on St. Simons." Cottonseeds rarely appear as a named cargo on the coasters. One of the few mentions of cottonseeds was the "10 casks" brought into Savannah by the sloop *Dirigo* from Darien in January 1828. On February 16, 1852, two vessels arrived in Savannah carrying cottonseed, a rather unusual occurrence. One was the sloop *Julia Ann* with "500 baskets" of cottonseed from "Rogers Plantation." The other vessel was the schooner *American Coin*, carrying fourteen hundred bushels of cottonseed from the town of Sunbury.[61]

## Sugarcane

Sugarcane was grown in Georgia in the eighteenth century, but its commercial cultivation was stimulated when a high tariff on imported sugar was instituted during the War of 1812. This tariff improved domestic markets and encouraged Georgia coastal planters to take up its cultivation. Though these tariffs were reduced after the war, sugar prices remained high into the early 1830s.[62]

Thomas Spalding of Sapelo Island was one of the main growers of sugarcane along the Georgia coast and one of its most ardent proponents, primarily as a supplement to cotton cultivation. Spalding was growing sugarcane by 1805 and constructed sugar works for grinding the cane and processing the juice into syrup, molasses, and sugar. Spalding reportedly earned $12,500 from his sugar crop of 1815. That year, the *Georgia Journal* reported, "It will not be long before this most valuable article [sugar] will become one of the stable commodities of our state. We shall be no longer dependent on the European colonies in the West Indies for Sugar, rum and molasses." The newspaper suggested that the cultivation of sugar in the southern part of the state might become more profitable than cotton. This was not to be, but by 1828, there were about two hundred plantations growing sugarcane in the vicinity

[61] James Gould's sale of cottonseed was printed in the *Brunswick Advocate*, January 11, 1838; the shipments of cottonseed appeared in the *Daily Georgian*, January 28, 1828, and February 16, 1852.

[62] Crook and O'Grady, "Spalding's Sugar Works," 9–13; Sullivan, *Early Days*, 109.

of Savannah and on the lower Georgia coastal plain. Not all of this production was for the commercial market; much was for home consumption, often as molasses and syrup.[63]

Syrup and molasses are the products of boiling the juice from grinding sugarcane. The machinery to grind cane to produce juice was simple and inexpensive, so most growers would have owned or had convenient access to this equipment. The most desired product was granular sugar itself, but sugar production required greater expertise, technical skill, and capital outlay than the production of syrup and molasses. Consequently, many cane growers along the Georgia coast boiled their sugar juice to produce molasses and syrup but did not make sugar. When syrup was carried by the coasters, it was typically identified as "Georgia syrup" to distinguish it from syrup that might be coming from Florida or even from the West Indies. Advertisements in Savannah newspapers make this distinction and imply that syrup from Georgia was more desirable than syrup from elsewhere, as in the 1852 advertisement by G. M. Willett & Co., which noted it had for sale "GEORGIA SYRUP, a superior article."[64]

The peak years for sugar production on the Georgia coast were 1829 and 1830. The decline of sugarcane cultivation after 1830 resulted from lower import duties on sugar in 1832, deflated foreign markets, and rising rice and cotton prices. The price of sugar from 1832 to 1840 averaged about six cents per pound, as opposed to as much as sixteen cents under the earlier tariff protection. The decline of the sugar industry in Georgia was such that by 1850, the state produced only 846 hogshead of sugar.[65]

Sugar, syrup, and molasses were shipped in wooden barrels or "hogsheads." These hogsheads held about seven hundred to a thousand pounds of packed sugar. The term "barrel" is sometimes used to refer to the sugar containers carried by the coasters. This might be a generic term that includes hogsheads, but sometimes it may have referred to a smaller cask with a capacity of between thirty and forty gallons, as opposed to the sixty-three or so gallons in a true hogshead. When the sloop *B. S. Newcomb* sailed into Savannah on January 15, 1846, she was carrying "9 hhds [hogsheads] sugar" and "8 bbls [barrels] molasses" from the Turtle River. In some instances, the barrels used to hold sugar products were delivered to the planter aboard coasting vessels. In December 1848, the sloop *Science* carried "80 bbls empty for syrup" to

[63] Gray, *History of Agriculture*, 2:748; Sullivan, *Early Days*, 111.
[64] *Daily Georgian*, January 31, 1852.
[65] Gray, *History of Agriculture*, 2:744, 748; Sullivan, *Early Days*, 111–12.

one of Richard Arnold's plantations on the Ogeechee River. Some planters had coopers skilled in making the water-tight barrels needed for carrying syrup or molasses; others found it easier and cheaper to obtain the barrels from Savannah.[66]

There are periodic accounts of arriving shipments of sugarcane products into Savannah and Charleston, but often these make no mention of quantities. When quantities are given, the amounts of sugar products carried were relatively small. Among the largest cargoes reported in the Savannah newspapers were the forty hogsheads of sugar carried from Darien by the sloop *Dirigo* on February 27, 1828; the fifty-six barrels of syrup brought into the city from Sunbury by the schooner *Sarah* in December 1843; and the thirty barrels of "Georgia syrup" brought into Savannah in December 1852 by Captain John Grovenstein aboard his schooner *Company*. More typical were shipments of less than twenty barrels. The small amounts of sugar products carried as individual loads indicate the relatively limited sugarcane cultivation along the coast. By the early 1830s, steamers running along the Georgia coast were carrying some of the region's sugar products, lessening that available to the coasters.

The transport of sugar products was seasonal. Most of the syrup was carried between November and March while shipments of sugar and molasses tended to by slightly later, between December and April, because of their longer processing time. In 1823, the published freight rate for shipping a hogshead of sugar between Darien and Savannah was two dollars, while freight on a "barrel" of sugar was fifty cents.[67]

## Timber Products

Georgia's forest and timber resources became important exports from the state's coastal ports in the nineteenth century. Georgia was the primary lumber-producing state in the South through much of the antebellum period. The state had huge coastal and inland resources of longleaf yellow pine, bald cypress, and live oak. Longleaf pine (*Pinus palustris*), also known as the Georgia pine or Georgia pitch pine, was the most important tree in the state's lumber industry. Growing to heights of eighty to one hundred feet, longleaf pine is fine-grained, strong, and durable, making it particularly important in ship construction. Bald cypress (*Taxodium distichum*), found along the freshwater

[66] *Daily Georgian*, January 15, 1846 (arrival of the *B. S. Newcomb*); Hoffman and Hoffman, *North by South*, 131 (report of sloop *Science*).

[67] Kayser & Co., *Commercial Directory*, 50.

swamps and rivers of the coastal plain, was used for various building purposes and producing shingles. The wood of the bald cypress is light, straight-grained, and not susceptible to rot. Live oak (*Quercus virginiana*) was an important timber resource in the eighteenth and early nineteenth centuries. Confined mainly to the sandy soils of the coast, its timber is tough and durable and was highly prized for ship's knees and frames.[68]

Timber and lumber were among the earliest exports from the colony of Georgia. Immediately after the establishment of Savannah, timber was being felled and rafted down the Savannah River to the city, where it was sawn into lumber for local consumption and, increasingly, for export. By 1738, just a few years after the settlement of Savannah, the firm of Robert Williams & Company was shipping lumber out of the city. Lumber became an increasingly important export during the latter years of the eighteenth century. In the 1790s, the firm of J. L. Lawrence & William Spencer was involved in purchasing, storing, and selling lumber brought from the interior into Savannah. They loaded lumber on ships bound for domestic locations, mainly in the Northeast, as well as foreign ports. The number of merchants handling timber and lumber in Savannah increased over time, particularly after the introduction of steam-powered sawmills. In 1817, the Chatham Mill, reportedly the first steam sawmill in Savannah, opened, and within a decade, several other steam-driven sawmills were constructed. The Vale Royal Mill, which opened in the 1840s, was one of the largest sawmills in the nation, sawing up to twenty-five thousand feet of lumber per day. By 1850, these mills made Savannah the most important exporter of lumber on the lower Atlantic coast.[69]

In 1810, lumber exports out of Savannah were valued at $23,559; by 1819 exports for the first three-quarters of the year were worth $65,000. In these early years, a considerable quantity of this lumber was shipped to markets in the West Indies. With an increase in the number of steam mills in the city, lumber exports expanded dramatically. Beginning in the late 1830s, there was a shift from West Indies markets, and more lumber began to be shipped directly to England and Canada. By the mid-1840s, more than "two hundred vessels of all sizes" were engaged in the lumber trade out of the city, and exports exceeded eighteen million board feet. In 1855–1856, exports rose to almost thirty-five million feet. These lumber vessels consisted mainly of brigs

[68] Eisterhold, "Savannah: Lumber Center," 526; Wood, *Live Oaking*, **[page?]**.

[69] Eisterhold, "Savannah: Lumber Center," 529, 537; M. Granger, *Savannah River Plantations*, 175.

and schooners fitted with hatches, or "ports," near the waterline at their bows through which lumber was loaded. Reportedly, a vessel of one hundred tons burden could carry about a hundred thousand feet of lumber. Cypress shingles and barrel staves were also important products of the Georgia timber trade and significant Savannah exports. In 1843, the city exported 5,175,000 cypress shingles and 66,000 oak staves.[70]

Other towns on the coast were heavily involved in the sawing and shipping of lumber, but Darien was the most important. The Altamaha River provided the town access to the huge pine and cypress timber resources in inland sections of the state. In the eighteenth century, sawmills around Darien were tidal-powered, and output was low. In 1817, a steam-powered sawmill was erected at Upper Bluff, just west of Darien, and it soon began to provide sawn lumber for export. The Upper Bluff mill included a rice mill featuring four gangs of crank pestles capable of milling four tierces of rice per hour. Three years later, another steam sawmill, the Eastern Steam Saw Mill, was in operation at Darien. Soon, large numbers of ships were visiting Darien and leaving with lumber bound for distant ports.[71] Although Darien continued as an exporter of timber, it was not until after the Civil War that the Darien timber industry blossomed.

Other lumber ports on the coast included St. Marys, on the St. Marys River, which had a steam sawmill by 1805, and Jeffersonton and Burnt Fort, located inland on the Satilla River, both of which became local centers for sawing and shipping lumber.

Although the state's timbering industry was large and important, the coasting vessels sailing in the local Georgia trade in the antebellum period were never heavily involved in the transport of the region's sawn lumber, because, even from an early date, most of the timber and lumber was carried directly away from sawmills and ports on the Georgia coast to distant destinations. This is reflected in newspaper notices of the period, such as one appearing in the June 27, 1820, Savannah *Daily Georgian*: "WANTED—to load at the Darien Steam Saw Mill, for Northern ports—Two or more Vessels capable of receiving PLANK on board of at least 35 feet in length. Immediate dispatch will be given. Apply at Mr. John McNish's or to Wm. Scarbrough."

[70] "Commerce of Savannah," *DeBow's Review*, May 1847, 3:402, 404; Eisterhold, "Savannah: Lumber Center," 528, 539.

[71] Vessel arrival information taken from the *Daily Georgian*, August 2, August 31, and October 3, 1842; April 13, 1843; and May 1, 1845.

There was little incentive to ship sawn lumber into Savannah first and then transship it. It simply would have been uneconomical to go through the process of unloading, storing, and then reloading sawn lumber in Savannah. Coasting vessels did transport some amounts of sawn lumber into Savannah, but much of this was likely for local use. Occasionally coasting ships coming into Savannah with lumber are known to have been bound elsewhere. For example, when the coasting sloop *Bolivar* arrived in Savannah from Darien with a cargo of "lumber" in June 1828, the *Daily Georgian* did note the vessel was "bound for New York."

Because sawn lumber was not a major item carried into Savannah and other Georgia ports, it is not often named in newspaper shipping lists. Even when arrivals of lumber are reported, quantities are rarely mentioned. The *Daily Georgian* noted the arrival of the sloop *Conductor* from Cumberland Island into Savannah on June 10, 1828, with "lumber to master." How much lumber may have been aboard is unknown; it was not assigned to a Savannah factor or merchant but was to be disposed of by Captain Nye of the *Conductor*. A few accounts were more specific. The Savannah papers reported on the arrival of the sloop *Huntress* in Charleston on July 15, 1828, with a load of "live oak timber" from "East Florida" and in June 1830 the arrival of the sloop *Conductor* in Savannah from St. Marys with "live oak" consigned to "F. Willink." This was Henry Frederick Willink, who operated one of the largest shipyards in Savannah, where the live oak timber was to be used.[72]

While coasting vessels were not carrying large quantities of sawn lumber into Savannah or Charleston for export, they did transport considerable amounts from sawmills to individual plantations and towns. Records of the movement of this lumber in the coasting trade is found in the few newspapers that exist for the smaller ports and scattered among extant plantation record books and store accounts. The *Brunswick Advocate* reported numerous shipments of lumber into that city aboard local coasters during the years it was published, 1837 to 1839. This lumber was coming from the sawmilling centers of Jeffersonton and Burnt Fort on the Satilla River and would be used for construction in and near Brunswick. The accounts of Charles C. Jones include a number of payments to coasting vessels for the receipt of lumber at his plantations in Liberty County. In October 1836, Jones paid $14.88 in freight charges "to sloop Macon" for the receipt of 4,043 feet of "boards" at his Maybank Plantation. In February 1840, a payment of $12.82 was made to Captain Leonard Bolles of the sloop *Eagle* for transport of "Lumber for Negro

[72] *Daily Savannah Republican*, June 17, 1830; *Daily Georgian*, July 15, 1828.

Houses" to Carlawater Plantation. The boards on the *Eagle* were shipped to Carlawater by the firm R & W King, one of the Savannah factors dealing extensively with coastal planters.[73]

In a few cases, lumber seems to have been an important cargo for some vessels, at least for part of the year. In the mid-1840s, the schooner *Vesta* carried lumber or timber into Savannah several times, including one cargo of live oak from St. Simons Island. Captain Charles Stevens carried lumber into Savannah at least three times in 1842 aboard his sloop *Splendid.* The newspaper accounts note only that Stevens had "lumber" aboard and that he was sailing from the Satilla River or the town of Jeffersonton. Stevens carried lumber during the summer and fall, when the more important coastal cargoes, rice and cotton, were not normally shipped.[74]

How much lumber an individual coasting vessel could carry is difficult to determine. The largest cargo of lumber carried by the schooner *Vesta* was just over thirty-two thousand board feet. For the schooner *Joseph*, it was thirty-one thousand feet; for the schooner *William D. Jenkins*, sixty-two thousand feet, and for the sloop *Virginia*, it was thirty thousand feet. It is unlikely these ships were transporting maximum loads of lumber because their loads were considerably less than the one thousand board feet per ton of burden reported for vessels designed specifically to carry lumber.[75]

In addition to sawn lumber, the coasting traders carried other timber products. The most important were naval stores, but other items included shingles, barrel staves, and "wood." Shingles, normally made from cypress, and staves, like sawn lumber, appear to have been commonly shipped directly from mills to distant ports, reducing their importance as cargoes in the local coasting trade. These items only occasionally appear as cargoes of the coasters sailing into Savannah. Charles Stevens, with his sloop *America*, carried one thousand "staves" into Savannah in April 1850. The *America* was coming from the Satilla River and also had on board turpentine and rosin, both forest products commonly shipped out of the Satilla River area.[76]

Not infrequently, Savannah newspapers record coasting vessels coming into port carrying "wood." In a few instances, the papers report a cargo of

[73] *Brunswick Advocate*, June 8, 1837–June 1, 1839; "Maybank Plantation Accounts," October 31, 1836, and "Carlawater Plantation Accounts," February 12, 1840, Charles C. Jones Plantation Books, Howard-Tilton Library, Tulane University.

[74] *Daily Georgian*, August 2, August 31, and October 3, 1842; April 13, 1843; and May 1, 1845 (vessel arrival information).

[75] Eisterhold, "Savannah: Lumber Center," 528.

[76] *Daily Georgian*, April 7, 1850.

"cord wood." Amounts are rarely given, an exception being the "31 cords light wood" brought into Savannah by the schooner *Elias Reed* from the Turtle River on April 4, 1854.[77] When wood is a cargo on the coasting vessels, there is often the notation "wood to master." The notation "to master" meant that the wood either belonged to or had been assigned to the vessel's master to dispose of and was not consigned to a specific factor or merchant.

## Naval Stores

The term "naval stores" originally referred to ship's stores in general, but by the eighteenth century, it came to more specifically mean tar, pitch, rosin, and turpentine. All were derived from the sap or oleoresins of various species of conifers. In Georgia, this tree was generally the longleaf yellow pine (*Pinus palustris*), the same tree that was so important in lumber production. Tar was essential in the manufacture of rope needed to rig sailing vessels, and it was used both as a coating and a lubricant. Pitch, a derivative of tar, was painted on the sides and bottoms of ships to act as a preservative and protectorate. Turpentine, derived from distilling pine sap, had a variety of uses as a preservative and a mixing agent with paints.

The production of naval stores in the American colonies was minimal before the passage of the Naval Stores Bounty Act in 1705. Britain's loss of naval stores from Finland and Sweden after Russia overran the two countries stimulated the act's passage. The production of naval stores had a slow beginning in the American colonies, but by 1725, they were providing all England needed. Most of these stores came from the Southern colonies. The early production of naval stores in Georgia tended to be supplementary to indigo and rice cultivation. The enslaved laborers who grew rice and indigo also worked to produce lumber and naval stores such that both were supplemental to agricultural pursuits.[78]

Georgia did not begin to commercially produce naval stores until the 1750s, and tar was the most important of these items. Between 1755 and 1765, Georgia exported only 187 barrels of turpentine, 106 barrels of pitch, and 2,125 barrels of tar. During the colonial period, these products were of minimal importance relative to lumber, staves, and shingles. The naval stores industry in Georgia did not become truly important until the middle of the nineteenth century. An article on the "Commerce of Savannah" in the May 1847 issue of *DeBow's Review* makes no mention of naval stores, despite

[77] Ibid., April 4, 1854.

[78] Herndon, "Naval Stores," 426, 428.

including a long discussion on the value and importance of the longleaf pine as a Georgia export.[79]

In the nineteenth century, the term turpentine specifically referred to the sap of pine trees although the term was also used to refer to the distilled product of the sap, more commonly called "spirits of turpentine." The sap is extracted by cutting incisions at the base of the trunk; sap drained from the incisions into a container attached to the side of the tree. Trees were usually cut in January and February, and the sap began to run by mid-March, increasing as the weather warmed. The sap could be shipped as is, or it could be distilled into oil or spirits of turpentine, usually in a simple still set up in the woods where the sap was gathered. The sap was boiled in a large metal tank, and the vapor was taken off through a metal tube passed through water to cool and condense the vapor into spirits of turpentine. The residue left from boiling the sap was rosin.[80]

The production of tar and pitch from longleaf pines was a winter activity that used the limbs and parts unfit for lumber. To produce tar, a circular floor of clay declining slightly toward the center was constructed. A narrow ditch or pipe of wood led from the bottom of the floor to a lower pit, where barrels were placed to receive the tar. Pinewood was cut into short pieces and stacked in a pyramid on the prepared floor. The wood pile was covered with a layer of pine boughs, and the pile was set afire. The fire was allowed to burn slowly, forcing the tar to flow from the wood to the prepared floor and out through the ditch or pipe into the waiting barrels. In this type of rude kiln, the average yield was one barrel of tar to one cord of wood. Pitch was produced by heating tar in large kettles and boiling off the moisture.[81]

Accounts of coasting vessels carrying naval stores seldom appear in Savannah newspapers until the late 1840s. In 1830, when more sailing coasters were working in the trade than at any other time, just a single shipment of "tar and rosin" into Savannah was reported. This shipment came from St. Augustine, Florida, aboard the sloop *Import*. After this shipment, no report of naval stores coming into Savannah aboard a sailing coaster appears until the November 8, 1847, arrival of the sloop *Liberty* with a hundred barrels of spirits of turpentine. The *Liberty* was sailing from Harris Neck, on the South Newport River in McIntosh County, and the naval stores she carried may

[79] Gray, *History of Agriculture*, 1:102; Herndon, "Naval Stores," 430; "Commerce of Savannah," *DeBow's Review*, May 1847, 3:397–407.

[80] Herndon, "Naval Stores," 428–29.

[81] Gray, *History of Agriculture*, 1:160; Herndon, "Naval Stores," 429–30.

have been shipped by William J. King, a prominent planter of Harris Neck and one of the *Liberty*'s owners. After 1847, naval stores became a more commonly listed cargo of the coasters, shipped in both barrels (capacity about forty gallons) and hogsheads (capacity sixty-three gallons). In 1848, eight shipments of naval stores arrived in Savannah aboard six different vessels. The sloop *Liberty* arrived three times in Savannah in 1848 with naval stores, comprising ninety-four barrels of spirits of turpentine, eighty-seven barrels of turpentine, and forty-five barrels of rosin. The eighty-seven barrels of turpentine represented nondistilled sap, often referred to as "soft turpentine" or "crude turpentine." The transport of naval stores into Savannah by sailing coasters peaked in 1850, when the *Daily Georgian* reported twenty-seven shipments representing at least fifty-one hundred barrels of turpentine, spirits of turpentine, and rosin. The number of shipments by sailing coaster declined after 1850, with ten or fewer arrivals annually for most of the following decade. Steamers, which were running regularly along the coast by the 1840s, may have been carrying some naval stores, but these products' high flammability may have discouraged their shipment by steamboat.[82]

The amount of naval stores that could be carried by a coaster is difficult to determine. In 1850, the sloop *Liberty* sailed into Savannah five times with naval stores. The quantities carried ranged from thirty-four barrels of rosin and spirit of turpentine carried into the city on April 2, to 129 barrels carried on October 11. The naval stores carried by the *Liberty* were small compared to the five hundred barrels transported on individual voyages by the larger schooners *Company* and *Elias Reed* that year. Unlike cotton and rice, the transport of naval stores does not appear to have been strongly seasonal. In 1850, for instance, when twenty-seven shipments came into the city, they arrived in every month except December. The most reported arrivals occurred in September, with five. In other years, shipments were similarly spread throughout the year.[83]

Some naval stores carried into Savannah were used locally, but much of it was transshipped overseas or to Northeastern American ports. Freight information on shipping naval stores is not widely available. In 1850, spirits of turpentine brought about thirty-two cents per gallon in Savannah, or about

[82] *Daily Savannah Republican*, November 24, 1830 (sloop *Import*); *Daily Georgian* for the years 1847 and 1848 (*Liberty* cargoes); Sullivan, *Early Days*, 248–50 (William J. King); *Daily Georgian*, the *Daily Morning News*, and the *Savannah Daily Republican* 1847–1861 (naval stores shipments into Savannah).

[83] *Daily Georgian*, 1850.

$12.80 per forty-gallon barrel. Soft turpentine, tar, and rosin were considerably less valuable, bringing two to two and a half dollars per barrel. The freight rates on naval stores were correspondingly low. Charles C. Jones paid thirty-eight cents freight on a barrel of tar in 1836 and fifty cents in 1840. Fifty cents per barrel or hogshead was probably a typical rate from 1847 to 1861.[84]

## Miscellaneous Commodities

In addition to the cargoes discussed above, the coasting ships carried smaller amounts of other commodities from the coastal area into Savannah and Charleston. These included items such as hides, lime, moss, oranges, potatoes, "ground nuts" (peanuts), and corn. The amount and value of these goods were insignificant over most of the year; however, during the summer, when little or no rice or cotton was shipped, some items were important cargoes.

Large quantities of corn, flour, peas, potatoes, and other foodstuffs arrived at Savannah aboard sailing ships, but little came from the nearby coastal region. Corn, for example, was grown on many of the coastal plantations, but it was mainly for local consumption or animal fodder and was not an important export crop. Plantations often ran short of corn and had to import it, usually from Savannah or Charleston. Charles C. Jones notes several shipments of corn to his family's plantations in Liberty County. In 1840, he recorded a payment for "150 bushels corn for plantation, summer of 1838" to Carlawater Plantation and, in 1858, a shipment of eighty bushels to Maybank Plantation aboard the sloop *Orange.* In the 1850s, Hugh Frazer Grant shipped corn from Savannah to his Elizafield Plantation at the mouth of the Altamaha River, and the James Hamilton estate accounts mention numerous shipments of corn from Charleston to Hopeton Plantation, also on the lower Altamaha. Most of the corn shipped out of Savannah and Charleston appears to have come into the two cities from North Carolina or the Chesapeake Bay area.[85]

The sailing coasters occasionally carried corn into Savannah from tidewater Georgia, South Carolina, or Florida. In 1820, Savannah newspapers list

[84] "Carlawater Plantation Accounts," October 31, 1836, and February 15, 1840, Charles C. Jones Plantation Books, Howard-Tilton Library, Tulane University; "Commercial" columns, *Daily Georgian*, 1850 (1850 naval store prices).

[85] "Carlawater Plantation Accounts," February 15, 1840, and "Maybank Plantation Accounts," July 14, 1858, Charles C. Jones Plantation Books, Howard-Tilton Library, Tulane University; House, *Planter Management*, 170–250; Glynn County, Georgia, estate of James Hamilton, Wills and Appraisements, Book F, 1851–1854.

three coasting vessels arriving with corn. One of these, the sloop *Suffolk*, was carrying "corn blades" from Darien, which are believed to refer to corn stalks with leaves attached. In later years, the number of arriving vessels carrying corn from coastal Georgia and South Carolina was similarly small. An exception was 1852, when the Savannah papers list nine local coasters arriving with corn. These corn shipments all originated at two Liberty County locations, Riceboro and Sunbury.[86]

Commonly, the papers list "corn" as part of the cargo, with no amounts given. When quantities are provided, they are usually less than a few hundred bushels. Exceptions would be the thirteen hundred bushels brought into Savannah in November 1839 by the sloop *Swallow* and the corn carried in 1852 by the schooner *Northern Belle*. The *Swallow*'s unusually large cargo of corn originated on Sapelo Island, probably from Thomas Spalding's productive island plantation. In 1852, the *Northern Belle,* under her owner and master Charles Thompson, sailed into Savannah with nine hundred bushels of corn on August 28, eight hundred bushels on October 24, and two hundred bushels on December 28. All three shipments originated in Riceboro on the North Newport River.[87]

Other food crops, such as potatoes, peas, and oats, rarely appear as cargoes of the coasters into Savannah. Typically, these crops were shipped by the bushel or "hamper," such as the twenty hampers of potatoes carried from Riceboro by the schooner *Fort George Packet* in April 1845 and the 269 bushels of peas brought by the schooner *Levant* in January 1860.[88]

Oranges were another of the minor commodities carried by coasters. The Georgia coast is at the northern limit of orange cultivation. Shipments of oranges normally originated in Florida or at Georgia locations south of Sapelo Island, where conditions were most favorable for their cultivation. Between November 1827 and November 1828, the *Daily Georgian* listed twenty-three shipments of oranges arriving aboard sailing coasters. These originated at places such as St. Simons, Jekyll, and Cumberland islands; St. Marys in Georgia; as well as St. Augustine and "St. Johns" in Florida. In December 1827, the Savannah merchants Jenny & Douglas advertised they had "30,000 sweet oranges for sale on board the sloop Conductor, Capt. Nye, at Anciaux's Wharf." This captain, Barnabas Nye, had sailed the *Conductor* into Savannah a few days earlier from Cumberland Island. An idea of the number of oranges

[86] *Daily Georgian*, 1820 and 1852.
[87] Ibid., 1839 and 1852.
[88] Ibid., April 12, 1845, and January 12, 1860.

grown on some coastal plantations is seen in an advertisement in the *Daily Georgian* on November 23, 1827, in which the firm of A. G. Miller offered for sale two hundred thousand "sweet oranges, now on the trees at Dungeness plantation, south end of Cumberland Isl."[89]

When newspapers list shipments of oranges, the number aboard is typically noted; no mention of containers for oranges has been found. Oranges were more commonly listed as cargoes of coasters before the 1840s; after that, they rarely are named. By this time, the fruit was more commonly shipped aboard steamboats, providing faster and more dependable service than sailing vessels.

Occasionally, orange juice is listed in newspapers as a cargo, such as the January 8, 1820, report of the arrival of the sloop *Union* at Savannah from Sapelo Island carrying unspecified quantities of cotton and orange juice. The orange juice almost certainly originated at the Sapelo plantation of Thomas Spalding, who experimented with the cultivation of various crops, including oranges. This shipment from Sapelo represents one of the few instances where cargoes of oranges originate north of St. Simons Island. On June 11, 1830, the *Daily Savannah Georgian* reported the arrival of the sloop *Leader* with a shipment of "orange trees" from St. Marys. This is the only account found of trees themselves being shipped.

On rare occasions, other citrus fruits are mentioned as cargoes. The sloop *Bridgeport* arrived in Savannah in November 1819 from Darien with a cargo that included lemons; in January 1810, the sloop *Maria* transported unreported quantities of oranges and lime juice into Savannah from St. Augustine; and two months later, the sloop *Packet* arrived with lime juice coming from "Amelia," a reference to Amelia Island just north of the mouth of St. Johns River in Florida.[90]

Infrequently, coasters carried unspecified fruit into Savannah. When fruit is mentioned, the origin of the vessels is most commonly a Florida location, such as St. Augustine or "St. Johns, East Florida." "Ground nuts," or peanuts, are another item that sometimes appeared as a cargo. In 1827, three ships sailing from Sunbury and Riceboro carried unspecified amounts of "ground nuts" into Savannah. On January 5, 1852, the sloop *Virginia* brought 120 bushels of "ground nuts" into the city from Darien. Peanuts had been grown in small amounts along the Southeastern coast since the seventeenth century, and it is believed that enslaved Africans introduced both the plant

[89] Ibid., November 27, December 8, and December 10, 1827.

[90] Ibid, November 27, 1819, and January 16 and March 27, 1810.

and knowledge of its cultivation. Peanuts were used for food and their oil, but during the early decades of the nineteenth century, peanuts were more commonly used as food for hogs and for soil improvement.[91]

Among the more unusual food items carried by coasting vessels was turtle. Turtles typically appear as a cargo item no more than two or three times a year, and after about 1830, they may have been more commonly carried aboard steamers rather than sailing coasters. All turtle shipments recorded originate in Florida, and the type of turtle most commonly carried was the green turtle (*Chelonia mydas*), a large marine species highly sought after as food. Turtles were shipped live, as an 1835 Savannah advertisement noted: "GREEN TURTLE—Just received and for sale at the City Hotel forty Green Turtles, weighing from 18 to 60 lbs, in fine order."[92]

One of the vessels carrying turtles into Savannah experienced one of the sea dangers occasionally faced by coasting ships. On July 24, 1820, the *Daily Georgian* reported the arrival of the schooner *Mary McKay* in Savannah from "Cape Florida" with an unspecified amount of "turtle" and noted that the ship had been "attacked by pirates off coast of Fla."

Another minor cargo carried by the Georgia coasters was animal hides. Normally, the type of hide is not identified, but most were likely cow hides, such as the one hundred "cow hides" carried by the schooner *Northern Belle* from Riceboro to Savannah in March 1852. Usually, the quantity of hides carried is not reported. Exceptions are the 530 hides brought into Savannah from Centerville by the sloop *Catherine Chard* in April 1852 and the "500 hides and beeswax" carried by the schooner *James & Augustus* from the Turtle River in April of the following year. Another cattle product rarely mentioned is tallow. How tallow was packaged for shipment is not reported, but it was likely stored in casks or barrels.[93]

Occasionally skins of other animals are identified as cargoes, such as the "6 bundles deer skins" aboard the schooner *Elias Reed* sailing from the Turtle River in April 1854. In February 1830, the schooner *Betsey Maria* arrived in Savannah from St. Marys with Sea Island cotton, beeswax, and "otter skins."[94]

Other rare cargo items carried into Savannah were moss, lime, and beeswax. "Moss" refers to Spanish moss (*Tillandsia usneoides*), the epiphytic plant

[91] *Daily Georgian*, 1827 and January 5, 1852 (vessel arrivals); Gray, *History of Agriculture*, 1:194, 2:828 (peanuts as a Southern crop).

[92] *Daily Savannah Republican*, November 12, 1835.

[93] *Daily Georgian*, March 29 and April 12, 1852, and April 19, 1853 (transport of hides).

[94] *Daily Georgian*, April 4, 1854; *Daily Savannah Republican*, February 16, 1830.

that grows in abundance in trees along the southeast Atlantic coast. Moss was used primarily as stuffing for mattresses and upholstered furniture. It was commonly packed and shipped in bales, as indicated by a cargo of twenty "bales" of moss carried by the sloop *Splendid* from Darien in 1851. In 1852, two sloops, the *Julie* and the *Splendid*, carried lime into Savannah, both shipments originating at Isle of Hope just southeast of Savannah. Lime was typically produced by burning oyster shells dug from the numerous prehistoric Indian shell middens in the coastal region. Beeswax was another of the rare cargo items transported in the Georgia trade. Some of the beeswax was shipped in casks, such as the "2 casks beeswax" aboard the schooner *Elias Reed* when she sailed into Savannah from Brunswick in May 1850. On this particular voyage, the *Elias Reed* was laden with a mixed cargo often carried by coasters during the late spring and summer when cotton and rice shipments were low. In addition to the beeswax, the *Elias Reed* had onboard 70 bags of cotton, 230 hides, 100 bushels of rough rice, 5 oak timbers, and 424 "pieces cedar posts."[95]

On rare occasions, a coasting vessel arrived in Savannah from a "wrecking voyage," normally referring to a trip from the east coast of Florida with salvage from a wrecked or stranded ship. The Savannah *Daily Georgian* of May 9, 1820, reported the schooner *Choctaw* had departed on a "wrecking voyage." Three weeks later, on June 2, 1820, the newspaper reported the return of the *Choctaw* from the "coast of Florida" with a "full load of mahogany," no doubt the cargo recovered from a wrecked vessel. The following year, the sloop *Hiram* sailed into Savannah from the coast of Florida with "sugar, piano fortes, oil cloth and rigging" salvaged from the wreck of the schooner *Richmond Packet*.

The coasters occasionally carried other unusual and unique cargo items. Among these was "one house frame" brought into Savannah by Captain Charles Porquet from St. Simons Island aboard the sloop *Maria* in March 1820. On November 9, 1858, the *Daily Georgian* reported the arrival of Captain Charles Stevens and his schooner *Northern Belle* from White Oak Creek in Camden County with a cargo of three hundred barrels of turpentine and "5 iron safes." In the 1850s, the sloop *Liberty* made several deliveries of "medicine" to South End Plantation on Ossabaw Island, mainly for treating the enslaved workers there.[96]

[95] *Daily Georgian*, May 13, 1850, February 6, 1851, and May 24 and August 5, 1852.

[96] Ibid., March 16, 1820, and November 9, 1858. The Ossabaw Island deliveries are reported in A. Thompson and Horton, "A Short Summary," 183.

Before the 1820s, coasters sailing into Savannah from the southern Georgia port town of St. Marys occasionally carried salt, nails, "cotton bagging," cheese, etc. The "bagging" referred to the inexpensive hemp or burlap-like cloth used to make bags to hold pressed cotton. These types of cargoes were not produced in St. Marys but were coming from Spanish Florida or from overseas, and they reflect the town's position at the time as a border settlement and port of entry where a range of goods were received for transshipment elsewhere. Before 1815, vessels often arrived in Savannah directly from Spanish Florida, and occasionally they carried salt. One such arrival was the schooner *Kitty Ann*, which sailed into Savannah on January 23, 1810, from Amelia Island with three thousand bushels of salt.[97]

## The Coasting Cargoes out of Savannah

The crops and other "products of the country" carried by the coasters represent cargoes for only one "leg" of the coasting trade—the movement of goods away from plantations and small towns into the ports of Savannah and Charleston. When departing Savannah and Charleston, these vessels carried entirely different types of cargoes: items required or desired by coastal residents that were not produced or available locally such as foodstuffs, clothing, building supplies and materials, farming tools, machinery, and the like. Unfortunately, the Savannah and Charleston newspapers in the first six decades of the nineteenth century rarely named these outbound cargoes, so there is no comprehensive list as there is for the inbound shipments. Some information on these outbound cargoes can be obtained from other sources, such as the few newspapers published in small ports receiving these shipments and from existing plantation and merchant account books and records. These sources were examined to understand the types and quantities of material that made up this outbound leg of the local coasting trade.

Between November 1820 and November 1821, the *Darien Gazette* reports the arrivals of coasters from Savannah and Charleston, as well as from other smaller regional ports. During this year, the newspaper reported 127 coaster arrivals from Savannah, representing 73 percent of all arrivals from locations south of Charleston. These reports illustrate Darien's (and other small Georgia ports') dependence on Savannah as a source of goods and the importance of Savannah to regional trade and commerce, even at this early date. Over this same year, only seventeen arrivals are reported from

[97] *Republican & Savannah Evening Ledger*, January 23, 1810.

Charleston, and even smaller numbers of coasters sailed into Darien from St. Marys, St. Augustine, and other coastal communities. The goods received from Savannah and Charleston were similar and consisted of salt, corn, bacon, flour, "groceries," and whiskey, plus numerous shipments identified only as "assorted cargo," "sundries," or "dry goods," as well as furniture, bricks, lime, nails, coal, and hay. Similarly, the *Brunswick Advocate* shows that between 1837 and 1839, more coasters sailed into the small port of Brunswick from Savannah than from any other port. These vessels were laden with cargoes identical to those carried into Darien in the previous decade and are often described only as "merchandise" or "sundry merchandise." Most were destined for merchants or planters in and near Brunswick.[98] The *Darien Gazette* and the *Brunswick Advocate* were only published for a few years, but the types of goods imported from Savannah and Charleston in these years reflect the entire period from 1800 to 1861.

Occasionally, newspaper advertisements provided comprehensive listings of the goods received aboard local coasters. In the April 7, 1855, issue of the *Daily Morning News*, the firm of Hooker & Burnett, located at the community of Mount Pleasant outside of the town of Brunswick, ran an advertisement naming merchandise recently received by the sloop *America* from Savannah:

> HOOKER & BURNETT, Mt. Pleasant, Ga. — Have just received by sloop America, a large assortment of Dry Goods, Ready Made Clothing, Hats, Caps, Boots, Shoes, Saddles, Bridles, Hardware, Crockery Ware, Tin and Wood Ware, Medicines, Perfumeries, Fancy Goods, Et., Etc.
>
> Also
> 20 bbls [barrels] Flour
> 10 bbls Biscuit & Pilot Bread
> 10 bbls Sugar
> 5 bbls Molasses
> 3 bbls Syrup
> 5 hhds [hogsheads] Bacon Sides and Shoulders
> 10 Bags Covvee
> 20 bbls Brandy, Gin and Whiskey
> 10 boxes Soap
> 10 Boxes Tobacco

[98] *Brunswick Advocate*, 1837–1839; *Darien Gazette*, 1820–1821.

25 Sacks Salt

Port, Fulton Market Beef, Sugar Cured Hams, Butter, Cheese, Lard, Herrings, Sardines, Pickles, Jellies, Catsup, Blackberry, Cherry and Ginger Brandy, Cordials, Madeira and Port Wine, Champagne, Porter, Scotch, Ale, Scheidam Schnapps, etc. Also 300 bushels Baltimore Corn, 10 bundles Northern Hay

For sale at Savannah Prices and Figures, Hooker & Burnet

It appears that most or all of the goods mentioned in this advertisement came on the *America*. Samuel Hooker, one of the partners in Hooker & Burnett at Mount Pleasant, purchased the sloop *America* in early 1855, and on March 6, 1855, he advertised in the Savannah *Daily Morning News* that the sloop "will take freight at Winkler's Wharf at West Broad Street for Brunswick and Turtle River, until Wednesday Evening when she will sail." It appears that it was on this voyage out of Savannah that the *America* carried the goods described in the April 7 advertisement.[99]

Some of the goods received by merchants in small towns like Brunswick were then delivered to local planters. However, when possible, coasters carried these goods directly to individual plantation landings. Specific information on the range and variety of the merchandise carried by coasters to individual plantations is found in extant plantation records. Examples include the records of the James Hamilton estate, administered by James Hamilton Couper of Glynn County. The estate had accounts with several Savannah merchants, but its main business was conducted with the Charleston firm of Mitchell & Mure, later the Robert Mure & Co., the factor that purchased most of the estate's rice and cotton crops. The main holdings in the Hamilton estate were Hopeton Plantation on the lower Altamaha River and Hamilton Plantation on St. Simons Island. Account records provide information on the varieties and amounts of goods carried by coasting vessels from Charleston to these plantations. In addition, these records list the prices of goods, freight costs, and other expenses. Figure 4.4 presents a transcription of an invoice listing a shipment made by the schooner *Hopeton* to Hopeton Plantation in November 1854. This shipment contained rather typical items delivered to the coastal plantations: hardware and equipment, some foodstuffs (corn and salt), various

[99] Pearson, "Georgia Coasting Trade," 490 (*America* ownership); *Daily Morning News*, March 6, 1855 (advertisement of *America*'s departure from Winkler Wharf).

medicinal products, and articles destined for enslaved persons on the plantation (blankets and shoes).[100]

Hopeton Plantation was one of the larger rice plantations along the Georgia coast. Smaller properties received fewer but similar types of goods. The Charles C. Jones records for properties in Liberty County list goods received by sailing coaster from various Savannah merchants. On October 31, 1836, Jones recorded payment to the sloop *Macon* for freight on the following items shipped to Carlawater Plantation:

| | |
|---|---|
| 5000 ft boards | 18.75 |
| 2 Bales goods $1.50, 4 pcs bagging $1 | 2.50 |
| 1 canne [?] $1.50, Mill Stone $1 | 2.50 |
| 1 barrel lime 50¢, Barrel tar 38¢ | .88 |
| 3084 ft boards $3.75 | 11.63 |

For May 1838, the accounts noted the following freight costs on items received from R. & W. King & Company in Savannah:

| | |
|---|---|
| To Freight one Hogshead of Bacon per Sloop Macon, as per Bill up to 1st Inst. Paid | 1.00 |
| To Freight, 1 Plough-50¢. 12 Hoes 24¢. 5 Pts 25¢ And Sack of Salt $2.75, & Ten bushels of Salt At 50¢ $5.00 pd Capt. Boles by Cap Jones, May 12th 1838. Pd by Mr. Law | 8.74 |

This particular entry records specific payments to a coasting captain for the transport of goods. In this instance, a one-dollar payment was made to the sloop *Macon* as freight on a hogshead of bacon, and there is a notation of five dollars paid earlier to a "Capt. Boles." John L. Grovenstein was master and part owner of the sloop *Macon* at the time, and the captain named "Boles," probably Leonard Bolles, was one of the several Northern men sailing in the local trade.[101]

[100] Glynn County, Georgia, estate of James Hamilton, Wills and Appraisements, Book F, 178; see House, *Planter Management*, for similar information for Elizafield Plantation at the mouth of the Altamaha River.

[101] "Carlawater Plantation Accounts," October 31, 1836, and May 25, 1838, Charles C. Jones Plantation Books, Howard-Tilton Library, Tulane University.

## Passengers

The coasting sloops and schooners sailing along the lower Atlantic coast often carried passengers because they provided a convenient means of travel in a region where overland journeys were difficult, particularly before the advent of steamboat service in the 1820s. Steamboats offered more reliable schedules than did the sailing vessels, and the passenger accommodations on steamers tended to be better than those provided on small trading sloops and schooners. Accounts of passenger travel and the conditions found aboard the small sailing coasters are rare, but newspapers often included a notation on the transport of passengers in the same sections listing cargoes, with no indication as to how many or who they might have been.

By 1820, Savannah newspapers were beginning to publish information on the number of passengers arriving on particular vessels and, occasionally, their names. Even so, it is apparent that this information was not collected for all vessels. The regular packets sailing between Charleston and Savannah in 1820, such as the sloops *Volant* and *Delight*, each of which sailed between the two ports approximately once a month, must have carried passengers on almost every trip even though the Savannah newspapers only list passengers for a few arrivals. During this time, the sloop *James*, under Captain James Vincent, a resident of Camden County, often arrived in Savannah with coastal produce and passengers from Florida and the lower Georgia coast. Examples include his March 1820 arrival from Amelia Island with 140 bales of cotton and three passengers and his May arrival from Fernandina with three passengers and "15 US troops for Charleston." Nothing is known about the passenger accommodations on the *James*, but it is likely to have been crowded on its May voyage with eighteen passengers and soldiers, plus a crew of two or three. Even on the March trip, the passengers had to share space with the 140 bales of cotton, which took up much of the main deck space and, possibly, portions of the hold.[102]

Occasionally, the names of passengers were listed in the newspapers, such as when the December 13, 1827, issue of the *Daily Georgian* reported the arrival of the schooner *Columbia* from the Satilla and Turtle rivers with 122 bags of cotton and three passengers, "Mr. Gould, Miss Gould and Miss Harris." The "Mr. Gould" mentioned was almost certainly James Gould, planter and lighthouse keeper on St. Simons Island, traveling with a daughter and a relative of his wife, Jane Harris Gould. Gould and his party probably

[102] *Daily Georgian*, March 24 and May 31, 1820.

boarded the *Columbia* at a landing on St. Simons, meaning the schooner made at least one stop after departing the Satilla or Turtle rivers. The Savannah and Charleston newspapers seem to have listed only where coasters picked up their principal cargoes as the place of origin for voyages. However, vessels often made stops en route to pick up passengers and, possibly, additional cargo, after departing their port of origin.[103]

By the late 1820s, steamboats were traveling along the coast between Charleston and Florida, and their accommodations, speed, and reliable schedules made them more desirable to travelers than sailing vessels. Consequently, accounts of passengers aboard sailing coasters in the newspapers began to decrease by 1830, and sailing coasters seem to have rarely carried passengers after 1840.

Many coasters advertised that they carried passengers as well as freight. A typical example is the advertisement for the sloop *Argo* appearing in the July 20, 1837, *Brunswick Advocate*:

For Savannah
The Sloop ARGO, Capt. Wendell, will sail
For the above port on Saturday Next. For
Freight or passage, apply at this office
or to the Captain on board

Some coasters specifically touted their passenger accommodations, especially for a long voyage. Typical of these is the May 1, 1828, advertisement in the *Daily Georgian* for the sloop *America* when she was returning to the Northeast in the spring after spending the rice- and cotton-shipping season in Georgia:

For New York
The fine Sailing Sloop
AMERICA
S.C. Luce, Master
Will have immediate dispatch
For freight or passage, having good accommodations,
Apply to Capt. L. on board at Jones Upper
Wharf or to Hall, Shafer & Tupper

[103] Information on James Gould from Vanstory, *Georgia's Land*, 161–62.

The precise nature of the accommodations for passengers on the coasters is largely unknown. A cabin and berths could be provided for just a few passengers, but the small amount of cabin space available on most coasters meant that some passengers slept on deck or even in the hold. The variable accommodations are indicated by the occasional mention of passengers traveling in "steerage." Accommodations may not have been a serious concern on most coasting voyages because the trips were short, often less than a day and rarely more than three or four days. On the longer voyages, such as those to or from Northeastern ports, berths might have been temporarily added to accommodate greater numbers of travelers. In 1821, an advertisement for the sloop *Bridgeport*, which was departing Darien for New York, noted that only passengers would be carried on the voyage and the sloop had "roomy accommodations, with state rooms."[104]

In early January 1810, the sloop *Eliza* arrived in Savannah from St. Marys with a cargo consisting of "cotton bagging, salt, and negroes." Twenty years later, in 1830, the sloop *Albert* carried fifty-seven enslaved persons into Savannah from Riceboro bound "for Beaufort, South Carolina." Enslaved people were certainly carried aboard coasters between plantations and towns along the Southern Atlantic coast in great numbers, but these accounts represent some of the few mentions of this activity in newspapers. Other records occasionally mention the transport of the enslaved by the sailing coasters. On December 26, 1844, Roswell King Jr. noted in his journal that he "shipped Abner Hannah Sue & Ann to Savannah for sale by sloop Eagle, Grovenstine, Master."[105]

When Congress prohibited the importation of enslaved people into the United States after 1808, the legislation included a provision requiring all vessels of forty tons or greater carrying persons already held as slaves in the United States to file two passenger manifests. One manifest had to be filed at the port of origin and the other at the destination. Both outbound and inbound slave manifests are extant for Savannah for most years between 1808 and 1860, and these show that many sailing coasters periodically transported enslaved people. For example, thirty-two inbound slave manifests were filed at Savannah in 1827, and many of the filing vessels were Georgia coasters. These included the schooners *Orange*, M. H. Hubbard master, and *Betsey*

[104] *Darien Gazette*, June 9, 1821.

[105] *Republican & Savannah Evening Ledger*, January 18, 1810; *Daily Savannah Republican*, February 15, 1830; "Roswell King Diary," December 26, 1844, Hill Memorial Library, Louisiana State University.

*Maria*, Jeremiah Bowen, master, each of which made several voyages from St. Marys transporting enslaved people while the sloop *America*, Stephen Luce, master, and sloop *William*, Elisha Luce, master, each made one voyage from Darien with bondspeople aboard.[106]

The outbound manifests show similar activity. Forty-one outbound slave manifests were filed with the customhouse in Savannah during the first six months of 1827. Among the vessels transporting enslaved people out of the city that year were several Georgia coasters, including the *America* and her master Stephen Luce, bound for Darien; Captain Job Delano and his sloop *Falcon*, sailing to the Satilla River; and Captain John Turner, sailing for Beaufort with the sloop *Cynthia*.[107] Captains Luce and Delano were both from Massachusetts, revealing that Northern captains were involved in the local transport of enslaved people even though some of these Northern men are known to have harbored antislavery sentiments. One suspects that the accommodations provided to the enslaved aboard these ships were minimal, with individuals probably spending the entire voyage on deck or in the vessel's hold. Fortunately, most of these slave-carrying voyages were of short duration.

[106] Beckwith and Lewis, *Savannah Slave Manifests*, 8–9.
[107] Ibid., 104.

# Chapter 5

## "Well Calculated for a Coaster": The Ships in the Georgia Coasting Trade

On January 21, 1803, a notice appeared in the Savannah newspaper *Columbian Museum & Savannah Advertiser* stating that the sloop *Patty* was for sale. The paper reported that the *Patty* was "of an easy draft of water and well calculated for a Coaster." This sloop is believed to be the 48 3/95-ton *Patty* carrying local plantation produce and merchandise in and out of Savannah during the first decade of the nineteenth century. In addition to her shallow or "easy" draft, register documents show the *Patty* was typical of the "coasters" sailing in the local Georgia trade over the next sixty years. Although the *Patty* worked in Georgia from 1800 to about 1810, the only official vessel documents found for her are registers issued at the port of New Haven, Connecticut.[1] The Connecticut registers indicate the *Patty* was built in Southold, New York, in 1797, and by 1799 was homeported in New Haven. The sloop measured fifty-three feet, three inches long; eighteen feet wide; and six feet deep, with a burden of 48 3/95 tons. She had one mast, one deck, and a "square stern." The *Patty*'s mast and sail plan, if typical of the coasting sloops of the period, would have consisted of a gaff-rigged mainsail and one, or possibly two, staysails, or jibs, set forward of the single mast, as shown in Figure 1.2.

Extant enrollment and register documents, bills of sales, newspaper advertisements, and other similar sources provide basic information on the Georgia coasting ships. The majority were small, single-masted sloops, like the *Patty*, and two-masted schooners of less than one hundred tons burden and measuring from forty to seventy-five feet long. Very rarely, a vessel described as a "pettiauger" appears in records. The term "pettiauger" dates from the early Colonial period and is adapted from the Spanish term "periagua." As discussed in Chapter 1, a pettiauger referred to a long, narrow vessel that

[1] Records for the sloop *Patty* are found in "Connecticut Ship Database, 1789–1939," hereinafter cited as "Connecticut Ship Database."

used a single, large hollowed-out log as a hull. Often, hulls were enlarged by splitting the log into two pieces and inserting a plank or planks between the two and by adding planks to the sides. A pettiauger could be either sloop- or schooner-rigged.[2]

Descriptive terms such as "sloop" and "schooner" are insufficient to provide an idea of what these vessels looked like. A small number of paintings and drawings and some photographs are available, but drafted plans (known as "draughts") that provide specific information on a ship's lines are uncommon for the small sailing vessels working in American coasting trades before the Civil War. Most of these ships were constructed without complete plans or the use of models; the builders relied on their skill, experience, and a few basic measurements. Howard Chapelle has discussed the difficulties in finding detailed information on small early-American sailing vessels, particularly before 1830. There was a tendency to produce draughts for craft involved in special activities, such as pilot boats, and later authors have concentrated on these more specialized vessels to the neglect of the typical coasters that made up most of the commercial vessels working along America's coasts. In addition to documentary sources, a few archaeological examples of nineteenth-century vessels from coastal Georgia and South Carolina provide construction details not found elsewhere. Recognizing the limitations, the following discussions provide information on the appearance and manner of build of the ships sailing in the Georgia coasting trade during the first six decades of the nineteenth century.[3]

The exact number of sailing ships involved in the Georgia coasting trade over the period of concern is unknown. No official compendiums of coasting vessels working in a local trade exist, as these craft rarely had to interact with or report to governmental officers or agencies during their everyday activities. Coasting vessels had to obtain official documents such as licenses or enrollments and, occasionally, registers, but many in the Georgia trade obtained these papers at distant ports, so these documents may not reveal their Georgia activities. As noted in the Introduction, vessels with burdens greater than twenty tons involved in foreign trade had to obtain registers while similar-sized vessels engaged in domestic coastal or fisheries trades were required to obtain enrollments and licenses. Smaller commercial vessels, between five and twenty tons in burden, had to obtain yearly licenses. Large numbers of licenses

[2] Fleetwood, *Tidecraft* (1995), provides a comprehensive discussion on the use of the pettiauger along the Southeast coast.

[3] Chapelle, *American Sailing Ships*; Chapelle, *Search for Speed.*

were issued to coasters at Georgia ports in the years before the Civil War, but few exist today, having been destroyed, lost, or misplaced over the intervening years. Large numbers of vessel enrollments and registers for Georgia customs districts exist at the National Archives and Records Administration for the years since 1815, but for many years, particularly those before 1830, numerous documents are missing, fragmentary, or faded and are difficult or impossible to read. Consolidated lists of enrolled vessels for Georgia ports and adjacent areas of South Carolina and Florida, known as "Master Abstracts of Enrollments," are available for the years after 1815. Although these lists provide less information than an enrollment document, they do record a vessel's rig (i.e., its masting and sail plan), burden in tons, the name of the master, the date and place of issuance of previous ship's documents, and the reason for the issuance of new ones. Neither the individual enrollments nor the master lists state specifically that a particular vessel sailed in the coasting trade although it is reasonable to assume that most of the small schooners and sloops enrolled in Georgia, lower South Carolina, and upper Florida worked to some extent in this trade.[4]

The most complete listing of vessels sailing in the Georgia trade is not found in governmental records; rather, it comes from information on vessel arrivals and departures published in port newspapers. As noted in the introduction, the most useful newspapers in this study are those published in Savannah because of that city's central role in the Georgia coasting trade. Shipping lists published in newspapers at smaller port towns, such as Darien and

[4] Most pre-1815 vessel documents submitted to the Treasury Department were apparently destroyed when the British burned Washington in 1814. The extant original individual certificates of enrollment and registry for 1815 and later and the "Master Abstracts of Enrollments," 1815–1911, for various ports are found in RG 41, Records Relating to Vessel Documentation, BMIN, NARA. The enrollments issued at Savannah that are of interest to this study are found in "Enrollments Savannah" and the "Master Abstracts" for that port. A small number of Savannah enrollments, registers, and licenses for the years 1793 to 1800 exist in a set of records titled "French Spoliation Claims, District of Savannah, Registers, Enrollments, and Licenses, 1793–1800." These French Spoliation Claims are found in RG 36, Bureau of Customs, NARA, Washington, DC, and are hereinafter cited as "French Spoliation Claims." In addition, the enrollment and register documents from a number of ports have been abstracted and published. Most of this work was done in the late 1930s and early 1940s as a part of the survey of Federal Archives undertaken by the Work Projects Administration. For the Atlantic coast, most of the ports included in these works are located in the northeast. See Compiled Vessel Documents in the bibliography for a complete listing of the published vessel documents examined for this study.

Brunswick, occasionally provide the names of coasters not found in Savannah papers. Similar lists published in Charleston newspapers are important because many of the coasters carrying Georgia produce sailed there, particularly during the first two decades of the nineteenth century. In addition to the lists of arriving and departing vessels, newspapers also contain advertisements, notices of sale, and the like, which provide information on coasting vessels not found elsewhere.

Relying on the various sources described, particularly the Savannah newspapers and enrollment and register documents, a list of 652 sailing vessels involved in the Georgia coasting trade between 1800 and 1861 has been compiled and is presented as appendix A. This listing of 652 ships is incomplete because so many contemporary records are missing. This list does, however, include most of the coasters working in the Georgia trade from 1800 to 1861.

The names of these ships in themselves are intriguing. They reflect the personal preferences of owners and builders, as well as more general ideas on what were considered appropriate ship names at the time they were built. The Georgia coasters were named after well-known and lesser-known persons, such as the sloops *George Washington*, *Governor Shelby*, *Little Bette*, and *Howard & James* and the schooners *Robert Habersham* and *Mary Ann*; after places or natural features, such as schooners *Altamaha*, *Fort George Packet*, and *Savannah* and the sloops *Frederica* and *Eutaw*; and after patriotic themes, like the sloops *Union*, *Liberty*, and *Independence*. Some names reflected economic topics, as did the schooner *American Coin* and sloop *Fair Trader*; many were named after animals, such as the sloops *Rice Bird*, *Humbird*, and *Leopard* and the schooners *Young Eagle* and *Alligator*, while others took on classical names, such as the sloops *Neptune* and *Argo* and the schooner *Idonius*. Some names were more whimsical, including the schooners *Young Sea Horse*, *Vexation*, *Drunkard's Return*, and *Due Bill* and the sloops *Saucy Jack* and *Jack-O-Lantern*.

A sloop is a sailing vessel with a single mast; a schooner is fitted with two or more masts. Using the information available, 327 (49.9 percent) of the Georgia coasters are identified as schooners and 328 (50.1 percent) as sloops. This represents 655 ships rather than 652 because three of the identified coasters were converted from sloops to schooners during their years sailing in the Georgia trade and are included in the total twice. By the early years of the nineteenth century, these two types of sailing vessels had come to dominate the coasting trade in Georgia and in America as a whole. The identification of the masting and sail plan, or "rig," of a vessel, either sloop or schooner, relies on information contained in vessel documents, newspapers, and other records. In a few instances, these records provide conflicting information, in which cases vessel documents are relied on, if available. In a small number of cases, it is difficult to distinguish between vessels of the same or similar name, and a few duplications probably exist in the list.

It is possible that vessels with the same name listed as both a sloop and a schooner in various sources could be the same ship that has been misidentified or had been converted to a different rig. Several instances have been found where a vessel was "altered," but the nature of that alteration is not always identified. These alterations may reflect changes to a vessel's hull, its rig, or both. Only five cases have been found that specifically document a change in rig, and in every instance, these record the conversion of a single-masted sloop into a two-masted schooner, the rig that became increasingly popular during the early years of the 1800s.

The numbers of sloops and schooners working as Georgia coasters between 1800 and 1861 were roughly equivalent. However, over these sixty-one years, the relative number of each type of vessel changed dramatically. Using information provided in Savannah newspaper shipping lists, Figure 5.1 presents information on the numbers of coasting sloops and schooners sailing into Savannah at approximately ten-year intervals. The number of sloops increased dramatically between 1800 and 1830, and, in every year but 1800, sloops outnumbered schooners carrying coastal products into Savannah during the first three decades of the nineteenth century. The number of coasters identified in 1800 is so small that the differences between the numbers of sloops and schooners may not be meaningful. In 1800, the Georgia coasting trade was in its early stages of development, but sloops came to dominate that trade over the next thirty years of expansion and maturity. After 1830, the total number of sailing vessels declined because of the increasing participation of steamers and the expanding railroad network, both of which transported cargoes formerly carried by coasters. As the number of sailing vessels declined, so did the participation of sloops. By 1839, the numbers of coasting sloops and schooners sailing into Savannah were equivalent, and after that year, the number of sloops declined significantly while the number of schooners rose. In 1861, thirty-one sailing coasters arrived in Savannah, 80 percent of which were schooners. Even though the number of schooners working each year did not change dramatically after about 1810, they became relatively more important after 1830 due to the significant decline in the number of sloops participating in the trade.

## The Coasting Sloop in Georgia

The sloop is a single-masted vessel fitted with "fore-and-aft" sails. A fore-and-aft sail is triangular and is "set" or stretched parallel to the length of a ship's hull. The fore-and-aft sail contrasts with the "square" sail, which is set perpendicular to the length of the hull. The sloop was brought from Europe to America during the early years of settlement and became one of the most common classes of vessels built and used in America. The English term "sloop" is derived from the Dutch "sloep," equivalent to a vessel known as a "chalopue" in French or "chalupa" in Spanish. Before the eighteenth century, the term "shallop" was sometimes used interchangeably with "sloop" to describe single-masted vessels with fore-and-aft sails. The English term "sloop" in America appeared in the 1630s in reference to Dutch vessels in New Amsterdam, and by the 1650s, it was used in English documents in Massachusetts. The term began to appear in records of the Southern colonies (e.g., Maryland and Virginia) about the same time. Sloops were being constructed in America by the third quarter of the seventeenth century, and small sloops of twenty-five to seventy tons burden are the most common vessel in early colonial shipping and shipbuilding documents. These early sloops were typically fitted with a single mast, a bowsprit, and fore-and-aft sails, including a boomed and gaffed mainsail and an unboomed headsail, or jib, fitted to the stay, extending from the mast top to the forward end of the bowsprit, as shown in Figure 5.2. Many of the earliest sloops had open hulls with no deck. Soon sloops were built with a partial or complete deck to protect cargo and crew, which was particularly important when these vessels began to sail in ocean waters. By the eighteenth century, the sloop was normally considered a single-decked vessel although some were constructed with raised quarterdecks at their sterns.[5]

The development and expansion of the West Indies trade after 1640 spurred American shipbuilding, concentrated in New England but conducted in all the colonies. Sloops came to dominate the West Indies trade. Although principally characterized by their fore-and-aft sails, by the eighteenth century, sloops involved in long-distance trades were often fitted with a square topsail (sometimes two) and a square "course," a sail set beneath the topsail that extended down to the main deck, in addition to the fore-and-aft mainsail and jib or jibs.

[5] Baker, *Sloops & Shallops*, 47, 150; Chapelle, *American Sailing Ships*, 11.

The square topsail and course generally fell out of use in America by the early years of the nineteenth century although they might occasionally be used on long sea voyages. A small, fore-and-aft sail above the gaffed mainsail, known as a gaff-topsail, came into general use in the mid-1700s.[6]

Sloops, with their fore-and-aft sails, exhibit sailing characteristics particularly adapted to coastal, harbor, and river settings. Sloops can run closer to the wind than square-sailed vessels and are well adapted to sailing in confined waters or along sinuous channels. Fore-and-aft rigged vessels require fewer crew members than do square-rigged ships of similar size. A further consideration for the builders and owners of sloops was that they were less expensive to build than two-masted schooners of the same size. However, the sloop rig, with its single mast, is ill-suited to large vessels because of the large size of the mast, boom, and main sail needed to efficiently drive larger craft. In very large sloops, the single boom and mainsail required to propel the ship become too big and too heavy for men to handle, plus the width and depth of the hull have to be increased significantly to support the large mast. Shipbuilders found the use of the schooner rig, with two or more masts, allowed them to keep booms and sails to a manageable size. They also could increase the length and capacity of a vessel's hull without significantly expanding the beam and depth, as would be needed to support a single, very large mast and sail. Long, narrow vessels are faster than wide vessels; a schooner rig also allows a smaller crew to handle more sail than is possible on a sloop of the same size.[7]

The sloop was a popular vessel in South Carolina's early history, and many were built in Bermuda, Barbados, or elsewhere in the West Indies. Island-built sloops remained popular in South Carolina until they were supplanted by sloops built in the Northern colonies of Massachusetts, Rhode Island, and Connecticut and, to a lesser extent, by locally built vessels.

By the end of the eighteenth century, many sloops were in use in the American coasting trade, particularly where short voyages were typical.[8] This was true in the Georgia coasting trade after 1800, where many sloops confined their activities to the area between the Savannah and the St. Johns rivers. Even after schooners became popular for coasting as well as sea voyages, sloops continued to be employed in long-distance trips. Many of the sloops used in the Georgia coasting trade were sailed or owned by residents of Northern states,

[6] Baker, *Sloops & Shallops*, 109–10.
[7] Chapelle, *American Sailing Ships*, 298; Fleetwood, *Tidecraft*, 39.
[8] Chapelle, *American Sailing Ships*, 298.

and these vessels commonly made two or more trips a year between those states and Georgia. Some sloops made long overseas voyages while working in the Georgia trade; however, these occur rarely after 1810.

By the beginning of the nineteenth century, several types of sloops were being built in the United States, expressing slight differences in construction or style. The variations in these types were related to how the ships were used, local sailing conditions, and regional style preferences. Sloops were often fitted with centerboards that could be extended through the bottom of the hull to act as a keel to reduce side drift and improve a vessel's sailing abilities into the wind. The centerboard was positioned inside a long, narrow, watertight box or "trunk" built in the bottom of the hull. Typically, the centerboard was attached at its forward end by a pin extending through the trunk and on which the board pivoted when the stern end was lowered beneath the hull.[9]

Sloops built expressly for sea-going or long-distance coastal sailing tended to have slightly deeper drafts and a "sharper" or more V-shaped hull than those that confined their activities to coastal areas and rivers. The construction of these "well modeled" sloops was concentrated in Northeastern states; few are reported to have been built in the South.[10]

Few detailed descriptions of the sloops sailing in the Georgia coasting trade in the nineteenth century exist.[11] There are, however, a variety of records that provide basic descriptive information for many of the vessels working in Georgia. The most useful are enrollment and register documents, as discussed previously. These documents provide information on physical characteristics of vessels, the ownership, place of build, name of master, and where previous documents were issued. An enrollment document for a typical Georgia coasting sloop, the *Conductor*, is presented as Figure 5.3 to demonstrate the types of information found in these documents.

This enrollment (or "enrolment" as it was commonly spelled in the first half of the nineteenth century) for the *Conductor* was Enrollment No. 10, issued at the "Port of Savannah" on May 29, 1837.[12] The document notes that the owner of the *Conductor* was Robert W. Pooler, a resident of Savannah, and that Alexander Wilson was "master" (captain). The owner and the master swore under oath that they were citizens of the United States, a requirement

[9] Baker, *Sloops & Shallops*, 139; Chapelle, *American Sailing Ships*, 299.

[10] Chapelle, *American Sailing Ships*, 298.

[11] Fleetwood's discussions of sloops on the Southeastern coast concentrate on colonial-era craft or those built locally (*Tidecraft*, 45–66).

[12] Enrollment No. 10, Port of Savannah, sloop *Conductor*, May 29, 1837.

under the act for enrolling and registering vessels in the American coasting trade. The enrollment notes the *Conductor* was "of Savannah," meaning its homeport was that city. This was a "Permanent" enrollment issued to a vessel that belonged to the Customs District of Savannah, was homeported there, and was the residence of the owner. A "Temporary" enrollment or register was granted to a vessel that was homeported in another customs district or whose owner resided in another district. Temporary documents were issued when a vessel's status changed during a voyage and allowed a vessel to proceed on her voyage. It remained in force until surrendered for a permanent document upon reaching her homeport.[13]

The enrollment for the *Conductor* goes on to note that the sloop was "Built at Rochester in the State of Massachusetts in the Year Eighteen Hundred and Twenty Five (1825) as appears by her Enrolment No. 18 issued at this Port on the 8th August 1834 and now surrendered on change of Property."

This paragraph provides information on when and where the *Conductor* was built, when and where the previous vessel document was issued, and why the old document was surrendered and the present one issued. In this instance, the last document issued to the *Conductor* had been an enrollment at Savannah ("this Port") in August 1834. This paragraph also reveals the vessel had been sold ("change of Property") and the reason for the issuance of a new enrollment. Elsewhere the enrollment indicates that a single individual, Robert Pooler, became the owner of the sloop on or about May 29, 1837.

The information presented in this paragraph is particularly important in following the life and working history of a vessel. For instance, because the *Conductor* had been issued enrollments rather than registers at Savannah in 1834 and 1837, the presumption is the sloop was working in the coasting trade during these three years. Pre-1834 information on the *Conductor* can be found in the referenced earlier enrollment. Many nineteenth-century Georgia vessel documents are missing, but the August 1834 enrollment for the *Conductor* is extant. In fact, each of the four enrollment documents issued to the *Conductor* between October 1828 and May 1837 have been located. These enrollments show the sloop was issued a register at New Bedford, Massachusetts, in September 1828, which was surrendered for an enrollment at St. Marys, Georgia, on October 21, 1828. When the *Conductor* was enrolled at St. Marys in 1828, her owners were Richard A. Hill, a prominent local coasting captain, and Robert Stafford (of Camden County). Stafford owned a

[13] *Ship Registers of New Bedford, 1796–1850*, iii.

cotton plantation on Cumberland Island and was one of the county's most prominent citizens. No new vessel documents were issued to the *Conductor* until January 1834, when the sloop was enrolled at Savannah. This enrollment shows the sloop was sold to Savannah resident William Williams on or about January 22, 1834. When Williams purchased the sloop, Patrick Ramsbottom was master. Less than seven months later, on August 8, 1834, a new enrollment was issued to the *Conductor* at Savannah. This document shows that Ramsbottom bought the sloop from William Williams and retained ownership of the *Conductor* until selling her to Robert Pooler in May 1837.[14]

These four documents provide information on the use and ownership of the *Conductor* over ten years. During this time, the sloop was homeported at the Georgia ports of St. Marys and Savannah and was issued only enrollments, indicating she was used in the coasting trade. Shipping information published in Savannah newspapers confirming this activity shows the *Conductor* was actively involved in local trade during this period.

The May 1837 enrollment also contains descriptive information, noting that the sloop had one deck and one mast and measured fifty-seven feet, seven inches long, had a breadth of nineteen feet, seven inches, and a depth of five feet, three inches. She had a burden of "Forty eight and 49/95 Tons" and is described as a sloop with "a square stern, No Galleries and a Billet head." This last description means the *Conductor* had a flat or "transom" stern, no projecting galleries in the aft cabin, and a plain, uncarved stem post ("Billet head") at the bow. The owner, Robert Pooler, appeared before Savannah customs officials to verify the information in the enrollment. He also had to post a bond as required in the congressional act for documenting and regulating vessels in the coasting trade. Unfortunately, pre-Civil War Bonds for Enrollments for Savannah have been lost, so the amount of Robert Pooler's bond is unknown. Records from other ports show that bonds for small sloops and schooners ranged from about one hundred dollars to several hundred dollars.[15]

Registers and enrollments were issued in triplicate. One copy went to the master to be kept aboard his vessel, one copy was kept by the collector at the port of issuance, and the third was sent to the Office of the Register of the Treasury in Washington.[16]

[14] Enrollment No. 18, Port of Savannah, sloop *Conductor*, August 8, 1834, and Enrollment No. 2, Port of Savannah, sloop *Conductor*, January 22, 1834.

[15] "Bonds for Enrollments, Port of Mobile, May 16, 1853–November 14, 1856," RG 41, Records Relating to Vessel Documentation, BMIN, NARA.

[16] Stein, *American Maritime Documents*, 74, 135; "Vessel Documents."

Tonnage, or "burden," is probably the most common measurement in documents relating to American coasting vessels. A vessel's burden is a measurement of its internal space and is not specifically related to weight, as is commonly believed. As early as the thirteenth century, a vessel's capacity was measured by the number of "tuns," a large barrel with a capacity of about 250 gallons, it could carry. By the sixteenth century, the tun had become a standard unit of measurement for ships, representing about 100 cubic feet of space and about 2,240 pounds of weight.[17] Tonnage is important, because, as a measure of internal space and, thus carrying capacity, it is the basis upon which import duties, bonds, port fees, and other taxes are assessed. In the United States, the method of determining tonnage used during the first half of the nineteenth century was established in 1789 and continued in use until 1864. The method required measurements of the length, beam (breadth), and depth of hold of a vessel. The formula was ([length - 3/5 beam] x beam x depth), the product of which was divided by 95, the number of cubic feet contained in a "ton" of burden and the reason all tonnages before 1865 were expressed in 95ths. Because the length of the keel was impossible to measure in a floating vessel, the estimated keel length was derived by subtracting 3/5 of the vessel's breadth or beam from the full "length" of the vessel. In enrollments, this length, sometimes referred to as "length on deck," was the straight line distance measured just above the main deck from the forward side of the stem (at the bow) to the after side of the sternpost.[18]

The beam or breadth measured the broadest part of the hull. The depth, or depth of hold, of a single decked vessel was the measurement from the underside of the deck planking to the top of the "ceiling" (flooring) planking in the hold.[19]

The *Conductor*'s burden, 48 49/95 tons, was about average for the sloops engaged in the Georgia coasting trade. The burden is known for 171 of the sloops sailing in the Georgia trade and this information is displayed in Figure 5.4. These range from 10 70/95 tons for the tiny *Thomas Butler King* to the 102 13/95 tons for the Charleston-built *Delight*. Very few Georgia coasters had burdens under twenty tons. This is because boats smaller than this could not have transported cargoes of sufficient size to make them economically feasible.

[17] Fleetwood, *Tidecraft*, 60.

[18] For information on computation of tonnage see Lyman, "Register Tonnage," and Gibson and Gibson, *Dictionary of Transports*, xxxi–xxxiii.

[19] Lyman, "Register Tonnage," 226.

The average burden of these 171 sloops was 53 11/95 tons. Most ranged from thirty to seventy tons and only thirteen had burdens greater than eighty tons. Over the first half of the nineteenth century, larger sloops entered the Georgia trade, and average burdens increased slightly from a low of 49 3/95 tons in 1820 to a high of 58 67/95 in 1850.

Dimensions other than burden are available for many of the sloops in the Georgia trade. These dimensions are those typically provided in enrollment documents; length, breadth and depth of hold. The lengths of 140 sloops are known and these ranged from twenty-seven to seventy feet and averaged fifty-four feet, four inches. Only seven sloops were less than forty feet long, and most were forty to sixty-five feet in length. The beam, or breadth, for 137 sloops is known and ranged from eleven feet, eight inches, for the *Thomas Butler King*, to twenty-three feet for the Charleston-built *Swallow*.[20] The average breadth was eighteen feet, three inches, and most vessels had breadths ranging from seventeen to twenty-one feet. The depth of hold for 137 sloops ranged from four feet to eight feet, six inches, with an average of six feet, three inches. Most (71.5 percent) of these vessels had depths between five and seven feet.

The average sloop in the Georgia trade, with a depth of hold of six feet, three inches would have drawn less than seven to eight feet or so of water. These shallow drafts were particularly adaptable to the water conditions commonly faced by coasters along the Georgia coast. Seventeen sloops had depths of holds of five feet or less, and the place of build is known for sixteen of these. Interestingly, only one of the seventeen sloops with depths of holds of five feet or less was constructed in Massachusetts, where more than half of the sloops working in the Georgia trade were built. Four of these very shallow draft and generally small vessels were built in Georgia or South Carolina, and nine were constructed at Middle Atlantic locations between Pennsylvania and North Carolina.

Although many of the sloops in the Georgia trade were quite small, a few were large vessels. The seventy-foot-long *Catherine Chard*, built in 1837 at Cows Bay, New York, appears to have been the longest sloop sailing in the Georgia trade. Although long, the *Catherine Chard*'s hull was only twenty-one feet, ten inches wide, and it was quite shallow at only six feet deep. This

[20] Register (no number given), Port of Newport, Rhode Island, sloop *Thomas Butler King*, September 30, 1837, listed in *Ship Registers and Enrollments of Newport*; Enrollment No. 18, Port of Charleston, sloop *Swallow*, April 8, 1854. Although specifically identified as a "sloop," some records state the *Swallow* had two masts.

vessel could have neither supported a tall mast nor carried a great expanse of sail on such a shallow and narrow hull. Further, the shallow hull probably required the use of a centerboard although no mention of one is found in the records that relate to the sloop. The *Catherine Chard* appears commonly in the Savannah newspapers of the 1850s, bringing in rice and cotton, often from the Satilla River and other areas on the South Georgia coast under the command of captain and part owner Lewis P. Wiggins.[21]

The shape of the hulls of the Georgia coasting sloops is difficult to determine due to a lack of plans or descriptions. It is suspected that most of the sloops sailing in Georgia were flat-bottomed with bluff bows and "full-built" hulls that did not narrow sharply with depth. These hulls were amenable to cargo carrying and to the shallow waters of the tidal creeks and rivers of the Southeastern coast.

The archaeological remains of several sailing vessels thought to have been used in the Southern coasting trade in the late eighteenth or nineteenth centuries have been discovered that provide some information on hull shape. One of these is the Clydesdale Plantation sloop, excavated in the Back River near Savannah in 1992. Believed to have been abandoned between 1780 and 1820, the Clydesdale sloop measured forty-three feet, nine inches long; fifteen feet, five inches in beam and six feet, three inches deep. The Clydesdale sloop had a flat, or transom, stern, and enrollment documents of Georgia coasting sloops show that the flat stern, often described as "square," was most common. The burden of the vessel was estimated at twenty to twenty-five tons, and because the local woods live oak and yellow pine were used in construction, the authors suggest the boat was built locally.[22]

Another archaeological vessel identified as a possible coaster is the Malcolm Boat, discovered in the Ashley River about twelve miles from Charleston. Excavated in 1992, this boat measured 41 feet, 10.25 inches, long, with a beam of about 11 feet, 9.25 inches, and an estimated depth of hold of 4 feet, 11 inches. It was round hulled, keeled and had a transom or flat stern. It was

[21] Captain Wiggins's ownership of the *Catherine Chard* is found in several enrollments issued at Savannah, including Enrollment No. 16, Port of Savannah, sloop *Catherine Chard*, December 1, 1849, and Enrollment No. 5, Port of Savannah, sloop *Catherine Chard*, June 8, 1852.

[22] Amer and Hocker, "Comparative Analysis," 299–300. Relying on the measurements of the Clydesdale sloop, her burden may have been closer to thirty-five tons rather than the twenty to twenty-five suggested by the authors.

made from locally available live oak, white oak, yellow pine, and cypress. The sloop's displacement was estimated to be about twenty-four tons.[23]

The Malcolm Boat was fitted with a single mast, but the authors suggest the vessel had been converted from a two-masted rig sometime in its career.

It seems that the flat bottoms and shallow drafts of these vessels would allow them to drift to leeward when the wind was abeam or when sailing to windward. This could have made them inefficient when sailing in the narrow, sinuous tidal sounds and rivers along the Southeastern coast. However, on the Clydesdale Boat, at least, the relatively flat bottom seems to have been offset through the use of a deep keel that provided added lateral resistance, improving sailing performance.[24]

Although centerboards are rarely mentioned for the Georgia coasting vessels in contemporary documents, they were certainly well known and in use on coasting vessels elsewhere by 1820. Many of the sloops used in the Georgia trade were constructed at Northern or Chesapeake area locations where the centerboard was common. It is believed that a number of the sloops working in Georgia were fitted with them, at least after about 1830, despite their lack of mention in records of the period.[25]

Despite their rare mention in historical documents, centerboards have been recorded on archaeological remains of nineteenth-century vessels found in the region. The remnants of two nineteenth-century centerboard ships were found along the south shore of Hutchinson Island, opposite the city of Savannah. One of these, designated Vessel 15, consisted only of a nineteen-foot-long segment of keel with the centerboard trunk still attached. The trunk was centered on the keel that had been slotted to accept the centerboard. This boat is believed to have been abandoned sometime after about 1853.[26]

The other vessel found in the same vicinity was more complete and consisted of much of the bottom hull of a centerboard vessel. The almost intact keel of this ship, identified as Vessel 6, was made of sweet gum and measured about eighty-five feet long. The original beam of the vessel was difficult to determine, but drawings suggest a breadth of at least fifteen feet. The depth of the hull could not be measured. The centerboard trunk for Vessel 6 was partially intact and was positioned to the starboard side of the keel. The authors suggest that Vessel 6 was originally schooner-rigged and fitted with two

[23] Amer, "Malcolm Boat," 81.

[24] Amer and Hocker, "Comparative Analysis," 299.

[25] Chapelle, *Search for Speed*, 264, 280.

[26] Panamerican Consultants, *Archaeological Data Recovery*, 45–47.

masts but later converted to a single-masted sloop. A shield-style, five-cent piece minted between 1867 and 1883 was found inside of the repositioned mast step, indicating the rig conversion took place after 1867 although the vessel is believed to have been constructed before the Civil War. Available evidence suggests that Vessel 6 was abandoned by 1882. Like the Malcolm Boat, this conversion of a schooner to a sloop contradicts the information on conversions of Georgia coasters recorded in extant vessel documents.[27]

Little written information on the masting and sail plans of nineteenth-century Georgia coasting sloops is available. However, there is no reason to believe they differed greatly from what is known for the typical sloops of the period. The early sloops were generally fitted with a bowsprit; a single mast with a gaff-rigged mainsail; one or sometimes two headsails (staysails) forward of the mast, both of which are known as "jibs"; often a square topsail; and a square "course," a sail set beneath the topsail (see Figure 5.2).

Most authorities state the square course and topsails fell out of use by about 1800 and were replaced by a triangular, fore-and-aft gaff topsail fitted above the mainsail.[28] However, Thomas Sanders, a sailmaker in Wareham, Massachusetts, left records of repair work he did in the 1840s on several sloops working along the Georgia coast that reveal the continued use of square sails. For example, in 1842, Sanders repaired the "Gaff top" for the sloop *Eleanor*, and in later years, he repaired the *Eleanor*'s "Square sail" and made the sloop "a jib with bonnet."[29] It would appear from these records that the *Eleanor* was fitted with a square sail or "course" as late as 1842.

Advertisements of the sale of sloops appearing in Georgia newspapers occasionally provide information on the sails carried. One of the more complete descriptions of a coaster's equipment is found in an advertisement of the sheriff's sale of the sloop *B. F. Sherwood* appearing in the March 26, 1846, issue of the *Daily Georgian*:

> The sloop B.F. Sherwood, as she now lies moored at Thunderbolt, with standing and running rigging, one main, one jib and one square sail, one chain and anchor, two pieces of Hawsers, one Carboose and cooking utensils, one brass and one wooden compass, three mattresses, four pillow and pillow cases, four sheets, one signal lantern, one large and one signal flag, two water casks, one yawl,

[27] Mid-Atlantic Technology, *Archaeological Data Recovery*, 22, 57.

[28] Chapelle, *American Sailing Ships*, 299.

[29] Records for the sloop *Eleanor*, for 1842, 1844, and 1846 in "Thomas L. Sanders Accounts, 1839–1881," Manuscript Collection, Peabody Essex Library.

> one chart, one lead and line, and one lot of carpenter's tools. Levied on the property of Edmund Bird, to satisfy a fi fa issued out of the Hon. The Court of Common Pleas and Oyer and Terminer for the city of Savannah, in favor of Montgomery Cumming vs. Edmund Bird. Property pointed out by plaintiff.
>
> Elisha Willy, Sheriff

The *B. F. Sherwood*, like the *Eleanor*, carried a square sail, and several other accounts have been found indicating that sloops working in Georgia were often fitted with square sails well after 1840. These sails might have been used principally in long distance voyages, but their relatively common mention suggests that square sails were in use on American coasting sloops later than many have suggested.[30]

The use of small boats on coasters, such as the "yawl" mentioned in the sale notice for the *B. F. Sherwood*, was common. In 1820, a notice appeared in the *Darien Gazette* offering a five-dollar reward for the return of a "yawl boat" stolen from alongside the coasting sloop *Neptune*. Forty years later, Henry J. Dickerson & Son, the largest stevedoring firm in Savannah, offered a fifteen-dollar reward for a boat lost from the coasting schooner *Blooming Youth*. The lost boat was described as "a new port built boat" that measured twenty feet long and was painted black with a yellow and white stripe. Some small yawls may have been fitted with a removable mast and sail, but for the most part, they were row boats. These boats were either carried aboard or towed behind coasters and were used to travel ashore as well as to transfer cargo and, when necessary, to pull a coaster against a tide or river current.[31]

The sale notice for the *B. F. Sherwood* also provides some information on the accommodations and navigation equipment found on the coasting sloops. It appears this sloop had at least three bunks ("three mattresses") or possibly four (four pillows, pillowcases, and sheets) and had some type of cooking hearth or stove (the "Carboose" or, more appropriately, "caboose") and cooking utensils. The navigation equipment included two compasses, a signal lantern and flags, a lead line, and a chart. These are typical of the navigation aids used aboard the Georgia coasters. The presence of three or four bunks

[30] See Chapelle, *American Sailing Ships*, for information on the use of square sails on American sloops.

[31] Fleetwood, *Tidecraft*, 63–64, describes these small boats, and Silva, *Early Reminiscences*, 20, describes using row boats to tow coasters against the current on the St Marys River. The notices for lost boats are from the *Darien Gazette*, November 4, 1820, and the *Savannah Daily Morning News*, April 1, 1861.

suggests a crew of this size, presumably a master and two or three crewmen. The *B. F. Sherwood*, with a length of forty-six feet, four inches and a burden of 38 72/95 tons, was smaller than the average Georgia coasting sloop. However, small crews such as this were typical of most coasters although larger crews may have been aboard for the longer voyages made periodically between Georgia and their homeports in Northern states.

By 1800, rules and guidelines had been developed by builders for masting and sparing vessels of varying sizes. These rules were not strictly followed, and builders often used their judgment and experience in determining the sizes of masts and spars. Sloops built in the Chesapeake area commonly had little (or sometimes no) standing rigging, which supported the main mast. This might consist of a single rope, known as a "shroud," extending from each side of the vessel to near the top of the mast. Sloops sailing in open ocean waters or on long-distance voyages required more substantial standing rigging, typically two to four shrouds per side, plus a headstay running from near the top of the mast forward to the bowsprit.[32]

Howard Chapelle notes that it was typical for American builders to use multiples of a vessel's beam to obtain mast lengths. For example, in the Chesapeake Bay region, the total length of a schooner's mainmast was commonly 3.1 to 3.5 times the vessel's beam. This ratio seems to have been common for shipbuilders. In *Elements of Mastmaking, Sailmaking and Rigging* (1794), author David Steel noted that British builders followed the same approach, stating that the length of a sloop's "Mast and topmast in one" should equal 3.75 times the breadth of the vessel, slightly greater than the proportion Chapelle provided for American vessels. This may reflect that he (Chapelle) was not including the topmast with the mainmast, as Steel does.[33]

If the rules for masting given by Chapelle and Steel hold true for sloops sailing in the Georgia trade, the mast and topmast together on the typical coasting sloop with a beam of about eighteen feet, three inches would have been about sixty-eight feet, five inches long. Typically, the mainmast was approximately 72 percent of this total length, meaning it would measure forty-nine or fifty feet long. The topmast was about 34 percent of the total length and in this example would have been twenty-three or twenty-four feet long. Approximately 85 to 90 percent of the topmast extended above the mainmast; the lower portion overlapped the mainmast where it was attached. The mainmast extended through the main deck to the bottom of the ship's hold. This

[32] Steel, *Elements of Mastmaking*, 185, 279.

[33] Chapelle, *Search for Speed*, 214; Steel, *Elements of Mastmaking*, 47, 193.

means the mainmast and topmast together rose about sixty-two feet above the deck on the typical Georgia coasting sloop, considering an average depth of hold of six feet, three inches.

Although sloops of the period were often fitted with a topmast, some of the Georgia sloops, such as the four-foot-deep *Liberty* and *Fair Trader*, had hulls too shallow to support a topmast. One of the few accounts of the dimensions of masts for a regionally built sloop suggests it did not have a topmast. This was the sloop *Harriet*, launched on May 10, 1808, near Charleston, South Carolina. The *Harriet* is described as a thirty-five-ton sloop with a keel length of thirty-eight feet, a beam of fifteen feet, and a depth of hold of five feet, eight inches. The sloop's mast measured fifty feet long, the boom forty-two feet long, and the "gaff" seventeen feet long.[34]

Available draughts and illustrations of sloops of the first half of the nineteenth century indicate the mast was stepped well forward in the hull. In describing the forty-nine-foot, nine-inch-long sloop *Mayflower*, built at Pembroke, Massachusetts, in 1828, William Baker notes the mast would have been stepped three-tenths of the vessel's length from the bow "following the rules."[35] However, drawings of other nineteenth-century sloops show the mast was often stepped a little farther forward, about 20 to 25 percent of the vessel's length from the bow.

The typical sloop sailing in the Georgia trade had a single main deck. The use of a "quarterdeck" (an elevated deck toward the stern of the vessel) on coasting sloops was going out of style in the late eighteenth century, and none of the enrollments or registers found for sloops sailing in Georgia between 1800 and 1861 mention a quarterdeck.

By and large, the sloops in the Georgia trade were plainly built vessels, constructed with little in the way of ornamentation or elaboration. By 1800, most of these ships were built with "plain" heads, meaning that the bow stem was plain and lacked ornamentation. Vessel documents reveal that a few sloops had "fiddle" or "scroll" heads where the top of the bow stem had been carved with a scroll design, as seen in the sloop pictured in Figure 5.5. The general lack of ornamentation on these working vessels is further reflected in the fact that few were fitted with figureheads. Vessel documents examined reveal only thirteen instances of figureheads on the sloops working in Georgia. Six sloops were fitted with the figurehead of a woman, identified as a

[34] Information on the *Harriet* is found in "Sloop Book," Baker Family Papers, South Carolina Historical Society.

[35] Baker, *Sloops & Shallops*, 150.

"woman's bust head," "female bust head" or "woman figurehead." These were the sloops *Halcyon*, built at Newport, Rhode Island, in 1806; the *Delight*, built in Charleston in 1811; the *Lady Washington*, built at Portsmouth, Rhode Island, in 1814; the *John Chevalier*, built in St. Marys, Georgia, in 1817; the *Angel*, built at Rochester, Massachusetts, in 1828; and the *Catherine Chard*, built at Cows Bay, New York, in 1837. A "man figurehead" or "man's bust head" was identified on three sloops: the *Niagara*, built in New York City in 1817; the *Chauncey*, built in Haddam, Connecticut, in 1818; and the *Jay*, built at Lyme, Connecticut, in 1795. The sloop *Express*, built at Newport, Rhode Island, in 1820, had an eagle as a figurehead while the sloops *Volant*, built at Barnstable, Massachusetts, in 1817, and the *Mary Howard*, built at Wareham, Massachusetts, in 1827, were both fitted with undescribed figureheads. The enrollment document for the *Mill Maid*, a sloop of 25 35/95 tons burden built in Savannah in 1828, states she had an "Alligator" head. This refers not to a figurehead but to a bow that was narrow, flat, or blunt. Figureheads were never common on the Georgia sloops, but they were most often found on those built before 1830.

These coasting sloops were mainly cargo carriers, and the few descriptions of working sloops indicate that cargo, such as bags of cotton and barrels and hogsheads of rice, sugar, or molasses, were often carried on deck, suggesting that decks were open, with few deck structures and a minimal amount of clutter. Baker provides a proposed deck plan for the previously mentioned sloop *Mayflower*, built in 1823 at Pembroke, Massachusetts, which was similar to what we know of the typical Georgia sloop (Figure 5.6).[36]

Baker's reconstruction of the *Mayflower* shows a small, low cabin extending through the main deck near the stern, with an entranceway from the stern. Baker refers to this cabin as a "cuddy," but it is more accurately described as a "trunk cabin" because it extends through a deck. The cabin's interior was high enough to provide reasonable headroom, but the portion extending above the main deck, generally known as the "trunk," was low so it would not interfere with the swinging of the boom. These cabins were commonly fitted with skylights or small windows to provide light to the interior.

On the *Mayflower* plans, a large cargo hatch occupies the central portion of the main deck between the cabin and the mast (Figure 5.6). Just forward of the mast is a smaller hatch to provide access into the forepeak of the vessel. The bowsprit is stepped into a single "bit post" with an adjacent log windlass or winch. The windlass, used for raising the anchor or hauling in lines and

[36] Ibid., Figure 49.

ropes for other purposes, was turned with handspikes. One or two simple hollow log pumps for removing water from the hull would have been aboard. These might have been positioned adjacent to the mast, although for the *Mayflower*, Baker suggests they were positioned at the forward end of the cabin. The bulwark, the portion of the sides extending above and surrounding the main deck, would have typically been low and simple, what is sometimes referred to as a "log rail."

The stern posts on most coasting sloops were straight, with a slight amount of rake aft, and steering was done with a tiller attached to the head of the rudder post, as shown in Figure 5.6. Some larger sloops might have been fitted with wheels for steering, but the tiller was more common.

As revealed in enrollment documents, the flat stern depicted for the *Mayflower* was typical for the Georgia sloops. The term "square" is used in these records to describe the sterns of most of the sloops. Only two of the Georgia sloops are described as having "counter" sterns, meaning the main deck extended aft of the sternpost, producing an overhanging stern. This overhang typically consisted of cabin space. The two Georgia sloops with counter sterns were the small, thirty-nine-foot *Eutaw*, built at Charleston, South Carolina, in 1835, and the *Visitor*, a sixty-three-foot-long sloop built at Greenwich, New Jersey, in 1833. The *John Chevalier*, a sixty-two-foot sloop built in St. Marys, Georgia, in 1817, was described as having a "square stern with a square tuck." The term "tuck" may have meant the same thing as an overhang.[37]

Round sterns were not common on the coasting sloops, and only one of the Georgia sloops is described as having a "round" stern. This was the *Mathews*, a forty-one-foot vessel built on the shores of Chesapeake Bay in Mathews County, Virginia, in 1817. Enrollment documents for the *Catherine Chard*, the longest sloop known to have worked in the trade, state she had a "round tuck," possibly referring to a round and projecting stern.[38]

The coasting sloops commonly carried passengers, particularly before the arrival of steamboats in the 1820s. Before the 1840s, newspaper advertisements for the sloops often note that a vessel was available for "freight or

[37] Chapelle, *Search for Speed*, 318. Stern descriptions are from Enrollment No. 1, Port of Savannah, sloop *Eutaw*, January 10, 1842; Enrollment No. 5, Port of Savannah, sloop *John Chevalier*, March 9, 1819; Enrollment No. 13, Port of Savannah, sloop *Visitor*, August 30, 1838, Registers and Certificates of Enrollment, BMIN, NARA.

[38] Stern descriptions from Enrollment No. 10, Port of Brunswick, sloop *Mathews*, December 29, 1823; Enrollment No. 16, Port of Savannah, sloop *Catherine Chard*, December 1, 1849, Registers and Certificates of Enrollment, BMIN, NARA.

passage." Those sloops sailing between Georgia and Northern ports at the beginning or end of the trading season often carried passengers, meaning accommodations had to be available for what could be a voyage of two weeks or more. Some sloops must have had cabins sufficiently large for several passengers, but it's unlikely these facilities offered many amenities or a great deal of comfort. As noted in chapter 3, some vessels did promote their passenger accommodations; the *Bolivar* advertised in 1828 that she had "excellent accommodations" for passengers, and the sloop *Bridgeport* advertised in 1821 that she had "roomy accommodations, with state rooms."[39]

Cooking facilities were required on the sloops for the crew and any passengers aboard. These consisted of a simple brick hearth or a small iron stove, likely in the aft cabin, where it could also provide heat. In February 1836, a "cooking stove" was purchased for the sloop *Mill Maid* in Savannah at a cost of ten dollars. In 1817, while at Savannah, the Rochester, Massachusetts, sloop *Harmony* expended six dollars for "mending coobose," meaning the "caboose," a term that referred to the cooking area, or more specifically, the cooking stove itself. Later, additional expenditures were made "for repairing [the] chimney" on the *Harmony*.[40]

Information on the cost of building a coasting sloop or schooner during the first six decades of the nineteenth century is scarce. Much of the available cost information is for larger vessels. While that information is generally useful, most authorities indicate that the cost per ton of building a small vessel was somewhat higher than for a larger one. In addition, the cost varied from locale to locale depending on labor rates and the availability and price of materials. The general practice was for a builder to charge for constructing the hull, masts, and spars. The sails, rigging, and outfitting were additional expenses.[41]

Geoffrey Footner reports that it cost twenty-six dollars per ton to build the 259 83/95-ton schooner *Patapsco* that was launched at Fells Point, Maryland, on the Chesapeake Bay in 1812. This cost did not include the sails and rigging. The *Patapsco* was much larger than the typical Georgia coaster, meaning the per-ton cost for a vessel in the forty- to seventy-ton range would have been higher. In the 1880s, Henry Hall reported that the cost of building

[39] *Daily Georgian*, June 8, 1828; *Darien Gazette*, June 9, 1821.

[40] The cost of the stove for the *Mill Maid* is found in Unidentified Cash Book, February 26, 1836, Georgia Historical Society. The repairs on the *Harmony* are recorded in "Paul Wing Accounts," October 23, 1817, and January 3, 1818.

[41] Hall, *Ship-Building*, 14.

a three-hundred-ton wooden ship in the United States ranged from forty-five to sixty dollars per ton of burden and that this cost was about the same as it had been in 1847.[42]

A few deeds of sale from courthouse records of port towns and some estate appraisements provide the purchase cost or estimated value of vessels sailing in the Georgia trade. In addition, a set of documents known as "Conveyances of Enrolled Vessels" that record sales of vessels are available for Savannah from 1850 to 1856. Of course, the age and condition of any given vessel affected its sale amount. This is particularly relevant for the Georgia coasters because many of the ships used in the trade were old when local buyers purchased them. Other factors, such as the buyer's relationship to the seller and general economic and market conditions, influenced the sale price. These factors make generalizations difficult, but these various records provide insight into a principal expense for the owners of coasting ships.

Several sloops are included in conveyance and bill of sale records available for Savannah. Most were old when sold, so few sale prices reflect the original cost to build the vessel. One sale that may approximate original construction costs concerned the tiny 23 67/95-ton sloop *Carrier Dove*, built in Stonington, Connecticut, in 1853. On February 8, 1854, Jeremiah Wilcox of Connecticut, the builder, sold the *Carrier Dove* to Savannah residents John Makin and William Davidson for six hundred dollars. This comes to $25.35 per registered ton for a vessel that could not have been much more than a year old. This cost seems low, given the average building cost of forty-five to sixty dollars per ton provided by Hall; however, neither the condition of the vessel nor the circumstances of the sale are known.[43]

Other sloops for which sale information is available were considerably older than the *Carrier Dove* when sold. In March 1851, Captain Charles Stevens of St. Simons Island sold his sloop *Splendid* to William Morrell of Savannah for $850. Several years later, in February 1855, Stevens sold his sloop *America* to S. A. Hooker of Brunswick for $1,600. The 38 40/95-ton *Splendid* was eighteen years old when Stevens sold her, while the 56 23/95-ton *America* was twenty-nine years old. Despite the age disparity, the sloops sold for about the same amount per ton, the *Splendid* for $22.13 per ton, the *America for* $28.45. In 1851, a half interest in the seven-year-old *B. S. Newcomb* sold for

[42] Footner, *Tidewater Triumph*, 112, 279; Hall, *Ship-Building*, 87.

[43] "Conveyances of Enrolled Vessels, Savannah," sloop *Carrier Dove*, February 8, 1854. John Makin and William Davidson appear in the 1850 Chatham County census found in Otto, *1850 Census Chatham County*, 17.

$15.76 per ton, and a year later, a one-quarter interest in the sloop sold for $12.95 per ton. In 1852, the fifteen-year-old *Catherine Chard* sold for $1,650, or $25.52 per ton, while the following year, a one-quarter interest in the sloop sold for $350.00 or $21.66 per ton. The sales of the *B. S. Newcomb* and *Catherine Chard* show a modest amount of depreciation over a year.[44]

The selling price of one Georgia sloop differed considerably from the norm. This was the locally built *Liberty*, constructed in McIntosh County in 1846. In 1852, William King of McIntosh County sold the six-year-old *Liberty* to Phineas Kollock for just $200, or $9.34 per ton. Two years later, Christy Holverson, a Savannah coasting captain, purchased the now eight-year-old *Liberty* for the modest sum of $100, representing a cost of only $4.67 per ton of burden. The *Liberty* might have been in poor condition at the time of both sales; however, the low price may reflect the fact that the *Liberty* was a poor candidate for a coaster. The *Liberty* had a very narrow (fifteen feet) and shallow (four feet) hull, meaning she was limited in the cargo she could carry and may have been able to carry only a small amount of sail without a danger of foundering. Considering her narrow, shallow hull, the *Liberty* may have been a log-built "pettiauger," or at least constructed in that style.[45]

## The Coasting Schooner in Georgia

The schooner is most commonly defined as a fore-and-aft rigged vessel with more than one mast. The origin of the term schooner and the vessel itself has long been debated among maritime historians. The most commonly accepted belief is that the schooner rig appeared in the Netherlands by the early seventeenth century. The rig was soon introduced into England and then brought to America before 1700. Some suggest the term "schooner" originated in the Dutch settlements of New England in the seventeenth century; however, one unverified account notes the word comes from "scoon," a slang term for a fast vessel first used at a launching in Gloucester, Massachusetts, in 1713. This oft-repeated story cannot be confirmed, but after this date, mentions and depictions of schooners become increasingly common. For its first century of

[44] "Conveyances of Enrolled Vessels, Savannah," sloop *Splendid*, March 24, 1851; sloop *America*, February 9, 1855; sloop *B. S. Newcomb*, July 3, 1851, and September 25, 1852; and sloop *Catherine Chard*, June 1, 1852, and June 21, 1853.

[45] "Conveyances of Enrolled Vessels, Savannah," sloop *Liberty*, March 25, 1852, and April 27, 1854.

use, the schooner was a two-masted vessel, but schooners with three or more masts entered production in the last decades of the eighteenth century.[46]

The schooner became popular as builders learned that the fore-and-aft rig with two masts allowed them to increase the length and capacity of a vessel without increasing the beam and depth, as would be needed to support a single, very large mast and sail on a sloop. As noted previously, long, narrow vessels were faster than wide vessels. A schooner rig also allowed a smaller crew to handle more sail than possible on a sloop of similar size.[47]

Although fore-and-aft sails were most common on schooners, many also carried square sails on their foremast, and a few had them on the mainmast. Particularly popular were one or two square sails carried at the top of the foremast, the so-called "topsail schooner." Also, like sloops, many schooners carried a large, square mainsail on the foremast, known as a "course." The "course" was most typical of early schooners, but the smaller, square topsails remained in common use well into the nineteenth century.[48]

Although the schooner rig originated in Europe, it was in America that it became most popular and reached its epitome of design. Because of their speed, schooners were popular as privateers during the American Revolution, especially the "topsail" schooner that could carry a large amount of sail for its size. By the end of the eighteenth century, small- and medium-sized schooners had become the most popular ships working in the coastwise trade of the United States. Larger versions were prevalent in overseas trade, including the slave trade. Among the larger versions was the "Baltimore clipper schooner," a vessel that came into existence by the mid-eighteenth century and by 1815 was reportedly carrying nine-tenths of the foreign trade of the United States. The rapid spread of the schooner design and its tremendous increase in numbers during the latter half of the eighteenth century are testaments to its commercial viability. By 1790, the two-masted schooner was the "national rig" of the United States. It became the most popular vessel in the coastwise trade before 1800 and remained so until the Civil War. It was so popular that the term "coaster" was often considered synonymous with "schooner."[49]

[46] Historical information on the origins of the schooner is derived mainly from Chapelle, *Search for Speed*, 11; Hahn, *Colonial Schooner*, 14–15; and MacGregor, *Merchant Sailing Ships*, 92.

[47] Chapelle, *American Sailing Ships*, 298; Fleetwood, *Tidecraft*, 39; Greenhill and Manning, *Schooner Bertha L. Downs*, 8.

[48] Chapelle, *American Sailing Ships*, 11.

[49] Davis, *American Sailing Ships*, 25; Chapelle, *American Sailing Ships*, 132–33, 220–21; Taylor, *Transportation Revolution*, 108.

As with other wooden vessels, the centers for schooner construction in the United States were in the Northeast, particularly New England. Soon, construction spread along the Atlantic coast to South Carolina and Georgia. The popularity of the schooner and its widespread construction meant that a number of regional types were developed to suit specific sailing and trade conditions.

By the early 1800s, most of the schooners involved in the American coasting trade had full or bluff bows and more rounded hulls than those schooners built expressly for speed and long-distance, deep-water sailing. As with sloops, cargo capacity and a shallow draft to allow travel on shallow and confined coastal waters were more important than speed for schooners sailing in the Georgia trade.

Two-masted schooners dominated the American coasting trade. The very earliest three-masted schooners seem to have been built around 1800, but it was not until the 1820s or 1830s that they appear in any numbers, and then almost exclusively in overseas trade. The dominance of the two-masted schooner in the coasting trade during the first half of the nineteenth century is expressed by the fact that every one of the schooners working in the Georgia coasting trade between 1800 and 1861 that can be identified was a two-masted vessel. Not a single three-masted schooner has been found in the records examined. It was not until after the Civil War that the three-masted schooner became popular in the coastwise trade, particularly for hauling lumber.[50]

The popularity of schooners increased during the second half of the eighteenth century, and by 1800, they appear more commonly in Savannah newspaper accounts of vessel arrivals and departures than do sloops. In 1800, the Savannah paper, the *Georgia Gazette*, lists numerous schooners arriving in Savannah from regional ports, as well as from Northern locales such as New Bedford, Boston, and Philadelphia, and overseas locations like Jamaica, St. Kitts, the Bahamas, Barbados, and Bermuda. However, as discussed previously, and as shown in Figure 5.1, despite the national dominance of schooners in the American coastwise trade by around 1800, in the local coasting trade out of Savannah, yearly arrivals by sloops typically outnumbered schooners through 1830.

Tonnage measurements have been found for 153 of the 327 schooners identified as Georgia traders between 1800 and 1861. These burdens range from 18 54/95 for the North Carolina-built *Eliza & Emma* to the 196 80/95

[50] Chapelle, *American Sailing Ships*, 258.

tons of the *Paugassett*, built at Wilmington, Delaware, in 1856. As shown in Figure 5.4, the average tonnage for these 153 schooners is 70 53/95 tons. This average burden is about 30 percent greater than the fifty-three tons average burden of sloops in the Georgia trade, and, as a class, the Georgia coasting schooners were larger than sloops. For example, only two sloops working on the Georgia coast had burdens greater than one hundred tons, while twenty-eight of the schooners did. Although there were Georgia schooners with burdens greater than one hundred tons, most ranged from about forty tons burden to ninety-five tons.[51]

Very large schooners, like the 196 80/95-ton *Paugassett* and the 175 50/95-ton *Alexander Blue*, were too big to serve many of the shallow water landings on the coast, and they mainly sailed in interregional coastwise trade between Savannah and Northeastern ports or in overseas trades. Few of those schooners with burdens over one hundred tons were regular participants in the local trade. Many appear once or twice in newspaper accounts sailing into Savannah with coastal produce from regional ports like Darien or Brunswick. These large schooners likely were placed into service carrying local commodities only as a matter of convenience between sailings to Northern or overseas ports.

The tiny *Eliza & Emma* and *Thomas Spalding* were the smallest schooners for which information on burden is available. The twenty-one-ton *Thomas Spalding* was enrolled at the port of Savannah in 1856. This enrollment notes the *Thomas Spalding* was built in 1852 "at Satilla," presumably referring to the Satilla River. The ship was named for the planter Thomas Spalding of Sapelo Island. Mention of the *Thomas Spalding* has been found twice in Savannah newspaper shipping lists. Both sailings occurred in 1856. In June, she sailed from "Sapelo" with sixteen bales of cotton and in August from Charleston, with no cargo listed.[52]

Information on the 18 54/95-ton *Eliza & Emma* is contained in an 1838 Chatham County deed recording the sale of a half share of the schooner by William Thomas to Joseph E. Silveria for $235.80. Joseph Silveria was a Savannah resident who held ownership in several coasters, including the sloops

[51] Chatham County, Georgia, burden information on the schooner *Eliza & Emma* is from deed of sale, schooner *Eliza & Emma*, William Thomas to Joseph E. Silveria, December 12, 1838, Deed Book 2W, 265–66; burden of the *Paugassett* is from "Connecticut Ship Database," schooner *Paugassett*.

[52] Enrollment No. 3, Port of Savannah, schooner *Thomas Spalding*, March 28, 1856, Registers and Certificates of Enrollment, BMIN, NARA; *Savannah Daily Republican*, June 10 and August 5, 1856.

*Bolivar* and *Eutaw* and the schooner *Marion*. The 1838 deed notes the *Eliza & Emma* was built in Smithville, North Carolina, in 1833, had been issued a register at Charleston in 1837, and was now (1838) being issued a license in Savannah.[53]

Another of the small schooners in the Georgia trade was the 29 25/95-ton *Fort George Packet*, built at Fort George Island, Territory of Florida, in 1825. She was forty-one feet, two inches long, fifteen feet, eleven inches wide, and had a depth of hold of four feet, nine inches. This schooner was active in the coasting trade from the 1830s to 1862, the year the *Fort George Packet* became one of several old coasters sunk by Confederate authorities as obstructions in the Savannah River. For part of her career, the *Fort George Packet* was owned by long-time coasting captain and shipowner Daniel Reddick of Camden County, and the schooner was a principal carrier of goods between St. Marys and Savannah. The *Fort George Packet* was a very narrow schooner, and Fleetwood suggests she was built on the "old periagua style."[54]

The average burden of schooners enrolled at Savannah between 1815 and 1861 ranged from a low of 45 60/95 tons in 1840 to a high of 71 80/95 tons in 1860. Over this period, the average burden of schooners was greater than that of sloops, but there was a considerable overlap in the sizes of the two types of vessels.

Dimensions other than burden have been obtained for many of the schooners working in Georgia. Lengths are known for 113 of these vessels; they ranged from thirty-five feet, three inches to 108 feet, 9.5 inches and averaged sixty-one feet, one inch. Most schooners were between fifty and seventy-five feet long; only nine had lengths greater than seventy-five feet. Breadth measurements are available for 112 schooners, and these average nineteen feet and range from the twelve feet, five inches of the *Eliza & Emma* to the twenty-eight feet for the very large *Paugassett*.[55]

The depth of holds of Georgia coasting schooners averaged six feet, seven inches, and ranged from four feet, three inches, to the nine feet, four inches, of the sixty-five-foot-long schooner *Savannah*. Of the 113 schooners

[53] Chatham County, Georgia, deed of sale, schooner *Eliza & Emma*, December 12, 1838, Deed Book 2W, 265–66.

[54] Enrollment No. 3, Port of St. Marys, schooner *Fort George Packet*, March 7, 1825; Fleetwood, *Tidecraft*, 132. Documents relating to the scuttling of the *Fort George Packet* and other vessels in the Savannah River during the Civil War are in of the "Vessel Papers."

[55] Chatham County, Georgia, deed of sale, William Thompson to Joseph Silveria, schooner *Eliza & Emma*, December 12, 1838, Deed Book 2W, 265–66; "Connecticut Ship Database," schooner *Paugassett*.

with known depths of hold, the four shallowest were all built in North Carolina. These were the *Emma & Eliza*, built at Smithville in 1833; the *E. B. Hackburn*, built in Craven County in 1853; the *Young Eagle*, constructed in 1830 at Wilmington; and the *Sarah Potter*, built at an unknown North Carolina location in 1816.[56]

As with sloops, a lack of information, particularly in the form of actual draughts, makes it difficult to determine the precise shape of the hull of these schooners. One of the few sources of information on the hull shape of nineteenth-century coasters is found in marine survey and classification documents made after 1857 at the port of New York. These records were produced for insurance purposes and contain information on hull shape and draft, as well as other building details, including the types of wood and fasteners used in construction.[57] In these records, twenty-five schooners that sailed in the Georgia trade from 1857 to 1861 have been identified. The Georgia schooners included in these survey records tend to be the larger vessels, but the information given would apply to the coasting schooners as a group. Table 5.1 presents information from these survey documents for several of these schooners to give an idea of the types of information contained in these records.

The hull shape of twenty-three of the twenty-five schooners is described as "full bodied" or "full modeled" (see Table 5.1). A full-modeled hull has a rounded or bluff bow and generally a flat bottom. These are hulls built to carry cargo, not for speed. Two of these schooners, the *Florida* and the *J. Truman*, had medium-bodied hulls, meaning they were somewhat "sharper" and had more V-shaped hulls than full-bodied vessels, characteristics that generally produced faster sailers.

The marine survey documents show that fourteen of the twenty-five schooners were fitted with centerboards (see Table 5.1). These records are among the few contemporary accounts that mention the use of centerboards on Georgia coasters. Official vessel documents rarely mention centerboards,

[56] Dimensions of the *Savannah* from Enrollment No. 6, Port of Savannah, schooner *Savannah*, February 27, 1815, however, although enrolled at the port of Savannah, this schooner may not have sailed in the local trade. Dimensions of the *E. B. Hackburn* are from deed of sale, Patrick Doyle to Fred Krenson and Rufus Hawks, November 2, 1860, Chatham County Deed Book 3T, 556–57; dimensions of the *Young Eagle* are from Enrollment No. 3, Port of Beaufort, schooner *Young Eagle*, November 13, 1843; and those of the *Sarah Potter* are from Enrollment No. [illegible], Port of Savannah, schooner *Sarah Potter*, July 20, 1819, Registers and Certificates of Enrollment, BMIN, NARA.

[57] Board of Underwriters, *New York Marine Register*.

and it appears they were noted only in certain situations, such as during a survey and inspection. The fact that so many of the Georgia schooners listed in these survey documents had centerboards suggests that many of the other schooners, as well as sloops working in the local trade, were similarly equipped.

As with sloops, by about 1800 accepted rules and guidelines had developed for masting and sparing schooners of varying sizes. For example, in the Chesapeake region, the total length of a schooner's mainmast was commonly 3.1 to 3.5 times the beam. Usually, the foremast was a couple of feet shorter than the mainmast, but it was greater in diameter because it supported the largest amount of sail—the foresail, headsails, topsails, and course. The length of the main boom was typically 2 to 2.5 times the vessel's beam, while the bowsprit length was about 1.2 times the beam. How strictly these "rules" were followed by builders outside of the Chesapeake area is unknown, but they do enable estimates of the sizes of masts and booms on the schooners in the Georgia trade.[58]

Larger schooners typically carried topmasts on both the foremast and mainmast. However, many schooners, particularly smaller ones, had no fore topmast, meaning they could not carry a gaff topsail on the foremast. Presumably, some of the Georgia coasting schooners, like the sloops, carried no topmasts at all. This would have been particularly true of the very narrow and shallow-hulled vessels that could not safely support tall masts or carry a great deal of sail. A painting by Alice Ravenel Huger Smith titled *Loading the Rice Schooner* depicts a small coasting schooner being loaded with rice in South Carolina. In what was certainly a typical scene during the rice shipping season, a line of enslaved women are carrying the rice to the coaster in baskets balanced on their heads. The schooner in this painting carries no topmasts.[59]

Little specific information is available on the sail plan of the Georgia schooners, but presumably, it differed little from that of other vessels working along the Atlantic coast. This means they carried three basic sails: a staysail or jib, a gaff-rigged foresail, and a gaff-rigged mainsail (see Figure 5.2). The mainsail would have been boomed and, often, hooped to the main mast although the foresail was often boomless, especially during the early 1800s. Many schooners carried smaller sails in addition to this basic set, most commonly a flying jib and gaffed topsails on the mainmast and sometimes on the

[58] Chapelle, *Search for Speed*, 214.

[59] MacGregor, *Schooner*, 72. A. R. H. Smith's painting *Loading the Rice Schooner* hangs in the Gibbes Art Gallery, Charleston, South Carolina.

foremast. In 1840, Massachusetts sailmaker Thomas L. Sanders charged $59.55 for "making main sail & Foresail & Jib & fly jib" for the Georgia coasting schooner *Roswell King*. The following year, Sanders made a "gafftop sail" for the *Roswell King* that used two and one-third yards of canvas, one and three-quarters pounds of twine, and three thimbles.[60]

Square sails remained in use on some Georgia coasters well into the nineteenth century as revealed in a notice of sale of the schooner *Olive* that appeared in the January 4, 1839, issue of the *Brunswick (Georgia) Advocate*:

> "For Sale—the schr. OLIVE, with all her tackle and apparel, containing of an excellent gauging of rigging, two fine and nearly new chain cables and anchor and a complete suit of sails, including topsail, top gallant sail and square sail. WA Howard, Brunswick."

From the advertisement, it appears the *Olive* carried three square sails, assuming the "topsail" and "top gallant" sails were small square sails carried on the foremast. If so, the "square sail" likely refers to a "course," the large square sail carried on the lower portion of the foremast.

On March 5, 1859, *Harper's Weekly* published a drawing of some of the colonial ruins at the old town of Frederica on St. Simons Island. In the background can be seen the upper portion of a two-masted schooner. The ship is shown with a gaffed, fore-and-aft sail on the mainmast and a large square sail at the top of the foremast. At the time, coasting captain Charles Stevens resided at Frederica and moored his schooner *Northern Belle* there. It is unknown if this illustration depicts the *Northern Belle*, but it does show the use of square sails on schooners sailing along the Georgia coast as late as 1859.[61]

By the end of the eighteenth century, coasting schooners, like sloops, tended to be plainly built vessels, with a single, "flush" deck surrounded by a low bulwark, often only a plank on edge or "log rail." The raised quarterdeck had generally disappeared by 1800. Mention of a quarterdeck has been found for only one of the Georgia schooners, the *Water Witch*, built in Savannah in 1816. An 1816 enrollment notes the *Water Witch* had a square stern and a "high quarterdeck." The *Alexander Blue*, built at Port Jefferson, New York, in 1856, had a "half poop," which might have resembled a quarterdeck, and rare mentions of "break to deck" occur, suggesting something other than a flush

[60] Chapelle, *American Sailing Ships*, 227; "Thomas L. Sanders Accounts, 1839–1881," November 1840, October 1841.

[61] *Harper's Weekly*, March 5, 1859. Information on Charles Stevens's residency at Frederica and activities as a coasting captain is found in Pearson, *Charles Steven*.

deck (see Table 5.1). Another Georgia-built schooner, the *Flora*, constructed at the town of Sunbury in 1821, was fitted with "quarter railing," which may indicate that the railing around the *Flora*'s stern was higher than the more commonly used log rail.[62]

Many schooners had a low trunk cabin toward the stern, as revealed in the New York marine survey records (see Table 5.1). These cabins generally provided enough space to accommodate the master and crew, although some of the larger schooners may have had separate crew's quarters in the forward part of the ship. Schooners often carried passengers, particularly in the years before 1830, before regular steamboat service began. Typically, when passengers are mentioned aboard schooners, the numbers are small, usually only two or three persons, but, occasionally, schooners did carry larger numbers.

Most schooners had a main hatch positioned between the two masts that provided access to the cargo hold. Large schooners may have had additional hatches.

Typically, the stern post on coasting schooners was straight, with some amount of rake aft, as found on most sloops. Available documents indicate that most of the Georgia schooners, like the sloops, had "flat" or "square" sterns. Some schooners, particularly the larger ones, may have had wheels for steering, but tillers were more common.

Schooners working in Georgia after 1800 were plain vessels and, like sloops, were built with little ornamentation. When a specific reference is made to the decoration of a schooner's stem, the term most commonly used is "billet head," or, less often, "plain head." As with sloops, this referred to a plain, undecorated stem post. A few schooners had "fiddle" or "scroll" heads, suggesting the top of the stem was carved with a scroll, or a scroll carving was fit forward of the stem under the bowsprit.

Only nine of the Georgia schooners had figureheads. The schooners *Northern Belle*, *Mary A. Rowland*, and *Minerva* all had female figureheads. Five schooners, the *Margaret Ann Howard*, *Sarah*, *Paugassett*, *Sarah Jayne*, and *John W. Anderson* are reported as simply having "figureheads." The schooner *Challenge* was fitted with an "eagle figurehead."

[62] Enrollment No. 9, Port of Savannah, schooner *Water Witch*, March 26, 1816; Enrollment No. 3, Port of Sunbury, schooner *Flora*, December 8, 1822, Registers and Certificates of Enrollment, BMIN, NARA. Information on the *Alexander Blue* is from Board of Underwriters, *New York Marine Register*, 186.

## Where the Georgia Coasters Were Built

If America had one great advantage in shipbuilding over Europe, it was the plentiful and cheap wood supply. Consequently, American-built schooners could cost one-quarter the amount of a similar vessel built in England or France although by the 1840s, this cost advantage began to disappear with the introduction of iron hulls. Various kinds of wood, dependent on regional supplies, were used to construct American schooners. Before about 1812, Chesapeake area builders used white oak, mulberry, pine, sassafras, chestnut, and cedar. Farther north, and in New England, builders used oak, hackmatack, birch, white pine, and fir. Chapelle notes that after the War of 1812, oak and cedar became common, except in New England, where pine, fir, and hackmatack remained popular. Southern builders used live oak and yellow pine, woods that, by the early years of the nineteenth century, were being shipped North in considerable amounts for use in shipbuilding.[63] On occasion, newspaper advertisements mention the types of woods used in Georgia coasters. On November 25, 1820, a notice of the pending auction of the schooner *Choctaw* appeared in the *Daily Georgian*, stating, "The fast sailing schooner CHOCTAW—as she now lies at Hall & Hoyt's Wharf, burthern about 300 barrels, 2 years old, built of live oak and red cedar, has lately undergone a thorough repair." No information has been found on the place or date of build of the *Choctaw*, but the use of live oak and red cedar may suggest a Southern location.

The New York marine survey documents discussed previously identify the types of wood used for the frames (ribs) and the hull planking of ships. As seen in Table 5.1, most vessels used white oak or a combination of white oak and chestnut or white oak and pine.

Rarely, the hulls of coasters were sheathed with metal to protect them from destructive marine organisms. Sheathing, or "metaling," a ship was an expensive process not worth the cost for most of the vessels working in the Georgia trade. However, at least three of the Georgia schooners were sheathed. These included the *Blooming Youth*, which was metaled with zinc in 1854, and the *Challenge*, which was sheathed with the same metal in 1859. The third coaster known to have been sheathed was the schooner *Governor*

[63] Taylor, *Transportation Revolution*, 127; Chapelle, *American Sailing Ships*, 246.

*White*, reported to have a "copper bottom" in an 1803 advertisement appearing in the *Columbian Museum & Savannah Advertiser*.[64]

A review of vessel documents and other records for the ships sailing in the Georgia trade reveals that most were built in New England, which dominated the construction of wooden ships throughout much of the history of the United States. Slightly over half of the 259 Georgia coasters for which place of build is known were constructed in the Northeast between Maine and Connecticut (Table 5.2). Seventy-two, or 28 percent, of these ships were built along the middle Atlantic seaboard, between New York and North Carolina. In contrast, just forty-one, or 16 percent, were constructed in South Carolina, Georgia, and Florida (Figure 5.7).[65]

As a group, most of the Georgia coasters were constructed in Northeastern states; however, there was considerable disparity in the number of sloops built there versus the number of schooners. As shown in Figure 5.7, one hundred, or approximately 74 percent, of all sloops were constructed in the Northeast, but only forty-seven, or 38 percent, of the schooners were built there. A higher proportion of the schooners working in Georgia were built in Middle Atlantic states (i.e., 42 percent) than in the Northeast or the South. This difference not only reflects the popularity of schooner-rigged vessels in the Middle Atlantic states but also is an expression of time. Ship construction was more widespread and better developed in the Northeast during the very early nineteenth century than elsewhere in the country. This was a period when sloops were popular and being built in large numbers. As schooners gained popularity, they were built in increasing numbers in the Northeast, as well as at other locations along the Atlantic coast. Sloops were more common than schooners in the Georgia trade in the years before 1830, so it is inevitable that many of these sloops were built in the Northeast, where the construction of both sloops and schooners was concentrated. This is borne out by the fact that the average year of build for the 138 Georgia coasting sloops for which this information is known is 1818. On the other hand, the average year of build for the 124 schooners is 1831.

During the colonial period, ships built in the West Indies, Bermuda, or the Bahamas were often used in South Carolina and Georgia, but these were

[64] Board of Underwriters, *New York Marine Register*, 197; Board of American Lloyds, *American Lloyd's*, 370; *Columbian Museum & Savannah Advertiser*, February 4, 1803.

[65] "Enrollments Savannah" and "Master Abstracts of Enrollments 1815–1911," for all Georgia customs districts and for Beaufort, South Carolina, RG 41, BMIN, NARA.

almost entirely replaced by American-built ships by the end of the eighteenth century.[66] A small number of these foreign-built sloops and schooners continued to work out of Savannah in the final years of the eighteenth century and the early years of the nineteenth. For example, on July 27, 1786, the Savannah newspaper *Georgia Gazette* contained the following notice of sale: "To be sold, the sloop Ann and Sally, built in the Bahamas 10 months ago of the best mahogany and cedar. Ralph Deposs, Vendue master."

Thirty-five years later, on February 28, 1820, the *Daily Georgian* published a notice of sale of the sloop *George*, stating that the sloop was two years old, ten tons "burthen" and was built "at Nassau of cedar and mahogany." It is unknown if such a tiny sloop was used as a commercial vessel, but it is the last mention of a Caribbean-built vessel found in the Savannah newspapers.

### *The Ships of Sippican*

By 1800, most vessels sailing in the coasting trade along the South Atlantic coast were American-built, and, as vessel documents for Georgia coasters show, most of these were constructed in New England, particularly in Massachusetts and Connecticut. The large number of coasters built in New England is not surprising, given the importance and long history of shipbuilding in that region. What is somewhat remarkable regarding the Georgia coasters is the large number constructed along the southern Massachusetts coast at locations bordering Buzzards Bay, such as the towns of Rochester, Freetown, and Wareham. These three towns alone built sixty-three of the Georgia coasters, representing nearly one-quarter of those for which a place of build is known. This small area in Massachusetts played an important role in the coasting trade in Georgia through the first half of the nineteenth century. This area not only produced many of the ships sailing in the Georgia trade but also was home to captains and crews who manned the Georgia coasters.

During its early years of settlement, the region bordering Buzzards Bay was known as "Sippican," a name reportedly used by Native American residents. Old Sippican includes the towns of Rochester, Marion, Wareham, and Mattapoisett (Figure 5.8). Although not specifically falling within what was considered "Old Sippican," the nearby towns of Fairhaven, Freetown, and Dartmouth were closely associated with and figured importantly in shipbuilding and the American coasting trade. Shipbuilding and seafaring became important in this region at an early date. By the mid-eighteenth century, the

[66] *Fleetwood*, Tidecraft, 61.

area's shipyards were producing many sloops and schooners for the coasting and West Indies trades and larger vessels for whaling.[67]

Many early immigrants to Sippican came from Scituate and Marshfield, south of Boston, where shipbuilding had been pursued for over a quarter century. It was not long before some gave up their attempts to farm the land around Buzzards Bay and turned to building ships. By about 1650, a vessel had been constructed on Buzzards Bay to be used for trade with the Dutch in New Amsterdam. Ships would be built along the bay for the next two centuries.[68]

The town of Rochester in Sippican was established along the Mattapoisett River, and four hundred to five hundred ships were constructed there before the late 1800s. Among the early family names of the Rochester area were Briggs, Hammond, Holmes, Meigs, Barstow, Bolles, Dexter, Nye, Pease, Luce, Tabor, Wing, and Blankinship. These names appear repeatedly as masters, crew, and owners of sloops and schooners working in the Georgia trade after 1800. Many of these same men also were involved in shipbuilding. At one time, as many as nine shipyards in Rochester were building sloops, schooners, brigs, and ships to be used as coasters or whalers.

Several of the early Rochester shipyards were located in the Sippican village at the upper end of Sippican Harbor. Shipbuilding remained an important activity there, but it was limited to small, shallow-draft sloops and schooners because of the shallowness of Sippican Harbor. Leading shipbuilders in Sippican, renamed Marion in 1852, during the first half of the nineteenth century included William Clarke, Edward Sherman, John Delano, and members of the Wing family.[69]

Rochester's importance in the construction of small sloops and schooners is evident from the fact that at least forty-six of the coasters working in the Georgia trade between 1800 and 1861 were built there. More Georgia coasters were constructed at Rochester than at any other single location.

The decline of shipbuilding in Rochester was due to several factors. For one, after the 1830s, steamboats began to take over the coasting trade in many areas, reducing the need for the small, sailing ships that were a significant portion of the Rochester production. Additionally, a nationwide economic panic in 1857 seriously impaired many of the businessmen of the Rochester

[67] Historical and shipbuilding information in "Old Sippican" comes from Ryder, *Lands of Sippican*.

[68] Leonard, *Mattapoisett and Old Rochester*, 279–80.

[69] Rosbe, *Maritime Marion*, 23.

area, including shipbuilders, and they never fully recovered. Finally, the decline of the whale fisheries after the discovery of petroleum ended the market for Rochester-built whalers.[70]

While the names of builders of many of the larger whaling vessels have been preserved, fewer records exist for those who built the smaller sloops and schooners even though these were often the same individuals. In 1912, eighty-year-old Charles Henry Delano, a member of a prominent family of shipowners and captains in Rochester, made a list of Rochester vessels and their builders. A few Georgia coasters are in this list, including the sloop *Angel* and schooners *Hopeton*, *Home*, and *Roswell King* (all built by William Clark) and the schooners *Ocean Queen* and *John Fraser* (built by Edward Sherman).[71]

In addition to shipbuilding, the Sippican region became an important salt producer in the early nineteenth century. Windmills were used to pump water out of Buzzards Bay into shallow vats, where it was left to evaporate in the sun. In some years, as many as twenty-thousand bushels of salt were produced in Rochester, and in 1806 the community of Sippican was producing more salt than any other settlement in the township. This salt became an important commodity of the ships sailing out of Rochester, and much was transported by coasters to ports along the Atlantic coast, including Savannah and Charleston. It is suspected the transport of salt first brought some of the Sippican men and their ships to Southern ports in Georgia and South Carolina. A severe hurricane in 1815 destroyed much of the Sippican salt industry, ending most of the area's trade in salt to Southern ports.[72]

Wareham, that portion of Rochester located near the head of Buzzards Bay, was settled soon after 1680 and was incorporated as a town in 1739. Several seafaring families from Wareham became prominent in the Georgia coasting trade. These family names included Bates, Hathaway, Savory, Sturtevant, Briggs, Gibbs, Hammond, Nye, Perry, and Swift. The Wareham area became an important producer of iron beginning in the early 1820s. Iron ore obtained from the area's swamps was used to produce nails and other important items for the shipbuilders of Rochester.[73]

[70] Leonard, *Mattapoisett and Old Rochester*, 285.

[71] Ryder, *Lands of Sippican*, 93.

[72] Rosbe, *Maritime Marion*, 18.

[73] Leonard, *Mattapoisett and Old Rochester*, 299; Ryder, *Lands of Sippican*, 88.

*Southern-Built Coasters*

Although the Northeast, and the Sippican area specifically, were important for the ships sailing in the Georgia trade, a small number of Georgia coasters were built in Southern states (see Table 5.2). The principal shipbuilding centers in these states were Savannah and Charleston. During the first half of the eighteenth century, shipbuilding was a modest industry in South Carolina, but production declined by 1800. In the late 1840s, Charleston had five shipyards involved mainly in ship repair although several coasting vessels were launched from the city during the first half of the nineteenth century.[74] Among the Charleston-built ships working in the Georgia trade were the sloop *Delight*, launched in 1811, which sailed principally as a Charleston-to-Savannah packet, and the schooner *Lucretia*, built in 1812, which sailed for many years between Charleston and Darien and other locations on the lower Georgia coast.

Shipbuilding in Georgia was never an important industry, particularly when compared to Northeastern states. Fleetwood comments on the lack of this industry in the state, reporting on an 1814 account about shipwrights, noting there were "not more than Six or Seven Carpenters in the whole State of Georgia."[75] Some small ships and numerous "plantation boats" were built in Georgia during the colonial period, but records on these are rare. Some local vessels were built by men held in bondage. In 1809, Major Pierce Butler, referring to enslaved laborers on his plantation at the mouth of the Altamaha River, wrote that his enslaved "ship carpenters have built me two Sea Vessels without any white person directing them."[76]

Savannah was the most important shipbuilding center in Georgia, but even there it was a modest industry, and information on ship construction in the city during the first decade of the nineteenth century is negligible. None of the Georgia coasters built before 1810 for which the place of build is known were constructed in Savannah. The identity of only five Savannah-built coasters, three schooners and two sloops, has been determined, but there certainly were others.

The most active shipyard in Savannah before the Civil War was the Willink Yard, founded by Henry Frederick Willink and later operated by his son, Henry Jr. Few documents identify the vessels they built, but Willink and his son are named as the original owners of several coasters constructed in

[74] Coker, *Charleston's Maritime Heritage*, 193.

[75] Fleetwood, *Tidecraft*, 107.

[76] Bell, *Major Butler's Legacy*, 116–17.

Savannah that they probably built. Henry Sr., a native of Hanover, Germany, was in Savannah by 1828, the year the sloop *Mill Maid* was reportedly built by "Fred Willink."[77]

Over the next three decades, the Willinks, father and son, expanded their ship repair and building facilities at Savannah. By 1853, the Willink yard, located at the downriver end of the Savannah riverfront, advertised a "Marine Railway" and was "prepared to take up vessels of any size" and could "repair, clean, pant or caulk" and build vessels of "any sizes or descriptions."[78] Enrollment documents show the Willinks built the schooner *John R. Wilder* in 1854, and they are listed as the original owners of the sloop *Alpha*, launched in 1833, and the schooners *C. A. L. Lamar*, built in Savannah in 1855, and *H. F. Willink, Jr.*, built in the city in 1856, suggesting they also constructed these ships. The *John R. Wilder* and *C. A. L. Lamar* were built as Savannah pilot boats.[79]

Another Savannah shipbuilder, Jeremiah Corwin, advertised in 1841 that he had taken over the wharf used by Mr. H. Segure and would commence as a "Ship Wright, Spar Making and boat building." In 1844, the *Daily Georgian* announced that "J. Corwin—Shipwright" had completed his ways for hauling out vessels. In 1842, Corwin built the schooner *George B. Cumming* and, in 1844, the *George W. Behn*, both Savannah pilot boats.[80]

Another Savannah shipyard was the Krenson & Hawkes yard. The partners in this firm were Rufus P. Hawkes and Frederick Krenson, the latter, like Henry Willink, a native of Germany. Krenson had settled in Savannah by 1850, and in 1858, he advertised himself as a "shipwright—caulker and sparmaker" who had a large stock of live oak, "white timber," and cypress planks on hand in addition to a "floating derrick for raising heavy timber."[81]

The coasters relied on shipyards like Willinks and Krenson & Hawkes to undertake major repairs on their vessels, but others in Savannah were involved in ship repair and outfitting. These included ship carpenters, sailmakers, and chandlery businesses. The 1850 federal census for Savannah lists

[77] Enrollment No. 1, Port of Savannah, sloop *Mill Maid*, January 22, 1834.

[78] *Daily Georgian*, August 11, 1853.

[79] Enrollment No. 1, Port of Savannah, sloop *Alpha*, January 29, 1836; Enrollment No. 1, Port of Savannah, schooner *C. A. L. Lamar*, October 18, 1855; Enrollment No. 13, Port of Savannah, schooner *H. F. Willink Jr.*, August 29, 1856, Registers and Certificates of Enrollment, BMIN, NARA.

[80] Advertisements for Jeremiah Corwin are from the *Daily Georgian*, August 24, 1841, and April 24, 1844. Information on the launch of the two pilot boats is from the *Daily Georgian*, June 20 and November 8, 1842.

[81] *Daily Morning News*, March 19, 1858.

thirty-two ship carpenters, four sailmakers, and one caulker. Few of these men were native-born Georgians; most were immigrants from other states or foreign countries. While most were White, three were identified as "mulattos" and one as a "free black."[82] Although few in number, free Black men seem to have been continuously involved in shipbuilding trades in Savannah throughout the first half of the nineteenth century. A list of "Free Persons of Color" appearing in the August 21, 1838, *Daily Georgian* included three ship carpenters: Paul Carmon (age twenty), Hannibal Roe (forty), and Joseph Summers (twenty-seven).

Some tradesmen worked for the major yards, but others operated independently. In 1847, J. Griffen & Co. advertised in the *Daily Georgian* that they had taken "the sail-loft at Exchange Dock, and are prepared to carry on the business in its various branches." In 1853, the paper ran an advertisement by M. Amorous noting that he had resumed his business of "Sail Making."[83]

The largest facilities for shipbuilding and repair along the Southeast coast were in Savannah and Charleston, but small shipyards were scattered along the coast between Charleston and Florida. Some smaller locales were building coasting vessels at an early date. For example, the schooner *York* was constructed on St. Simons Island in 1801, and the schooner *Experiment* was built there in 1807. In October 1805, Francis Young advertised that he was building a ship at St. Marys. The advertisement noted that the vessel was:

> forty-eight feet keel, fifty-seven feet on deck, seventeen and a half feet beam, seven and a half feet hold, will measure about sixty-four tons. Live Oak and Red Cedar frame, and planked with the best yellow pine, can be launched in a month from this time.

The identity of this ship is unknown, but its size was typical of the coasting sloops and schooners of the period. Young went on to note in his advertisement that he would contract "to build at St. Mary's vessels of any size and description that may be required."[84]

Several other coasters were built at St. Marys in later years. These included the sloops *Eliza* and *John Chevalier* and the schooners *Ellen* and *Betsey Maria.* Who built these ships is unknown, but the 1850 federal census for

[82] Otto, *1850 Census Chatham County.*

[83] *Daily Georgian,* November 6, 1847, and July 8, 1853.

[84] *Charleston City Gazette,* October 17, 1805.

Camden County, including St. Marys, listed three men as ship carpenters, indicating the existence of ship repair or shipbuilding in the town.[85]

Similarly, the small port town of Brunswick in Glynn County boasted facilities for ship repair and construction. In the 1850 census, two men in the county were identified as ship carpenters, and one, Charles Flanders, a native of Maine, as a "ship wright."[86] Three coasters, the schooners *Elias Reed*, *Satilla*, and *William D. Jenkins*, are known to have been built in Brunswick, but there were certainly others.

## The Age of the Georgia Coasters

In spring 1862, the schooner *Fort George Packet* was one of sixteen coasting ships and pilot boats acquired by Confederate authorities and sunk as obstructions in the Savannah River.[87] The schooner was thirty-six years old, having been built in 1825 at Fort George Island, Territory of Florida. The *Fort George Packet* worked in the coasting trade longer than most vessels, but several others were quite old when they ended their careers. For example, among the other coasters sunk as obstructions in 1862 were the twenty-five-year-old sloop *Catherine Chard*, as well as the sloop *Splendid* and schooner *Levant*, both twenty-nine years old. Eleven of this group of scuttled ships had sailed in the Georgia trade; their average age when sunk was 20.5 years. One would think the ships chosen as river obstructions would have been older vessels or those in disrepair. However, examining the age of ships sailing in the Georgia trade indicates that many were as old as the group selected for scuttling.

Information on the average age of sloops and schooners working in the Georgia trade for various years is presented in Figure 5.9. The ships included in Figure 5.9 are those for which year of build information is available. Two trends are obvious in these data. First, for every year shown, the average age of sloops was greater than the average age of schooners, and, second, there was a considerable increase in the average age of all vessels in the coasting fleet over time. Furthermore, this increase occurs principally after 1835. Before that year, the average age for both sloops and schooners was less than about ten years, although sloops did tend to be slightly older than schooners. In 1809–1810, the year of build is known for only eleven of the approximately seventy different vessels sailing into Savannah from local ports. This number

[85] Otto, *1850 Census Camden County*.

[86] Ibid., 3.

[87] Records of the seizure and scuttling of the *Fort George Packet* are in "Vessel Papers," file F-22.

includes just three schooners and eight sloops, and their average age was less than eight years. The year of build is known for thirty-three of the sixty-two local coasters sailing in 1825. This includes eight schooners and twenty-five sloops; their average age was about ten years. The youngest coaster sailing in 1825 was the two-year-old sloop *St. Marys* while the oldest appears to have been the *Nancy*, a sloop constructed at Providence, Rhode Island, forty years earlier.[88]

The year of build is known for twenty-five sloops and nine schooners sailing into Savannah as coasters in 1835. The youngest were three schooners and a single sloop, all just one year old. These were the schooners *Ellen*, *Florida*, and *President* and the sloop *Stranger*, all constructed in 1834. The average age for the coasters sailing in 1835 was 8.7 years, slightly younger than those sailing ten years earlier.

The average age of the sailing vessels working in Georgia remained stable for the first thirty-five or so years of the nineteenth century. Subsequently, however, the coasting fleet became increasingly older, as seen in Figure 5.9. In 1845, the average age of thirteen coasting sloops sailing into Savannah was 14.5 years while that of fourteen schooners was 10.4 years. More than one-third of the coasters sailing into Savannah in 1845 were over fifteen years old. The introduction of steamboats into the Georgia coasting trade in the 1820s and their expanding importance in transporting cargo and passengers were important factors behind the increasing age of the sailing ships remaining in the trade. It was becoming economically less feasible to construct or enter new sailing ships into the local coasting trade, where they had to compete against the speed and reliability provided by steam vessels. Many sailing vessels did continue to work, and some new ones were built, but as a group, coasting captains and owners kept their ships working for as long as possible, or, when they did replace them, they tended to purchase older, less expensive vessels.

In 1861, forty-three different coasting vessels sailed into Savannah from local ports. The year of build is known for twenty-six of these ships: five sloops and twenty-one schooners. The average age of all twenty-six vessels was 16.5 years. The youngest coasters sailing in 1861 were the schooners *Alexander Blue* and *Paugassett*, both of which were five years old. The oldest was the thirty-six-year-old *Fort George Packet*. As seen in earlier years, sloops tended to be older than schooners. The average age of the five sloops sailing in 1861 was 23.6 years while that of the twenty-one schooners was 14.8 years. The fact

[88] Enrollment No. 9, Port of Savannah, sloop *St. Marys*, January 1, 1827; Enrollment No. 4, Port of Savannah, sloop *Nancy*, February 20, 1800.

that sloops were consistently older than schooners reflects the increasing popularity of schooners throughout the nineteenth century. As fewer and fewer sloops were built relative to schooners, the average age of those sloops that continued to sail increased relative to that of schooners.

Were the sloops and schooners working in the Georgia coasting trade unique types of vessels relative to coasters built or working in other areas? The available information does not suggest this. First, most coasters were built outside the area, and many sailed in other coasting trades. Even among those coasters built in Georgia, South Carolina, and Florida, most were not different from those constructed elsewhere. A few, such as the schooner *Experiment*, built at St. Simons Island in 1807, were very narrow and may have been constructed along the lines of log pettiaugers ("periaguas"), making them unique.

Regardless of place of build, the Georgia coasting ships tended to incorporate those characteristics that made them amenable for use in the waters found along the Southeastern coast. The characteristic that was most critical and most obvious was shallow draft.

The Georgia coasters tended to be small. The average burden of sloops was 53 11/95 tons, and that of schooners was 70 53/95 tons. The small burden of coasters goes hand-in-hand with shallow draft. Owners and captains in the Georgia trade selected vessels that would accommodate the water conditions along the coast, regardless of where that ship was built. The 1803 report that the sloop *Patty* was "of an easy draft of water and well calculated for a Coaster" would characterize the Georgia coasting fleet for the first six decades of the nineteenth century.[89]

[89] *Columbian Museum & Savannah Advertiser*, January 21, 1803.

# Chapter 6

## "Apply to the Master Onboard": The Men in the Georgia Coasting Trade

The coasting trade in Georgia was a complex economic system involving a range of participants. Three groups of individuals were key actors, each performing a specific and vital role in the coastal trading system. These were 1) the masters and crewmen who manned the coasters and carried the coastal cargoes; 2) the planters and merchants who provided the cargoes; and 3) the factors and commission merchants in Savannah and Charleston who received and handled the cargoes shipped to market by the planters. Shipowners make up a fourth group, but many of the owners of Georgia coasters were members of one of the three groups named here. From the perspective of this study, the central figures in the trade were the ship captains, such as Charles Stevens, and the crews who sailed the coasters. At one end of their voyage, these captains dealt with planters shipping their goods to market or with local merchants receiving wares from the port centers of Savannah and Charleston. At the other end of their voyage, they dealt with the factors and commission merchants who received rice, cotton, and other produce from individual planters and shipped supplies back to planters and merchants on a vessel's outbound voyage.

Others were directly or indirectly involved in the coasting trade, but their roles were peripheral to the three groups listed here. For instance, there were those who did the laborious work of loading and unloading the coasting vessels. Most of these workers were enslaved Black laborers, but in Savannah and Charleston, some factors hired White dock workers or "stevedore" companies, such as that owned by Henry J. Dickerson in Savannah from the 1840s to the 1860s. Most of those involved in this work were men, but not all. Alice Huger Smith recalled the "long lines of men and women" filling the holds of schooners with basketloads of rice at her family's South Carolina plantation. There were also businessmen in Savannah and Charleston, who, while not directly involved in shipping and receiving coastal cargoes, provided goods to planters and small town storeowners. These men supplied the merchandise carried aboard coasters, but these goods were often shipped by or through one of the larger factorage firms. This was particularly true when the merchandise was going to individual planters. Others associated with the coasting trade were shipyards, which built or repaired coasting vessels; ship's chandlers, which provided supplies and material needed by the coasters; and banks, which furnished the capital required by the other participants.[1]

Almost nothing has been written about the men who sailed the Southern coasters, the shipmasters and crewmen. They are the persons of greatest interest in this study, but few records of their lives exist. Most of these men occupied the middle and lower levels of society, groups history has tended to neglect, especially in the South. Drawing on various sources, this chapter deals mainly with the Georgia coasting captains and crewmen, expanding our minimal understanding of their work and their lives. Other participants, such as factors and planters, are touched on as necessary to elucidate their involvement with the coasting trade.

## The Georgia Coasting Captains

On March 15, 1838, the *Brunswick Advocate*, the short-lived newspaper published in the Georgia port town, reported on the arrival of three sloops and one schooner in its "Marine Intelligence" column. These entries were:

> Schr. *Mohawk*, Parker [master], Baltimore, with 3000 bushels corn and 24 bundles hay to Nightingale & Couper.
>
> Sloop *Bolivar*, Richardson, Savannah, Mdze. [Merchandise] To Joseph Bancroft and hay to Nightingale & Couper
>
> Sloops *America*, Burr, and *Argo*, Taylor, Savannah, Mdze.

These were typical newspaper shipping entries, including only the last name of the master, the port of origin, the cargo carried, and the merchant, factor, or other person receiving it. Rarely was the first name or initial of the master included, often making it difficult to identify that person, particularly if it was a common name. Despite this shortcoming, newspaper shipping lists provide our most comprehensive source of information on the names of those who commanded the ships sailing in America's coasting trades.

[1] Walker, "Henry James Dickerson"; A. R. H. Smith, *Carolina Rice Plantation*, 62.

The records examined for this study show these ship captains achieved varied degrees of economic success. Many, especially those residing in the South, ended their lives in meager financial circumstances although a small number did achieve modest wealth and prosperity. Few attained the elite social status and financial success that many coastal planters and Savannah merchants did.[2] Because of their modest circumstances, the lives of the great majority of coasting captains received little note or attention in contemporary accounts and documents. As a result, the identity of many and the lives of most remain obscure or unknown. However, by using other sources of information in addition to newspaper shipping lists, including census records, vessel documents, courthouse records of various sorts, and family histories, it is possible to identify many of these men and gain an expanded understanding of the lives of a few of them.

For example, in the case of the four masters named in the March 15, 1838, issue of the *Brunswick Advocate*, other sources enable a positive identification of two of these men. The captains who can be positively identified are Edmund Richardson of the *Bolivar* and Leonard Burr of the *America*. The master of the *Argo* may have been William C. Taylor, while nothing is known about Captain Parker of the *Mohawk*.

Edmund Richardson and Leonard Burr are both named in enrollments for their respective vessels, the sloops *Bolivar* and *America*, and their names appear in other newspaper shipping lists as well as in official documents for other vessels they captained or owned. A variety of records, such as census, land, and marriage documents, as well as published and online family histories

[2] The planter society that developed on the Sea Islands of Georgia and South Carolina has attracted a considerable amount of attention from scholars as well as popular writers. Two popular overviews of the history of the Georgia Sea Islands include Vanstory, *Georgia's Land*, and Lovell, *Golden Isles*, which, together, mention most of the important figures occupying the islands and the adjacent mainland during the nineteenth century. Kemble's famous *Journal of a Residence* and Leigh's *Ten Years on a Georgia Plantation* provide information on the Butler family's cotton and rice plantations on St. Simons Island and along the Altamaha River. Other discussions on the coastal planter society are found in Coulter, *Thomas Spalding*, and M. Granger, *Savannah River Plantations*. A. R. H. Smith's *Carolina Rice Plantation* and Rosengarten's *Tombee* present specific information concerning coastal planters and factors while Woodman's *King Cotton* provides a thorough discussion of the Southern cotton factorage system. Bell's *Major Butler's Legacy* is a more recent examination of the Butler family, including its activities in coastal Georgia. Other discussions on the coastal planter society and the workings of the coastal plantations are found in Clifton, *Life and Labor*; Hoffman and Hoffman, *North by South*; Sullivan, *All Under Bank*; Stewart, *What Nature Suffers*; and Bullard, *Robert Stafford*.

and genealogies, provide additional information on captains Richardson and Burr. Edmund Richardson, a native of Massachusetts, was a resident of St. Marys, Georgia, from 1811 until his death on November 11, 1850. He was active in the St. Marys-Savannah coasting trade between 1820 and 1842, and he may have first sailed in this trade as early as 1809. Leonard Burr was a native and resident of Freetown, Massachusetts, who sailed at least three different ships along the Georgia coast between 1828 and the early 1850s. Captain Burr was one of the many captains from the "Sippican" area bordering Buzzards Bay in southern Massachusetts who sailed in the Georgia trade during the first six decades of the nineteenth century. What we know about the captains named in this single day's shipping list is typical of what has been found for the entire sixty-one-year period of interest.[3]

The exact number of men who worked as masters of sailing ships in the Georgia coasting trade between 1800 and 1861 is unknown. The available records are too incomplete to develop a full listing, just as they are for the coasting vessels themselves. Newspaper shipping lists and vessel documents have produced a list of 875 names of "masters" of sailing coasters in Georgia between 1800 and 1861. This listing contains many duplicates for a variety of reasons; most often, it's impossible to distinguish between individuals with the same surname, the way newspaper shipping lists typically identify shipmasters. This is particularly true when common names such as Davis, Moore, Smith, and White are involved. Another reason is that masters' names often appear in different documents with different spellings, further making identification difficult. Despite these difficulties, out of this list of 875 captains, 471 men are specifically identified in that both a surname and a given name or initials are known. Appendix B provides a listing of the 471 identified captains.

Some men regularly involved in the coasting trade appear as master of only a single vessel, often sailing this ship for several years. James H. Pitcher, of Rochester, Massachusetts, appears as the master of the sloop *Mariner* sailing in the Georgia trade between 1827 and 1841. However, most captains who worked in the trade for more than five or six years commanded two,

[3] Enrollment No. 38, Port of Fall River, sloop *America*, August 3, 1836; Enrollment No. 1, Port of St. Marys, sloop *Bolivar*, March 9, 1836, Registers and Certificates of Enrollment, BMIN, NARA, and Works Progress Administration, Ship Registers of Dighton-Fall River. Information on Captain Edmund Richardson is found in US Census Bureau, "US Censuses of Population, Camden County, 1820 through 1850"; Fields, "The Crypt." Information on Leonard Burr comes principally from Savannah newspaper shipping lists.

three, or more different coasters over time. Leonard Bolles, one of the Sippican captains, sailed as master of at least seven sloops and three schooners between 1819 and 1861. John L. Grovenstein, a native of Camden County, captained three sloops and two schooners between 1835 and 1861, and Daniel Reddick, of Savannah, was master of four sloops and two schooners during his years as a captain between 1838 and 1852.[4]

Successful captains of a coasting ship needed a variety of skills and personal attributes. These included competency in handling a sailing vessel, knowledge of the sailing routes and conditions in the region, and some degree of business understanding and acumen. Some skills were gained through experience; others were the product of individual personality. The skills needed to operate a sailing ship, such as handling sails and lines, were best learned through working aboard a vessel. Presumably, most of the captains began their careers as crewmembers and remained in this position for some period before taking a command. Sylvanus Staples, of Taunton, Massachusetts, began working on coasting vessels when he was ten years old, and became the master of a sloop when he was eighteen.[5]

Some sailing skills may not have been as important to the coasting captain as they were to the deep-sea sailor. For instance, the ability to use navigation instruments, such as the quadrant or sextant, was not critical to the local coasting captains because the typical sailing routes were along the inland passage or just offshore within sight of land. Only a few references to such navigation instruments appear in relation to the Georgia coasters. Captain Steele of the schooner *Harriet* was robbed of his "quadrant, books, charts, &c." by the British while sailing along the South Carolina coast in August 1813. In January 1818, Captain Paul Wing purchased a "log glass" (a small hourglass for measuring the speed of a vessel) for the sloop *Harmony* during a voyage from Massachusetts to Georgia.[6] Most local captains did not need quadrants or log glasses. Rather, they would have only carried items such as a compass and lead line for measuring directions and water depths, both critical in local navigation. When the coasting sloop *B. F. Newcomb* was advertised for auction in Savannah in 1846, she carried "one brass and one wooden compass" as well as "one lead and line," in addition to a chart.[7] Sophisticated

[4] The trading activities of these men come from shipping lists in the Savannah newspapers *Daily Georgian*, *Daily Savannah Republican*, and *Daily Morning News*, and the Darien newspaper *Darien Gazette*.

[5] Beers & Co., *Representative Men*, 104.

[6] *Republican & Savannah Evening Ledger*, August 21, 1813; "Paul Wing Accounts."

[7] *Daily Georgian*, March 26, 1846.

navigation instruments, such as quadrants, were needed only on those coasters sailing the long voyage between Georgia and Northern ports, or on the few that made the occasional trip to the Bahamas or the Caribbean.

More important to the Georgia captains than competency in the use of navigation instruments was knowledge of the visual landmarks designating the locations of various sounds, rivers, and inlets; the depths and current conditions of these water bodies on any given tide; and the locations of sandbars, mud flats, and shell banks. The extreme tidal range along the coast from Charleston to the St. Johns River in Florida meant that rivers and creeks easily traveled by a shallow-draft coaster at high tide might be entirely dry at low tide. For landings situated on larger streams, such as Darien on the Altamaha River or St. Marys on the St. Marys River, the tide was not a serious concern. However, many of the landings visited by coasters were on streams that were difficult or impossible to travel during low tide. Reasonably accurate navigation charts were available by the 1840s, and navigation aids, such as buoys and lighthouses, were in place even earlier, but learning the landmarks, sailing routes, and tidal conditions required firsthand experience, best gained by working as a crewmember.

The coasting captains also had to be able to maintain basic records accounting for the cargoes they carried. On any given voyage, shipmasters were responsible for hundreds if not thousands of dollars of cargo, often from several different shippers. Captains certainly had to keep track of the goods they carried in terms of quantity and ownership, meaning some types of records were kept. It may be that captains who sailed their vessels strictly as employees were not required to maintain their own records; their employers or the shippers of cargoes could have kept the necessary accounts. On the other hand, captains who owned their vessels and acted as independent businessmen had to maintain some sort of accounts of cargoes, payments, fees, etc., meaning they had to be literate and have some basic understanding of financial record keeping. Unfortunately, business records maintained by local coasting captains are extremely rare. A few accounts kept by New England captains who worked seasonally in the Georgia trade survive, but similar records kept by masters residing in Southern states are almost nonexistent. Accounts exist for the sloops *Mill Maid* and *Conductor* and the schooner *Fox* while working in the Savannah River from 1834 to 1836, and a few pages of shipping accounts kept by Charles Stevens of St. Simons Island have survived. The Stevens accounts cover part of the years 1859 and 1860 when he sailed his schooner *Northern Belle*. These records are minimal and not organized in the standard double entry credit (income) and debit format typically used in basic accounting. Captain Stevens's accounts consist of single-line entries giving a date, the articles transported, who the shipper or receiver was, the amounts involved and, often, whether the account was settled (a typical entry: "June 8, 1859, 2 Hhs [hogsheads] salt to Hull and Scarlett, $2.50, Settled"). Apparently, the cost of shipping these two hogsheads was paid. Other entries in the accounts do not include the word "settled," suggesting they remained unpaid, and a few entries specifically indicate that some amount was still due. Figure 6.1 presents an entry from Charles Stevens's accounts dated July 10, 1860. This entry records the freight costs on the delivery of several items to a "Mrs. Riley," including four boxes weighing a total of 1108 pounds, one barrel of glass, one tub, and one "Rocking Chair." "Mrs. Riley" might have been Elizabeth Riley, who in 1860, lived in Glynn County near the community of Bethel, one of the communities visited frequently by Charles Stevens with the *Northern Belle*. Some of Captain Stevens's entries list costs incurred for "drayage" and "wharfage," reflecting amounts the captain had paid or owed to someone else. Drayage referred to transporting the cargo to his vessel and wharfage represents costs charged by a wharf owner for use of his facilities or for holding the cargo in storage until Captain Stevens arrived. It may be that Stevens's very simple, even crude, account keeping was typical of the records maintained by many coasting captains. Even in this simple form, the captains had to keep track of several types of expenses for every shipment if he had any hope of making his business profitable. At the very least, captains needed basic mathematical skills.[8]

Georgia coasting captains commonly became masters of ships in their early to mid-twenties. Some boys certainly began to work on coasting ships in their teens, or even younger, but it took several years to acquire the sailing experience and absorb the intimate knowledge of regional geography needed to serve as a ship's captain. Further, some level of maturity, acquired with age, would have been necessary. Charles Stevens seems to have been rather typical, and he first appears in Savannah newspapers as master of a coasting sloop in 1841 when he was twenty-five years old. He had been in Georgia for several years before 1841, possibly working as a member of the crew and then mate on a coaster.[9] Other men first appear as masters in newspaper shipping lists

[8] Unidentified Cash Book, Georgia Historical Society. The Charles Stevens accounts are in the Frewin, Stevens, Taylor Family Papers, Georgia Historical Society. Elizabeth Riley is listed in US Census Bureau, "US Census of Population, Glynn County, 1860," 237.

[9] Pearson, *Charles Stevens*, 490–506.

at a similar age. Among these were Michael Peck, who was twenty-four; Antonio Lawrence, twenty-six; William Watson, twenty-five; Daniel Reddick, twenty-six; Luke Christie twenty-four; John L. Grovenstein, twenty-seven; John Lightbourn, twenty-three; Thomas Snow, twenty-six; Stephen Williams, twenty-two; and Barron Hadley, also twenty-two. John Delano, a native of Rochester, Massachusetts, and the captain and owner of the coasting sloop *Reformation*, was twenty-three when he died in Savannah in 1814.[10] A few of the captains began at slightly younger ages. Barnabas Nye, another Rochester, Massachusetts, native, appears as a captain on the Georgia coast when he was only twenty-one, as did Peter Thompson of Savannah. Dennis Pacetty, long-time coasting captain from St. Marys, was twenty years old when he was first named as a master of a coasting sloop.

Life on a sailing coaster, as with life at sea in general, was hard and physically demanding, meaning masters and crewmen alike needed basic physical abilities, strength, and stamina. Individual coaster voyages tended to be short, typically lasting no more than three or four days, but the captains and crews spent much of their time on deck exposed to the elements. Although coasters sailed year-round, their activity was most intense during the rice- and cotton-shipping season, extending from October to April, encompassing the coldest months of the year. Even though the climate along the Southeastern coast is mild, freezing weather does occur, and its effects are intensified by winds from weather fronts that cross the area. During the summer, the coasters were exposed to extremely hot and humid weather, frequent rains, and, occasionally, hurricanes.[11] These weather discomforts, together with the strenuous physical labor involved in raising and lowering sails, hauling lines, and shifting cargo would have taken their toll on the men aboard the coasters. Although the captains and crew had to be physically fit, at least one local master with a known physical disability could command a coaster. Captain James Frewin, resident of Frederica on St. Simons Island, served as master of the schooner *Maria* in the 1820s even though he had a wooden leg.[12]

The arduous and often harsh conditions found aboard ships might have forced many captains to give up their profession at young ages, but the age at which men left the trade varied considerably. A review of the ages of captains given in the 1850 and 1860 censuses for Georgia coastal counties indicates that most were between twenty-five and forty-five years old although a few

[10] Georgia Historical Society, *Register of Deaths*, 3:83.
[11] Wilkes et al., *Soil Survey*, 67.
[12] Pearson, *Charles Stevens*, 12.

were older. Charles Stevens was serving as a captain at the age of forty-nine, when the Civil War ended his career; David Pidge died in Savannah in 1828 at the age of fifty-one, just two weeks after he sailed the sloop *Mill Maid* into Savannah; and Charles Porquet was sailing as captain of the sloop *Dolphin* two months before his death in Savannah in 1829 at the age of fifty-five. In 1865, Charles Thompson of Savannah was described as "one of the oldest coasters in the trade" when he was fifty-four years old. Edmund Richardson of Camden County seems to have made his last voyage as a coasting captain when he was about sixty. It does appear that few coasting captains were active past their mid-fifties. Of course, this was during a period when the average life expectancy of White American males was about forty years.[13]

In common with all maritime trades, work in the coasting trade could be dangerous. The captains and crews aboard the coasters faced the perils of weather as well as hazards from the simple act of working on the water and aboard a ship. Scattered accounts of these dangers appear in reports of accidents, sinkings, and diseases that befell the ships and men in the trade. Drowning, always a danger when working aboard ship, was a particular hazard in the nineteenth century because many people, even sailors, couldn't swim. Newspapers occasionally mentioned the specific circumstances of drownings, such as the January 28, 1862, report in the *Savannah Daily News* that Captain Domingo Galleo accidentally fell overboard from his schooner *Eliza Ann* and drowned in the Savannah River. Other newspaper accounts report only that a person was "lost at sea," which may or may not have been from drowning while others describe accidents and some sinkings of coasters but do not always report on injuries to those onboard. In 1825, the *Darien Gazette* reported that Elias Boles of Rochester, Massachusetts, was lost overboard from the schooner *Volutia* "in a gale of wind off Cape Hatterass." The *Volutia* (or *Volucia*) was a regular trader along the Georgia coast in the early 1820s, and she was apparently coming South when Elias Boles was lost. Death records for the city of Savannah include accounts of those who died aboard ships; in many instances, these persons were working aboard coasters. Some men died from accidents, such as forty-six-year-old Peter James, a

[13] Information on Charles Stevens from Pearson, *Charles Steven*; information on Charles Thompson from R. M. Myers, *Children of Pride*, 1701; and information on others from Georgia Historical Society, *Register of Deaths*, 4:224, 239. The estimate of Edmund Richardson's age comes from the *Daily Georgian*, June 8, 1838. Life expectancy information is for Massachusetts males at birth, which was 38.3 years, from US Census Bureau, "Life Expectancy."

native of Italy, who in 1809 was killed by a falling block aboard the coasting sloop *Polly* while the vessel was docked at the Savannah waterfront.[14]

Savannah death records are also replete with the names of coasting captains and crewmen who died from the range of diseases that were a danger to everyone in the first half of the nineteenth century, many of which were especially ubiquitous in port cities. The graves of "seaman" who died from a wide range of causes fill the city's cemeteries. The prevalence of diseases, coupled with the generally ineffective medical practices of the period, meant that many coasting captains and crewmen died young or were incapacitated and left the trade. Among the coasters struck by diseases in Savannah were John Delano, captain of the sloop *Reformation*, who died in 1814 of "Bilious Fever"; Timothy Driscoll, who died in 1815 of "Pleurisy"; Peter Caesar (or Cesar), who died in 1817 of "Consumption"; Stephen Waddington, who succumbed to "Dropsy" in 1815; Peter Worthington, who died in 1844 of "Apoplexy"; and Savannah native coasting captain, and river pilot, James Dent, who died of "Cholera Morbus" in 1850.[15]

Most young men with average physical and mental abilities could acquire the practical sailing skills needed to handle a coaster. To be truly successful, captains also had to possess the personality traits and social skills needed to deal at a personal as well as business level with those who shipped or received the cargoes. This was particularly true of those captains who owned their ships and worked as independent entrepreneurs. These men had to develop and maintain relationships with planters, merchants, and factors to insure they were called on when cargoes needed to be carried. Few records exist to shed light on how these relationships were made or how permanent they might have been. Some captains were certainly more successful than others, and some must have been unable to please everyone they served. This latter circumstance seems true of Captain Shadrack Smith of the schooner *Mary Adams*. On July 17, 1828, a letter appeared in the *Daily Georgian* from "several citizens of Glynn County" to their factors in Savannah asking them "not to ship any articles again on board the schooner Adams, as her Captain passed by our landings, and has not delivered our freight...." The Glynn County residents asked that factors in Savannah "give a preference to such vessels as are owned by Citizens of Georgia or Carolina, and earnestly hope that our

[14] *Darien Gazette*, February 2, 1825; Georgia Historical Society, *Register of Deaths* 2:61.

[15] Death information found in Georgia Historical Society, *Register of Deaths*, vols. 2–6.

Legislature will tax Northern Coasters…." Captain Smith may have failed to stop at these planters' landings, but he seems to have had a good reason because the same issue of the *Daily Georgian* reported that the *Mary Adams* had been seized by authorities when caught smuggling goods in the St. Marys River.

In 1841, Roswell King Jr., then living on Colonels Island in Liberty County, Georgia, complained that the captain of the sloop *Swallow* delivered a cargo of lathing and unabashedly asked to be paid $20 in freight, as much as the laths had cost. On this same voyage, King noted the *Swallow* charged only $2.50 to deliver corn to someone at Harris Neck, a delivery that involved sailing "*10 miles out of his way*." In the 1850s, prominent rice planter Charles Manigault employed a Captain Wallace to carry rice from his lands on Argyle Island in the Savannah River to his factor in Charleston, but he was not always pleased with the service. Wallace had apparently failed to stop at Argyle Island on a previous trip, and Manigault wrote his son Louis to "talk strong" to the captain when he next arrived and tell him "if you don't come back immediately here for the next load you need not come at all. I will give it to another." Concerning Captain Wallace, and coasting captains in general, Charles Manigault wrote, "You cant [*sic*] depend on these people. They have always 2 or 3 strings to their bow."[16]

Other accounts mention similar dissatisfaction with coasting captains, particularly when they failed to arrive at a specified time. In December 1844, Anna Page King of Retreat Plantation on St. Simons Island, in a letter to her husband, Thomas Butler King, wrote, "That man Chevalair has not yet landed our goods. This is truly provoking." She was referring to a member of the Chevalier family of St. Marys, several of whom were involved in the coasting trade.[17] It was often difficult for the sailing coasters to maintain precise schedules because they were so dependent upon weather, wind, and tide. Before the advent of steamboat service, coastal residents were used to and probably more accepting of these vagaries in schedules. However, once people became accustomed to the predictability of service provided by steamboats in the 1820s, they became less tolerant of the delays and missed deadlines inevitably associated with sailing vessels.

The statement about "Northern Coasters" in the 1828 *Daily Georgian* is a reference to the many captains who brought their ships down from

[16] "Roswell King Diary," August 4, 1841, Hill Memorial Library, Louisiana State University; Clifton, *Life and Labor*, 231, 233.

[17] Anna Page King's comments are found in Pavich-Lindsay, *Anna*, 30–31.

Northeastern ports every year to work in the coasting trade out of Savannah and Charleston. These men came mainly from ports in Connecticut, Rhode Island, and, particularly, the Buzzards Bay area of Massachusetts, and were involved in carrying coastal cargoes during the fall, winter, and spring. In some years, these Northern men dominated the local Georgia trade, representing close to 50 percent of the captains working. There are occasional references to conflicts between these Northern men and Southern merchants, planters, and shippers, and there must have been some antagonism between these Northern captains and those residing in the South. In 1858, Charles Manigault, in complaining about the non-arrival of the schooner *Minx*, wrote that she was "a Yankee (I think) come South to make money." Manigault also wrote that when he employed the sloop *Bee*, he was given assurance that she was owned by a "Charleston concern" and not a Northerner. He noted that the *Bee*'s owner (a Charleston factor) indicated, "Yankeys ought not to have so much," apparently referring to a share of the Charleston coasting trade.[18]

Despite Charles Manigault's stated antagonism against Northern captains, Charleston and Savannah factors and coastal planters relied heavily on their services and ships from 1800 to 1861. This reliance was made necessary by the lack of sufficient Southern masters and ships to carry cargo during the peak shipping seasons. The numbers of Northern captains in the Georgia trade decreased after the 1840s, but the type of discord described by Charles Manigault continued over time as sectional differences intensified, ultimately leading to the Civil War.

## The Ownership of Georgia Coasters

Ownership in a ship, even a small coasting vessel, was an important form of capital investment in the United States before the Civil War. In the South, investment could be made in land and slaves, but along the Georgia coast, the initial capital required to develop a successful cotton or rice plantation was beyond the means of most men. Ship ownership provided opportunities for those with modest amounts of capital to invest. Even so, there was no long history or strong cultural incentive to participate in ship ownership in the South, as was the case in Northern states, so the total number of Southern residents involved in the ownership of Georgia coasting ships was small throughout the nineteenth century.

[18] Clifton, *Life and Labor*, 263, 264.

The pattern of ownership of Georgia coasters took many forms, including ownership by a single individual, ownership by a small group of four or five individuals, ownership by a larger group, often a dozen or more persons, and ownership by business firms. These categories of ownership are similar to what Robert Albion has described for vessels documented at the port of New York in 1850.[19] In the case of the Georgia coasters, the men serving as masters were also commonly owners. Out of the list of 471 specifically identified masters, 169, or approximately 36 percent, held ownership in one or more Georgia coasters. This number is certainly an underestimate because so many documents providing ownership information for Georgia vessels are missing.

Approximately 90 percent of the 169 identified captain-owners held ownership in the vessel they commanded. A small number also had ownership in other coasters. Over the sixty-one-year period of interest, a captain might own several vessels, either in sequence or simultaneously. Georgia captains could be the sole owner of the vessel they sailed, or they could share ownership with others. Vessel documents indicate that sole ownership by masters was common for those captains living in Georgia, South Carolina, or Florida. A captain who was the sole owner of his vessel acted as an independent entrepreneur although records show they often carried cargoes for a specific factor or merchant over the course of a season or longer, suggesting business alignments of some sort. When Southern masters owned a share in a vessel, ownership was usually spread between a small group, typically no more than four or five individuals. These ownership groups often consisted of coasting captains. Such was the case of the schooner *Edmund & Francis*, which in 1850 was owned equally by Stewart Austin, Domingo Galleo, Henry J. Dickerson, and Simon Santini. All were Savannah residents who worked as masters of coasters even though Henry Dickerson, by this time, had given up the captaincy of coasters and was more fully involved in ship ownership and stevedoring.[20]

Some local captains shared ownership in vessels with other categories of individuals who had economic interests in the coasting trade, including both planters and merchants. In the late 1850s, long-time captain John Grovenstein owned the schooner *Elias Reed* in partnership with Hugh Frazier Grant, a prominent planter in Glynn County. The *Elias Reed* often transported rice from Grant's Elizafield Plantation on the Altamaha River.

[19] Albion, "Early Nineteenth-Century Shipowning," 1–11.

[20] "Conveyances of Enrolled Vessels, Savannah," schooner *Edmund & Francis*, November 26, 1850.

Similarly, Robert Stafford, a planter from Cumberland Island, Georgia, shared ownership in at least two coasting vessels with the men who served as masters. In the 1830s and 1840s, he and Captain Samuel Flood of St. Marys were the owners of the schooner *Ellen*, and, in the 1820s, Stafford shared ownership in the sloop *Conductor* with her captain, Richard Hill, also of St. Marys. In the 1840s, brothers John and Pierce Butler, owners of plantations in Glynn County, shared ownership of the schooner *Roswell King* with several residents of Rochester, Massachusetts; at least two co-owners, Frederick Bolles and Pelag Blankinship, served as masters of the schooner.[21]

Lewis P. Wiggins, native of Riga, Russia, and captain of several Georgia coasters in the two decades before the Civil War, was master of the sloop *Catherine Chard*, as well as a part owner with Boston & Gumby, a Savannah factorage firm (Figure 6.2). Newspaper listings of arrivals of the *Catherine Chard* indicate that Captain Wiggins traded almost exclusively with locations along the lower Georgia coast, such as the Satilla River and the communities of St. Marys, Centerville and Jeffersonton. The Boston & Gumby firm had business ties with several merchants and planters along the lower coast, and they used the *Catherine Chard* to serve their clientele in that area. Boston & Gumby are often named as consignee of cargoes carried into Savannah by the *Catherine Chard*, but other merchants also received goods via the sloop. Obviously, Captain Wiggins was not bound to transport goods exclusively for his co-owners.[22]

Many of the Georgia coasters were sailed and owned by Northern men. As with the captains living in the South, it was common for masters of Northern vessels to hold ownership in the coaster they commanded. However, it was less common for a Northern captain to be the sole owner of his vessel than it was for a Southern captain. This was particularly true for the captains and vessels from the Sippican area. When the master of one of these Northern ships was an owner, it was typically with a group of other men. This group sometimes consisted of fewer than four or five individuals, as was frequently the case for the Southern-owned coasters. However, it was common for the Sippican ships to be owned by groups, sometimes of ten or more people, an arrangement almost never seen for Southern-owned ships. In 1826, for

[21] Enrollment No. 5, Port of Savannah, schooner *Elias Reed*, June 17, 1857; Enrollment No. 6, port of St. Marys, schooner *Ellen* December 13, 1834; Enrollment No. 7, Port of St. Marys, sloop *Conductor*, October 21, 1828; *Enrollments of New Bedford, 1841–1939*, schooner *Roswell King*.

[22] Enrollment 5, Port of Savannah, sloop *Catherine Chard*, June 8, 1852.

example, twenty-four men, including her master Stephen C. Luce, owned the sloop *America*. Most of the *America*'s owners were residents of Rochester, Massachusetts.[23]

Some of the owners of the *America* were ship captains, while others were merchants or ship builders. In addition, several of these owners held shares in other coasters working in Georgia, and it is likely they owned interests in vessels sailing in other trades. These Sippican men were in the business of ship ownership, in addition to the coasting trade itself. The dispersal of ownership among a large number of investors helped limit the loss of any one individual in the event a voyage was unprofitable or a vessel was lost. Albion noted a similar pattern of large numbers of men sharing the ownership of the small coasters enrolled in New York in 1850. The rareness of large group ownership among Southern owners reflects the lack of a tradition of ship ownership in the region, as well as the absence of an established system that would support or encourage capital investment in the ownership of ships.[24]

Most commonly, multiple owners of a Georgia coaster resided in the same town or location. Occasionally, however, ownership was spread among men living in distant communities. In 1828, ownership of the newly built sloop *Angel* was shared by a group of Rochester, Massachusetts, men, including her master Elisha Luce, and by several residents of Darien, Georgia, including Zephaniah Kingsley, Philip Yonge, Bayard Hand, Henry Yonge, and the firm of P. R. Yonge & Son. Undoubtedly, the *Angel* was intended to transport merchandise for her Northern and Southern owners. Two years later, the *Angel* sailed into Savannah thirteen times, and every voyage originated at Darien. The *Angel*'s cargoes consisted almost entirely of upland cotton, shipments probably arranged by her Darien merchant owners. This shared North-South ownership seemed to work for the *Angel* because ten years later, in 1838, the sloop was jointly owned by prominent coasting captains Stephen C. and Elisha Luce of Rochester and Savannah merchants Robert Hutchison and Andrew Low. An advantage in having owners in different ports was that these men could act as local agents, looking out for the interests of the owners.[25]

How active merchants such as P. R. Yonge & Son might have been in the operations of the vessels they owned is unknown. Albion indicates that

[23] Enrollment No. 61, Port of New Bedford, sloop *America*, October 10, 1826.

[24] Albion, "Early Nineteenth-Century Shipowning," 9.

[25] *Enrollments of New Bedford, 1808–1840*, sloop *Angel*; *Daily Savannah Republican*, 1830.

this type of ownership was generally "passive," meaning the merchant used the ship for transporting their goods but made few decisions on its day-to-day operations.[26] However, it would be in their economic self-interest for merchants to make local arrangements to secure cargoes for a vessel in which they shared ownership.

Although many Georgia coasters were owned in part or entirely by their captains, not all were. A few Georgia and South Carolina planters owned vessels outright. For example, Major Pierce Butler of Philadelphia, prominent St. Simons Island planter, was the sole owner of the schooner *Experiment* before his death in 1822.[27] In most instances, the plantation owner employed his vessel to service his own plantation, but it almost certainly carried goods for other customers as a business venture.

Merchants owned some Georgia coasters and hired captains to sail them. In the 1840s, Andrew Low Sr. and Jr., partners in the prominent Savannah firm Andrew Low & Co., owned the sloop *America*. During their ownership, they employed several masters for the sloop. Members of the Habersham family of Savannah, both as individuals and as a company, owned several coasters sailed by various hired captains. These ships included the sloop *George* and the schooners *Robert Habersham*, *Hamilton*, *Savannah*, and *Science*. When the sloop *Fashion* was enrolled in Savannah in 1856, the Savannah mercantile firm of Bell & Prentiss and city resident William H. Gladding were listed as owners. Gladding was apparently an employee of the United States Revenue Service at Savannah.[28]

When a coasting captain owned his vessel outright, his income derived from the freights received in the operation of the vessel. Presumably, a similar

[26] Albion, "Early Nineteenth-Century Shipowning," 2.

[27] Enrollment No. 16, Port of Darien/Brunswick, schooner *Experiment*, November 18, 1821; Enrollment No. 8, Port of Darien/Brunswick, schooner *Experiment*, October 11, 1823; Enrollment No. 3, Port of Darien/Brunswick, schooner *Experiment*, July 28, 1828; Bell, *Major Butler's Legacy*, 107.

[28] The Low's ownership of the *America* is recorded in Enrollment No. 3, Port of Savannah, June 1, 1843. Habersham family ownership of various vessels is contained in several vessel documents, including Enrollment No. [illegible], Port of Savannah, schooner *Robert Habersham*, May 21, 1834; Enrollment No. 7, Port of Savannah, sloop *George*, May 2, 1836; Enrollment No. [illegible], Port of Savannah, schooner *Hamilton*, August 30. 1838; Enrollment 13, Port of Savannah, schooner *Science*, September 10, 1840; and Enrollment No. 19, Port of Savannah, schooner *Savannah*, October 6, 1845. Information on ownership of the sloop *Fashion* is in Enrollment No. 3, Port of Savannah, December 2, 1856. William Gladding's occupation from Otto, *1850 Census Chatham County*, 79.

situation occurred when the captain was a part owner. When a captain was in the employ of others and held no ownership in the coaster he commanded, the manner in which he obtained his income is uncertain. Records relating to the business arrangement established in these circumstances for Georgia coasters are rare, but it appears these non-owning captains worked in the employ of the shipowner, either on a set salary or on a share of the total freight income received by the vessel. In some maritime trades, such as whaling, pay was typically on a share basis, with the income derived from a ship's activities divided among the owner(s), captain, and crew following a set formula. However, on vessels transporting cargo, such as coasters, it appears that set wages were most common for the crew and the captain, if he was not an owner. Among the few documents relating to hired coasting captains in Georgia concern Savannah resident Patrick Doyle, who in the 1850s, was hired by the Savannah factors Reed & Tison to serve as master of their schooner *Elias Reed*. Captain Doyle had no ownership in the schooner, and he was hired at a set wage of forty dollars per month. Hiring captains on set wages seems to have been common in the local Georgia trade.[29]

The Southern captains often resided in the areas where their trading activities were concentrated. For example, a small group of ship captains living in Camden County consistently sailed into Savannah and Charleston from ports in that county such as St. Marys, the Big Satilla River, Cumberland Island, and the inland towns of Centerville and Jeffersonton, or from nearby "East Florida" and Amelia Island. These men rarely traded with other areas along the Southeastern coast, and newspaper entries on vessel arrivals in Savannah indicate these local men often dominated the St. Marys-Savannah trade with few outsiders participating.

Some of the captains residing in Savannah concentrated their activities in the region near the city, carrying most of their cargoes from the nearby Savannah and Ogeechee rivers and landings in between. By 1810, the town of Riceboro, located on the North Newport River in Liberty County, was listed as a place of origin of coasting vessels, reflecting the expansion of cotton and rice agriculture in that area. After 1810, several captains sailed almost exclusively from Riceboro or nearby landings in Bryan and Liberty counties, such as Sunbury and St. Catherines Island. In the early 1820s, Nicholas Sallowich, apparently a resident of Savannah, sailed his sloop *Union* exclusively between Savannah and the communities of Sunbury, Hardwick, and

[29] Daggett, *Fifty Years of Fortitude*; "Accounts for the Schooner *Elias Reed*," Gordon Family Papers, Southern Historical Collection, University of North Carolina.

Riceboro. Thomas Snow, one of the few captains living in Liberty County, typically arrived in Savannah from locations in or near that county, and, with his schooner *Sarah*, he was the only captain to sail regularly from the old colonial town of Sunbury during the late 1840s and early 1850s.[30]

Coasting captains not only served particular regions but also carried goods for specific plantations or planters. The schooner *Hopeton* seems to have carried most, if not all, of the production from Hopeton and Hamilton plantations in Glynn County to the Charleston factor Robert Mure & Company in the 1850s. However, the two plantations received goods from merchants in both Charleston and Savannah aboard several ships in addition to the *Hopeton*.[31]

## The Southern Captains in the Georgia Coasting Trade

The captains who sailed in the Georgia trade comprised two main groups, those who resided along the South Atlantic seaboard in Georgia, South Carolina, and Florida and those who were residents of Northern states who sailed South to work for part of the year. Throughout the first six decades of the nineteenth century, most of the Southern captains lived in or near Savannah, the local trade's principal port and population center. A small number lived at other locations along the Georgia coast, such as St. Marys in Camden County and St. Simons Island and Brunswick in Glynn County. A few captains resided on the lower South Carolina coast and at locations on the northeastern Florida coast. Other captains resided in Charleston, but these individuals were not commonly involved in trade into Savannah or along the Georgia coast after the 1820s.

The principal difference between the Southern and Northern captains, in terms of their trading patterns, is that Southerners tended to sail in the local trade all year long. The Northern captains typically sailed in the Georgia trade only during the rice and cotton shipping seasons, from September through April. However, even the Southern captains decreased their activities in the summer when rice and cotton were unavailable as cargo. Differences other than sailing patterns existed between the two groups. Unlike the Northern captains, those living in the South often became sole owners of their vessels and more often worked as independent businessmen. In addition, a large

[30] The activities of captains trading with the central Georgia coast are derived from shipping lists published in Savannah newspapers. See entire list in the bibliography.

[31] Glynn County, Georgia, estate of James Hamilton, Wills and Appraisements, Book F.

proportion of the captains residing in the South were immigrants while most Northern captains were native-born Americans. This difference in nativity reflects the deep and strong maritime heritage found in many Northern states, a heritage far less pronounced in the South. In addition to foreigners, several Georgia captains residing in the South were transplants from Northern states.

The men who regularly sailed as captains in the Georgia trade were a small group whose numbers changed considerably over the years. Figure 6.3 presents information on the total number of coasting captains sailing into Savannah for various years from local ports other than Charleston. After 1820, Savannah newspapers eliminated departure information for the local coasters, so Figure 6.3 incorporates only arrival data. Many of these arriving captains made only a single voyage into Savannah, so Figure 6.3 also shows the numbers of captains making three or more arrivals into Savannah annually. These are the masters regularly employed in the local trade.

In 1800, only fifteen different captains sailed into Savannah from local ports. Savannah newspapers in 1800 published complete information on departing vessels, including the names of masters and the ports of destination. From these records, nineteen additional captains can be identified sailing out of Savannah to local ports. Only one of these men, a Captain Guillenet of the schooner *Friendship*, made more than three departures, all to St. Marys. In 1800, only one captain made more than three voyages into Savannah. This was Captain Rudolph, believed to be Robert Rudolph, who sailed the schooner *Alective* into the city four times from St. Marys.

As seen in Figure 6.3, the total number of captains sailing into Savannah from local ports increased through 1820, when approximately eighty-seven different masters are identified. After that, the number of arriving captains declines, with thirty-three different captains sailing into the city in 1860. However, over this sixty-year span, there were normally fewer than twenty-five or thirty captains regularly involved in the trade in any given year. As few as fifteen or twenty of these regular traders might be Southern residents. If these men sailed in the trade for any length of time, they had many occasions to meet, given they were sailing into the same ports, often carrying cargoes for the same factors and planters. The 1850 and 1860 censuses for Chatham County show that coasting captains often lived together, particularly if they were natives of the same country. Further, many of the Southern captains shared ownership in the same ships and sold vessels among themselves. A variety of other interactions occurred among this small fraternity of Southern captains. For example, the children of some Southern captains married men who were captains or became captains. Records also show that captains sold land to other captains. In the early nineteenth century, Sarah Frewin, wife of Captain James Frewin of St. Simons Island, purchased property at the town of Frederica from two other coasting captains, James A. D. Lawrence and Charles Porquet. Captain Charles Stevens named one of his daughters Elizabeth Watson, almost certainly after William Watson, a fellow Dane and coasting captain, reflecting the close relationship between the two men. Public documents reveal that Southern captains loaned money to one another and served each other in various legal transactions, such as executors of wills and administrators of estates. These and other similar interactions must have been common, but most have gone unrecorded. The Southern coasting captains were a small band of men bound together by their occupation, a bond strengthened by their status as immigrants. Their occupation and their nativity set them apart from broader Southern society, and their modest economic status certainly excluded them from the society of the coastal planters and merchants with whom they interacted on a regular basis.[32]

Similar interactions must have occurred between the Southern captains and those from the North. We know, for instance, that individuals from these two groups owned coasters in common, and vessels were sold between the two, occurring most often when Northern men sold their ships to Southerners. Friendships must have developed between many Northern and Southern captains. However, there were tensions between the two groups beyond the personal level, as evidenced in the occasional complaints against Northern captains by Southern planters and shippers. Despite the fact that there were an insufficient number of Southern-owned coasters to handle the local trade during the peak of the rice and cotton shipping season, many considered Northern captains outsiders, infringing on the business of Southern captains and shipowners. In addition, there was also slavery, the great divider that existed between the South and the North and an underlying cause for intensifying sectional differences throughout the first half of the nineteenth century. How views on slavery may have specifically affected personal relationships between Southern and Northern captains is unknown, but an obvious expression of it was found in the composition of the crews of the coasters themselves. All available evidence suggests that enslaved African Americans commonly worked as crewmen aboard Southern-owned coasters, and it is suspected that these enslaved men comprised a significant proportion of the crewmen sailing

[32] Otto, *1850 Chatham County Census*; Georgia Historical Society, *1860 Census*; Sarah Frewin's property purchase found in Frewin, Stevens, Taylor Family Papers, Georgia Historical Society; Pearson, *Charles Stevens*.

in the Southern coasting fleet. This was not the case for the Northern coasters, which were manned by Whites and, possibly, a small number of free Blacks.

Who were the men who commanded the Georgia coasters? We know the names of many, but very little of the lives of most. The names of many captains appear in the federal censuses, and these records provide a starting point for identifying these men. Those censuses compiled after 1840 are of particular value because they include information on occupation, a help in the positive identification of a particular captain. The 1840 census included categories for individuals employed in the "Navigation of canals, lakes, rivers" and "Navigation of the ocean" while the 1850 and 1860 censuses included specific identifications of "Occupation." To gather information on Southern captains, the federal census records for the coastal counties of lower South Carolina (Colleton and Beaufort counties), coastal Georgia (Chatham, Bryan, Liberty, McIntosh, Glynn, and Camden counties) and northeastern Florida (Nassau, Duval, and St. Johns counties) were examined. These are the counties where most of the local captains resided.[33] The censuses have enabled the identification of many of the coasting captains. Additional published and private records have been used to expand an understanding of the lives of a small number of these captains, as discussed in following sections.

### *The Captains of Savannah and Chatham County*

Savannah was the population and trading center of coastal Georgia during the first six decades of the nineteenth century, and it was home to many coasting captains and shipowners. Between 1800 and 1861, more individuals working in maritime trades of various sorts resided in Chatham County, which includes the city of Savannah, than any other county in the state. Detailed information on most of the Savannah captains is scarce, particularly for the years before about 1820. Listings in newspaper shipping columns and official vessel documents, such as enrollments, registers and conveyances, provide basic information on a number of these men, such as their full names and places of residence, and shed light on their sailing activities and vessel ownership. Federal census records also contain information on those involved in the

[33] The online versions of the 1840, 1850, and 1860 federal censuses of population for all of these counties were examined at Ancestry.com. These online records were supplemented with the 1850 population census records for the Georgia coastal counties published privately by Otto, and the 1860 population census of Chatham County compiled and published by the Georgia Historical Society.

coasting trade, and public documents, such as wills, deeds, and estate records elaborate on the lives of a few of these men.

For those Savannah captains sailing before the 1820s, typically only their names and the ships they commanded are known, derived from newspaper shipping lists and surviving vessel enrollments and registers. Several of the Georgia captains sailing at the beginning of the nineteenth century are included in a 1799 tax list of Chatham County residents that enumerates occupations and names. Among the occupations recorded are fifteen "mariners," one "sailor," and nine "pilots." Several men identified as "mariners" in this tax list also appear in newspaper shipping columns or official vessel documents as masters of coasting vessels. These include George Hillary, master of the sloop *Friendship*, sailing in the Sunbury-Savannah trade in 1800, and Thomas Keen, commander of the schooner *Savannah Packet*, working in the Charleston-Savannah trade from 1800 to 1802. Among the nine pilots named in the 1799 tax list is William Pindar, Jr., master of the coasting schooner *Industry* in 1801.[34]

Other Savannah captains sailing before 1820 were men such as William Brown, master of the sloops *Republican* and *Rising Sun*; Stephen Waddington, master of Major Pierce Butler's schooner *Experiment*; Peter Caesar of the sloop *Diana*; John Emerus of the schooner *Two Friends*; and Elisha Hopkins of the schooner *Vexation*. Savannah newspapers name Timothy Driscoll as master of the schooner *James Madison* in 1814, and when he died in Savannah of pleurisy in 1815, it was noted that he was "Captain and owner of a coasting sloop." Bermuda native Samuel Lightbourn seems to have settled in Savannah in about 1797, the year he swore an oath of citizenship in the city. Over the next twenty years, Captain Lightbourn served as master of the sloops *William*, *Good Intent*, *President*, and *Three Sisters*, often sailing between Savannah and the old colonial town of Hardwick. German native, Henry F. Willink, who later established the preeminent shipbuilding and repair yard in Savannah, was master of the sloop *Henry* in 1813, sailing between St. Marys and Savannah. Nicholas Sallowich, who first appears in newspaper shipping lists in 1813 as master of the sloop *Polly*, continued sailing as a captain of coasters to about 1835. Among the more active Savannah captains of this early period were

[34] "General Tax Return of Chatham County, for the Year 1799," Georgia Archives, Georgia's Virtual Vault; Hagy, *People and Professions*, 101.

David Pidge and Noah B. Sisson, each of whom served as master and owner of several coasters during their careers, which extended well past 1820.[35]

During the first two decades of the nineteenth century, many of the captains living in Savannah were immigrants. These men had to take oaths of naturalization, a requirement to serve as master or to own a vessel employed in the American coasting trade. George Hillary took an oath of naturalization in Savannah in 1793, and William Pindar Jr. in 1806, although these oaths do not name the native country of either man. Captains Stephen Waddington and Timothy Driscoll were natives of Ireland; Henry Willink was from Germany. John Emerus was a native of Sweden, and Peter Caesar and Nicholas Sallowich were from Italy. Two of the most active captains of the period, David Pidge and Noah B. Sisson, were originally from Northeastern states, Pidge from Massachusetts and Sisson from Rhode Island.[36]

In general, more information is available on those captains sailing after 1830. Among the Savannah captains of this period were William Austin, John Arnaud, Jeremiah Boar, Luke Christie, Henry J. Dickerson, Patrick Doyle, Domingo Galleo, John Grovenstein, Antonio Lawrence, Joseph Raffile, John Russell, Noah B. Sisson, Charles Thompson, Peter Thompson, William Watson, Lewis Wiggins, John Williams, Alexander Wilson, and William Worthington. The names of these men appear in newspaper shipping lists and in other records, including the federal censuses. The 1840 federal census of Chatham County identifies 201 individuals employed in "Navigation of the ocean" and forty in "Navigation of canals, lakes, rivers" (Table 6.1).[37] Most of these men lived in the city of Savannah. Some of the 201 individuals named in the "Navigation of the ocean" occupation, the category most likely to include those sailing in the coasting trade, may have worked as crewmembers aboard coasters. Others, however, were sailors employed on ocean-going sailing ships or steamers, not on local coasters. Not all 241 individuals listed with these occupations are named in the census; some are just enumerated by age, sex, and occupation because the 1840 census typically provides names only for the head of a household, not everyone residing in that household. In several instances, a single household may include several persons assigned one of the

[35] Georgia Historical Society, *Register of Deaths* 3:48, 113; Hemperley, "Federal Naturalization Oaths," 472.

[36] Otto, "Oaths of Allegiance," 6, 11; Georgia Historical Society, *Register of Deaths* 2:83, 3:113, and 5:223; Hemperley, "Federal Naturalization Oaths," 459, 464; Otto, *1850 Census Chatham County*, 105; information on David Pidge and Noah Sisson is from Sprague, *Genealogies*, 6124, 6129; and Harrison, "Noah B. Sisson."

[37] U.S. Department of State, *Compendium of the Sixth Census*, Table 2.

"navigator" occupations, but only the household head is named. For example, three persons identified as "Navigators of the ocean" resided in the household of William Canuet, a known coasting captain. One of these was William Canuet himself, identified as a male between thirty and forty years old. The names of the other two "navigators" residing in the household are not provided.[38]

Approximately two dozen of the men assigned the two navigator occupations in the 1840 Chatham County census can be equated with captains whose names appear in newspaper shipping lists or vessel documents. These persons are listed in Table 6.2. Some of the other 241 men engaged in navigation that year certainly served as masters of coasting vessels, but they cannot be positively identified. It is unknown why some men were assigned the occupation "navigation of the ocean" while others were assigned "navigation of canals, lakes and rivers."

The 1850 and 1860 federal censuses of population provide specific designations for occupations, making it easier to identify those engaged in the coasting trade versus other maritime activities. The 1850 census lists 140 persons in Chatham County with occupations related in some way to marine work. Among the occupations enumerated are "boat builder," "boatman," "capt. of steamer," "mariner," "customs inspector," "sail maker," "seaman," and "ship carpenter." One occupation category that specifically identifies those working in the coasting trade is the word "coaster" itself. Eight Chatham County men were assigned this occupation in 1850. However, when the names of individuals in the 1850 Chatham County census are compared against the names of known masters of coasting vessels derived from shipping lists and vessel documents, it is apparent that some coasting captains were assigned the occupations of "mariner" or "seaman" (Table 6.3). Other men assigned these occupations cannot be identified as captains and may have worked as crewmen.[39]

The 1850 Chatham County census lists thirty-four men employed as "pilots." These men were primarily involved in piloting vessels into and out of the port of Savannah. A few do appear as masters or owners of coasting ships. One of these pilots, Alexander Johnston, was the sole owner of the coasting schooner *Joseph* in 1841 and a part owner in 1844–1845. Alexander Johnston occasionally served as master of the *Joseph* on coasting voyages. Another pilot, John Fleetwood, was the master of the sloop *Eutaw* in 1841, a vessel that in

[38] US Census Bureau, "US Census of Population, Chatham County, 1840."
[39] Otto, *1850 Census Chatham County.*

later years and under other masters often sailed into Savannah carrying coastal crops.[40]

A few of the men listed in Table 6.3 had once served as coasting captains but seem to have left that occupation by 1850. Among these was Francis Hernandez, who served as a master up to about 1845, but is identified as a "ship carpenter" in 1850. By 1860, Hernandez was working as a Savannah pilot.

The 1850 census reveals that a number of Chatham County residents were involved in ship construction, repair, and outfitting. These included four "sail makers," one "caulker," thirty-two "ship carpenters," and one "boat builder." The large number of ship carpenters reflects the amount of vessel repair and construction in Savannah at the time. Much of this activity was associated with work on those sloops and schooners sailing as coasters.

The 1860 federal census of population for Chatham County enumerates a large number of persons with occupations related to marine and riverine activities. These include occupations such as "boatman," "fisherman," "caulker," "mariner," "flat hand," "pilot," "master shipbuilder," "navigator," "shipsmith," "ship carpenter," "steamship waiter," "sail maker," and even "diver," of which there were twenty-one living in the county in 1860. Many of these occupations can be reliably associated with steam-powered vessels, fishing, stevedoring, etc. Others can be specifically associated with the coasting trade, such as "capt. of coasting vessel," "capt. of schooner," "capt. of river sloop," "mariner on coasting vessel," and, possibly, "mariner sloop."[41] This census identifies seven individuals as captains of coasting vessels, and all are named in newspaper shipping lists as masters of coasters sailing into or out of Savannah. The 1860 census identifies ten men as mariners on coasting vessels whose names do not appear in newspaper shipping lists as masters. These men likely served as crewmembers, meaning this census is one of the few documents providing any information at all on the men who crewed the coasters.

The 1860 Chatham County census identifies 257 "mariners." Many of these men were sailors on ocean-going sailing ships or steamers, some were crewmembers on coasters, and a few were captains of coasting ships. These known captains are Peter Collins, Patrick Doyle, Antonio Lawrence, William Phillips, William Postell, William Watson, and Stephen Williams (see Table 6.3). Some of these men were quite active in the Georgia trade. In the 1850s,

[40] Enrollment No. 6, Port of Savannah, schooner *Joseph*, April 20, 1841; Unnumbered enrollment, Port of Charleston, schooner *Joseph*, January 18, 1844; *Daily Georgian* April 7, 1845.

[41] Georgia Historical Society, *1860 Census*.

Patrick Doyle sailed as master of the sloop *Carrier Dove* and schooners *E. B. Hackburn* and *American Coin*; William Watson was captain of at least five sailing coasters between 1843 and 1860. Stephen Williams was captain of at least four, and Antonio Lawrence of five. Additionally, several of these men held ownership in one or more coasting vessels. It is unknown why they are identified as "mariners" rather than "captains" in this census.

We have only the barest outline of the lives of most of the Georgia coasting captains during the first six decades of the nineteenth century, including those from Savannah and Chatham County. Newspaper shipping lists and vessel documents provide information on their sailing activities, but outside of this, little is known of the lives of most. Even this admittedly scanty information, when combined with other records, such as public documents and private papers, expands our knowledge of the lives of a few of these men. Relying on these types of records, the careers of three Savannah captains are briefly discussed to provide a general idea of their lives and activities. These men are captains David Pidge, Henry James Dickerson, and Domingo Galleo.

Captain David Pidge of Savannah is representative of the Georgia captains in that what is known about his life derives principally from records relating directly to the coasting trade. He appears often in newspaper shipping lists and vessel documents from 1815 to 1828 and was one of the most active Savannah captains in the first three decades of the nineteenth century. The earliest record of David Pidge's residency in Savannah and his involvement in the coasting trade is a June 9, 1815, Savannah enrollment for the Massachusetts-built sloop *Flora*. In this enrollment, Pidge is named the master and sole owner and is identified as a "Savannah mariner."[42]

However, David Pidge was not a native of Georgia. He was born in 1777 in the town of Rehoboth in Bristol County, Massachusetts, and moved to Savannah by 1815 after living for a short while in Rhode Island and South Carolina. Most New England captains operated their ships in Georgia for part of the year, but a few, such as David Pidge, ultimately settled in Georgia, drawn initially to the area through their coasting activities. David Pidge appears in the 1820 Chatham County census as "Captn. Pidge," aged twenty-six to forty-five and engaged in "Commerce."[43]

[42] Enrollment No. 40, Port of Savannah, sloop *Flora*, June 9, 1815.

[43] US Census Bureau, "US Census of Population, Chatham County, 1820"; Georgia Historical Society, *Register of Deaths* 4:12.

Captain Pidge was thirty-eight when he enrolled the sloop *Flora* in Savannah in 1815. This was older than average for Georgia coasting captains, suggesting he was sailing during earlier years, possibly when living in Rhode Island and South Carolina. After 1815, David Pidge served as master and owner of several coasters sailing out of Savannah. These included the schooners *Due Bill* (built originally as a Jeffersonian gunboat), *Patsey*, *Industry*, *Three Sisters*, and *Experiment* and the sloops *Hermit* and *Cynthia*. In the 1827 and 1828 *Daily Georgian*, Pidge's name appears a few times as master of the sloops *Union* and *Mill Maid*; however, it is unknown if he ever held any ownership in these ships. Before 1820, Captain Pidge mostly sailed between Charleston and Savannah, but after 1825, most of his activity was confined to the area along the lower Savannah River, often transporting rice from local rice mills. During his career, David Pidge served as master of at least ten coasting ships and probably held ownership in most of these. In the years just before his death, Captain Pidge was operating at least three coasting vessels simultaneously.[44]

Considering the number of ships Captain Pidge owned, entirely or partially, it appears he was reasonably successful in the trade. He received an additional economic boost when named a winner in the Georgia land lottery of 1820 and received a 202-acre tract of land located in present-day Clayton County, near the city of Atlanta.[45]

On November 10, 1828, David Pidge sailed the sloop *Mill Maid* into Savannah from an unreported location with eighty tierces of clean rice to the factor Robert Habersham. This was the last time Captain Pidge's name appeared in the newspapers and likely represented his final voyage as he succumbed to a "fever" just two weeks later, on November 25, 1828. City death

[44] David Pidge is listed as an owner or master in Enrollment No. 5, "Master Abstracts," Port of St. Marys, sloop *Hermit*, April 7, 1815; Enrollment No. 9, "Master Abstracts," Port of Savannah, sloop *Cynthia*, March 14, 1815; Enrollment No. 16, Port of Savannah, sloop *Cynthia*, June 26, 1816; Enrollment No. [illegible], Port of Savannah, schooner *Due Bill*, March [illegible], 1816; Enrollment No. 2, Port of Savannah, schooner *Patsey*, January 8, 1818; Enrollment No. 23, "Master Abstracts," Port of Savannah, schooner *Patsey*, November 2, 1819; Enrollment No. 13, "Master Abstracts," Port of Savannah, schooner *Patsey*, March 15, 1820; Enrollment No. 2, "Master Abstracts," Port of Savannah schooner *Industry*, January 17, 1820; Enrollment No. 7, "Master Abstracts," Port of Savannah, schooner *Three Sisters*, June 6, 1825; and Enrollment No. 11, "Master Abstracts," Port of Savannah, schooner *Experiment*, April 11, 1827. His name appears as master of the sloops *Mill Maid* and *Union* in the *Daily Georgian* for 1827–1828.

[45] Ancestors Unlimited, "Land District 13."

records report he was a fifty-one-year-old "Seaman" and resided on East Broad Street in the city's Washington Ward.[46]

David Pidge made a will on November 23, 1828, two days before his death. He left his estate to his son Benjamin, including about four hundred dollars in property that formerly belonged to his deceased wife, Eliza. He also had a tract of land in South Carolina known as "Pine Island," household furniture, a hundred dollars in cash, and two enslaved persons, a woman named Dapling and young boy named Ned. Captain Pidge also left his son items related to his life on the water. These consisted of "two flats, also two Boats, also a quantity of old sails and anchors." David Pidge named two men closely associated with the coasting trade as executors of his estate, Henry F. Willink, owner of the largest shipyard in Savannah, and Captain Wright White, a Savannah pilot. Like David Pidge, White was a native of New England (Connecticut) who moved to Savannah as a young man. Records show that other coasting captains often chose men closely associated with the trade to act on their behalf in various legal matters, such as executors or trustees of their estates. For the Southern captains, these actions reflect the close ties among the small group of men who shared this maritime occupation.[47]

Another of the Savannah captains was Henry James Dickerson. Born in Somerset County, Maryland, in 1813, he reportedly arrived in Savannah as a crewman aboard a small sloop from Baltimore in 1832 at age nineteen. Henry Dickerson ultimately achieved greater financial success than any other Georgia captain. He seems to have begun his career as a captain in about 1838 when he appears as master of the sloop *Georgia* transporting cotton and rice from Darien to Savannah. Over the next six years, Dickerson also served as captain of the sloops *America* (1843) and *Mary* (1843) and the schooner *Levant* (1844). By 1845, Dickerson had given up life as a master of vessels and become involved in the ownership of coasters and the operation of a "stevedore" company in Savannah. In his stevedoring activities, Dickerson was engaged in the loading and unloading of ships, including his own, and the movement of cargoes through the docks at Savannah.

Between 1838 and 1861, Dickerson held ownership in at least eighteen different sailing vessels and steamboats working in the local trade. In addition, in 1857, along with Joshua Eddows, he acquired the steam tug *Uncle Sam* that

[46] *Daily Georgian*, November 10, 1828; Georgia Historical Society, *Register of Deaths* 4:224.

[47] Chatham County, Georgia, will of David Pidge, November 11, 1828, Will Book G, 69.

he used in his stevedoring work in Savannah.[48] On September 1, 1859, the *Daily Morning News* announced the launching of Henry Dickerson's steamer, the *Robert Habersham*, one of the few steamboats designed to transport rough rice. Dickerson may have owned other vessels, but incomplete records make this impossible to determine. With these eighteen or so ships, Henry Dickerson was more heavily involved in the ownership of Georgia coasting ships than any other state resident in the two decades preceding the Civil War. During the 1850s, Dickerson simultaneously held ownership in as many as ten or twelve coasting vessels. In this practice of multiple ship ownership, he followed an economic model more common in the North than the South.

Henry J. Dickerson achieved considerable wealth in his stevedoring and shipping businesses. The 1860 census indicates the combined value of his real estate and personal property was $101,000. He died on June 25, 1883, and was buried in an elaborate family vault in Savannah's Laurel Grove Cemetery. At the time of his death, he was one of Savannah's wealthiest residents. In this, Henry Dickerson was certainly atypical of the small company of Southern coasting captains, most of whom ended their lives in modest financial circumstances.[49]

More typical of the Savannah coasting captains was Domingo Galleo (or Gallio), who served as a master of coasting vessels from the early 1840s to 1862 and held ownership in at least two, the sloop *Visitor* and the schooner *Eliza Ann*. As with most of the Georgia captains, little is known about Captain Galleo's life beyond the information contained in newspaper shipping lists, a few vessel documents, and other public records. Despite numerous listings in Savannah newspapers as a master of coasters, no person who can be identified as Domingo Galleo appears in any census record for Chatham County although it is believed he lived in Savannah. On January 28, 1862, the *Savannah Daily News* carried the following account of Galleo's death:

> DROWNED – Saturday night last, as DOMINGO GALLEO, master of the schooner Eliza Ann was going aboard his vessel, lying near the central Cotton Press, he accidentally fell into the Savannah river and was drowned. Deceased was a native of Italy, and for the past thirty years has been engaged in the coasting trade.

[48] *Daily Morning News*, April 13, 1857.

[49] "US Census of Population, Chatham County, 1860"; Chatham County, Georgia, estate of H. J. Dickerson Extrs., 1883–1901, Estate Accounts File 321; Walker, "Henry James Dickerson."

This notice, and other records, reveal that Galleo was an immigrant, a status he shared with a great many of the captains living in Georgia.[50]

The earliest mention of Galleo as a captain appears in 1843 newspaper shipping lists naming him master of the New Jersey-built sloop *Visitor*. In October of that year, Galleo appeared at the Custom House in Savannah and swore that he was master of the *Visitor*, having replaced the former captain named "Joseph Raffale." Joseph "Raffale" was Joseph Raffile, a fellow Italian and Savannah resident who sailed in the Georgia trade from the 1830s to the 1850s. Galleo acquired part ownership of the *Visitor* in 1844 and served as her master until the mid-1850s. In 1855, he purchased a three-quarters share in the Charleston-built schooner *Eliza Ann* and remained master and part owner until his drowning in 1862.[51]

Although Domingo Galleo seems to have sailed as master of only the *Visitor* and the *Eliza Ann*, he did own shares in two other coasters. One was the schooner *Edmund & Francis* that he owned in conjunction with three other Savannah men, who were all coasting captains: Stewart Austin, Henry Dickerson, and Simon Santini. For a short period, Captain Galleo was a part owner of the Baltimore-built schooner *John W. Anderson*, in conjunction with Henry J. Dickerson.[52]

Domingo Galleo left no will. An inventory and appraisement of his estate after his death shows he owned little other than his three-quarters share in the schooner *Eliza Ann*.[53] The inventory listed the following property:

| | |
|---|---|
| 3/4 part of Schooner Eliza Ann | $900.00 |
| 1 Chest and contents | 20.00 |
| 1 trunk & contents | 8.00 |
| Beds & Bedding & two pair Boots | 10.00 |

[50] Filby, *Passenger and Immigration*, 29.

[51] Record of Domingo Galleo as a ship's master is found in an October 28, 1843, endorsement on the reverse of Enrollment No. 3, Port of Savannah, sloop *Visitor*, February 2, 1843; Galleo as owner and master is found in Enrollment No. 9, Port of Savannah, sloop *Visitor*, November 1, 1844; Enrollment No. 10, Port of Savannah, sloop *Visitor*, July 1, 1847; Enrollment No. 6, Port of Savannah, schooner *Eliza Ann*, September 27, 1855. Joseph Raffile appears in Otto, *1850 Census Chatham County*, 45; and Georgia Historical Society, *1860 Census*, 84.

[52] "Conveyances of Enrolled Vessels, Savannah," schooner *Edmund & Francis*, November 26, 1850; "Conveyances of Enrolled Vessels, Savannah," schooner *John W. Anderson*, March 12, 1853.

[53] Chatham County, Georgia, estate of Domingo Galleo, Inventory and Appraisement, Estate Accounts File G-178.

| | |
|---|---:|
| Cash found on Body | 230.00 |
| One Silver watch | 20.00 |
| Money deposited in Planters Bank | 171.00 |
| Money found in Chest | 20.00 |
| One due bill on Wm. Morrill | 100.00 |
| Due of John Monigan of Schr Eliza Ann | 95.00 |
| Due bill James Hunter | 100.00 |
| One Shaving Box & contents | .50 |
| | $1674.50 |

Not mentioned in the inventory is the enslaved woman Hannah, whom Galleo had purchased in 1850 from John Foster for $367.00. Hannah was forty-eight years old in 1850, and if not sold by 1862, she may have been deceased.[54] The lack of residential property and household furnishings in this inventory raises the possibility that Galleo resided in a boardinghouse or aboard his schooner.

Domingo Galleo's debtors cannot all be identified, but the "John Monigan" in the estate inventory may have been a mate or crewmember of the *Eliza Ann* or had performed work on the schooner. Mathias Amorous was named the executor of Galleo's estate. Amorous is listed in the 1850 Chatham County census as a thirty-nine-year-old "boatman" from the island of "Monorca," and he had been associated with Galleo for a number of years.[55] In April 1862, Mathias Amorous petitioned the Chatham County Court of Ordinary to sell Galleo's share in the *Eliza Ann* for the benefit of the estate. That same month, the Confederate government acquired the *Eliza Ann* for twenty-five hundred dollars and sank her as an obstruction on the lower Savannah River.[56]

Domingo Galleo was broadly representative of many of the coasting captains residing in the South from about 1830 to the Civil War. He was an immigrant to America who achieved little financial prosperity. What wealth he did have was principally in his ownership of coasting vessels.

[54] Chatham County, Georgia, deed of sale, John Foster to Domingo Galleo, December 24, 1850, Deed Book 3H, 36.

[55] Georgia Historical Society, *1860 Census*, 262; Otto, *1850 Census Chatham County*, 2.

[56] Chatham County, Georgia, petition to sell interest in the schooner *Eliza Ann*, April 9, 1862, Estate Accounts File G-178; "Vessel Papers," File E-28.

### *The Captains of Glynn County, Brunswick, and St. Simons Island*

Some of the resident Georgia captains lived outside of Savannah, particularly in Glynn County on the central Georgia coast and Camden County on the lower coast. Some of the captains in Glynn County lived on the mainland in or near the port town of Brunswick, others lived on St. Simons Island. Coasting captains dwelling in Glynn County included McGregor Baisden, James Frewin, John D. Denson, Joseph DuBignon, William Gatchett, James A. D. Lawrence, E. B. Learned, John J. Morgan, Richard Owens, Fenn Peck, Michael Peck, Anthony Shaddock, and Charles Stevens. Several of these men lived at the old town of Frederica on St. Simons. These included McGregor Baisden, James Frewin, Fenn and Michael Peck (St. Simons, possibly Frederica), Charles Porquet, Anthony Shaddock (St. Simons, possibly Frederica), and Charles Stevens.

Several of these men's lives have been examined using various extant historical records. One of them, James Frewin, a native of England, settled at Frederica about 1817, having spent more than fifteen years serving in the British Royal Navy. Frewin was born near London in 1780 and entered the Royal Navy when he was sixteen. He sailed aboard several British naval ships until 1815, almost the entire period of the Napoleonic Wars. Eventually rising to the rank of "quarter gunner," he served in the Far East and Indian Ocean and ended his naval career aboard the frigate HMS *Seahorse* in the War of 1812. He lost a leg during his service, possibly at the Battle of Lake Borgne in Louisiana when the *Seahorse* was participating in the failed British attempt to capture New Orleans in December 1814.

Frewin established a modest farm and mercantile store at Frederica before 1820 and served as master and owner of the coasting schooner *Maria*. Captain Frewin had the *Maria* at the docks in Savannah when it sank along with several other ships in the great hurricane of September 1826. Frewin was among a handful of island residents who refused to leave their homes when federal forces occupied St. Simons during the Civil War. At the time of his death in December 1863, Captain Frewin's family was refugeed on the mainland, and his burial at Christ Church near his home was undertaken by sailors from the United States bark *Fernandina*, one of the ships in the federal blockading squadron stationed in St. Simons Sound. One can only surmise that these American seamen knew of James Frewin's long service in the Royal Navy and were providing him an honorable burial as a fellow sailor.[57]

[57] Pearson, "Captain James Frewin," 2001.

Charles Porquet, a native of France and resident of Frederica for a short time, was sailing along the Georgia coast as early as 1812 as master of the sloop *Concord*. Captain Porquet later served as owner and master of the sloop *Maria* and remained in the trade through the 1820s as master of the schooner *Experiment* and sloops *Favorite*, *Mill Maid*, and *Dolphin*. He seems to have resided at Frederica for a short time before 1820. In November 1828, Charles Porquet died in Savannah from "Dropsey." Captain Porquet's fortunes declined near the end of his life. At the time of his death, he was an occupant of the Poor House & Hospital in Savannah.[58]

The 1840 federal census identifies only six persons in Glynn County employed in "Navigation of the ocean" and eight whose occupation was "Navigation of canals, lakes, rivers" (see Table 6.1).[59] The Glynn County census for 1850 classifies two persons as "coasting captains" and three as "seaman" (see Table 6.3). The two coasting captains were Charles Stevens of Frederica and Richard Owens of Brunswick. The three seamen were J. L. Fraser, John D. Denson, and William H. Spaulding. John Denson (or Dennison) resided in Brunswick and was captain of the schooner *Columbia* and sloop *Alamode* in the late 1840s.[60]

Richard Owens, one of the men identified as a "coasting captain" in the 1850 census, was a native of Wales who arrived in Georgia by August 1845, the month his name first appears in Savannah newspaper shipping lists as master of the schooner *Joseph* and the sloop *B. S. Newcomb*. Owens worked as the "hired" master of the *B. S. Newcomb* from 1845 through 1848 when the sloop was owned by prominent Savannah merchant Elias Reed. In 1849, Owens became captain of the new schooner *Elias Reed*, launched in Brunswick that year.

On September 16, 1850, Captain Owens sailed the *Elias Reed* into Savannah from the Turtle River with a cargo of cotton, turpentine, and rosin. This was his last voyage; he died shortly thereafter, and on October 23, 1850, an "Inventory and Appraisement" of his estate was made and recorded in Glynn County Superior Court. Richard Owens's entire estate was valued at $4,322.50, of which 98 percent was incorporated in his ownership of two vessels. One was an unnamed schooner identified as "Vessel on Stocks" that was

[58] *Republican & Savannah Evening Ledger*, November 2, 1812; *Daily Georgian*, 1820, 1821, 1827–1828; *Savannah Republican*, 1825; Enrollment No. 10, Port of Savannah, sloop *Maria*, March 15, 1815, Registers and Certificates of Enrollment, BMIN, NARA; Georgia Historical Society, *Register of Deaths* 4:239.

[59] US Census Bureau, "US Census of Population, Glynn County, 1840."

[60] Otto, *1850 Census Glynn County*, 7.

assigned a value of twenty-five hundred dollars, and the other was his half ownership in the schooner *Elias Reed*, described as "equipped & running" and valued at $1,750. At the time of his death, Owens was in debt to the Brunswick shipwright Charles M. Flanders for one thousand dollars. Presumably, this debt related to the "Vessel on Stocks" that was apparently under construction. Richard Owens's widow Martha sold the two vessels to settle the estate, but they brought less than their appraised values. Gustavas Friedlander, a Glynn County merchant, purchased the half share of the *Elias Reed* for $1,350 while Joseph V. Connerat, a wholesale grocer in Savannah, bought the schooner under construction for $1,324.[61]

Despite his brief career in the coasting trade, just five years, Richard Owens achieved some degree of success, becoming a half owner of the schooner *Elias Reed* and, at the time of his death, having another schooner under construction in Brunswick. Based on the appraisement made of his estate, Owens's investment in the two ships was about four thousand dollars, no mean sum at the time.

In addition to the captains, several other men living in Glynn County in 1850 were associated with coasting vessels. These included two men identified in the 1850 census as "ship carpenters" and one as a "ship wright." Outside of Savannah and Chatham County, Glynn and Camden counties were the only two coastal Georgia counties with resident shipwrights and ship carpenters in 1850. They were few, but their presence indicates that ship repair and construction occurred in these two counties, principally at the port towns of Brunswick and St. Marys.[62]

The 1860 census for Glynn County lists fourteen people with ship-or marine-related occupations. These included the occupations of "ship carpenter," "mariner," "pilot" and "seaman." The census identifies four individuals as "mariners," and two of these, Charles Stevens and John Q. A. Butler, worked in the coasting trade (see Table 6.3). Seven individuals are identified as "ship carpenters" and one as a "ship smith."[63]

Charles Stevens, named as a "coasting captain" in the 1850 census and a "mariner" in the 1860 census for Glynn County, served in the Georgia coasting trade from the 1830s until the Civil War. Stevens was born in Denmark in 1816 and immigrated to the United States at the age of eighteen. On

[61] Glynn County, Georgia, estate of Richard Owens, Wills and Appraisements, Book F, 187, 268–69.

[62] Otto, *1850 Census Glynn County*, 3.

[63] US Census Bureau, "US Census of Population, Glynn County, 1860."

December 7, 1840, he signed an Aliens Declaration in Savannah renouncing his allegiance to "every foreign Prince, Potentate, State or Sovereignty whatever: and particularly to the King of Denmark whose subject I am." He likely worked as a crewman or mate on a coaster between his arrival in Savannah in 1836 and his first appearance as a master in 1841. In October 1843, Charles Stevens married Sarah Dorothy Hay at Frederica on St. Simons Island. Sarah Hay was the niece of the coasting captain James Frewin, and she resided at Frederica with Frewin and his wife.[64]

Charles Stevens settled at Frederica and operated his coasting business there for the next two decades. He also acquired land at Frederica, eventually gaining possession of the old fort, most of the colonial town site, and several hundred acres of surrounding land, including considerable property fronting the Frederica River. Although Stevens seems to have always maintained a primary economic interest in his shipping business, he did develop a small cotton plantation on his St. Simons property and achieved modest wealth in his endeavors.

The earliest mention of Charles Stevens as master of a Georgia coaster is in a December 3, 1841, advertisement in the *Daily Georgian* noting that the sloop *Splendid*, with Stevens as master, would receive freight for "Darien, Brunswick, Turtle River, St. Simons and Jeffersonton," and shippers should "Apply to the captain on board." Charles Stevens would continue in the coasting trade as captain and vessel owner until the Civil War, when he was captured by Union forces and sent as a prisoner of war to Fort Delaware, New Jersey.[65]

In November 1843, Charles Stevens, as sole owner and master, enrolled the sloop *Splendid* at the port of Savannah. Captain Stevens bought the sloop from Henry J. Dickerson, and it appears he worked as master of the *Splendid* in Dickerson's employ for two years before the purchase. Stevens sailed the *Splendid* in the Georgia trade for another eight years. In 1847, he purchased a second ship from Henry Dickerson, the fifty-seven-foot sloop *America*. Captain Stevens's purchase of the *America* and his operation of both sloops for the next few years suggests some level of success in the coasting business.[66]

In May 1855, about three months after selling the sloop *America* for

[64] The biographical information on Charles Stevens presented here is from Pearson, "Georgia Coasting Trade," and Pearson, *Charles Stevens*.

[65] Pearson, *Charles Stevens*.

[66] Enrollment No. 14, Port of Savannah, sloop *Splendid*, November 23, 1843; Enrollment No. 18, Port of Savannah, sloop *America*, May 1, 1847; "Conveyances of Enrolled Vessels, Savannah," sloop *America*, February 9, 1855.

sixteen hundred dollars, Charles Stevens purchased the schooner *Northern Belle* for three thousand dollars from Captain Charles Thompson. He enrolled the ship at Savannah as sole owner and master. Unlike most Georgia coasters, plain vessels with little ornamentation, the sixty-two-foot *Northern Belle* was fitted with a "woman's head" at her bow. With a burden of almost seventy-eight tons, she had a considerably greater cargo capacity than the two sloops Stevens previously owned. The *Northern Belle* was nine years old and possibly in some disrepair when purchased because, soon after his purchase, Captain Stevens bought new sails, repaired the schooner's "water ways," and had her "thoroughly and solidly overhauled."[67]

On May 9, 1855, the day after Charles Stevens purchased the *Northern Belle*, the Savannah *Daily Morning News* ran the following notice: "Capt. Charles Stevens, having purchased the schooner NORTHERN BELLE, will continue to run her regularly between this Port and Darien, St. Simons, Turtle River, Big and Little Satilla Rivers."

Charles Stevens sailed the *Northern Belle* until the Civil War, serving locations along the central and lower Georgia coast. With the Union blockade along the coast during the first year of the war, Stevens and his family evacuated St. Simons and settled on the mainland. He and other coasting captains attempted to keep their ships working, but by winter 1861, this task was becoming increasingly dangerous. Captain Stevens's last voyage into Savannah occurred in late October 1861 when he sailed the *Northern Belle* from the Altamaha River with thirty-five hundred bushels of rough rice consigned to Robert Habersham & Sons. By this time, Stevens had changed the name of his schooner to *Southern Belle*. Unable to keep his ship working, he moved her up the Altamaha River above Darien to hide her from federal blockading forces. He was unsuccessful, and in June 1862, Union forces captured the *Southern Belle* and took her to St. Simons Island. Through mishandling, the schooner sank in the Frederica River shortly after her arrival at St. Simons.[68]

On April 5, 1860, Charles Stevens acquired half ownership of the eighty-two-ton schooner *Florida* from her owner, John Q. A. Butler, another Glynn County resident. The document of sale notes that Stevens paid only two hundred dollars for his half share.[69] Stevens apparently never sailed the *Florida*

[67] "Conveyance Savannah," schooner *Northern Belle*, May 8, 1855; Enrollment No. 4, Port of Savannah, schooner *Northern Belle*, May 8, 1855; Pearson, *Charles Stevens*, 42, 44.

[68] *Daily Morning News*, October 28, 1861; Pearson, *Charles Stevens*, 55–56.

[69] Glynn County, Georgia, deed of sale, John Q. A. Butler to Charles Stevens, April 5, 1860, Deed Book N, 177–78.

into Savannah, and there is no record of the vessel working in the local trade after 1857. Either the *Florida* was involved in some other trade after 1857, or she was in poor condition and not sailing at all, a possible reason for the low price of two hundred dollars Stevens paid in 1860. Charles Stevens and John Butler kept the *Florida* for just two months, selling her in June 1860 to three residents of Jacksonville, Florida.[70]

At the outbreak of the Civil War, Stevens was forty-five years old and did not enter Confederate service early in the war. However, by 1863, he was a member of the Glynn County Militia Company, and in 1864 he became a member of Company I, 29th Battalion, Georgia Cavalry. On December 22, 1864, Charles Stevens and several others serving picket duty on the Altamaha River were captured by federal troops from the blockade vessel USS *Ethan Allen*. These prisoners were carried north to Philadelphia, and some of them, including Charles Stevens, were transferred to the prison at Fort Delaware in the Delaware River. On February 1, 1865, only three weeks after his incarceration at Fort Delaware, Charles Stevens died in the prison hospital. Two notations on his record indicate that he died of "Laryngitis" and "Inflammation of Larynx," no doubt an infection brought on by exposure to the cold and dampness of the island prison.[71]

Charles Stevens's family returned to St. Simons after the war ended. They received word that he had died in prison but were provided no other information. In 1873, Charles Stevens's widow, Sarah, filed a federal government claim for four thousand dollars for the loss of the schooner *Northern Belle*. In that appeal, she stressed that she and her husband had opposed secession and the dissolution of the Union. The federal government rejected her claim.[72]

Charles Stevens's life came to a tragic end, but his career in the Georgia coasting trade had been quite successful. Within three years of his first appearance as a master of a coasting ship, he purchased the vessel he commanded, the sloop *Splendid*. In 1847, he bought his second ship, the *America*. Over the next several years, Stevens served as master of the *America* and employed others to command the *Splendid*. In addition to his coasting business, he acquired land on St. Simons Island and established a modest farm. By

[70] Register No. 1, Port of Savannah, schooner *Florida*, April 5, 1860; Register No. 15, Port of Savannah, schooner *Florida*, June 27, 1860.

[71] Pearson, *Charles Stevens*, 54–62.

[72] Ibid., 62–63; "Petition by Mrs. Sarah D. Stevens," RG 233, Records of the U.S. House of Representatives, Southern Claims Commission, NARA.

1860, he owned 624 acres of land worth $2,500, and his property was valued at $11,250. Much of his wealth was contained in the sixteen enslaved laborers he owned, some of whom farmed his two hundred acres of "improved" land while others worked aboard his schooner *Northern Belle*. Few of the other Southern coasting captains achieved the economic success of Charles Stevens, certainly a measure of his personal abilities and resolve.[73]

*The Captains of Camden County and St. Marys*

Several coasting captains and shipowners resided in Camden County and its principal town, St. Marys. Camden County was second only to Savannah in the number of resident coasting captains. This stemmed in part from St. Marys's importance as an entrepôt to Spanish Florida before 1820. The town served as a convenient port of call for vessels trading with Florida and was a center for smuggling and the illicit movement of goods into and out of Georgia.

Among the Camden County citizens involved in the coasting trade in the early years of the nineteenth century were members of the Chevalier and Hubbard families. Those in the Hubbard family included Elihu Hubbard and his sons, Malatiah Haley and John. The father, Elihu, was born in 1759 in Woodstock, Connecticut, and moved to Camden County by 1792, the year he acquired a lot in the town of St. Marys. Elihu Hubbard made his living in the coastwise trade as master and owner of the sloop *Two Sisters* in the years before 1800. When Elihu died in 1812, he bequeathed to his two sons his share of the sloop "'Two Sisters' to be held by his friend, Frances Chevellier."[74]

Elihu Hubbard's sons, Malatiah (or Meletiah) and John, participated in the coasting trade in the 1820s and 1830s. Malatiah served as captain of the schooner *Orange* and the St. Marys-built schooner *Betsey Maria*, both of which were regular Savannah-St. Marys traders. It appears that Malatiah Hubbard ended his career on coasters by 1840, and by 1850, he had moved his family to Newton County, Georgia, where served as a Methodist minister. He died in Morgan County, Georgia, in March 1861.[75]

[73] US Census Bureau, "US Census of Population, Glynn County, 1860," and "US Census Slave Schedule, Glynn County, 1860."

[74] Huxford, *Pioneers*, 201–202; Camden County, Georgia, deed of sale Thomas Steel to Elihu Hubbard, September 24, 1792, Camden County Deed Book A, 81–82; Camden County, Georgia, will of Elihu Hubbard, Book A, May 21, 1806; sailings of the sloop *Two Sisters* are found in the *Republican & Savannah Evening Ledger*, 1807–1808.

[75] US Census Bureau, "US Censuses of Population, Camden County, 1820, 1830 and 1840"; Helene, "Helene Family Tree"; *Daily Georgian*, 1827–1828; Enrollment No.

Malatiah brother, John, was born in St. Marys in 1803 and resided in the town until his death in May 1850. An 1820 St. Marys's enrollment document names John Hubbard as master of the forty-one-foot sloop *Mathews* even though he was only seventeen or eighteen. He remained captain of the *Mathews* until at least 1827. Subsequently, Captain Hubbard served on steamboats operating along the Georgia coast.[76]

Another of the Camden County captains was John Chevalier of St. Marys, who served as master and owner of several coasting ships from 1799 to 1830. John Chevalier was a native of France, born sometime between 1760 and 1770. He reportedly came to the United States in 1794, possibly to St. Marys, where he acquired land by 1797.[77] John Chevalier may have started working on a coaster soon after his arrival in the United States in 1794. He was certainly involved in the coasting trade by 1799 when he was named captain of the sloop *Betsey*, sailing between St. Marys and Savannah. The following year, he was one of two captains of the schooner *Alective*, sailing in the St. Marys-Savannah trade. The other captain was Robert Rudolph, another Camden County mariner and Chevalier's close associate. For several years, captains Chevalier and Rudolph dominated the sea-borne trade between St. Marys and Savannah with the *Alective*.[78]

By 1805, Chevalier was captain of the sloop *Polly*, probably named after his wife, Polly Ann Sleigh. On May 28, 1805, the *Georgia Republican & State Intelligencer* published the following advertisement:

7000 HARD BRICKS — For Sale —
on board the sloop POLLY, John Chevalier
master, from New York.
Sloop will sail for St. Marys on Thursday

10, Port of St. Marys, schooner *Betsey Maria*, March 17, 1829; Enrollment No. 13, Port of St. Marys, schooner *Betsey Maria*, October 16, 1829; *Daily Savannah Republican*, 1830.

[76] Hibbard, *Genealogy*; Enrollment No. 6, Port of St. Marys, sloop *Mathews*, June 13, 1820; Enrollment No. 14, Port of Savannah, sloop *Mathews*, June 7, 1827; *Daily Georgian*, 1820, 1827–1828; *Savannah Republican*, 1825; *Daily Savannah Republican*, 1830.

[77] S. J. Thompson, *People of Camden*, 39; US Census Bureau, "US Census of Population, Camden County, 1830"; Camden County, Georgia, deed of sale, James Macomb to John Savalier [Chevalier], April 15, 1797, Deed Book B, 373.

[78] The sailing of the *Betsey* is reported in the *Columbian Museum & Savannah Advertiser*, January 29, 1799; sailings of the schooner *Alective* are found in the *Georgia Gazette*, 1799–1800 and *Georgia Republican & State Intelligencer*, 1805.

This advertisement shows that Captain Chevalier had sailed northward, returning to Savannah with merchandise for sale. This practice of sailing North in the spring or summer, after the bulk of the cotton and rice crops were carried to market, was undertaken by other Georgia coasting captains.

After 1805, Chevalier sailed as master of several other Georgia coasters; including the *Julia Ann*, variously identified as a schooner and a "galliott," and the sloops *John Sleigh*, *Dosoris*, and *John Chevalier*, launched at St. Marys in 1817. On May 16, 1821, the following advertisement appeared in the *Daily Georgian*:

For Sale
The beautiful and staunch built sloop
JOHN CHEVALIER,
Built of the best materials, and well cal-
culated for a Packet in almost any
Trade

It is unknown if the sale was successful, but John Chevalier continued as the sloop's captain into 1825. In 1827, he acquired the new fifty-two-foot sloop *Leader*, built that year in Bridgeport, Connecticut. With the *Leader*, Chevalier continued his active role in the coasting trade, principally sailing between St. Marys and Savannah and to lower coast locations such as the Satilla River, the town of Jeffersonton, and Jekyll Island.[79]

In addition to his coasting activities, Captain Chevalier was involved in agriculture. On October 25, 1830, the Savannah mercantile company Palmes & Lee advertised in the *Daily Savannah Republican* that they had for sale "5000 [oranges] Fresh from the Grove of Capt. Chevalier, carefully picked and in fine order for transportation."

John Chevalier gave up the coasting trade around 1831. By then, he had worked at sea for more than thirty years and was somewhere between sixty and seventy years old—too old to undertake the strenuous work associated with operating a coasting trader. In 1832, Captain Chevalier conveyed a considerable amount of property to Louis Dufour, who was acting as trustee for

[79] Information on vessels that John Chevalier served as master on, or owner of, is found in the following vessel documents: Enrollment No. 3, "Master Abstracts," Port of St. Marys, sloop *John Sleigh*, April 3, 1815; Enrollment No. 8, "Master Abstracts," Port of St. Marys, sloop *Dosoris*, May 12, 1815; Enrollment No. 9, Port of Savannah, sloop *John Chevalier*, July 19, 1825; Register No. 248, Port of New York, sloop *Leader*, August 3, 1827; and Enrollment No. 3, Port of St. Marys, sloop *Leader*, August 27, 1828.

a woman named Elizabeth Brewer and her four children. Louis Dufour, like John Chevalier, was a native of France who immigrated to Florida and later moved to St. Marys where he was a merchant. In Chevalier's 1832 deed of trust, Elizabeth Brewer is identified as a "colored woman," and other records identify her as a mulatto and "free person of color." John Chevalier and Elizabeth Brewer lived together and had four children, all born as free persons of color. Captain Chevalier's close relationship with Elizabeth Brewer was highly unusual for the period. The property given to Elizabeth included a lot in St. Marys and several small marsh islands near Jekyll Island, known as Latham's Hammocks, which Chevalier acquired in 1829. In addition, Elizabeth Brewer received their household goods, plus items of personal property, including "all his silver," the latter described as "already in possession of said Elizabeth." John Chevalier died shortly after executing this trust deed.[80]

At least one other member of the Chevalier family, Francis, became a coasting captain. It is unknown if Francis was John Chevalier's son or a nephew. Francis Chevalier was born in Camden County in 1810 and died there in 1844. By 1835, Francis was sailing as master of the sloop *Virginia*, transporting goods between Camden County and Savannah. The sixty-foot *Virginia* sailed in the Georgia trade until she was scuttled as an obstruction in the Savannah River during the Civil War.[81]

The 1840 census for Camden County lists six persons employed in "Navigation of the ocean" (see Table 6.1). These were Robert Day, John Stotesbury, and four unnamed individuals living in the household of Samuel Flood. No other record of Robert Day working in the coasting trade has been found, but John Stotesbury worked as a captain of coasters along the Georgia coast from 1813 to the mid-1830s. During his career, Stotesbury sailed as master of several ships, including the sloop *Good Intent* and the schooners *Trimmer* and *Collector*. He was an owner of the *Good Intent* and the sloop *Bolivar*. Stotesbury worked for the Customs Office in St. Marys in the 1820s when he is identified as the "Surveyor" on enrollment documents issued there. In 1822, Stotesbury purchased and enslaved Sarah, Charles, Samuel, Mary, and Amelia in a judgment against the estate of David Lewis. The two men who witnessed the sale document, Jeramiah Bowen and Malatiah H. Hubbard, were

[80] Camden County, Georgia, deed of trust, John Chevalier to Louis Dufour, July 14, 1832, Deed Book M, 70; Huxford, *Pioneers*, 132; US Census Bureau, "US Census of Population, Camden County, 1840."

[81] Fields, "The Crypt"; *Daily Savannah Republican*, 1835; Enrollment No. 2, Port of St. Marys, sloop *Virginia*, September 25, 1840; "Vessel Papers," File V-13.

both coasting captains. This is another expression of the intertwined relationships that existed among the small group of captains sailing in the Georgia trade. John Stotesbury last sailed around 1834 and died at St. Marys in 1847.[82]

The 1850 census for Camden County lists two men with the occupation of "capt. sloop" (see Table 6.3). Both men, William Laen (sometimes Lane) and Dennis Pacetty, appear in other records as coasting captains. The census also listed one "pilot," Sylvanus Church; one "seaman," seventy-two-year-old Edmund Richardson; and one "mariner," Peter Carb (Corb), who were all masters or owners of coasting ships. The occupation of three Camden County residents in 1850 was "ship carpenter," reflecting the presence of ship repair and construction facilities.[83]

By 1860, the Camden County census identifies a single resident as a coasting captain. This was John Louis Grovenstein, whose occupation is given as "sea captain" in the census. John Grovenstein was a native of St. Marys; he sailed in the Georgia trade for at least thirty years, from 1831 to 1861. Grovenstein was born in Camden County on September 25, 1811 and died in St. Marys in 1882. Before 1830, he lived in St. Marys, but subsequently moved his residence several times along the Georgia coast, apparently related to his trading activities. In the mid-1830s, he was living near Riceboro and by 1850 was a resident of Savannah. The federal census for that year gives Grovenstein's occupation as "coaster" (see Table 6.3). By 1860, he was living in St. Marys. In 1880, two years before his death in St. Marys, the Camden County census lists sixty-nine-year-old John Grovenstein as a "steamboat captain."[84]

John Grovenstein began his career as a coasting captain in 1831, the year he appears as master of the *Flora* when that schooner was enrolled at Savannah. The *Flora* was a small, thirty-two-ton vessel built at Sunbury, Georgia, in 1821. Grovenstein was twenty years old in 1831, and the *Flora* probably represented his first command. By 1835, newspapers name him as captain of the sloop *Ann* and later of the sloops *Jackson* and *Macon*. With these vessels, Captain Grovenstein was involved mainly in transporting cargoes between

[82] Enrollment No. [illegible], Port of Savannah, sloop *Good Intent*, [illegible], 1816; Enrollment No. 1, Port of St. Marys, sloop *Bolivar*, June 6, 1834; Fields, "Slave Deed Abstracts," 65; FamilySearch, "John Stotesbury."

[83] Otto, *1850 Census Camden County*.

[84] Fields, "The Crypt," burial record for John L. Grovenstein; US Census Bureau, "US Census of Population, McIntosh County, 1840," and "US Censuses of Population, Camden County, 1860 and 1880"; Otto, *1850 Census Chatham County*, 40.

Savannah, Riceboro, and other nearby locations. According to the 1840 census, he and his family were living in McIntosh County near Riceboro.[85]

In 1840, Grovenstein served briefly as master of the sloop *George Washington* before becoming master and part owner of the sloop *Eagle* in 1841. The other owners of the *Eagle* were Leonard and Charlton Bolles, residents of Rochester, Massachusetts, where the sixty-foot sloop was built in 1827. The Bolles brothers were prominent Rochester shipowners, and Leonard was one of the Sippican captains sailing in the Georgia trade for many years. In 1843, the Bolles brothers sold their share in the *Eagle* to Savannah merchant Henry Brigham, leaving him and John Grovenstein as owners. Between 1841 and 1850, Grovenstein, as master of the *Eagle*, sailed into and out of Riceboro and other areas on the central Georgia coast, such as the Ogeechee River and Darien.[86]

In 1843, Captain Grovenstein purchased a half share of the eighteen-year-old schooner *Fort George Packet* from Daniel Reddick. Grovenstein held complete or partial ownership in the *Fort George Packet* until 1854, when she was sold to two coasting captains, John Russell and Thomas Room.[87]

On December 23, 1848, John Grovenstein enrolled the Macon-built sloop *B. S. Newcomb* in Savannah as her sole owner. He owned the sloop for three years, selling her in 1851 to two Savannah residents, Stephen Williams, another coasting captain, and Albert L. DeLorge, a grocer. After 1841, John Grovenstein served as master of his sloop *Eagle* while others sailed his two other ships, the *B. S. Newcomb* and *Fort George Packet*, apparently under hire. In 1851, Grovenstein purchased a partial interest in the schooner *Company* and served as her master until 1855, when he and Henry Brigham, the other owner, sold her to Henry J. Dickerson for twelve hundred dollars.[88]

[85] Enrollment No. 3, Port of Sunbury, schooner *Flora*, November 8, 1822; US Census Bureau, "US Census of Population, McIntosh County, 1840."

[86] Enrollment No. 24, Port of Savannah, sloop *Eagle*, November 24, 1841; Enrollment No. 16, Port of Savannah, sloop *Eagle*, December 16, 1843; *Daily Georgian*, 1841–1850.

[87] Enrollment No. 17, Port of Savannah, schooner *Fort George Packet*, December 28, 1843; Enrollment No. 1, Port of Savannah, schooner *Fort George Packet*, July 6, 1854.

[88] Enrollment No. 24, Port of Savannah, sloop *B. S. Newcomb*, December 23, 1848; "Conveyances of Enrolled Vessels, Savannah," sloop *B. S. Newcomb*, July 3, 1851; "Conveyances of Enrolled Vessels, Savannah," schooner *Company*, January 22, 1851; Enrollment No. 3, Port of Savannah, schooner *Company*, Port of Savannah, March 22, 1851; "Conveyances of Enrolled Vessels, Savannah," schooner *Company*, July 3, 1855; Otto, *1850 Census Chatham County*, 10; Georgia Historical Society, *1860 Census*, 36.

For almost twenty years, from the late 1830s to the late 1850s, six ships, the *George Washington*, *Macon*, *Eagle*, *B. S. Newcomb*, *Fort George Packet*, and *Company* handled a major portion of the sailing trade between Savannah and Riceboro. John Grovenstein was the sole or partial owner of each of these vessels.

In 1857, Grovenstein became master and owner of a half-share of the schooner *Elias Reed*. The other owner of the schooner was Hugh Fraser Grant of Glynn County. Records of Grant's Elizafield Plantation show the *Elias Reed* commonly carried Grant's rice crop to market in Savannah and transported supplies back to the plantation. By 1858, Grovenstein had moved back to St. Marys. He continued to sail as master of the *Elias Reed* until the Civil War.[89]

The federal blockade along the Southeastern coast during the early years of the Civil War ended Grovenstein's trading activities, but the *Elias Reed* continued to sail during the war. In April 1862, the schooner transported the "St. Marys Volunteers" from St. Marys to Brunswick, and the following year, she attempted, but failed, to run the blockade. On November 6, 1863, the USS *Octorara* captured the schooner in the Bahamas on her way from St. Marys to Nassau. Aboard were eight bales of Sea Island cotton, seventy-six barrels of turpentine, and 365 barrels of resin. John Grovenstein was not aboard the *Elias Reed* when she was captured.[90]

Over his career, John Grovenstein acquired ownership in at least seven sloops and schooners, a measure of some level of prosperity in the trade. Despite this success, the few records available indicate Grovenstein's financial worth was modest, at least after 1850. He was living in Savannah that year, and census records show he owned no real estate. The slave schedule reveals he did own a single enslaved person, a nine-year-old boy. A year later, John Grovenstein owned an enslaved man named Robert, "aged about thirty seven years." Robert's age suggests he may have worked aboard Captain Grovenstein's ships. The 1850 census does not list the value of Grovenstein's personal property, which would have included his ownership in several coasting vessels. In 1860, now a resident of St. Marys, Grovenstein's personal property was valued at $1,130 and his real estate at $1,500. His part ownership in the

[89] Enrollment No. 5, Port of Savannah, schooner *Elias Reed*, June 17, 1857; Enrollment 2, Port of St. Marys, schooner *Elias Reed*, June 17, 1858; House, *Planter Management*.

[90] "Vessel Papers," File E-159; *War of the Rebellion: Official Records of the Union and Confederate Navies*, ser. 1, 1:533–34, hereinafter cited *ORN*.

*Elias Reed* likely accounted for most of his personal property evaluation. John Grovenstein was one a handful of Georgia captains who continued working in the coasting trade after the Civil War although aboard steamboats. In 1870, the value of his real estate was one hundred dollars, and the value of his personal property was the same. Despite a thirty-year career as a ship captain and ownership in at least seven coasting ships, John Grovenstein had little money to show for it at the end of his life.[91]

*Other Georgia Captains*

Only a small number of coasting captains or shipowners resided outside of Savannah, St. Marys, or Glynn County. One of these was Claude Phillip Leset, who lived in Liberty County in the 1830s. Variously identified as "Lucette," "Lisett," "Lissette," or "Lisset," in addition to Leset, he was a native of France who arrived in the United States at Providence, Rhode Island, in 1800. By 1820, Leset had moved to Savannah, and by 1825 he was master of the sloop *Albert*, sailing in the Georgia trade. Captain Leset was forty-four years old in 1825, older than usual for men beginning their careers as coasting captains. He first appears in the 1830 census for Liberty County as "C. P. Lucette," residing with his wife, Elizabeth, and four children.[92] As captain of the *Albert*, Claude Leset was more active than most of the captains, at least in terms of voyages into Savannah. In 1830, the *Daily Savannah Republican* reported that Leset made fifteen arrivals into Savannah with the *Albert*, more than any other local coasting captain. On February 15, 1830, the *Albert* sailed into Savannah from Riceboro with thirty-one bales of cotton and "57 slaves for Beaufort, SC." The newspaper noted this normally short voyage from Riceboro took three days, which must have been an uncomfortable journey for the Black captives, crowded together on the fifty-two-foot sloop for so long in the February cold. Captain Leset apparently delivered the human cargo to Beaufort, accounting for the *Albert*'s arrival in Savannah from there just five days later.

Claude Leset remained master of the *Albert* until 1831, when he became captain of the fifty-seven-foot *Jackson*, a sloop owned by Ralph and William King, proprietors of R & W King, one of the prominent Savannah

[91] Otto, *1850 Census Chatham County*, 40; US Census Bureau, "US Censuses of Population, Camden County, 1860 and 1870," and "US Census Slave Schedule, Camden County, 1860"; Chatham County, Georgia, deed of sale, Robert, John L. Grovenstein to Scranton, Johnston & Company, February 25, 1851, Deed Book 3H, 172.

[92] Hemperley, "Federal Naturalization Oaths," 472; US Census Bureau, "US Census of Population, Liberty County, 1830."

commission merchant firms. By 1835, Leset was captain of the sixty-foot sloop *Mary Cumming*. The *Mary Cumming*, like Leset's previous ship, the *Jackson*, was owned by a group of Savannah merchants. No records show that Claude Leset ever held ownership in a coaster, suggesting he sailed as a hired captain. With the *Mary Cumming*, Leset was an extremely active trader. In 1835, the *Daily Georgian* reported he sailed the *sloop* into Savannah fourteen times, more entries than any other coasting captain that year. With this sloop, Captain Leset frequently carried cargoes from the Ogeechee River area into Savannah, a shift from his earlier trading patterns with the *Albert*, when he more commonly sailed from Riceboro, where he lived. Claude Leset sailed the *Mary Cumming* until May 1836, when he appears as master of the sixty-two-foot sloop *Science*. The *Science* was Captain Leset's last command. He seems to have ended his coasting career about April 6, 1837. On that day, an endorsement was made on the back of the May 1836 enrollment for the *Science*, noting that Peter Worthington "having taken the oath required by law is at present Master of the within named vessel in lieu of C. P. Lisset late master." Claude Leset may have withdrawn from the coasting trade due to ill health because five months later, on September 17, 1837, he died in Savannah at the age of forty-eight.[93]

The 1840 federal census identifies no persons employed in "Navigation of the ocean" living in Bryan or Liberty counties, but it does list a John Buckley in Liberty County employed in the "Navigation of canals, lakes, rivers" (see Table 6.1). Buckley served as a captain of Georgia coasters for about ten years, from the mid-1830s through the mid-1840s. The ships he sailed included the sloops *Angelica*, *Two Friends*, and *Liberty*. The 1840 census identifies eleven persons in McIntosh County as "navigators of the ocean." None of these men can be equated with any known coasting captain (see Table 6.1). In 1840, ninety-seven men in McIntosh County were employed in the "Navigation of canals, lakes, rivers" (see Table 6.1). None can be equated with a known coasting captain, and, while some may have been crewmen aboard coasters, most probably worked on the steamers, flatboats, and poleboats, or lumber rafts then traveling in large numbers on the Altamaha River.[94]

In the 1850 census for Bryan County, two individuals are identified as "seaman"; neither can be identified as a coasting captain. Similarly, only two

[93] Enrollment No. 12, Port of Savannah, sloop *Jackson*, November 18, 1831; Enrollment No. 8, Port of Savannah, sloop *Science*, May 2, 1836; Spears, "Claudius Leset."

[94] US Census Bureau, "US Census of Population, Liberty, Bryan and McIntosh counties, 1840."

men in the 1850 Liberty County census were assigned maritime occupations; one identified as a "sailor" and one as a "sea farer." The "sea farer" was Thomas Snow, who commanded several coasters during the 1840s and 1850s, including the sloop *Splendid* and the schooners *Orange*, *Sarah*, and *Argo* (see Table 6.3). Like many of the Georgia captains, Snow was an immigrant, born in 1814 or 1815 on the island of St. Bartholomew, then a Swedish colony. He came to New York in 1830 and arrived in Savannah in 1835, possibly already working aboard a coaster. He took an oath of naturalization at Savannah in August 1840, making him legally eligible to own and command a ship in the coasting trade. Thomas Snow sailed as master of the forty-six-foot schooner *Sarah* from 1842 to the early 1850s, becoming a part owner by 1846. With the *Sarah*, Captain Snow was the principal trader with the town of Sunbury for over a decade.[95]

In McIntosh County in 1850, the only maritime occupation used by the census taker was "pilot." The five individuals assigned this occupation worked as pilots for vessels sailing into and out of Darien. At least one of these "pilots," George Russell, is named in Savannah newspapers as master of the coasting sloops *Georgia* and *Levant* in the 1840s and early 1850s.[96]

Only a single person listed in the 1860 census of McIntosh County had an occupation related to coasting and this was Stephen Blankenship, identified as a "mariner." Stephen Blankenship cannot be equated with any coasting captain or shipowner.[97]

## *The Florida Captains*

Three counties on the northeast coast of Florida, Nassau, Duval, and St. Johns, fall within the normal sphere of trade of the coasters sailing into and out of Savannah. The 1840 federal census identifies six men involved in "Navigation of the ocean" in Duval County and six in St. Johns County.

One of the persons named in the 1840 census of Duval County is Timothy Wightman, who was captain of the schooners *Decatur* and *Koret* in the late 1820s. Captain Wightman was a native of New London, Connecticut, and was master of the coasting sloop *Cherub* out of New London in 1822. By 1830, he was employed as a pilot, living in Duval County at the port town of

[95] Otto, *1850 Census Bryan County*, 5, 6; Otto *1850 Census Liberty County*, 2, 17; Hemperley "Federal Naturalization Oaths," 483; Enrollment No. [illegible], Port of Savannah, schooner *Sarah*, March 2, 1846.

[96] Otto, *1850 Census McIntosh County*.

[97] US Census Bureau, "US Census of Population, McIntosh County, Georgia, 1860."

Fernandina on Amelia Island. Captain Wightman made several voyages into Savannah and Charleston with the *Koret* and the *Decatur* in 1827 and 1828. His voyages originated at Florida locations, from the St. Johns River to as far south as Key West. The two schooners sailed by Wightman may have been pilot boats that he occasionally used to transport coastal cargoes as an adjunct to his principal occupation.[98]

The 1850 federal census for the Northeast Florida counties lists several individuals involved in maritime trades of one sort or another. The 1850 Nassau County census names three persons in maritime occupations, all identified as "pilots," but none can be correlated with captains of Georgia coasters named in Savannah newspapers. Duval County had eighteen individuals in marine-related occupations that included categories such as "navigator," "sailor," "pilot," "ship builder," and "ship carpenter." None of the men assigned these occupations can be identified as coasting captains.[99]

The 1860 censuses for the northeastern Florida counties also list a number of men with marine-related occupations. Many were pilots, apparently serving on the St. Johns River and the entrance to St. Augustine. Only one person in this census can be identified as a captain sailing in the Georgia trade. This is Dennis Pacetty, listed as a fifty-one-year-old "sea captain" residing at Fort Clinch in Nassau County. Pacetty, born in St. Augustine in 1808, lived in St. Marys, Georgia, from before 1830 through the 1850s and captained several ships in the Georgia trade. These included the sloops *Bolivar* and *Independence* and the schooners *Betsey Maria* and *Orange*. Captain Pacetty lived in Florida only a short time, residing for most of his life in Camden County, where he died around 1880.[100]

### *The South Carolina Captains*

A few captains in the Georgia trade sailed almost exclusively between Savannah and the town of Beaufort or other locations on the lower South Carolina coast. Several of these men were residents of Beaufort or the surrounding area; others lived in Savannah but had trading ties with planters or merchants in South Carolina. Among the men sailing in the Beaufort-

[98] US Census Bureau, "US Census of Population, Duval County, Florida, 1840; P. Thompson, "Timothy Wightman III"; "Connecticut Ship Database," sloop *Cherub*.

[99] US Census Bureau, "US Censuses of Population, Nassau and Duval counties, Florida, 1850."

[100] US Census Bureau, "US Census of Population, Nassau County, Florida, 1860," and "US Censuses of Population, Camden County, 1830, 1840, 1870 and 1880"; Otto, *1850 Census Camden County*, 16; Mathews, "Pacetty and Bonelly Families."

Savannah trade in the early nineteenth century were Daniel Blythwood (schooners *Delaney* and *Mary & Eliza*, 1809–1813), Isaac Pierce (sloop *Mathews*, schooners *Flora* and *Isabella*, 1816–1830), Thomas McMillan (schooner *Isabella*, 1828) and John W. Turner (sloops *Lawrence, Cynthia*, and *Swallow*, and schooner *Idonius*, 1820-1835, 1820–1835).

The 1840 federal census identifies six individuals engaged in the "Navigation of the ocean," living in Beaufort County, South Carolina, while none appear in adjacent Colleton County. One of the "navigators" in Beaufort County was Thomas McMillan, who served as master of the schooner *Isabella* in the late 1820s. This schooner was built in 1815, apparently in Beaufort, and for fifteen years was regularly involved in the Beaufort-Savannah trade under several different masters. Farther north, in Charleston County, which includes the city of Charleston, three hundred men listed in the 1840 census were engaged in the "Navigation of oceans," greater than the number found in Chatham County and Savannah. It is unknown how many of these men worked on coasters; a large number were seamen serving aboard ocean-going ships or coastwise sailing and steampackets, not local coasting sloops and schooners.[101]

Surprisingly, the 1850 federal censuses for Beaufort and Colleton counties list no one who can be identified as a captain working in the Georgia trade. The same census for Charleston identifies many people associated with maritime trades. Those who might have worked in the coasting trade were assigned occupations of "mariner" or "master mariner," but few of these men were sailing coasters into Savannah or other Georgia locations. Among the few who did were William H. Bee, identified as a fifty-year-old "mariner," who occasionally sailed the schooners *Eliza Ann* and *Experiment* into Savannah, and twenty-eight-year-old "mariner" David S. Little, who served as captain of the sloop *Swallow* and schooner *Ellen* along the Georgia coast in the 1850s. By 1850, most of the captains residing in Charleston seem to have confined their activities to trade between that city and the adjacent coasts of South and North Carolina; few sailed to Georgia. The 1850 census for Charleston shows that a large proportion of those men assigned a maritime occupation were originally from Northern states or immigrants from foreign

[101] US Census Bureau, "US Censuses of Population, Charleston, Colleton and Beaufort counties, South Carolina, 1840"; Enrollment No. 3, "Master Abstracts," Port of Beaufort, schooner *Isabella*, November 20, 1815.

countries, just as it does for the city of Savannah.[102]

The 1860 federal census for Beaufort County lists one man, John Murry, as a "ship captain" who sailed in the Georgia trade. John Murry, a native of England, appears to be the "Captain Murry" (or "Murray") who regularly sailed between Beaufort and Savannah in the 1830s and 1840s as master of several coasters, including the schooners *Isabel* and *Science*. The 1860 census for Colleton County, South Carolina, also lists a single individual who was a coasting captain. This is Charles E. Anderson, a thirty-year-old "sailor" and native of Sweden who immigrated to Charleston in 1850. Anderson was master and part owner of the schooner *Margaret Ann Howard* in the mid-1850s and occasionally sailed into Savannah. Many mariners are listed in the 1860 Charleston census. However, as in the 1850 census, those thought to be associated with the coasting trade typically confined their activities to areas north of the Savannah River.[103]

### *The Nativity of the Southern Captains*

One fact that is apparent in federal census records is that most coasting captains residing in Georgia, South Carolina, and Florida were foreign-born or native to Northern states. This was particularly true after 1830. The 1850 and 1860 censuses list only four captains residing in Georgia who were native to the state. These were John L. Grovenstein, John L. Lightbourn, William R. Postell, and Daniel Reddick. Postell's Georgia nativity is questionable as he is identified as a native of South Carolina in the 1860 census (see Table 6.3). Many of the captains living in Georgia in 1860 were from the Scandinavian countries of Norway, Denmark, and Sweden. These included Charles Stevens, Thomas Snow, Stephen Williams, William Watson, Andrew Hanson, Peter Collins, Andrew Backman, and Peter Thompson. A smaller number were from the British Isles, Germany, Italy, or France. Other foreign-born captains who worked in earlier years or are not included in the 1850 or 1860 censuses were James Frewin (England), John Chevalier (France), Lewis Wiggins (Russia), William Laen (England), James Peterson (Denmark), Claude Phillip Leset (France), Domingo Galleo (Italy), and Anthony Shaddock (England).

[102] US Census Bureau, "US Censuses of Population for Charleston, Colleton and Beaufort counties, South Carolina, 1850."

[103] US Census Bureau, "US Censuses of Population for Charleston, Colleton and Beaufort counties, South Carolina, 1860"; Motes, *Migration*, 160.

As seen in Table 6.3, twenty-four men, or 56 percent of the forty-three known captains living in Georgia in 1850 and 1860, were foreign-born. A review of the captains living in lower South Carolina and Northeastern Florida shows a similarly high proportion of European immigrants. The seafaring heritage of these European countries, coupled with the great number of European immigrants coming into Southern states after 1830, partially accounts for this prevalence of foreign-born captains. It is also true that economic and cultural factors prevailing in the South in the nineteenth century limited the numbers of native Southerners who took up a sea-faring life. Agriculture was the economic base of the region, and heritage and economics directed young men in that direction. This was particularly true after the opening of huge tracts of western lands in the 1820s and 1830s attracting large numbers of men from Southeastern states, including Georgia and South Carolina, to travel west and pursue their futures as cotton planters.

A number of the Georgia captains listed in the 1850 and 1860 censuses were natives of Northeastern and Middle Atlantic states, such as Virginia, Pennsylvania, New York, Rhode Island, and, particularly, Massachusetts. These were states with a longer and stronger maritime heritage than the South, and many of the men from these states were involved in the coasting trade before settling in Georgia. These Northern-born men comprised 26 percent of the captains living in Georgia in 1850 and 1860, compared to only 19 percent who were natives of Southern states (see Table 6.3). Ship captains were not the only Northern men to move South; many merchants in Savannah and Charleston were from the North. The number of Northern men was so large in these urban centers that as early as 1810, a group in Savannah formed an association known as the "New England Society of Georgia."[104]

The captains from the South, as well as those from the North, were almost exclusively White men. Clement Sabatty, a resident of Savannah, appears to be one exception to this White brotherhood. Sabatty appears as master and managing owner of the sloop *Mathews* in an 1820 St. Marys enrollment. He is believed to be the sixty-six-year-old Clement "Sabattie" residing in Savannah in 1850. In the census for that year, Sabattie is identified as a "Mulatto" and native of "St. Domingo." Other than this possibility, no account of a non-White person serving as either captain or owner of a Georgia coasting vessel has been discovered. This contrasts with the crews on those

[104] *Republican & Savannah Evening Ledger*, March 1, 1810.

vessels owned and sailed by Southern captains that included many enslaved African Americans.[105]

## The Northern Captains in the Georgia Coasting Trade

### *The Captains from Sippican*

On June 13, 1814, John Delano, a native of Rochester, Massachusetts, died in Savannah of "Bilious Fever." Delano was twenty-three years old, and death records identified him as captain and owner of the coasting sloop *Reformation*.[106] Captain Delano was one of many captains sailing in the Georgia trade before the Civil War who were natives and residents of the Sippican area bordering Buzzards Bay in southern Massachusetts (see chapter 5 for the discussion of Sippican-built vessels). These men were so significant in the Georgia trade that they deserve special consideration.

In 1835, a compilation of ships owned in the Customs District of New Bedford was published that contained a "List of Coasting Vessels" owned in several towns, including the Sippican communities of Rochester, Fairhaven, and Wareham. This list, which includes the names of vessels and their "managing owners," notes that two schooners and three sloops were owned in Fairhaven, five schooners and twenty-two sloops in the town of Rochester, and one schooner and thirteen sloops in the town of Wareham. Most of these vessels sailed in the Georgia coasting trade in the first half of the nineteenth century. Men identified as manager-owners of these vessels include Silas Briggs, Job Blankinship, Leonard Dexter, Frederick B. Bolles, Theophilus Pitcher, Jr., Leonard Bolles, Elisha Luce, Richard Gurney, Joseph Wing, Joseph Allen, and Timothy Savory—names that appear time after time in newspaper shipping lists as masters of vessels carrying coastal produce into Savannah between 1800 and 1861.[107]

Exactly when the captains and shipowners from Sippican began their involvement in the Georgia coasting trade is unknown, but it was before 1800. In the late eighteenth century, Rochester men were heavily engaged in the West Indies trade and visited Charleston and Savannah. On their southward voyage, some carried salt and other merchandise from Sippican and returned home with rice and cotton. In 1800, three Sippican captains are named in

[105] Enrollment No. 6, "Master Abstracts," Port of St. Marys, sloop *Mathews*, June 30, 1820; Otto, *1850 Census Chatham County*, 7.

[106] Georgia Historical Society, *Register of Deaths* 3:83.

[107] "List of Shipping Owned in the District of New Bedford," The Mariners' Museum.

shipping lists published in the *Columbian Museum & Savannah Advertiser* sailing into Savannah from local ports. These were captains Dexter with the sloop *Eliza*, John Pierce with the schooner *Fox*, and Burr with the schooner *John*. These Sippican men represent 20 percent of the fifteen named captains sailing into Savannah from local ports that year (Figure 6.4). Each made just a single arrival into Savannah in 1800, and their voyages made up only a small percentage of the total number of local coaster arrivals into the city.[108]

It is obvious the Sippican men had a minor involvement in the Georgia trade at the start of the nineteenth century. However, at that time the trade itself was small and, as the volume of the local coasting trade grew over time, so did the participation by Sippican captains. In the year between April 1809 and April 1810, eleven Sippican captains are named in the shipping columns of the *Republican & Savannah Evening Ledger* sailing into the city from local ports. Among these Northern men were Captain Briggs of the sloop *Anna Matilda*, Captain Alden Wing of the sloop *Liberty*, Captain Allen of the schooner *Renown*, Captain Handy of the sloop *Packet*, Captain James Delano of the sloop *Reformation*, and Captain Bolles of the sloop *Sophia*. These Sippican natives represent 19 percent of the fifty-eight named coasting captains sailing into Savannah that year, and they made 15 percent of the ninety-eight arrivals of all local coasters.

It was in the years after the War of 1812, that Sippican captains and their ships became increasingly involved in the Georgia trade. In 1820, eighteen, or 21 percent, of the eighty-seven named captains sailing in the local trade into Savannah were from Sippican, accounting for one-third of the local coaster arrivals into the city. Six of the fifteen captains making more than five arrivals into Savannah in 1820 were Sippican masters (see Figure 6.4). In 1830, Sippican residents accounted for 31 percent of the named coasting masters sailing into Savannah and about 42 percent of the 479 coaster arrivals. That year, thirty-five coasting captains made five or more arrivals into Savannah, and at least twenty, well over half, were men from Sippican.

The increasing involvement of Sippican men in the Georgia trade after 1815 corresponded with an expansion in the trade itself and particularly with the rise in importance of the port of Darien in the local trade into the 1830s. Throughout this period, Sippican men dominated the Darien-to-Savannah trade, transporting the huge amounts of upland cotton carried down the Altamaha River from the Georgia interior. The participation of Sippican men in the Georgia trade peaked in the early 1830s. After that, the relative number

[108] Ryder, *Lands of Sippican*. 92.

of Sippican captains decreased, as did the total number of sailing vessels, as they were displaced by steamers. Additionally, the importance of Darien as a transshipment point for upland cotton declined as railroads began to carry this cotton directly into Savannah from the interior.

With the withdrawal of the Sippican men, the number of local captains increased. In 1839, although several Sippican men were among the most active of the coasting captains in voyages into Savannah, these Northern masters accounted for only 30 percent of the 122 coaster arrivals into the city and just 22 percent of the identified captains. By 1850, only four of the thirty-four or so named masters of local coasters can be identified as men from Sippican. Together, these captains made eleven voyages into the city, accounting for just 5 percent of the 225 coaster arrivals. Ten years later, in 1860, only three coasting captains arriving into Savannah were Sippican men. These captains made four voyages into the city, representing only 4 percent of the 108 arrivals by sailing coasters (see Figure 6.4).

As with the Southern captains, newspaper shipping lists, enrollment documents, censuses, and genealogical records enable the identification of many men from Sippican who sailed as Georgia captains. The Allen, Bates, Blankinship, Bolles, Briggs, Delano, Luce, Snow, and Wing families of Sippican were particularly heavily involved in the Georgia trade. The Blankinship family was a prominent maritime clan, and they appear as ship captains and shipowners from the early eighteenth century. James and Job Blankinship, possibly brothers, were born in Rochester in the 1790s, and both men served as masters and part owners of the sloop *Howard & James*, which carried local crops into Savannah in the 1820s. There were at least two Pelag Blankinships involved in the Georgia trade. One, born in Rochester in 1809, was both master and an owner of the schooner *Roswell King* in the 1830s. He appears in the 1840 federal census for Rochester engaged in "navigation of the ocean," but later moved to Savannah, where he worked as a steamer captain. Pelag Blankenship did not remain in Georgia; by 1860, he was back in Sippican, living in Marion. The federal census for that year identifies him as a "master mariner"; the occupation used in the census to identify men actively sailing as ships' captains.[109]

[109] *Enrollments of New Bedford, 1808–1840*, sloop *Howard & James*; information on Pelag Blankenship from New England Historic Genealogical Society, *Vital Records*, 41; Otto, *1850 Census Chatham County*, 7; US Census Bureau, "US Censuses of Population, Plymouth County, Massachusetts, 1840 and 1860."

In addition to serving as a ship captain, Pelag Blankinship was an owner of at least four schooners sailing in the Georgia trade after 1840. In these cases, he was one of a group of owners, each holding a small share in a vessel, a form of ship ownership common for Sippican residents.[110]

The Bolles (also often spelled "Boles" and "Bowles") family was another of the old Rochester seafaring families whose members were involved in the Georgia trade for more than sixty years. Ebenezer and Savory Bolles were among the first in the family to work in the Southern trade. In 1807, an Ebenezer Bolles was master of the new Rochester-built sloop *Sophia*, sailing between Savannah and ports like Sunbury and Riceboro. An Ebenezer Bolles, presumably this person, is named in several vessel documents of ships sailing in the Georgia trade, including the sloops *Sophia* (1810), *Riceborough* (1816), and *Harmony* (1815, 1818) and the schooner *Eclipse* (1823). By 1825, Ebenezer Bolles lived in Savannah, the year the *Savannah Republican* named him in "A Correct List of Pilots," for the port of Savannah.[111]

Another member of the Bolles family, Savory Bolles, a native of Rochester, was sailing along the Georgia coast by 1809 as master of the Rochester-built sloop *Polly & Betsey*. Captain Bolles died in 1826 at age fifty-one. Shortly before his death, he captained the sloop *Mercy* in the Georgia trade (Figure 6.5). At least three of Savory Bolles's sons became captains or owners of Georgia coasters. These were Leonard, Charlton, and Savory A. The eldest son, Leonard, was active in the Georgia trade for almost forty years. Beginning in 1819, when only nineteen years old, he appears as captain of as many as thirteen sloops and schooners carrying coastal produce along the Georgia coast. In addition, Leonard Bolles held ownership in at least six other ships working in the trade. Leonard Bolles served as a coasting captain into the late 1850s.[112]

Several members of the Briggs family of Rochester were active in the Georgia trade. These included Nathan Briggs, who served as master of at least seven Georgia coasters between about 1820 and 1845; Paul Briggs, who was master of the schooner *Columbia* in the 1820s; and Silas Briggs Jr., who sailed as captain of the schooners *Eagle* and *Harmony* between 1815 and the late

[110] *Enrollments of New Bedford, 1841–1939*.

[111] Information on the two mariners named Ebenezer Bolles from Bolles, *Genealogy*, 49–50; *Ship Registers of New Bedford, 1796–1850*; *Enrollments of New Bedford, 1808–1840*; *Savannah Republican*, January 26, 1825.

[112] Information on Savory Bolles from Bolles, *Genealogy*, 49; and *Enrollments of New Bedford, 1808–1840*, sloop *Polly & Betsey*. Information on Leonard Bolles from New England Historic Genealogical Society, *Vital Records*, 44; and US Census Bureau, "US Censuses of Population, Plymouth County, Massachusetts, 1850 and 1860."

1820s. Silas Briggs Jr. moved to Savannah by the mid-1820s, and the *Savannah Republican* of January 26, 1825, named him as one of the city's pilots.[113]

The Delanos were another of the Rochester families active in Georgia. James Delano and Harper Delano Jr., possibly cousins, were sailing along the Georgia coast before 1820. James served as master of the sloops *Good Hope* and *Reformation* between 1810 and 1816, while Harper was captain of the schooner *Elizabeth* in 1818. Harper Delano Jr. died at "Port au Prince WI" in May 1819; his brother Jabez Delano died "at sea" the same year. These notations in death records reflect the wide-ranging voyages of the Rochester captains and the hazards they faced in their chosen work.[114]

Members of the Luce family of southern Massachusetts appear commonly in records of the Georgia trade. The brothers Elisha and Stephen C. Luce were particularly active between 1820 and the mid-1850s. They shared ownership in several coasters, including the sloop *Angel* and the schooner *Altamaha*, and both served as masters of these and other vessels. Stephen C. Luce gave up working aboard ship in his later years but remained involved in the coasting trade as a merchant and shipowner.[115]

A few of the Georgia captains were from areas immediately adjacent to Sippican. This included the town of Fairhaven, whose seafaring family names included Allen, Egery, Delano, Taber, Nye, Eldridge, Church, Hammond, Hathaway, and Russell. Leonard Hammond, a resident of Fairhaven in 1850, sailed as captain of the sloops *Regulator* and *Stranger* along the Georgia coast from about 1825 to 1841.[116]

[113] *Ship Registers of New Bedford, 1796–1850*, schooner *Harmony*; *Enrollments of New Bedford, 1808–1840*, schooners *Columbia* and *Eagle*; Georgia Recorder Office, *Official Register*, Silas Briggs entry, 120.

[114] Sailing activities of James and Harper Delano come from the *Republican & Savannah Evening Ledger*, 1810; *Columbian Museum & Savannah Daily Gazette*, 1818; and *Enrollments of New Bedford, 1808–1840*, sloop *Good Hope*. The deaths of Harper and Jabez Delano are found in New England Historic Genealogical Society, *Vital Records*, 104, 372.

[115] Sailing activities and ship ownership information for the Luce captains come from several Savannah newspapers including the *Daily Georgian*, 1820, 1841–1854; *Savannah Republican*, 1825; and *Daily Savannah Republican*, 1830, 1835; as well as *Enrollments of New Bedford, 1808–1840*, sloop *Angel*; and *Enrollments of New Bedford, 1841–1939*, schooner *Altamaha*. Luce family information from New England Historic Genealogical Society, *Vital Records*, 207–208; 405; US Census Bureau, "US Census of Population, Plymouth County, Massachusetts, 1860," in which Stephen C. Luce is identified as a "Merchant."

[116] Tripp, "Story of Fairhaven," 1–8; US Census Bureau, "US Census of Population, Bristol County, Massachusetts, 1850."

Several of the Georgia captains were residents of Freetown, located in Bristol County just a few miles northwest of Rochester. Among these Freetown captains were Daniel C. Brown, Leonard Burr, Luther Cudworth, and Isaac Gifford. Captain Luther Cudworth sailed as master of at least three Georgia coasters in the 1840s and 1850s: the schooners *Carrier*, *Harriet Lewis*, and *President*. Leonard Burr of Freetown sailed in the Georgia trade from the late 1820s to the 1850s as master of at least three ships, the sloops *America* and *Trader* and the schooner *Company*, all built at Sippican towns.[117]

Very few of the Northern captains sailing in Georgia were from the whaling center of New Bedford, located immediately west of Sippican. New Bedford captains were commonly engaged in trade between Northern ports and Southern locations from Virginia to Savannah; however, few worked in the local Savannah trade. One who did was Joseph Howland, member of one of the prominent Quaker families of New Bedford. Captain Howland sailed the sloops *Bolivar*, *Rosetta*, and *Three Brothers* in the Georgia trade between 1818 and 1830. Joseph Howland was one of several Quaker captains sailing in Georgia; some were abolitionists or, at least, harbored antislavery sentiments. During his trading activities in Georgia, Captain Howland met Patrick Gibson, owner of a plantation on Creighton Island, near the town of Darien. Having become familiar with the antislavery movement in the town through Joseph Howland, Gibson desired to free enslaved persons he owned, and in 1834, he and four of his enslaved laborers traveled to New Bedford aboard a lumber schooner. Patrick Gibson died before legally freeing these individuals, two of whom were his daughters, as well as other slaves from Creighton Island, some of whom were carried to New Bedford aboard coasting ships operated by Northern captains. In a lawsuit arising over the ownership of these enslaved persons, Joseph Howland testified that Patrick Gibson told him that he "wished to free his slaves" and that he understood Gibson "brought his children to the north to free them and educate them." Ultimately, antislavery citizens in New Bedford prevented the Creighton Island slaves from being forcibly returned South, and several remained and settled there.[118]

Other Quaker mariners residing in New Bedford and Sippican included members of the Wing, Taber, and Delano families, some of whom served as masters and owners of Georgia coasters. We do not know their feelings about

[117] Information on Luther Cudworth from Kelly, "Luther Cudworth, Stamper Family." Leonard Burr's sailing activities are from Savannah newspapers, 1825–1855.

[118] Grover, *Fugitive's Gibraltar*, 150–51.

slavery, but they likely sympathized with the antislavery sentiments held by many Quakers.

Many Sippican men involved in the Georgia trade were also engaged in other maritime activities. They were associated with whaling, as ship captains, owners, and agents. Stephen C. Luce, active in the Georgia trade, was an owner of the whaler *Cossack* and served as agent for the *Shylock* while his brother Elishu was an agent for the whaling bark *Dryno*. The brothers also were part owners and served as masters of the sixty-one-foot schooner *Altamaha* that, in addition to sailing as a Georgia coaster, worked as a whaler. In 1855, the *Altamaha* left Sippican on a whaling voyage of only six months and nine days. The schooner returned with a cargo of 240 barrels of sperm oil and eight of blackfish oil, valued at $13,500. After paying off the crew and outfitting her for another voyage, her owners divided $8,000 in profits.[119]

The schooner *Admiral Blake*, another Georgia coaster from Sippican, made whaling voyages under Captain B. B. Handy, including a two-month trip that netted $11,000 worth of sperm and blackfish oil. The Sippican whalers seem to have commonly undertaken these short whaling voyages, called "plum pudding" voyages by the hard-bitten whalers of Nantucket and New Bedford. However, the whalers out of Sippican ranged around the entire world. Archelus Hammond, born in 1759 in Mattapoisett, the town just south of Rochester, is reported to have "struck" the first whale in the Pacific and introduced whaling to that ocean.[120]

Typically, the coasting captains from the Sippican area and other Northern states worked seasonally in the Georgia trade. Each year, their sloops and schooners arrived in Georgia in October or November, when shipments of new rice began. Most of these men continued sailing along the Southeastern coast until the spring or early summer. They departed for Northern ports between April and June after most of the season's rice and cotton had been shipped and before the onset of summer, which brought a variety of dangerous diseases striking down many in port towns such as Savannah and Charleston.

This seasonal activity of the Sippican captains is reflected in shipping lists and advertisements in Georgia newspapers. Notices of Sippican captains and their vessels arriving in Savannah from ports such as Rochester, New Bedford, Providence, or New York usually began to appear in city newspapers around October. Beginning in the spring, most Sippican captains began to sail North, returning home. These departures were commonly listed in

[119] Ryder, *Lands of Sippican*, 114–15, 122.
[120] Starbuck, *American Whale Fishery*; Leonard, *Mattapoisett and Old Rochester*, 295.

Savannah newspaper. Sometimes, advertisements for several Northern-bound ships appeared in newspapers on the same day or same week, suggesting they were sailing north together. These group sailings also occurred among coasters traveling South in the fall.

*Other Northern Captains*

Although most of the Northern captains working in the Georgia trade were from the region of Sippican, not all were. A few captains resided in other areas of Massachusetts, and several were from Rhode Island and Connecticut. It appears that the captains from Rhode Island and Connecticut participated in the Georgia trade mainly during the first two decades of the nineteenth century; most dropped out of the Southern trade before 1830. Jonathan Landon, of New Haven, was sailing the sloop *Patty* in the Georgia trade in 1800 and possibly later. Several members of the Bulkley (or Bukley) family of Connecticut sailed in the Southern coasting trade. One of these was Walter Bulkley, a resident of Fairfield, who sailed the sloop *Chauncey* along the Georgia coast in the 1820s. The brothers Ambrose and Jeremiah Burrows of Stonington, Connecticut, both appear to have served as master of the schooners *Jolly Sailor* and *Three Friends* in Georgia before 1810. William Grant, another resident of Stonington, was master of the sloop *Franklin* in 1805 and 1806, sailing into Savannah from Beaufort and St. Marys. Another Connecticut captain was Daniel Goff Phipps, who, between 1800 and 1816 sailed in the St. Marys–Charleston trade as master of the sloops *Cinderilla*, *Ruth*, and *Rising Planet*. John M. McColley, a resident of Derby, Connecticut, master and part owner of the sloop *Othello*, sailed in Georgia from about 1825 to 1830, after most Connecticut residents had given up the Southern trade. With the *Othello*, Captain McColley principally sailed into Savannah from locations in lower South Carolina. Other Connecticut captains who worked as Georgia coasters were Joel Chittendon, Jabez Stanton, Giles Savage, Sylvanus Osborne, and William Gorham.[121]

A few captains in the Georgia trade were residents of Rhode Island, including Nicholas Gardner of North Kingston and Jessie Martin and Thomas Prentice, both of Providence. Savannah newspapers show Jessie Martin was

[121] Jonathan Landon's residency is found in US Census Bureau "US Census of Population, New Haven County, Connecticut, 1790"; the residency and vessel ownership for Walter Bulkley, Ambrose Burrows, Jeremiah Burrows, William Grant, and John M. McColley are found in "Connecticut Ship Database"; information on Daniel Goff Phipps' service during the American Revolution is from Lincoln, *Naval Records*, 434.

captain of the sloop *Sally* from 1806 to 1808; Thomas Prentice sailed the schooner *Minerva* in 1806, and Nicholas Gardner sailed the sloop *Lady Washington* in 1816. The Rhode Island captains seem to have left the Georgia trade earlier than those from Connecticut.

Considering that several ships working in the trade were built in Virginia, New Jersey, New York, and Delaware, it's surprising that not one person residing in the Mid-Atlantic states south of Connecticut has been positively identified as a master in the local Georgia trade after 1800.

A few men from these Mid-Atlantic states did command vessels in the Georgia trade, but only after they took up residency in the South. Henry J. Dickerson, owner of many Georgia coasters, was a native of Baltimore, but he appears as a captain in the Georgia trade only after moving to Savannah in the late 1830s. Samuel F. Flood, who sailed the vessels *James* and *Ellen* in the St. Marys–Savannah trade from 1820 to the early 1840s, was a native of New York but moved to Camden County by 1805, before his name appears as a coasting captain.

## The Crews of the Georgia Coasters

Crews working aboard the Georgia coasters represent a largely unknown group. The few accounts dealing with the subject indicate the number of crewmen on coasters was small, generally consisting of the master and two to four men, one of whom might be classified as a mate. The smallest coasters might require only the master and a single crewman. In 1817, the crew of the sixty-three-foot sloop *Harmony* consisted of five people: the captain, a mate, and three crewmen. James S. Silva of St. Marys wrote that the crew aboard Captain Dennis Pacetty's sloop *Independence* in the 1840s "usually consisted of a white captain and colored cook and two deck hands." Occasionally, newspapers mentioned the number of crewmen aboard coasters, often when a vessel was involved in an accident, and particularly if someone was injured or died. When the schooner *Harriet* was captured by British sailors on the lower South Carolina coast in 1813, she was manned by a captain and four crewmen. The schooner *Susan*, sailing out of Charleston, was run down by the schooner *Justice* and sank in December 1827. The *Susan*'s captain, two men, and a boy were saved; one man drowned.[122]

[122] Information on the sloop *Harmony* is in "Paul Wing Accounts"; Silva, *Early Reminiscences*, 20; the capture of the *Harriet* is from the *Republican & Savannah Evening Ledger*, August 21, 1813, and the sinking of the *Susan* from the *Daily Georgian*, December 25, 1827.

The names of crewmen on coasters appear occasionally in newspapers, again commonly when reporting an accident. On February 3, 1821, the *Darien Gazette* reported the drowning of twenty-four-year-old George Harris, identified as a "seaman" from "the sloop Neptune, Capt. Drinkwater." Harris was a native of Springfield, Connecticut. The *Neptune* had been transporting cargo from Darien to Savannah. On June 29 of that same year, the *Daily Georgian* noted that Joseph Snow, the mate aboard the coasting sloop *Support*, was knocked overboard and drowned when the sloop was sailing from Savannah to New Bedford. On April 29, 1808, the Savannah newspaper *Public Intelligencer* contained a notice that John Cowing, master of the sloop *Olive*, was offering a reward for the return of sixteen-year-old Elias Davis, who had run away. Davis was described as an "indentured apprentice," and Captain Cowing was offering "six & a quarter cents" as a reward. Apparently, young Elias Davis was apprenticed as a crewman on the *Olive*.

Many sailors from coasters died of disease in Savannah and are identified by name in city death records. Most of these men were in their twenties and from Northern states. A small number were older, such as John Willis, a sixty-year-old native of Ireland who was identified as a "seaman belonging to the sloop Union" when he died of fever in 1817.[123]

The United States census records for the coastal counties of Georgia for the years 1850 and 1860 include the names of numerous individuals assigned occupations such as "sailor," "seaman," or "mariner," some of whom were probably crewmen on coasting ships. The 1860 census for Chatham County identifies thirteen men whose occupations were "mariner on coasting vessel," "mariner, coasting vessel," or "mariner, sloop." At least one of the men assigned these occupations, William Frazier, served as a captain of a Georgia coaster, but the others were probably crewmen. Two of these thirteen mariners were natives of the South; the others were from Northern states or England, a phenomenon mirroring what is seen among the captains of the coasters. Four of these mariners are identified in the census as "black" or "mulatto" (Table 6.4). These men were members of the community of 725 "Free Negroes" then residing in Chatham County.[124]

While there is little specific information on who the crewmen of Georgia coasters were, it appears that many of the men working aboard Southern-owned vessels were enslaved. When captured by British forces off the South Carolina coast in 1813, the crew aboard the schooner *Harriet* consisted of

[123] Georgia Historical Society, *Register of Deaths* 3:172.

[124] Georgia Historical Society, *1860 Census*.

Captain Steele and "4 black men." In 1810, a vessel under the command of a Captain Fowler swamped off the Georgia coast while sailing from St. Simons Island to St. Marys. Among the survivors were "3 negro sailors."[125] These accounts do not specifically state these men were enslaved, but it is presumed they were. Specific evidence that slaves were employed aboard Southern-owned coasters is scattered in federal census records. The 1840 census for Camden County records that the household of known coasting captain William Laen included three men engaged in "navigation of canals, lakes, rivers," a category that commonly included those working on coasters. Since William Laen himself was the only White male over five years old in the household, two of the men involved in navigation must have been among the five enslaved men between the ages of twenty-four and thirty-six listed in Laen's household. That same year, Francis Chevalier, another Camden County coasting captain, owned three enslaved men between ten and twenty-four years old; all employed in the "navigation of canals, lakes, rivers." A review of 1840 census records for the other coastal Georgia counties shows that several other coasting captains owned slaves who were employed in "navigation."[126]

When James Silva noted that the crew aboard Daniel Pacetty's sloop *Independence* consisted of a "colored cook and two deck hands," he indirectly reveals the two crewmen were Black by noting that Captain Pacetty was "always glad to have a well grown white boy to go with him as company and to assist in the steering." The deckhands aboard the *Independence* are likely to have been enslaved, but it is unknown if they were owned by Dennis Pacetty. Captain Pacetty enslaved a few laborers: none in 1840 and two in 1860, a sixty-year-old woman and a fifty-year-old man, who might have worked on Captain Pacetty's ship. Captain Charles Stevens of St. Simons Island owned five enslaved people in 1850 and sixteen in 1860. Two of the enslaved men owned by Stevens in 1850 were Robert, age twenty-six, and Abram, age thirty-three, both of an age to perform the work required to sail a coaster. In 1852, one of the enslaved men owned by Captain Stevens drowned. This man is thought to have been the man named Abram who may have been working aboard Steven's sloop *America*.[127]

[125] *Republican & Savannah Evening Ledger*, August 21, 1813; *Columbian Museum & Savannah Advertiser*, November 15, 1810.

[126] US Census Bureau, "US Census of Population, Camden County, 1840."

[127] Silva, *Early Reminiscences*, 2; US Census Bureau, "US Census Slave Schedule, Camden County, 1850"; Pearson, *Charles Stevens*, 37.

Occasionally, newspaper advertisements of the sale of slaves and notices of runaway slaves reveal information on men who may have worked on coasters. Although somewhat earlier than the period of interest, in 1786 Lachlan McIntosh advertised for sale in the *Georgia Gazette* an enslaved man named Fergus, who was "well known in this town to be the best pilot and boatman of any negro in the state." The February 1, 1804, issue of the *Columbian Museum & Savannah Advertiser* published a notice by Thomas Rigby of Brunswick concerning a runaway slave named Harry. The notice stated that Harry was "a good sailor and may perhaps attempt to get employment in that way." On April 17, 1813, the *Republican & Savannah Evening Ledger* contained an advertisement for an auction of "Seventeen NEGROES, in families." Among the individuals to be sold was a man described as "an excellent oarsman, and good pilot" who knew the inland navigation between Savannah and Charleston. In May 1805, the schooner *Franklin* took on a "negro pilot" at Frederica on St. Simons Island to guide the ship around to the eastern side of the island where timber was loaded.[128]

Although some of the Georgia captains may have engaged their own enslaved laborers aboard their ships, a review of federal slave census records for 1850 and 1860 shows that only fourteen of the forty-two captains residing in Georgia were slaveowners. In most instances, these fourteen captains owned a single or very few persons, often women or young boys unlikely to have worked aboard ships. John Louis Grovenstein, one of the most active of the Georgia captains, owned one nine-year-old boy in 1850 and no one in 1860. Captain William Watson of Savannah owned two enslaved girls in 1850 and a twelve-year-old boy in 1860. The twelve-year-old might have been old enough to work aboard the schooner that Watson then commanded, the *John W. Anderson*.[129]

The fact that most coasting captains residing in Georgia in 1850 and 1860 were not slaveowners might reflect their economic inability to own enslaved workers. However, cultural factors may also have been at work, and some of these men may have had an aversion to slave ownership. By 1850, most captains residing in Georgia were immigrants from states or countries where slavery had never existed or was outlawed, and they may have rejected the practice.

[128] Wood, *Robert Durfee's Journal*, 105, note 134.

[129] US Census Bureau, "US Censuses, Slave Schedules, Camden, Chatham, Glynn, Liberty and McIntosh counties, 1850 and 1860."

Although the captains may have been unable to provide slaves for work aboard locally owned coasters, shipowners, including factors and planters, possibly provided this labor. Many of these shipowners were prosperous men who owned many slaves, some of whom may have been forced to work aboard ships. Henry J. Dickerson, who held ownership in many coasters, was a large slaveholder after 1850. In 1850, he owned eighteen enslaved persons, ten of whom were males between the ages of twelve and thirty-five. In 1860, he held thirty-seven individuals in bondage, nineteen of them males between twenty and forty-five years old. Some of these young men probably worked in Dickerson's stevedoring business; others may have been put to work aboard the coasters and steamers he owned. Several of the factors and merchants in Savannah who owned coasting ships were also slaveholders. Robert Habersham, the largest of the rice factors in the mid-nineteenth century, held more than 250 persons in bondage in 1850. Most of these individuals worked on Habersham's plantations near Savannah, but some would have worked at his factorage house on the Savannah River, and others may have been forced to work on coasters he owned or that transported his cargoes.[130]

While many of the crew aboard Southern-owned coasters were enslaved men, not all were. Some White "mariners" listed in the 1850 and 1860 censuses for Georgia are believed to have been crewmen, possibly on locally owned vessels. The 1860 census for Savannah lists known coasting captain Charles Thompson as a "head of household," and living with him were three White "mariners." These men were John Anderson, Andrew Yaeger, and Stephen Williams, all natives of Denmark. One of these mariners, Stephen Williams, did serve as a coasting captain, but no record shows that either John Anderson or Andrew Yaeger did. It is likely these two men were employed as crewmen, possibly mates, on coasters, perhaps aboard Captain Thompson's schooners *Northern Belle* or *William Totten*. Similarly, in 1860, James Ferguson, identified as a twenty-seven-year-old "mariner" and native of Quebec, Canada, resided in the Savannah home of coasting captain Andrew Hanson. Captain Hanson owned the schooner *Fort George Packet*, and Ferguson may have worked aboard as a mate or crewman. There are several other references to White "mariners" residing in the households of coasting captains, and some probably reflect instances where crewmen were living with their captains. In

[130] US Census Bureau, "US Censuses, Slave Schedules, Chatham County, 1850 and 1860."

addition, it is believed that most men who became captains of coasting vessels began their careers as crewmembers.[131]

No record of an enslaved person as a master of a coaster has been found. However, experienced Black sailors certainly would have been able to, and probably occasionally did, perform the sailing duties of a captain. Some of these men acquired sufficient knowledge of the coastal sailing routes to be recognized as competent pilots, as was the "negro pilot" taken aboard the schooner *Franklin* at St. Simons Island in 1805. The several "Negro" and "mulatto" men identified in the 1860 Chatham County census as mariners aboard coasting vessels shows that a few free Blacks worked aboard coasters, but none can be correlated with any named captain. A list of "Free Persons of Color" appearing in the August 21, 1838, *Savannah Daily Georgian* identifies no mariner or sailor aboard a coaster but does name Henry Wall as a pilot and three men as ship carpenters, Paul Carmon, Hannibal Roe, and Joseph Summers.

The Northern coasters working seasonally in Georgia employed free men as crew. This is demonstrated in the common mention of these men in death and accident accounts and in the few other documents bearing on this topic. Financial accounts for the schooner *Harmony* provide information on what the crewmembers of that coaster were paid during her voyage along the Georgia coast in 1817–1818. John Caswell, the mate, received a total payment of $194.40 ($24 per month) for his eight months of service, extending from approximately September 25, 1817, to May 25, 1818. Crewmen on the *Harmony* were paid approximately $14 per month. These salaries seem low, but there was additional compensation in that the crew received free food on the voyage. Comparable records for men working aboard Southern-owned coasters have not been found, but White crew members likely received similar or slightly lower salaries.[132]

Records identifying the food eaten by the men serving aboard coasters are scarce. Accounts of the sloop *Harmony* on its voyage along the Southeastern coast in 1817–1818 record expenditures for beef (presumably brined or pickled), beets, potatoes, meal, and rum at the beginning of her voyage at Rochester, Massachusetts. This rather meager supply of food was supplemented during the sloop's voyage through purchases of a variety of foodstuffs. In Savannah, for instance, items such as potatoes, meat, pickles, sugar, bread, molasses, fish, eggs, sausages, butter, cheese, pepper, and mustard were

[131] Georgia Historical Society, *1860 Census.*

[132] "Paul Wing Accounts."

bought. These items probably reflect the typical fare for Northern coasters coming South to work. The food consumed on the Southern-owned coasters would have included some of these items, but regional foods like rice, bacon, hominy (grits), and fish would have been more commonly eaten on these local vessels.[133]

The crew members on the coasters suffered from various injuries and illnesses. The financial accounts of the *Harmony* include expenditures for medicines and medical services. These include payments for a doctor "to bleed" one of the crew and to purchase a purgative for him. These expenses were deducted from the man's pay. While in Savannah, one dollar was deducted from the pay of each member of the *Harmony*'s crew to pay for a "protection," presumably referring to a smallpox vaccination. No records of these types of expenditures for Southern-owned coasters have been found. Given that many of the crew aboard these Southern coasters were enslaved, these expenses would have been borne by the slaveowners.[134]

The sloop *Harmony*'s crew also had "hospittle [hospital] money" deducted from their pay at the end of their voyage. For those who worked the full eight months, hospital money amounted to $1.60, or twenty cents per month. This hospital money was related to the marine hospital fund established by Congress in 1798 to provide medical services for American merchant seamen in American ports. The fund was administered by the Treasury Department and financed by monthly deductions from seamen's wages. Although Paul Wing, the captain of the *Harmony*, submitted hospital fund monies to the government, many captains did not. As early as 1804, it was reported that only about three-fifths of the appropriate hospital fund monies were being paid. How many of the Georgia coasters complied with this requirement is unknown, but those that used enslaved labor aboard their coasters would not need to participate in the fund, at least for the crew members.[135]

## The Economic Success of the Coasting Captains

Specific information on the level of economic success achieved by the Georgia coasting captains is difficult to obtain. Many of the Georgia captains worked in the local trade for twenty years or more, suggesting it was an economically successful career. However, it is possible that some men were forced to remain in the trade out of economic necessity or because they lacked the skills to

[133] Ibid.
[134] Ibid.
[135] Ibid., Parscandola, "Public Health," 87.

pursue other types of work. Only a very few, such as Henry Dickerson of Savannah, achieved what can be considered a wealthy status. However, Dickerson's wealth came mostly from owning ships, stevedoring, and other business interests, not his service as a coasting captain. Charles Stevens achieved moderate financial success, partly from his shipping business and partly from the small plantation he developed at Frederica. In general, however, the captains residing in the South fared poorly relative to economic achievement. On the other hand, many of the Northern captains appear to have achieved modest financial success.

Details on the financial structure of the coasting trade, as it relates to the captains themselves, are unclear. Many captains, particularly those who owned the ship they sailed, relied on income from cargo freight charges. Some record of these charges for the pre-Civil War period exist, as do a few accounts and ledgers that record freight payments to specific captains or vessels. This documentation is fragmentary, making it impossible to develop a full measure of the income that an individual captain achieved on an annual basis. Newspaper shipping lists make it possible to track the voyages of some Georgia captains over long periods and estimate the quantity of plantation produce they carried into Savannah. From these figures, estimates of the freight charges the captain, or the shipowners, received can be projected for the period after about 1820, when some cargoes began to be identified in detail. While this inbound cargo information is available, there are no systematic or comprehensive records of what the coasters carried out of Savannah to plantations and communities along the coast. Except for occasional references in plantation accounts and other personal records, the amounts paid to captains for transporting these outbound cargoes are unknown. Thus, any estimate of the income the captains received is only a very rough assessment and, additionally, considers only the income derived from their voyages into Savannah transporting the "products of the country."

Despite this shortcoming, it is useful to examine this topic as another means of further illuminating the generally neglected lives of these men. Until comprehensive financial records are located, the newspaper accounts of cargo carried into Savannah—incomplete though they are—provide the best source for addressing the question of income. The sailing activities of Captain Charles Stevens have been followed in detail over the twenty-one years he sailed as a shipmaster in the Georgia trade. Information on the economic aspects of several years of Stevens's trading activities are presented here. These discussions shed light on issues related to the income Georgia captains received from transporting cargo and on more general aspects of the economics of the trade.[136]

The *Daily Georgian* reports that Captain Stevens brought his sloop *Splendid* into Savannah eight times in 1844. Stevens was the sole owner of the *Splendid* from 1843 until 1853, so he presumably received all the income from the ship's operations during these years. The cargo carried by the *Splendid* into the city during 1844 included 4,070 bushels of rough rice, 172 bales of cotton, an unspecified amount of rough rice on one voyage in March, an unspecified amount of wood from St. Simons Island, and, on one trip, also from St. Simons, an unspecified number of hides. Considering the prices these commodities were bringing in 1844, the cargo of rice was worth forty-six hundred dollars, and the Sea Island cotton was worth about twelve thousand dollars. The freight on rice during this period was approximately five cents per bushel, meaning the freight charge would have been around $203 for the rice transported over the year, not including the unknown amount carried on one voyage. The freight on cotton varied but was about one dollar per bale or bag, meaning that Stevens received approximately $172 for transporting cotton. The freight charges on the hides and wood are unknown. Thus, it can be estimated that in 1844, Stevens received at least $375 for the plantation products he transported into Savannah. In 1845, the *Splendid* again made eight sailings into Savannah. The cargo carried consisted of 282 bales of cotton and 3,100 bushels of rough rice. On one voyage in January, the newspaper records the *Splendid* arriving with seventy-one bales of cotton and an unspecified amount of rough rice. The Sea Island cotton carried during the year, at an average price of twenty cents per pound, was worth $16,920, and the rice, at ninety cents a bushel, brought about $2,790. Stevens would have received at least $437 in freight charges for transporting these goods.

By 1860, Stevens had acquired the schooner *Northern Belle*, and that year he carried into Savannah 281 bales of Sea Island cotton worth an estimated $30,292; 6,500 bushels of rough rice worth about $10,270; 200 bushels of corn; 872 recorded barrels of rosin and turpentine, and unknown amounts of rosin, hides, wood, and cedar posts. The estimated value of the known cargo carried by the *Northern Belle* into Savannah that year was $42,716, with cotton being the most valuable commodity. Captain Stevens received a minimum of $620 as freight on these goods.

The estimated freight incomes for these three years represent minimal amounts and do not include the freight payments Stevens received for

[136] Pearson, *Charles Stevens*; Pearson, "Georgia Coasting Trade."

miscellaneous items, such as corn, wood, and hides. These incomes don't include the freight on the non-enumerated rough rice carried on one voyage in 1844. Importantly, these estimated incomes relate only to cargo carried into Savannah; they do not incorporate freight charges on goods that Stevens carried out of Savannah or between smaller coastal ports and plantations where shipping activities were rarely documented. These incomplete cargo records exist only in scattered sources, such as extant plantation accounts. For example, on April 21, 1845, Hugh Fraser Grant notes a payment to "Capt. Stevens" of twenty dollars for delivery of one hundred bushels of corn and other supplies to Grant's Elizafield Plantation on the Altamaha River. In August 1855, the records of the estate of Dr. James Troup, also owner of a plantation on the Altamaha River, note payments to Charles Stevens of $87.83 for the delivery of "bricks and lime" and $40 for "freight on Machinery, 12,000 shingles and 3,000 bricks."[137] Charles Stevens and the other coasting captains made many similar deliveries over the course of every year, but the vast majority of these have gone unrecorded.

Despite their serious shortcomings, these examples provide information on the types and sources of income that Charles Stevens and the other coasting captains or shipowners received from operating their vessels. In the case of Charles Stevens, it is possible to state that he made more than $375 in 1844, more than $437 in 1845, and more than $620 in 1860. These amounts were for cargoes inbound to Savannah only. Still, based on what he was paid for transporting cargoes out from Savannah to the plantations of Hugh Grant and James Troup, his annual income from shipping was considerably greater than these sums. Even if the amounts given here are doubled, the incomes might seem low. However, these earnings need to be viewed in the context of the times. For example, the average annual income for a non-farm worker in Georgia in 1840 was $287, but the United States average was $399. Further, there are formulas for comparing wages and prices in the past to the present day. Particularly valuable is the information provided by the website Measuringworth.com, which presents a variety of ways to compare measures of monetary worth and value through time. Relying on the measure comparing "relative income" value, the $375 made by Captain Stevens in 1844 is equivalent to $266,000 in modern (2020) dollars; the $620 he made in 1860 is equivalent to $280,000 today. These represent significant annual incomes, and it must be reiterated that these revenues represent the minimum incomes that Stevens

[137] House, *Planter Management*, 186; Glynn County, Georgia, estate of James Troup, Wills and Appraisements, Book F, 348.

received. In this light, Captain Stevens apparently derived more than modest earnings from his trading activities. One reflection of his economic success is his purchase of several tracts of land at Frederica after 1850. This included three hundred acres from Christ Church in April 1851 and ninety-six acres from Theodore Pease in June 1860, for which he paid six hundred dollars.[138]

Although the income Stevens received from freight charges seems to have been reasonable, he and all the men who owned the ships they sailed faced a variety of expenses in operating their vessels. These included basic maintenance of a ship's hull, sails, and rigging, which could be considerable given the conditions under which coasting ships operated. Occasionally, ship captains or owners had to pay for repairs that could be undertaken only at a shipyard. Such repairs could be quite expensive. Ship captains also had to provide food for themselves and crew members.

Charles Stevens and the other coasting captains who owned their vessels were independent entrepreneurs who received their income from freight charges. Other captains were salaried employees working for shipowners, factors, and commission merchants. As noted previously, in the 1850s the Savannah factors Reed & Tison employed Savannah resident Patrick Doyle to serve as master of their schooner *Elias Reed* at a set wage of forty dollars per month. In 1834 and 1835, William Williams paid monthly wages to several men for work aboard the sloops *Mill Maid* and *Conductor* in transporting cargo and "lightering" work in the Savannah River.[139] Men who owned a share of the vessel they commanded may have received wages or a percentage of the income derived from freight charges. Unfortunately, available records do not shed light on these partial-ownership situations.

Vessel enrollment and register documents do show that many captains owned or held shares in coasting ships, meaning they had the funds or collateral to make these purchases. Other documents, such as bills of sale and official conveyance records, provide specific information on the amounts involved in these transactions. Charles Stevens sold his sloop *America* to the merchant Samuel A. Hooker for sixteen hundred dollars in February 1855, and three months later, he purchased the schooner *Northern Belle* from Captain Charles Thompson for three thousand dollars. A month after his sale of the *Northern Belle*, Captain Thompson paid New York owner David Laforge twenty-one hundred dollars for the schooner *William Totten*. In 1851, longtime captain

[138] Easterlin, "Interregional Differences," Appendix A; MeasuringWorth.com, "relative income" calculator; Pearson, *Charles Stevens*, 39.

[139] Unidentified Cash Book, Georgia Historical Society.

John L. Grovenstein, together with Savannah merchant Henry Brigham, paid W. D. Jenkins of Glynn County five hundred dollars for a one-quarter share in the schooner *Company*. In June 1855, Captain Antonio Lawrence bought the small, fifteen-ton schooner *Col. Long* from Alfred Haywood for $220.[140]

These examples of sales reveal that some captains accumulated the necessary capital to purchase vessels or shares in vessels. In some instances, these amounts were considerable. It is rarely recorded if these purchases involved payment in full at the time of the sale or if some type of scheduled payment was involved. In 1841, when Joseph Silveria sold the schooner *Marion* to Captain John Robbins in Savannah for twenty-three hundred dollars, the entire amount was paid at the time of the sale. Some captains relied on their ownership in vessels to secure loans. In 1839, Savannah coasting master John Russell used his sloop *Eliza* as collateral to secure a $1,225.25 loan from S. B. Hills. Russell was required to pay back the loan, with interest, in ninety days.[141]

For many Southern captains, ownership in a coasting vessel represented their greatest capital investment and often constituted a significant proportion of their net worth. When Domingo Galleo drowned in Savannah, his three-quarters share of the schooner *Eliza Ann* was valued at $900, representing more than half the value of his entire estate ($1,674.50). When Captain Richard Owens of Glynn County died in 1850, $4,250 of his entire $4,322.50 estate reflected his ownership in two vessels.[142]

Some information on the relative financial success of individual captains is available in the federal census for 1860, which provides information on the value of an individual's real estate separate from their personal estate. Table 6.5 presents information on the value of real estate and personal estates for several Georgia coasting captains. The list includes those men who were serving as masters in 1860, or in the immediately preceding years, and it is separated by those masters residing in the South and those residing in Northern states. The valuations given for several of the Savannah shipowners and

[140] "Conveyances of Enrolled Vessels, Savannah," sloop *America*, February 9, 1855, schooner *Northern Belle*, May 8, 1855, schooner *William Totten*, June 21, 1855, schooner *Company*, January 22, 1851, and schooner *Col. Long*, June 7, 1855.

[141] Chatham County, Georgia, deed of sale, Joseph E. Silveria to John Robbins, October 20, 1841, Deed Book 2Z, 392–92; ibid., deed of sale, John Russell to S. B. Hills, July 7, 1839, Deed Book 2W, 493–94.

[142] Chatham County, Georgia, estate of Domingo Galleo, Inventory and Appraisement, Estate Accounts File G-178; Glynn County, Georgia, estate of Richard Owens, Wills and Appraisements, Book F, 187, 268–69.

merchants most heavily involved in handling coastal commodities are provided for comparison.[143]

Significant differences exist among these groups. Many of the Southern captains owned no real estate, and the values of their personal estates were low. The lack of ownership of real property is expressed in the fact that several of the captains living in Savannah boarded with other men involved in the coasting trade. Some captains' ownership of a coasting vessel constituted a significant proportion of their personal estate. However, in some instances, the personal estate does not seem to include vessel ownership. Andrew Hanson is believed to have owned the schooner *Fort George Packet* in 1860, but his personal estate is valued at only two hundred dollars. Either Hanson's ship ownership is not included in the amount, or the schooner had an extremely low value.[144]

As a group, the coasting captains living in Georgia, South Carolina, and Florida were not wealthy; many seem to have owned no real estate, and the values of their personal estates were low. An exception is Charles Stevens of St. Simons Island. By 1860, he had acquired several hundred acres of land at the old town of Frederica, accounting for the high value of his real estate. Stevens's high personal estate value of $11,250, the highest among active captains living in the South, reflects his possession of sixteen enslaved persons that year, in addition to his ownership of the schooner *Northern Belle* and, possibly, his half ownership of the schooner *Florida*.

In contrast to the captains residing in the South, those living in the North typically owned real estate and personal estates of moderate to high value. By 1860, some of these men, such as Pelag Blankenship and Stephen Hadley, had given up commanding coasters but remained involved in the ownership of vessels. Regardless, as shown in Table 6.5, the Northern captains, as a group, were considerably wealthier than their Southern counterparts.

The list of shipowners and merchants living in Savannah provided in Table 6.5 includes those men most heavily involved in handling coasting produce and a few principal shipowners. Henry J. Dickerson was more involved in the ownership of coasters than any other Savannah or Georgia resident,

[143] US Census Bureau, "U. S. Censuses of Population, Camden, Chatham, Glynn, Liberty and McIntosh counties, Georgia, Beaufort County, South Carolina, Nassau County, Florida, and Bristol and Plymouth counties, Massachusetts, 1860."

[144] Enrollment No. 5, Port of Savannah, schooner *Fort George Packet*, October 13, 1858.

accounting for much of the value of his personal estate. Henry Brigham was a commission merchant, but he also owned several coasters. Henry Frederick Willink, who sailed as a coasting captain early in his career, became wealthy operating a shipyard. By 1860, his estate was valued at fifty-nine thousand dollars, and he held twenty-three people in bondage. Like Henry Dickerson, Willink's move into the upper economic class of Savannah occurred after and outside his tenure as a coasting captain.[145] The men in this list were among the wealthiest in Savannah, and their great wealth stands in contrast to that of the coasting captains; those men whom they depended on to move coastal crops and merchandise to and from market.

[145] Fraser, *Savannah in the Old South*, 273.

# Chapter 7

## "Georgians are madly devoted to cotton": The Georgia Coasting Trade from 1800 to 1830

In 1800, the Georgia coasting trade, although in existence for almost seventy years, was limited in the number of vessels involved and the quantity of goods carried. However, this trade was on the verge of rapid growth as the population along the coast from South Carolina to Florida increased and, most importantly, as the cultivation of Sea Island cotton and rice expanded. This growth trend would continue over the next thirty years, after which the sailing trade declined. This expansion in trade in the first three decades of the nineteenth century went hand in hand with the advance of Savannah as a mercantile center, reflected in Ebenezer Kellog's 1817 comment that "The Georgians are madly devoted to cotton."[1] One direct expression of this growth is the increasing number of coasters arriving at the port of Savannah after 1800. Figure 7.1 presents information on the number of local coaster arrivals into Savannah and the number of vessels involved for various years between 1800 and 1861. This information is derived from shipping lists published in several Savannah newspapers. For 1800, the *Columbian Museum & Savannah Advertiser* and *Georgia Gazette* list only twenty arrivals by fourteen ships from coastal ports and plantations between Beaufort, South Carolina, and the St. Johns River in Florida. By 1809–1810, sixty different sloops and schooners made ninety-eight trips into Savannah from this same area; by 1820, these numbers had risen to 213 voyages into the city by eighty-three vessels and, by 1830, reached 479 arrivals, again made by eighty-three different sailing ships. Over the first three decades of the nineteenth century, there was a six-fold increase in the number of coasters working in the trade, while the number of Savannah arrivals had increased by a factor of twenty-four. In addition, the

[1] Martin, "New Englander's Impression," 252.

number of regional ports of origin for coasters sailing into Savannah rose from four in 1800 to forty-six named locations in 1830.[2]

The year 1830 represented the peak of participation by sailing vessels in the Georgia coasting trade. After that year, the number of these ships in the trade declined. Although the coasting trade in Georgia grew significantly during the first thirty years of the nineteenth century, that growth did not simply correlate with regional population and agricultural expansion. National and international economic and political issues influenced the American coasting trade, and several had damaging impacts against which the trade struggled. Much of the Sea Island cotton and rice produced in Georgia, South Carolina, and Florida was carried directly to overseas ports, so the international demands or needs for these crops influenced their price in the market centers of Savannah and Charleston. These price trends influenced production that, in turn, affected the amounts available as cargo to coasting ships. For example, Sea Island cotton prices rose steadily to about 1818, encouraging coastal planters to increase cultivation, ultimately providing more cargo for coasters. However, after 1818, cotton prices dropped significantly during the nationwide economic depression known as the Panic of 1819. Rice prices also dropped, and these low prices curtailed the cultivation and shipment of these two critical cargoes.[3]

Before 1819, other events occurred to curtail the general growth trend in the Georgia coasting trade. Among these was the Embargo Act of 1807, which affected the coasting trade throughout the United States, not just in Georgia. This act prohibited all trade with foreign ports and was imposed by President Thomas Jefferson, principally in retaliation to British and French trade restrictions and harassment of American shipping. This act was unpopular and was replaced in 1809 by the Non-Intercourse Act, a more lenient statute that opened foreign trade with all countries except Britain and France. The Non-Intercourse Act of 1809 was no more acceptable than the Embargo Act and was repealed in 1810.[4]

Although in effect for less than four years, these two acts created hardships on those involved in the coasting trade in Georgia by closing off the primary overseas markets of the region's planters. The War of 1812 also had a detrimental effect on the coasting trade. British markets were closed to

[2] *Republican & Savannah Evening Ledger*, April 1809-April 1810; *Daily Georgian*, 1820; *Daily Savannah Republican*, 1830.

[3] Gray, *History of Agriculture*, 2:1030–1031.

[4] Sears, *Jefferson and the Embargo*.

Georgia crops, and British naval forces restricted and interrupted the activities of vessels all along the Southern Atlantic coast. This disruption is seen in a significant decrease in coaster arrivals at Savannah in 1813 (see Figure 7.1). That year, the *Republican & Savannah Evening Ledger* reported fifty-three arrivals of coasters into Savannah from regional coastal ports other than Charleston. This was half the volume of coaster traffic that existed just three years earlier, before the outbreak of hostilities.

The impacts of the Embargo and Non-Intercourse Acts and the War of 1812 were serious but not long-lasting. After 1814, the trade quickly revived, and by 1820, it had stabilized and was entering a period of expansion that, for the sailing coasters, continued for the next decade. Two events occurred around 1820 that had long-lasting effects on the Georgia coasters. One of these was the acquisition of Florida from the Spanish in 1821, which changed forever the nature of trade that had existed between Georgia and Spain. The other was the introduction of steamboats into the coasting trade in about 1820. Initially, steamboats had little impact on the sailing trade. It would not be until the 1830s, ten years after their first appearance on the Georgia coast, that steamboats made serious inroads into the activities of the sailing coasters.

In the following discussions, published newspaper listings of vessel arrivals and departures were used to assess the pattern and structure of the Georgia coasting trade during the first three decades of the nineteenth century. To do this, several years of Savannah newspapers were fully examined to obtain information on vessel arrivals and departures and, when available, cargo carried. These years are 1800, 1805, 1808, 1809–1810, 1813, 1820, 1821, 1825, 1827–1828, and 1830. In addition, newspapers from other years were scanned for coaster information, including shipping activities, advertisements, and sale notices. Also, several Charleston newspapers were examined, although not completely, and primarily to complement information provided by Savannah papers and to gather comparative data. In addition, the *Darien Gazette* for the years 1819 to 1824 was examined to obtain information on coaster activity at this small, local port.[5]

## The Georgia Coasting Trade in 1800

A full assessment of the nature and extent of the local sailing trade in Georgia in 1800 is fraught with difficulties because so few official records were maintained. Shipping lists printed in newspapers provide the most comprehensive

[5] The newspapers examined are named in the bibliography.

information on the activities of coasters. While these listings are invaluable for this study, they do have limitations. Newspapers in both Savannah and Charleston were publishing shipping information by 1800, but these listings omit some categories critical to a full understanding of the coasting trade. Typically, newspapers in 1800 printed only the type and name of the vessel, the name of the master and the port of origin for arriving ships, and the port of destination for departing vessels. Information on the cargoes carried by local vessels rarely appears in newspapers before about 1810 and is inconsistent until the 1820s. Thus, for the first two decades of the nineteenth century, the number and the identity of the vessels sailing into and out of Savannah in the local trade can be determined by looking at ports of origin and destination. However, it is impossible to identify the types or quantities of commodities moving in the trade from published shipping lists. Some Charleston newspapers name cargoes carried into that city for the period before 1820, but before about 1810, these papers do not regularly list vessels operating in the local trade.

Savannah newspapers record only a segment of the entire Georgia coasting trade for the years before about 1830 because many coastal planters were shipping their crops to Charleston. During the first decade of the nineteenth century, Charleston newspapers, like their Savannah counterparts, rarely named cargoes in their shipping news. This makes it impossible to establish how much of the Georgia production was carried into Charleston during these early years. But, based on the number of coasters sailing into Charleston from Georgia locations, it was considerable. Also, more vessels carried coastal commodities into Savannah than the sloops and schooners named in the newspapers. Some very small sailing vessels, often referred to as "plantation boats," were not recorded in newspaper listings, and "flats" or "poleboats," which often carried rice and cotton from nearby locations along the Savannah River, are rarely mentioned in newspapers before the 1820s.

Because the Savannah newspapers typically printed little information on coasting cargoes before 1815, it is impossible to correlate the increase in the number of coaster arrivals in Savannah with an increase in the quantity or value of cargo carried. However, the value and amounts of local produce (particularly rice and cotton) shipped on coasters did increase considerably, as evidenced in Savannah and Charleston export statistics and the few records existing on regional crop production.

At the start of the nineteenth century, the entire population of Chatham County, in which Savannah lies, was only 12,946, of whom almost two-thirds were enslaved African Americans. Chatham County contained about half of the total population residing along the Georgia coast, and Savannah was the political, social, and economic center of the state. Sea and river trade were critical to the city's economy, and on any given day, ships of all types and sizes might be unloading goods at the wharves and docks along the riverfront. Many of these vessels carried cargo from the Northeast as well as overseas ports. Some of the imported commodities aboard these vessels were marketed and consumed locally in Savannah. Others were reshipped from Savannah to merchants in small towns and to plantations scattered along the coast or carried inland along the state's major rivers and the few interior roads. Small sloops and schooners typically carried the merchandise from Savannah to coastal communities and plantations. These same ships returned to Savannah laden with those coastal commodities available for shipment.[6]

The first issue of the *Georgia Gazette* for 1800, published on January 2, reported in its "Marine List" column that fourteen vessels had "Entered Inward" and seven had "Cleared Out" since the last issue of the paper, published a week earlier (Figure 7.2). These ships were arriving from and departing for a variety of American, European, and Caribbean ports, and they typify the maritime activity at Savannah during this period. Eight schooners and sloops arrived from ports such as New York, Philadelphia, New Bedford, Boston, Charleston, and Jamaica while six schooners departed for similar ports. These sloops and schooners are the vessels sailing in the local coasting trade, but only one, the sloop *Patty*, sailed from one of the regional ports that were to become important in this trade. This was the town of Brunswick on the central Georgia coast.

Although Savannah newspapers of the period rarely identify the cargoes of arriving or departing vessels, Savannah merchants often advertised the receipt of goods by specific vessel, and these advertisements provide information on arriving cargoes. Typically, vessels from American ports in the Northeast or along the middle Atlantic coast carried general merchandise, building materials, and foodstuffs such as wheat, corn, or salt. In the January 2, 1800, issue of the *Georgia Gazette*, several merchants advertised merchandise recently received by ship. The firm of Macleod & Miller noted they had received "by the Cotton Planter from Charleston, and the Huntress from New York" a variety of foodstuffs, including walnuts, India mangoes, English cheese, and currant jelly, as well as "pewter plates and dishes." Benjamin Maurice & Company advertised it had on hand a variety of foods and dry goods received on

[6] Chatham County population from J. F. Smith, *Slavery and Rice*, Appendix A, Table A-5.

the brig *Mary* from New York. These items included cloth, such as "Fancy calicos," "Ginghams" and "Oznaburg flaxen," as well as rum, apple brandy, beef, and "Virginia pilot bread." On January 30, 1800, the firm of Taylor, Miller & Company reported in the *Gazette* that it had sugar and salt for sale "Received on consignment by the sloop Commerce, James Armstrong Master, from St. Bartholamews," one of the Leeward Islands in the Caribbean. Other newspaper notices and advertisements confirm that the main cargoes from the West Indies were sugar, rum, salt, coffee, and, sometimes, enslaved people. Ships coming from European ports carried an assortment of foodstuff, raw material, and manufactured items.

The only true "Georgia coaster" arriving in Savannah in the first week of 1800 was the sloop *Patty*, sailing from Brunswick under the command of Captain Jonathan Landon, of New Haven, Connecticut. The *Patty* appears a few other times in Savannah newspapers in 1800. The February 27, 1800, issue of the *Georgia Gazette* noted the *Patty*, with Landon as master, departed for Norfolk, Virginia. Nine months later, the November 27 issue of the *Gazette* reported the arrival of the *Patty* from New York. The master of the *Patty* on this voyage was Charles Chalker, who, like Captain Landon, was a resident of New Haven, Connecticut. Approximately a week later, on December 4, the *Georgia Gazette* reported the departure of the *Patty* for St. Marys, under the command of Captain Chalker.

The January 2, 1800, issue of the *Georgia Gazette* also reported that seven ships departed Savannah the previous week. All were schooners bound for the ports of New York, Boston, Havana, Charleston, Pasquotank in North Carolina, Cap Francois in Haiti, and Tortola in the Virgin Islands. It was not until January 9 that another local coasting vessel left Savannah, when the *Gazette* reported the departure of the schooner *Sally* for Sunbury in Liberty County, Georgia.

It is unreported what these two coasting vessels were carrying during the first weeks of 1800. However, Savannah newspapers in 1800 often contain notices by merchants stating they had local products, such as rice and cotton, for sale or that they wanted to buy these items. For example, on January 2, 1800, the firm of Meins & Mackay reported that it had "NEW RICE For sale, in whole and half barrels" while Johnston, Robertson & Company advertised that "Cash Will be Given for Best Quality Sea Island cotton." Coasters like the *Patty* and *Sally* carried some of these goods into Savannah.[7]

[7] *Georgia Gazette*, January 2, 1800.

The cargoes of those vessels departing Savannah, like those arriving, cannot be identified with certainty. However, those coasters leaving for local coastal ports such as Beaufort, Sunbury, Brunswick, and St. Marys typically carried manufactured goods, household items, building materials, foodstuffs and luxury items. Some of these goods were for merchants and stores in small coastal towns and some were destined for specific plantations. The *Darien Gazette,* published from 1818 to the mid-1820s, provides information on the goods carried into that small port town by coasters from Savannah. Between November 1820 and November 1821, coasters sailing from Savannah to Darien carried sugar, salt, mackerel, groceries, dry goods, flour, stationery, coal, corn, whiskey, gin, bricks, nails, hats, bacon, lime, furniture, and iron, and many carried goods identified only as "assorted cargo" or "sundries." Local coasters sailing in 1800 would have carried similar cargo.

A notion of the pattern and scope of the Georgia coasting trade at the beginning of the nineteenth century can be obtained by examining coaster arrivals and departures at Savannah for a full year. This information for 1800 has been collected from shipping information published in two Savannah newspapers, the *Georgia Gazette* and the *Columbian Museum & Savannah Advertiser*. The *Georgia Gazette* was issued every Thursday, and shipping information appeared in a column titled "Marine List." The *Columbian Museum & Savannah Advertiser* was issued twice a week, and shipping information was printed in a column titled "Marine Register." In almost all instances, the entries for vessel arrivals and departures in the two papers are identical. Ships are identified as local coasters if they are sloops or schooners arriving from or departing for coastal locations in lower South Carolina, Georgia, or Northeastern Florida. The shipping lists reveal that thirty-three different coasting vessels made seventy-two arrivals into Savannah in 1800. Eleven of these ships were sloops, and twenty-two were schooners. The number of monthly arrivals in Savannah is evenly distributed over the year; such regularity suggests that the transportation of coastal agricultural products was not the dominant factor in the schedule of coaster sailings. If it were, then coaster traffic would be closely tied to the harvest of these crops and be highest during the winter and spring, when the greatest quantities of rice and cotton were shipped to market. In 1800, many Georgia planters shipped their crops to Charleston. It would be several years before enough cotton and rice were shipped to Savannah to influence the seasonal pattern of coaster arrivals.

### *The Savannah-Charleston Trade in 1800*

The "Marine List" in the January 2, 1800, issue of the *Georgia Gazette* notes that one sloop, the *Henry*, arrived in Savannah from Charleston, and one schooner, the *Gosport*, departed for Charleston. These were two of the many vessels sailing in the most important of Savannah's sailing trades. The early waterborne trade between Savannah and Charleston often involved the transport of local agricultural commodities from Savannah to Charleston. However, it rarely included the movement of similar products from Charleston into Savannah. Thus, the Savannah-Charleston trade must be considered a unique facet of the local "coasting trade" as defined here. However, many of the ships, as well as the captains, participating in the trade between these two cities also transported, at one time or another, agricultural products from all along the Southeastern coast into Savannah and Charleston; therefore, it's appropriate to consider the sailing trade between these two ports at the opening of the nineteenth century.

In 1800, Savannah newspapers reported fifty-two arrivals of sloops and schooners from Charleston and forty departures for that city. This represents more than half of all coasting vessel arrivals in Savannah and almost half of the eighty-six departures. Twenty-nine sloops and schooners sailed between the two cities during the year, with at least one vessel arriving in Savannah from Charleston almost every week. Most of these ships made only one or two trips over the year, but several sailed as regular packets between the two cities. These were the schooners *Orange*, *Savannah Packet, Industry*, and *Cotton Planter* and the sloops *Rachel* and *Dove*. Each of these ships made several voyages between Charleston and Savannah in 1800.

Savannah and Charleston had maintained an active maritime trade since the establishment of Georgia. The survival and ultimate growth of the Georgia colony depended on this trade because so many of the imports into and exports from the colony passed through Charleston. As Savannah grew, an increasing number of ships sailed directly between there and overseas ports, but Charleston remained a principal trading partner, partly because so many planters and merchants in Georgia maintained ties with firms and institutions in the South Carolina capital. By the early nineteenth century, the small vessels sailing from Charleston to Savannah typically transported passengers, general merchandise, and foodstuffs, much of which was being transshipped through Charleston from Europe, the West Indies, or American ports along the middle and upper Atlantic coast. Coastal products, such as cotton or rice, were almost nonexistent in this trade. Passengers were particularly important because the water passage provided the most convenient mode of travel between the two cities. In 1805, one of the regular Charleston-Savannah packets was the Rhode Island-built sloop *Cordelia*, sailed by her captain William Gorham. It was reported that the *Cordelia*'s "accommodations for Passengers are in a style of elegance that cannot be surpassed by any vessel in port."[8]

The vessels sailing in the opposite direction, from Savannah to Charleston, carried passengers and rice, cotton, tobacco, and other products gathered in Savannah and reshipped to Charleston merchants or derived from plantations along the lower Savannah River. Little information on the cargoes carried by these Charleston-bound vessels is available for 1800, but by 1805, Charleston newspapers occasionally identified the cargo, most often cotton and tobacco, carried by coasters from Savannah. Presumably, similar cargoes were being carried in 1800. Although unreported, it is likely the cotton aboard these coasters from Savannah was Sea Island cotton. In 1800, Sea Island cotton production in Georgia was in its early stages of development, but it was a more abundant crop than upland, short-staple cotton. The amount of cotton exported from Savannah in 1800 is unrecorded. Charleston exported 25,157 bags, or over eight million pounds, of cotton in 1800–1801, all or most of which was Sea Island cotton. A considerable amount of the cotton exported from Charleston undoubtedly came from Savannah, carried aboard coasters sailing between the two ports, or from Georgia planters who shipped directly to Charleston. Rice is rarely mentioned as a cargo of the vessels sailing into Charleston from Savannah or other Georgia locations before 1805.[9]

In later years, newspapers reported that many coasters sailed from Savannah to Charleston "in ballast," meaning without cargo. The lack of exportable goods in Savannah during the early years of the nineteenth century likely forced many vessels to return empty to Charleston.

### *The Savannah-St. Marys Trade in 1800*

In 1800, next to Charleston, St. Marys, located on the southern Georgia border, carried on the most intensive coasting trade with Savannah (Figure 7.3). That year, St. Marys was the origin of ten coaster arrivals in Savannah and the destination of twenty-six departures (Tables 7.1 and 7.2). The cargo of these vessels is unreported, but those traveling to St. Marys carried a range of manufactured and household goods, foodstuffs, and luxury items. The cargoes carried from St. Marys must have included some of the Sea Island cotton

[8] *Charleston City Gazette* January 8, 1805.

[9] Charleston export figure from Gray, *History of Agriculture*, 2:680.

then grown in the area, as well as the more unusual crops of the region, such as oranges and limes. These ships also carried merchandise obtained legally or illegally from East Florida, then a foreign country.

The St. Marys area at the time was remote, and the population small. Only 946 Whites and 735 enslaved Blacks lived in Camden County in 1800; however, this was a significant increase over the 305 Whites and 84 enslaved persons living in the county ten years earlier. The almost nine-fold increase in the number of the enslaved, largely reflects the expansion of Sea Island cotton agriculture in the region.

From 1800 to 1820, the St. Marys region was a center for lawlessness due to its border position, isolated location, and Florida's unstable political situation. Smuggling was rampant, and contraband was brought from Florida across the St. Marys River or through the town of St. Marys. Much of this illegal trade originated at Amelia Island, on the Florida coast immediately south of the border. By 1807, the Spanish declared Amelia Island a free port of entry. Slaves became an important item of contraband after 1793, the year Georgia outlawed trafficking slaves from Florida or the West Indies. The importance of smuggling slaves out of Florida increased when the United States banned importing slaves in 1808. Efforts were made to suppress smuggling in the St. Marys region, but it did not end until 1821, when the United States acquired Florida.[10]

The seaborn trade between St. Marys and Savannah in 1800 was dominated by a single ship, the schooner *Alective*, and the two men who served as her captain, John Chevalier and Robert Rudolph. The *Alective* made twenty-two trips between Savannah and St. Marys in 1800. In addition, a vessel named *Friendship* sailed five times to and from St. Marys under captains named Brewster, Guillenet, and Hillary. The *Friendship* is identified as both a sloop and a schooner in the newspapers and may represent two vessels with the same name. The seven other vessels sailing between Savannah and St. Marys represent single listings of arrivals or departures.

Coasters sailed between St. Marys and Charleston in 1800, but this activity was rarely printed in Charleston newspapers. By 1805, the *Charleston City Gazette* was publishing the arrivals of some local coasters in its "Ship News" column, including those from St. Marys. Many of these were the same vessels trading between St. Marys and Savannah. The 1805 *City Gazette* occasionally printed information on cargoes, so some information on the local goods transported in the St. Marys-Charleston trade is available. Among the

[10] Bullard, *Robert Stafford*, 6; 272; Coulter, *Georgia*, 185.

coasters arriving in Charleston from St. Marys in 1805 were the schooner *Phoebe*, with a cargo of "cotton and cedar," reported on January 30; the sloop *George*, with "cedar and staves," on April 9; the sloop *Two Sisters*, with staves on July 1; the sloop *Akerly*, with cedar on April 23; the schooner *Industry*, with "naval stores and lumber" on October 17; and the sloop *Polly*, with cotton, on November 22. One of the regular coasters in the Charleston-St. Marys trade was the sloop *Ruth*, under Captain Daniel G. Phipps of New Haven, Connecticut. Several advertisements for the *Ruth* in the 1805 *City Gazette* stated she was "fast sailing," with "good accommodations" for passengers. Another of the regular traders between St. Marys and Charleston in 1805 was the sloop *Polly*, commanded by Camden County resident John Chevalier.[11]

The information on cargoes published in the 1805 *Charleston City Gazette* indicates that goods arriving from St. Marys consisted of forest products, such as cedar, lumber, staves, naval stores, and cotton. It is presumed similar cargoes were carried into Savannah.

Savannah maintained an active sailing trade with Spanish Florida in 1800. That year, Savannah newspapers reported six schooners sailed into the city from St. Augustine, the capital of East Florida, and there were five departures for that city (see Tables 7.1 and 7.2). Trade with St. Augustine at the time involved miscellaneous goods, such as cloth, rope, rum, salt, coffee, and, lesson commonly, food, such as oranges and turtles. Sea Island cotton was grown in Northeastern Florida at the time, but how much was carried by coasters into Savannah is unreported.

### *Savannah's Trade with Other Local Ports in 1800*

What most likely constituted the movement of plantation products in the coasting trade in 1800 is represented by the small number of vessels sailing into Savannah from regional ports other than Charleston, St. Marys, and St. Augustine. These vessels were few; only four arrivals into Savannah are identified in 1800. Two of these ships sailed from Beaufort, South Carolina. These were the schooner *Fox* and sloop *Harriet*, each making a single trip into Savannah (see Table 7.1). What these vessels carried is unidentified, but Beaufort was a center of rice production, and cotton cultivation was becoming important.

The *Fox* and *Harriet*, along with the sloops *Two Friends* and *Cinderilla*, sailed from Savannah to Beaufort in 1800. These ships carried merchandise

[11] Sailing information from *Charleston City Gazette*, 1805.

from Savannah to merchants and planters in Beaufort or Beaufort County and carried passengers between the two areas.[12]

Only two other coasters that might have carried local produce arrived in Savannah in 1800, and both sailed from Brunswick. These were the sloops *Patty* and *Eliza* (see Table 7.1). It is possible these sloops carried Sea Island cotton.[13]

It is apparent that few vessels entering Savannah in 1800 were involved in the movement of plantation produce. About fifty-seven million pounds of rice were exported from the United States in 1800, the majority grown in South Carolina. How much rice was produced in Georgia is unrecorded, but little of it seems to have been carried into Savannah aboard coasters. Many Georgia rice growers dealt with Charleston factors and shipped their rice there. Further, most of the rice produced in Georgia in 1800 was along the Savannah River, making it convenient to ship it to Savannah aboard small flats or plantation boats, whose arrivals were not reported in newspapers.[14]

By 1800, the cultivation of Sea Island cotton along the Georgia coast was well established and expanding. Export data for Savannah are sketchy for the first two decades of the nineteenth century, but in 1795, the entire cotton crop of the United States was thirty-five thousand bales, much of it Sea Island cotton grown in Georgia and most of it shipped through Charleston. These cotton exports, coupled with rice, made Charleston one of the greatest exporting points in America.[15]

In 1800, several vessels departed Savannah for small coastal ports soon to become important in the regional coastal trade. These included the towns of Beaufort, Brunswick, Sunbury, and Hardwick, as well as St. Simons Island (see Table 7.2). These coasters carried passengers and miscellaneous merchandise, food, and the like.

### *Georgia Coasting Ships and Captains in 1800*

Outside of those sailing to and from Charleston, St. Marys, and Florida, few ships sailed out of Savannah in the Georgia coasting trade in 1800. Those that can be identified are the sloops *Cinderilla*, *Eliza*, *Friendship*, *Fox*, *Harriet*, *Mary*, *Patty*, *Three Sisters*, *Two Friends*, and *William* and two schooners, the

[12] Sailing information between Savannah and Beaufort from the *Georgia Gazette*, 1800.

[13] Sailing information between Savannah and Brunswick from the *Georgia Gazette*, 1800.

[14] Rice export figures from Gray, *History of Agriculture*, 2:1030.

[15] U. Phillips, *History of Transportation*, 45, 57.

*Sally* and the *William and Sally*. The small, thirty-two-ton *Cinderilla* sailed in the Georgia trade until 1810 as a regular trader between Savannah and Beaufort. In 1795, the *Cinderilla* was registered in New Haven, Connecticut, with Daniel Goff Phipps as her master. Daniel Phipps served as the *Cinderilla*'s captain until at least 1800.[16]

In 1800, the sloop *Three Sisters* principally served the central Georgia coast, making three voyages to Brunswick and St. Simons Island. In 1801, young Reuben King obtained passage on the *Three Sisters* sailing from Savannah to Darien. King's account of the voyage represents one of the few descriptions of passenger travel aboard a coaster, and his was not a pleasant experience. He wrote that the *Three Sisters* departed Savannah in mid-February with fourteen passengers aboard bound to Darien along the "inland passage." In six days, the sloop could travel only to Sapelo Sound, less than sixty miles, where the frustrated King left the ship to continue his travels overland.[17]

The sloop *William* made a single voyage to Hardwick on the Ogeechee River in 1800. The master of the *William* was a Captain Lightbourn, believed to be Samuel Lightbourn, a native of Bermuda who had been sailing along the coast since the 1790s and remained in the trade to about 1820.[18]

Several of the coasters sailing into and out of Savannah in 1800 were owned and homeported at locations such as New Bedford and Somerset in Massachusetts; Portsmouth and Newport, Rhode Island; and New Haven, Connecticut. Few of these early vessels, or their masters, can be associated with the Sippican region that soon came to figure so prominently in the Georgia coasting trade.

## *Savannah Merchants in the Coasting Trade in 1800*

Much has been written about the lack of a developed mercantile and financial infrastructure in Savannah during the first twenty or thirty years of the nineteenth century, forcing many coastal planters in Georgia to trade with Charleston.[19] Although Charleston attracted a considerable amount of the Georgia coastal production, there was a small group of Savannah commission merchants and factors handling coastal produce in 1800. Among these was Joseph Machin, who advertised in January 1800 that he had "commenced the FACTORAGE BUSINESS" and had ample space for receiving produce "on

[16] "Connecticut Ship Database," sloop *Cinderilla*; 1795.

[17] King's voyage reported in Fleetwood *Tidecraft*, 99.

[18] Sullivan, *From Beautiful Zion*, 39–41, 50.

[19] See House, *Planter Management*, 57–82.

Mr. Clay's Wharf, and on Mr. William Wallace's Wharf at Yamacraw." That same month, two firms, Johnston, Robertson & Company and Taylor, Miller & Company, advertised they wanted to buy Sea Island cotton, and three months later, George Anderson, founder of what was to become one of the most important factorages in Savannah, advertised that he had "200 Tierces Rice" for sale.[20]

One of the oldest mercantile firms in Savannah was started by James Habersham in the 1740s. Joseph Habersham succeeded his father, James, and by 1803, was advertising that his firm was continuing in the "Factorage & Commission Business" in the building formerly used as the "custom house."[21]

The factors and commission merchants of Savannah were critical elements in the region's agricultural economy and were the key figures in marketing the crops. The factors served as a combination merchant, buyer, and banker for the planter, providing an outlet for plantation crops, a source of credit, and as a store for many of the finished and luxury goods plantation owners desired. In Savannah and Charleston, these merchants received plantation products carried by the coasters, commonly paying the freight charges directly to the coasting captains. The factor would then charge this payment, as well as other incurred costs, such as drayage, storage, or insurance, and his own commission for handling, storing, and selling the rice or cotton to the account of the plantation owner whose cargo was being received. Usually, these commission costs amounted to 2.5 percent of the gross sale price of the crop. The factor also arranged or handled the shipment of merchandise requested by individual planters.

The exact number of factors and commission merchants operating in Savannah in 1800 is unknown, but the number would increase rapidly over the next two decades. In 1820, the *Daily Georgian* listed eighty-three different individuals and firms receiving produce brought into Savannah aboard sailing coasters. By 1823, sixty-one individuals and firms specifically named as "factors" or "commission merchants" operated in the city. It is unknown how many of these handled coastal produce, but many certainly did.[22]

Savannah seems to have had adequate physical facilities for handling coasting cargoes by 1800. Many wharves and storehouses were located along

[20] *Columbian Museum & Savannah Advertiser*, January 2–3 and March 4, 1800.

[21] *Columbian Museum & Savannah Advertiser*, June 11, 1803; Shelnutt, "Robert Habersham," 4–5.

[22] See Woodman, *King Cotton*, and House, *Planter Management*, for discussions of the duties of factors and commission merchants and their relationships to planters; Kayser & Co., *Commercial Directory*, 38.

the riverfront. Names of wharves mentioned in newspaper notices and advertisements include Clay's, Clark's, Alger's, N. Anciaux's, Williamson's, Dennis's, Putnam's, Morel's, Belcher's, Watt's, and William Wallace's, located at Yamacraw at the upper end of the city. In 1801, Alexander Watt offered a "Wharf Lot" for lease that had a new "head and crane" for handling cargo and two storehouses. One was a three-story building "with piazzas," and the other was a two-story structure measuring seventy-four by twenty feet.[23]

At the start of the nineteenth century, Savannah also contained several businesses supplying ship chandlery and other stores and materials for vessels. However, the city was not an important shipbuilding center, and none of the Georgia coasters built before 1810 appear to have been constructed in Savannah.

## Slow Expansion of the Coasting Trade, 1800 to 1810

Savannah newspapers published between 1800 and 1810 show the number of local coasting vessels sailing in and out of the city remained low until 1807. Throughout these seven years, shipping in the port was dominated by vessels coming from or departing for European, West Indies, or Northeastern American ports. In January 1805, thirty-two ships, brigs, schooners, and sloops arrived from or departed for distant ports, such as Liverpool, Cadiz, Jamaica, St. Thomas, Barbados, New York, New London, and Norfolk.[24] During this same month, only seven arrivals and departures were reported for vessels sailing to or from local ports south of Charleston. During 1805, the *Georgia Republican & State Intelligencer* listed only ten arrivals and fourteen departures identified as local coasters, excluding vessels sailing between Savannah and Charleston (See Table 7.3).

In 1805, the regional sailing trade at Savannah continued to be dominated by traffic to and from Charleston, with several vessels working as regular traders between the ports. These included the schooners *Rising Sun* and *Carolina* and the sloops *Delight*, *Cordelia*, and *Julia*. Excluding ships sailing between Savannah and Charleston, the *Georgia Republican & State Intelligencer* for 1805 names eighteen vessels in the local coasting trade. Six of these were schooners, and twelve were sloops. Most made only a single voyage into or out of Savannah. As had been the case in 1800, excluding Charleston, the town of St. Marys was the most important regional trading partner with Savannah in 1805. Captain John Chevalier continued to be an important figure

[23] *Columbian Museum & Savannah Advertiser*, April 28, 1801.

[24] *Georgia Republican & State Intelligencer*, January 1805.

in the St. Marys-Savannah trade, although by 1805, he was sailing the sloop *Polly*, not the schooner *Alective*. St. Augustine also remained an important trading partner. The few other local ports visited by coasters in 1805 were Beaufort, Hardwick, Sunbury, Sapelo Island, and Darien.

The *Georgia Republican & State Intelligencer* lists a single vessel, the sloop *Experiment*, departing Savannah for Darien in 1805. At this time, a considerable portion of the Darien coasting trade was with Charleston, so the town is rarely mentioned in Savannah newspapers. This was true of other Georgia ports (and plantations) as well, and several of the coasting vessels and captains sailing between Savannah and local ports in 1805 were also sailing between those ports and Charleston.

On November 22, 1805, the *Charleston City Gazette* reported the arrival of Captain John Chevalier and his sloop *Polly* from St. Marys with a cargo of cotton, one of the earliest cargo identifications for a local coaster. The Savannah newspapers would not regularly include this information for another decade. As in 1800, the arrivals of coasters at Savannah from local ports in 1805 were evenly distributed over the entire year (see Table 7.3). There were no significant increases in arrivals during the cotton and rice shipping season, between November and April, suggesting that the transport of these crops did not dominate the activities of Georgia's small coasting fleet.

The several years after 1805 saw little change in Georgia's coasting trade. Savannah's sailing trade remained dominated by traffic to and from Charleston, with several vessels working as regular traders between the ports. St. Marys remained the second most important regional trading partner with Savannah, followed by St. Augustine and Beaufort. The few other local ports visited by coasters between 1805 and 1810 included Brunswick, Hardwick, Sunbury, Darien, Frederica, and several of the coastal islands: St. Simons, Sapelo, St. Catherines, Skidaway, and Jekyll.

The town of Hardwick on the lower Ogeechee River was a port of call between 1805 and 1810, but usually, less than a half dozen vessels sailed to and from the tiny settlement yearly. In January 1806, the sloop *Jane*, under command of a Captain Hammond, departed Savannah for "Ogeechee," the first time the name of the river itself, rather than Hardwick, is listed as a port of call.[25] Sunbury retained a slightly more important position than Hardwick, with as many as a dozen or so voyages originating there yearly through the 1850s.

[25] Ibid., January 7, 1806.

Brunswick and St. Simons Island, or sometimes the town of Frederica, occasionally appear as points of origin or destination for coasters, but before 1807, there are only a handful of sailings to and from the island each year. After 1807, the number of sailings to and from St. Simons increased due to the prosperous Sea Island cotton plantations established on the island.[26]

By 1810, several other Sea Islands visited by coasters out of Savannah were the locations of Sea Island cotton plantations. The sloop *Anubis*, under her master, Captain DuBignon, sailed to and from Jekyll Island several times in 1808 and 1809. The master of the *Anubis* seems have been Joseph DuBignon, a member of the family of exiled French aristocrats who then owned Jekyll.[27]

Before 1810, Sapelo Island appears in newspaper shipping lists a few times in relation to coasting vessels. At the time, the island was owned by Thomas Spalding and members of exiled French families. What these ships carried is unreported, but Sea Island cotton was likely the principal cargo. Thomas Spalding was experimenting with sugar production on Sapelo at this time and may have shipped sugar or syrup from the island.[28]

In 1805, only a single coaster sailed from Darien into Savannah, by 1808 this number had risen to four, and the following year to nine. By 1820, ninety-one coasting voyages into Savannah originated at Darien, more than any other local port, and the town on the Altamaha River retained this preeminent position in the local trade into the 1840s. Darien's growing importance as a coasting port after 1805 was partly due to the expansion of Sea Island cotton and rice cultivation in the region. The lower Altamaha River was well suited to tide flow agriculture, and several productive rice plantations were established there, but not until after 1820. The most important contributor to Darien's rise as a coasting port in the second decade of the nineteenth century was the increasing amount of upland cotton carried down the Altamaha River from the interior, where cotton cultivation expanded at an extraordinary rate after 1805. This cotton was gathered in Darien and loaded aboard sailing coasters for shipment to Charleston or Savannah.

In December 1807, Congress passed what was known as the Embargo Act of 1807, prohibiting all seaborne trade with foreign ports. This Act was urged by President Jefferson as retaliation for British and French trade restrictions and harassment of American shipping. The Embargo Act was not

[26] Vanstory, *Georgia's Land*, 130–59.
[27] Keber, *Seas of Gold*, 3, 146–82, 197, 208.
[28] Sullivan, *Early Days*, 85–87.

aimed directly at coasters operating along the Georgia and South Carolina coasts, because they did not normally sail to foreign ports. However, it affected them in several ways. For one, stopping rice and cotton exports produced a decline in prices, lessening production, and in turn, the amount available for coasters to carry. Additionally, the Embargo Act prohibited trade with Spanish Florida, an activity in which some Georgia coasters engaged. The act also added directly to the paperwork and potential cost of operating a coaster by requiring all coasting vessels to post a bond to the Collector of Customs double the value of the vessel and its cargo.[29]

Notice of the embargo began to appear in Savannah newspapers in early 1808. On January 8, the *Public Intelligencer* reported that no American ship could sail to a foreign port, but that any foreign vessels in American ports could leave. The paper noted that coasting vessels needed to post the new, increased bond with the Collector of Customs at their departure port. A month later, the paper carried a notice from the Customs Collector for Savannah stating that "All vessels owned by a citizen of the United States and sailing under the American flag, of whatever description they might be" were forbidden to sail to any foreign port, including boats under five tons such as "batteaux, canoes and flats."[30]

The Embargo Act failed to influence Britain or France, and it was extremely unpopular in the United States. Many ship captains and owners flaunted it. Some vessels out of Savannah and Charleston continued to sail to and from Florida ports illegally, and smuggling through St. Marys was active. Accounts of this illegal activity occasionally appeared in Savannah newspapers. For example, the seizure of the schooner *Maggy* by a United States gunboat for making an illegal voyage from Savannah to St. Augustine and Amelia Island, Florida, was reported. The sloop *Polly* was also involved in illegal activities and sold with her cargo of "sugars" at a marshal's sale in Savannah for "violation of the embargo laws."[31]

In March 1809, just three days before Jefferson left office, Congress repealed the Embargo Act and replaced it with the Non-Intercourse Act, opening foreign trade with all countries except Britain and France. This act was unpopular, failed in its objectives, and was allowed to expire in 1810 to be replaced by a law known as Macon's Bill No. 2. This bill lifted the embargo

[29] The requirements and implementation of the Embargo Act are reported in the Savannah newspaper *Public Intelligencer*, January 8, 1808.

[30] *Public Intelligencer*, February 5, 1808.

[31] *Public Intelligencer*, May 3 and October 11, 1808.

against both Britain and France but contained the provision that the embargo would remain lifted only for the country that first agreed to stop its harassments and seizures of American vessels. France agreed to this in writing, while Great Britain did nothing. In February 1811, President James Madison reinstated the embargo against Great Britain, one of the factors leading to the War of 1812.[32]

Although in effect for only a short time, the Embargo and Non-Intercourse acts damaged the American shipping industry and the businesses and communities tied to it. Between 1807 and 1808, the value of American exports dropped from $108 million to $22 million. With its great reliance on shipping, New England was particularly hard hit by the embargoes, which certainly affected the economies of Savannah and Charleston. To what extent the embargos affected the vessels sailing in the Georgia coasting trade is difficult to determine. The embargo was implemented at a time when the agricultural economy of the Southeastern coast was expanding, and it certainly ended coastal planters' ability to sell their crops to their most important foreign markets, meaning they may have shipped less to Charleston and Savannah. Small amounts of the coastal crops of rice and cotton were carried to ports in the Northeast, such as New York, Providence, and Philadelphia, and the prices paid during the embargo years were low. Even when planters shipped rice and cotton to Charleston and Savannah, the incomes they received were significantly less than in previous years, plus the crops might languish in the care of a factor for a considerable time before a buyer was found.

The embargo was implemented just at a time when Savannah was on the verge of an economic expansion that would ultimately enhance its position in the regional coastal trade relative to Charleston. In January 1808, an announcement for the sale of shares in the recently formed and incorporated "Planters Bank" in Savannah appeared in the *Public Intelligencer*. This was the first bank in Savannah and its formation reflected the growing availability of capital within the city, important to planters as well as other businessmen, including the owners and operators of coasting vessels.[33]

The newspapers of the period carried advertisements for the increasing numbers of merchants, stores, factors and commission merchants in the city. Among these were John P. Williams, a "Factorage & Commission business on Morel's Wharf," who would make "liberal advances" to planters who put their crops in his possession. Joseph Habersham, from his offices at Coffee-

[32] Sears, *Jefferson and the Embargo*.

[33] *Public Intelligencer*, January 6, 1808.

House Wharf, No. 10, advertised that he had "large and safe stores for the reception of COUNTRY PRODUCE, and will pay attention to the interest of those who may entrust him with their business." Other Savannah factors advertising in the local papers included Williamson & Morel, who published they were prepared "for the reception of produce" at Morel's Wharf; James & John I Gray; Mead & Jenkins; R. Richardson & Co.; and the firm of Bulloch & Glen, who opened their business in 1807 at Smith's Wharf.[34]

The city saw the growing numbers of commission merchants as a source of income and, in 1808, passed an ordinance placing a tax of fifty cents on every hundred dollars of "Goods, Wares and Merchandise" sold on commission that were not produced in Georgia. Recognizing the importance of the agricultural goods produced in South Carolina that were handled by Savannah commission merchants, particularly the rice and cotton crops from plantations along the lower Savannah River, the ordinance excluded "Rice, Cotton, Lumber, Corn, Tar and unmanufactured Tobacco" grown in that state from the tax.[35] The ordinance, while an effort to bring revenues into the city, was designed to protect and stimulate domestic production.

Despite the embargo, the coasting trade remained active. In 1808, the *Republican & Savannah Evening Ledger* reported the arrival of sixty coasting vessels from regional ports, except Charleston. St. Marys remained the principal port of origin, followed by Sunbury, Frederica, St. Simons Island, Darien, and Hardwick. Only a single vessel arrived from St. Augustine that year. This dramatic decrease in sailings between Savannah and Florida reflects the Embargo Act's prohibition against trade with foreign ports by American-owned vessels.

Although the embargo ended trade to Spanish Florida, it did not curtail the local trade to any great extent. The sixty arrivals of coasters from local ports reported by the *Republican & Savannah Evening Ledger* during the height of the embargo in 1808 represented six times more arrivals than had occurred just three years earlier. Only thirty-one, or about half of these coasters, arrived in Savannah during the principal rice and cotton shipping season, from November to April. In later years, beginning in the mid-1820s, 75 to 80 percent of all arrivals occurred during the shipping season, directly reflecting the importance of transporting these crops in the Georgia trade.

[34] *Georgia Republican & State Intelligencer*, January 7, 1806; *Republican & Savannah Evening Ledger*, January 22, 1807; *Republican & Savannah Evening Ledger*, October 15, 1807.

[35] *Public Intelligencer*, January 6, 1808.

## The Beginnings of Growth: The Georgia Coasting Trade in 1809–1810

By the end of the first decade of the nineteenth century, the Georgia coasting trade was taking on the form that would exist for the next half-century. Coasting ships were still sailing into Savannah from Charleston, reflecting the continued importance of trading ties between the two ports. However, Savannah, with a population of 5,212 in 1810, was growing increasingly important as a port, at Charleston's expense. The institution of the embargo in 1808 crippled Charleston's commerce, and it has been argued that this initiated the decline of Charleston as a seaport. By 1810, Savannah's reliance on Charleston as a source of manufactured goods and foodstuffs decreased, as more goods were being shipped directly into Savannah rather than through Charleston. This trend is seen in the decreasing number of coasters traveling between the two cities. In 1800, sloop and schooner arrivals in Savannah from Charleston constituted fully 72 percent of all the small vessel arrivals from local coastal ports. By 1809–1810, these voyages accounted for only 31 percent of arrivals.[36]

In addition to an increase in the number of coasters sailing into Savannah, the number of named locations from which they sailed grew from four in 1800 to thirteen in 1809–1810, excluding Charleston (Figure 7.4). The repeal of the Embargo Act in March 1809 and its replacement with the Non-Intercourse Act meant that coasters out of Savannah and Charleston could once again trade with Spanish ports in Florida.

The *Republican & Savannah Evening Ledger*, from April 1809 to April 1810, and its "Port of Savannah" column were used to examine the workings of the Georgia coasting trade at the end of the first decade of the nineteenth century. In this one year, the paper listed ninety-eight arrivals of sloops and schooners from thirteen different coastal locations south of Charleston (see Figure 7.4). These ninety-eight arrivals were made by fifty-nine different vessels, representing a 300 percent increase in the number of local ships sailing into Savannah since 1800. This increase was due mainly to Savannah's growing importance as a regional commercial center and to the renewal of trade with Spanish Florida. The increase was also due to the expansion of cotton and rice agriculture in the region and the transport of these crops into Savannah. However, even with the expansion of Savannah's mercantile

[36] Savannah population figures from Coulter, *Georgia*, 278; Coker, *Charleston's Maritime Heritage*, 154, writes on the effects of the embargoes on Charleston; sailing information from the *Georgia Gazette*, 1800; *Columbian Museum & Savannah Advertiser*, 1800 and *Republican & Savannah Evening Ledger*, 1810.

infrastructure, a considerable amount of the rice and cotton grown in the region was still being shipped to Charleston or to Florida.

Starting in January 1810, the *Republican & Savannah Evening Ledger* began to include information on cargo carried by arriving coasters and the name of the person or firm to whom cargo was consigned. Although the listing of cargoes appears to be complete only for the first four months of 1810, it represents some of the earliest specific information on cargoes of the Georgia coasters and the names of Savannah merchants receiving them. Most of those vessels coming into Savannah from Charleston were carrying the manufactured goods and foodstuffs that had always been carried on the Charleston-to-Savannah route. These cargoes included "merchandise," rope, glassware, coal, bricks, iron, cotton bagging, and various foods, such as wine, sugar, salt, tobacco, gin, and flour (Table 7.4). Most cargoes represent items transported from the Northeast or overseas and then transhipped to Savannah.[37]

### *Renewal of Trade with Florida*

With the repeal of the Embargo Act in 1809, trade with Spanish Florida was once again legal. The coasting captains quickly reentered this trade, and vessels from Savannah and Charleston began to sail to and from Florida in large numbers. On May 4, 1809, less than two months after the repeal, the sloop *Rosetta* arrived in Savannah from "Amelia," a reference to Amelia Island off the Northeastern Florida coast. Over the next twelve months, thirty-eight coaster voyages into Savannah originated in Florida; thirty-five of these came from Amelia Island. During this same period, the *Republican & Savannah Evening Ledger* listed thirteen coaster departures for Florida, all to Amelia Island. More coasters sailed between Savannah and Florida that year than between any other regional port except Charleston. Based on information available for the first four months of 1810, most of the ships sailing from Florida were in ballast (without cargo) (Table 7.5). Charleston newspapers of the period also show heavy coaster traffic between that city and Amelia Island, with most vessels returning from Florida in ballast.[38]

The trade between Savannah and Florida in 1810 mainly involved transporting cargoes into Spanish Florida, not in carrying goods out of that country. Specific information on the cargo carried by the vessels sailing to Florida

[37] *Republican & Savannah Evening Ledger*, January 23 and April 5, 1810.

[38] Savannah shipping data from the *Republican & Savannah Evening Ledger*, 1810 and Charleston data from the *Charleston City Gazette & Commercial Daily Advertiser*, January-March 1810.

is not included in the newspapers, but other sources reveal these ships were primarily laden with cotton, as were those sailing from Charleston. This cotton was taken to the thriving port at the northern end of Amelia Island, a town sometimes known as Louisa and renamed Fernandina in 1811.[39] There, the cotton was offloaded or transferred to foreign vessels for shipment overseas. Most of the ships receiving the cotton were British or were destined for British or French ports. This trade permitted American merchants and planters to conduct business with these nations indirectly, even though the Non-Intercourse Act prohibited American vessels from trading with them through the first five months of 1810. The volume of this illegal, or at least surreptitious, trade was huge. As previously noted, on January 16, 1810, the *Charleston City Gazette* reported that more than 150 British and American vessels were at Amelia Island transacting business for the shipment of cotton overseas. The amount of cotton carried to Florida reflects the increased crop production in the South, as well as the backlog in the hands of factors and planters since the imposition of the Embargo Act.

*Savannah's Trade with Other Local Ports 1809–1810.*

The number of regional locations listed as ports of origin for coasters coming into Savannah increased from four in 1800 to thirteen in 1809–1810. St. Marys remained the most important Georgia port of origin, with twenty-three arrivals reported in 1809–1810. St. Marys was also the destination for twelve coaster departures from Savannah, more than any other Georgia port. These thirty-five voyages between Savannah and St. Marys were exceeded only by the sailings between Savannah and Spanish East Florida. The cargoes carried to St. Marys are mostly unreported, but some information on those carried from St. Marys to Savannah in 1810 is provided in the *Republican & Savannah Evening Ledger*. As shown in Table 7.5, the newspaper listed five arriving coasters carrying salt, cheese, nails, and cotton bagging. The sloop *Eliza* had "negroes" on board when her arrival was reported on January 18. Most of these goods were obtained from Florida, possibly smuggled into St. Marys to avoid paying duties. The Black passengers on the *Eliza*, who were certainly in bondage, may also have come from Florida, meaning they were smuggled into Georgia, given that the importation of slaves into the United States had been outlawed in 1808.[40]

[39] Bullard, *Robert Stafford*, 65.
[40] Hoffman, *Florida's Frontiers*, 256.

In 1809–1810, coasting vessels were also sailing into Savannah from locations like Frederica on St. Simons Island, Darien on the Altamaha River, Hardwick on the Ogeechee River, and Sunbury on the Medway River, as well as from Sapelo Island, Riceboro, Brunswick, and Beaufort.

Darien was growing as a local mercantile center because of its advantageous position on the Altamaha River, which was already developing into an important avenue of transport for Georgia's interior cotton crop. In 1801, Darien contained only eight houses and three stores; by 1805, the population had grown to about 145 persons.[41] The growing commercial importance of Darien is expressed in the steady increase in coaster voyages from the town into Savannah—from none in 1800 and 1805, to four in 1808, to nine in 1809–1810.

By 1810, information on the merchants involved in handling goods carried by coasters was appearing, as shipping lists included the names of "consignees" of cargoes. This meant the cargo was destined for the named merchant because either the merchant had purchased the cargo or it was being shipped by a planter to a merchant to handle its sale. Among the Savannah merchants receiving or shipping goods on coasting vessels in 1810 were Andrew Low & Company, Small & McNish, J & J Carruthers, J. Meigs, R. Richardson & Co., and Joseph Habersham (see Table 7.5). The Small & McNish firm was heavily involved in the trade with Florida and St. Marys. One member of this company, John McNish, would soon form his own business, and by 1820, he was a leading dealer in Sea Island cotton, receiving more shipments by coasting vessel that year than any other merchant in Savannah.[42]

During the first four months of 1810, only four of the thirty-five coaster arrivals at Savannah from local ports carried Sea Island cotton and rice. These ships came from the small port towns of Beaufort, Sunbury, Hardwick, and Darien and represent the only sailings from these ports. All the voyages occurred between November and April, the main shipping season for rice and cotton. There were also ten coaster arrivals from Frederica, Riceboro, Brunswick, and Sapelo Island between November 1809 and April 1810. No cargo information is provided for any of these voyages, but given the time of year these sailings occurred, most, or all, of these ships probably carried coastal crops.

By the end of the first decade of the nineteenth century, the Georgia coasting trade was poised for growth and expansion and developing into the

[41] Sullivan, *Early Days*, 67, 70.

[42] Vanstory, *Georgia's Land*, 151–52; *Daily Georgian*, 1820.

system that would remain in effect for the next half-century. Trade with Florida rebounded with the repeal of the Embargo Act in 1809, and Savannah and Charleston merchants were heavily involved in the shipment of cotton to the Spanish colony. Much of this cotton was ultimately destined for British ports, locations that were still banned for American-owned vessels by the Non-Intercourse Act. St. Marys retained an important position in the local coasting trade, principally in the shipment of goods obtained from Florida, either legally or illegally. A few Georgia and South Carolina residents were sailing the ships in the local trade, but Northern men were beginning to bring their ships South to work for part of the year. Increasing quantities of cotton and rice were shipped into Savannah, and growing numbers of factors and merchants in the city handled these goods. Even though Charleston's importance as a regional port declined, many coastal Georgia planters continued to send their crops to factors there.

## The War of 1812

The growth seen in the Georgia coasting trade in 1810 was impeded by events related to the War of 1812. On June 18, 1812, Congress acceded to the request of President James Madison and declared war against Great Britain. War with Britain was based on four principal complaints: impressment of American seamen, violation of American neutrality and territorial waters, British refusal to permit neutral trade with France and her colonies, and British blockade of American ports.[43] Most of these complaints were long-standing and were factors behind the Embargo and Non-Intercourse Acts. These interferences in American rights produced widespread anti-British sentiment in the United States and helped elect pro-war James Madison in the election of 1810–1811.

Despite anti-British feelings, support for a war with Great Britain was not universal. Most of New England, as well as New York, New Jersey, and Delaware objected to the declaration of war or did not strongly support it. These attitudes stemmed largely from economic interests. These states were the centers of the American shipping business, as well as its nascent manufacturing base, and Great Britain and her colonies, having been reopened to American trade with the repeal of the Non-Intercourse Act in 1810, had again become important trading partners. Despite the occasional seizures and stoppages of ships by the Royal Navy, American merchants and shipowners

[43] Jacobs and Tucker, *War of 1812*, 12–13.

wanted to maintain this trade and were disappointed when James Madison reinstated the embargo against Great Britain in February 1811.

As early as 1807, British ships cruising along the Georgia coast had harassed American vessels. These relatively minor incidents escalated with the coming of the war. In early 1813, the British blockade of the New England coast was extended south to Florida.[44] This blockade was not airtight but did create a danger to ships trying to enter or leave Atlantic ports, and it curtailed the shipment of cotton and rice out of Charleston and Savannah to Northern and overseas ports. To complement their blockade, the British began to send small boat parties into the sounds and inland waterways to harass and disrupt coastal shipping. The declaration of war and suppression of overseas trade depressed Sea Island cotton prices, which dropped from thirty cents a pound before the declaration of war to seventeen cents in 1812 and to just thirteen cents in 1813.[45]

The Charleston and Savannah newspapers began to carry news of the war's impacts on coasting shipping in summer 1812. One expression was an increasing number of advertisements stressing that departing coasters would sail "by the inland passage." This was considered a safer option than sailing "outside" in the Atlantic, where vessels would be more vulnerable to capture by British ships.

The coasters sailing the inland passage began to travel in groups, often under the protection of United States gunboats. In December 1813, the gunboats stationed at St. Marys accompanied a convoy of thirty-six coasters north to Savannah. The size of this convoy suggests these vessels had waited in St. Marys until a large group was assembled. In some instances, gunboats escorted a few or even a single coaster on its voyage. On June 10, 1813, the sloops *Delight* and *Polly* arrived at Savannah from St. Marys under convoy of "Gun-vessel No. 63," and the following month the same sloop *Polly* arrived alone from St. Marys escorted by *Gunboat No. 24*.[46]

In addition to the Royal Navy, the coasters had to contend with foreign privateers visiting coastal waters. In April 1813, the *Columbian Museum & Savannah Advertiser* reported that two coasting schooners, the *John* and *Planter*, and an unnamed sloop were "cut out" by a privateer schooner while at anchor within the St. Helena bar in lower South Carolina. It was reported that the privateer came up "in disguise as an inland coaster." In light of the

[44] Ibid., 80.

[45] Gray, *History of Agriculture*, 2:738; Keber, *Seas of Gold*, 213.

[46] *Republican & Savannah Evening Ledger*, June 10 and July 13, 1813.

dangers, some of the Georgia coasters undertook to protect themselves. Among these was the *Maria*, which was identified as an "armed sloop" when advertised to take on freight or passengers at Savannah for St. Marys. What the *Maria*'s armament may have been is unreported.

The blockade and the attacks by the British Navy and marauding foreign privateers took their toll on the Georgia trade. Coasters continued to sail in 1813, despite the obvious dangers, but in reduced numbers. Between December 1812 and November 1813, the *Republican & Savannah Evening Ledger* reported only fifty-five coaster arrivals into the city from local ports other than Charleston (see Figure 7.1). This was half the number of arrivals reported in 1809–1810. Amelia Island, the most important port of origin for the Savannah coasters immediately before the war, was the origin of only eight arrivals in 1813.

It is apparent the shipping lists in the *Republican & Savannah Evening Ledger* contain some omissions of coaster arrivals and departures in 1813. These omissions continue and seem to increase over the next year, possibly a result of disruptions caused by the war. In 1814, the *Columbian Museum & Savannah Advertiser* did not even print a regular shipping list until March 21, when a column titled "Ship News" appeared. However, throughout 1814, the *Columbian Museum & Savannah Advertiser* printed advertisements and sale notices for coasting vessels, and it published accounts of coasters being captured, run aground, or burned by the British as well as warnings alerting coasting captains to the dangers of sailing. In June 1814, the paper reported the British captured several coasters in Buttermilk Sound, near Darien, some loaded with cotton and rice. These included the sloops *Good Intent*, *Polly*, and *Frederica*. Two months later, boats from the British frigate HMS *Lacedemonion* captured the coasting sloop *Hester* in Warsaw Sound, and her captain, Anthony Shaddock, was seized and taken to New Providence in the Bahamas. Later that year, there were reports that the British burned two coasters from "Amelia or St. Marys" in St. Catherines Sound and captured the sloop *Francis* and the schooners *Delight* and *George & Joseph* in Teakettle Creek, south of Savannah.[47]

The American gunboats did have some successes against the British along the South Carolina and Georgia coasts. In September 1814, two gunboats captured the British privateer *Fortune of War* in Sapelo Sound. At the time of her capture, the schooner was flying an American flag. In summer

[47] *Columbian Museum & Savannah Advertiser*, June 16, August 22, October 10, and December 12, 1814.

1814, a "Capt. Thompson" from Savannah successfully recaptured from the British the sloops *Good Intent*, *Maria*, and *Martha* and the schooner *Elizabeth & Jane*. It is unknown if Captain Thompson was a United States naval officer or commander of a privateer.[48]

In 1814, in addition to harassing coastal shipping, British landing parties went ashore at several locations along the Georgia coast. These parties damaged crops and property and took many slaves. In January 1815, British forces under the command of Admiral George Cockburn landed on Cumberland Island. They occupied the area around St. Marys, moving as far inland as the town of Coleraine on the St. Marys River. However, the British remained only a short time before returning to their ships.[49]

On December 24, 1814, before Admiral Cockburn had landed his men in South Georgia, commissioners from the United States and Great Britain signed a peace accord at Ghent in the Netherlands. Great Britain ratified the treaty within a few days, but it took almost two months before American commissioners reached Washington with copies of the treaty. There the Senate ratified it unanimously on February 17, 1815.[50]

With the war's end, the coasting trade in Georgia quickly returned to its pre-war levels. British markets were reopened to American goods, and cotton exports grew rapidly. In addition, New England mills required increasing amounts of Southern cotton. The number of American mills and their output had expanded considerably during the years of the Embargo and Non-Intercourse acts when the importation of foreign cloth was halted or restricted. After the war, prices for Sea Island cotton began to rise rapidly, reaching fifty-six cents per pound in 1817 before jumping to an average high of 63.2 cents in 1818, and occasionally reaching as much as eighty cents that year.[51] The combined needs of American and British mills for Southern cotton and the resulting high prices spurred planters to cultivate more cotton. There was little increase in the production of Sea Island cotton because it had reached the limits of its geographical range of cultivation. The great expansion in cotton production after the War of 1812 was in short-staple, upland cotton, whose cultivation rapidly extended westward across the South to the Mississippi River and beyond.

[48] Ibid., September 15, and June 16, 1814.

[49] Coulter, *Georgia*, 213; Keber, *Seas of Gold*, 214–16.

[50] Lord, *Dawn's Early Light*, 320, 338.

[51] Gray, *History of Agriculture*, 2:1030; Keber, *Seas of Gold*, 208–209, 234.

The demand for rice as food increased after the War of 1812, and exports grew. In 1813, during the height of the war, approximately 6.8 million pounds of rice were exported from the United States. This quantity increased in 1814 and reached 82.7 million pounds in 1815. Rice prices rose after the war, from about 3.3 cents per pound before the War of 1812, to 4.3 cents in 1815, to 6 cents per pound in 1817.[52] The increasing demand for rice, plus rising prices, led to an expansion in rice agriculture in South Carolina and Georgia. This expansion in the years immediately after the War of 1812 produced greater cargoes for the coasters.

One facet of the Georgia coasting trade that changed after the War of 1812 was the trade with Florida. During the war and the years of the embargo, the main market for American cotton, Great Britain, was closed to trade. To get around this restriction, considerable amounts of cotton were carried by coasters to Amelia Island and St. Augustine in Spanish Florida, where it could be sold into British hands. With the British market again open to Americans in 1815, this spurious cotton trade with Florida ended. The cargoes now carried to Florida out of Savannah and Charleston consisted of a wide range of foodstuffs and merchandise, including flour, beef, pork, tobacco, wine, nails, axes, candles, crockery, shoes, osnaburg, and "sundry articles."[53] Some of these goods were destined for residents of Florida, but much was loaded aboard other vessels at Amelia Island and shipped to overseas ports.

## The Panic of 1819

After the War of 1812, the coasting trade in Georgia rebounded quickly, echoing a period of economic growth seen nationwide. By 1818, Sea Island cotton was selling for as much as 63 cents per pound. However, the following year, 1819, the price tumbled to 55 cents per pound and declined to 26.7 cents in 1821. Like Sea Island cotton, rice prices dropped from an average of 4.6 cents per pound in 1818 to 2.8 cents in 1820.[54]

The decline in cotton and rice prices in 1819 and 1820 was related to a financial crisis known as the Panic of 1819, a nationwide economic depression stimulated by a dramatic fall in agricultural prices. This depression was America's first great economic crisis and in large measure, was related to the war's effects on the American economy. After the war, the pent-up demand for

[52] J. F. Smith, *Slavery and Rice*, 214; Gray, *History of Agriculture*, 2:1030; Keber, *Seas of Gold*, 214–15, *Charleston City Gazette & Commercial Daily Advertiser*, January 27, 1817.

[53] "Outward Foreign Manifests, Savannah" RG 36, Bureau of Customs, NARA.

[54] Gray, *History of Agriculture*, 2:697, 1030.

imports resulted in an influx of foreign goods into the United States. The prices for these goods dropped, putting many domestic manufacturers at a disadvantage. On the other hand, there was a great demand for American agricultural products in Europe, particularly in Britain, and the prices for these, especially cotton, rose as agricultural exports increased. The inflated cotton prices influenced planters to borrow increasing amounts of money to expand their production, especially by purchasing lands in the South and West as cotton cultivation spread in those regions. Banks issued large numbers of notes and extended increasing amounts of unsecured credit to planters and other borrowers. In fall 1818, the boom ended as banks began to recall their loans and limit their extension of credit. The severe contraction of the money supply led to a rapid drop in prices. At the same time, there were high crop productions in Europe, lessening the demand for American agricultural commodities and contributing to the precipitous drop in export prices paid for agricultural goods. All these factors resulted in an economic recession lasting into 1821.[55]

The drop in Sea Island cotton and rice prices had a negative effect on planters and merchants along the Southeastern Atlantic coast. Coker estimates that fewer than one-third of local Charleston merchants survived the crisis.[56] The panic's effect on Savannah merchants is unrecorded, but they may not have been as severely affected as those in Charleston. Savannah was undergoing a period of expansion and growth in 1819, and local merchants were beginning to capture some coastal trade that had formerly gone to Charleston. The events of 1819 quickened the already strengthening position of Savannah as an economic center and port relative to Charleston, including its standing in the regional coasting trade.

How the panic specifically affected the Georgia coasting trade is difficult to ascertain; however, coastal planters were certainly affected by the sharp drop in the prices of their two most important crops. Figures on rice and cotton production or exports for the period are fragmentary; however, Savannah did experience a slight decline in Sea Island cotton exports from 11,895 bales in 1819–1820 to 10,888 the following year. On the other hand, the export of upland cotton from Savannah increased significantly, rising from 134,528 bales in 1819–1820 to 157,099 in 1820–1821. This rise reflects the increasing amounts of upland cotton transported down the Savannah and Altamaha rivers, despite any effects the depression had on inland planters during these

[55] Rothbard, *Panic of 1819*, 6, 18–22.

[56] Coker, *Charleston's Maritime Heritage*, 174.

years. A lack of export data for 1818, immediately before the start of the depression, makes it difficult to assess its full regional impact.[57]

It appears the Panic did not significantly decrease the volume of coaster traffic at the port of Savannah. For 1820, the *Daily Georgian* listed 213 coaster arrivals into Savannah from regional ports, excluding Charleston. This is almost a fourfold increase over the fifty-five arrivals reported for 1812–1813. There were eighty-three individual vessels sailing in the Georgia trade in 1820, again a significant increase over the thirty ships participating in 1812–1813.

## Steamboats Come to the Coasting Trade

The participation of steamboats in the coastal trade was to have significant and, ultimately, detrimental consequences for the sailing coasters. However, competition from steamboats did not seriously affect the work of sailing vessels until about fifteen years after steamboats were first introduced to Georgia. The first commercially successful steamboat to operate in the state was the ninety-foot *Enterprise*, built in Savannah in 1816 by brothers Samuel and Charles Howard. In 1814, the State of Georgia granted Samuel Howard the exclusive rights to operate steam-propelled vessels in the state, with the provision that he have a boat operating within three years. The *Enterprise* was initially used to tow sailing vessels the long distance between the Savannah waterfront and the sea. This activity proved to be a great convenience to sailing ships negotiating this often difficult passage. The *Enterprise* was later placed in service on the Savannah River, pulling poleboats upriver and towing barges loaded with cotton downriver from the interior to Savannah.[58]

The *Enterprise* proved to be a financial success. In 1817, the Howard brothers organized the Steam Boat Company of Georgia and were soon operating several steamboats. These included the sidewheelers *Georgia* and *Carolina*, both built in Charleston in 1817.[59] These Steam Boat Company boats generally confined their operations to the Savannah River, transporting merchandise up the river and cotton downstream. In January 1819, the Howard steamboat *Altamaha* undertook its first voyage on the Altamaha River. By 1820, the steamboats *Samuel Howard* and *Ocmulgee* joined the Howard fleet and worked on the Savannah and Altamaha rivers.[60] These early Altamaha

[57] Kayser & Co., *Commercial Directory*, 12, 44.

[58] Goff, "Steamboat Period," 238; Fleetwood, *Tidecraft*, 94.

[59] Fleetwood, *Tidecraft*, 95; Mitchell, *Merchant Steam Vessels*, 30, 85.

[60] Sullivan, *Early Days*, 153; Mitchell, *Merchant Steam Vessels*, 161.

River steamers were primarily involved in towing barges and flats loaded with cotton from upstream landings on the Oconee and Ocmulgee rivers, down to Darien.

Initially, these steamers presented little competition to the sailing coasters because they were confined mainly to hauling cotton and merchandise on the Altamaha and Savannah rivers. They carried little in the way of commodities from the regional coastal ports and plantations into Savannah.

On June 15, 1820, the departure of the steamboat *Ocmulgee* from Savannah for Darien was reported in the *Daily Georgian*. The *Ocmulgee* returned to Savannah on June 22, carrying cotton, one of the first shipments of domestic produce carried along the Georgia coast by a steamboat. By 1821, advertisements announced that the steamer *Georgia* was available to carry freight and passengers between Savannah and Darien. After this, the amount of local produce carried by steamboats along the coast slowly increased. As late as 1825, steamboats were traveling along the coast south of Savannah only intermittently.

## The Georgia Coasting Trade in 1820

The workings of the Georgia coasting trade become clearer after about 1820 because Savannah newspapers began to include with some consistency the identity and quantity of the principal cargoes carried by coasters into Savannah and the consignees receiving them. With this type of information, it is possible to assess the quantity and value of cargo carried into the city by sailing coasters, and the activities of individual ships and masters become clearer. This level of detail is available only for ships and cargoes arriving into Savannah. Most years after 1820, newspapers published incomplete information on departing coasters and their cargoes.

In 1820, the *Daily Georgian* listed 213 coaster arrivals in Savannah from regional ports other than Charleston. These arrivals were made by eighty-three different vessels; fifty-eight sloops and twenty-five schooners. The preponderance of sloops was a continuation of the pattern seen during the previous two decades. In 1820, coasters sailed into Savannah from thirty-four named locations, a considerable increase over the thirteen named ten years earlier (Figure 7.5). Darien had become the most important port of origin for the coasters, with ninety-two sailings from there. The towns of St. Marys and Riceboro were distant seconds, with sixteen arrivals from each while fourteen coasters arrived from Beaufort and twelve from Sunbury. The position of St. Marys in the local coasting trade had declined significantly since 1815.

Similarly, the importance of Spanish Florida as a port of origin for Georgia coasters had fallen, with only fifteen arrivals from Florida in 1820. By 1820, coasters were arriving in Savannah from almost every one of the Sea Islands south of St. Helena Island in South Carolina, as well as from the region's main tidal rivers. Sea Island cotton agriculture was fully established on the islands, and rice agriculture was flourishing along some rivers and being established on others.

In 1820, the *Daily Georgian* provided cargo information on all but seventeen coasters arriving in Savannah. The remaining 196 arrivals constitute more than 90 percent of the sailings, a sufficient proportion to examine with reasonable accuracy the movement of coastal produce and the influence its transport had on coaster trading patterns. By 1820, the seasonal pattern of coaster sailings that would continue to the Civil War was established. That year, coasters arrived in Savannah every month, but 143 arrivals, representing 67 percent of the total, occurred between November and April, when the rice and cotton crops were shipped to market.

It is impossible to state exactly how many different masters brought coasting ships into Savannah in 1820 because several had the same or similar names. However, there appear to have been approximately eighty-seven different captains sailing that year. About half these men sailed into Savannah a single time. The names of many of the captains arriving once rarely or never appear again in Savannah shipping lists, suggesting these men never worked in the local trade again. Table 7.6 lists those forty-five or forty-six masters who made two or more arrivals into Savannah in 1820.

Nicholas Sallowich, a native of Italy residing in Savannah, was the most active coasting captain in 1820, making eleven voyages into the city with his Connecticut-built sloop *Union*. Eight of these trips were from Sunbury, and Captain Sallowich sailed as a principal trader to Sunbury, Riceboro, and Hardwick for more than two decades, from 1813 to the mid-1830s. Other active traders in 1820 included Edmund Richardson of St. Marys, who made ten voyages into Savannah as master of two ships, the sloops *Dosoris* and *Providence*. Charles Porquet, with his sloop *Maria*, and Allen Chase, with the sloop *Rosetta*, each made seven arrivals. Masters making six trips into Savannah in 1820 were John Chevalier and the sloop *John Chevalier*, James Deverger with the sloop *Favorite*, Joseph Howland with the sloop *Rosetta*, and John Anderson with the sloop *Governor Shelby*. Most of the captains making five or more arrivals into Savannah in 1820 resided locally; however, several Northern captains, especially those from the Sippican area, were active in the trade. These included Joseph Howland and Allen Chase, who shared captaincy of the sloop *Rosetta* in 1820, as well as Ephraim Allen, Joseph Allen, and Barnabas Nye, each of whom sailed into Savannah several times that year (see Table 7.6).

### *The Savannah-Darien Trade in 1820*

In 1820, Darien was the most important port of origin for coasters sailing into Savannah. The ninety-two sailings from the town represented 43 percent of all local coaster arrivals that year. This was a tremendous increase over the nine arrivals from Darien just ten years earlier. Darien's growth as a port was related almost entirely to its position on the Altamaha River, down which large quantities of inland cotton were shipped after the War of 1812. This cotton was collected in Darien, and most was placed aboard coasting vessels for shipment to factors in Savannah or Charleston.

Forty-seven different coasting ships made the ninety-two voyages into Savannah from Darien in 1820. About half of these ships made just one voyage, and only a few made more than four or five trips. The fifty-foot sloop *Rosetta* made twelve voyages from Darien to Savannah, over twice as many as any other coaster. Two men served as master of the *Rosetta* this year, Captain Joseph Howland and a Captain Chase, possibly Allen Chase, both from Sippican. In 1820, at least thirteen of the forty-five identified captains sailing from Darien into Savannah had Sippican surnames, demonstrating the importance of these Northern men in the Savannah-Darien trade.

The principal cargo carried by ships from Darien was cotton, and it was aboard seventy-one of the ninety-two arrivals in Savannah. Often, newspaper shipping lists distinguished between Sea Island cotton and upland cotton, but the *Daily Georgian* did not do so in 1820, and vessels are reported only as carrying a certain number of bales or bags of "cotton" or just "cotton" with no quantity given. The number of bales of cotton carried is provided for forty-six of the arrivals from Darien; the vast majority would have been upland cotton carried down the Altamaha River from the interior. Some small amount of Sea Island cotton was included in these shipments, but how much is impossible to determine. The amount of cotton carried by individual coasters from Darien ranged from 70 bales to the 340 bales carried by the sloop *Atlantic*. These 340 bales represented the largest number carried into Savannah on a single voyage by any vessel from any port in 1820. Most coasters carried between 150 and 250 bales on each trip.

The total amount of cotton enumerated in the shipping lists for coasters from Darien in 1820 was 8,572 bales. At 320 pounds per bale, the average weight of upland cotton bales at the time, these bales represented more than

2.74 million pounds of cotton. The average price for upland cotton in 1820 was 15.2 cents per pound, meaning the approximate value of the enumerated cotton carried by the coasters from Darien was $416,500. If some of this cotton included Sea Island cotton, which brought a higher price, then this value would be higher.[61]

Today, it is difficult to comprehend the worth of the various commodities carried by the coasters as it was perceived in 1820. The stated value of the various items shipped into Savannah that year have a much greater value in present-day currency, and it seems fitting to provide modern equivalencies to obtain a better appreciation of the economic standing and impact of the coasting trade at the time. Relying on computations provided in the website MeasuringWorth.com, it is estimated that the $410,400 worth of cotton shipped from Darien would have a modern-day (in 2020) value on the order of 9.4 million dollars. This figure represents a rough estimate, but it conveys an idea of the great monetary value of the cotton shipped through Darien and carried by the sailing coasters to Savannah.[62]

It also must be remembered that the quantity of cotton carried is given for only forty-seven of the seventy-one vessel arrivals transporting cotton from Darien. It is unknown why the newspapers did not provide the number of cotton bales or bushels of rough rice aboard every vessel carrying these crops. But this omission was not uncommon before the 1830s and may have been due to the inability of the individual collecting the information to connect with a person having specific knowledge of a vessel's cargo, such as the captain or a consignee. Regardless of the reasons for these omissions, there is no reason to believe that the quantities of cotton aboard these unrecorded arrivals significantly differed from those where specific amounts were listed. For 1820, we have no data on the quantities of cotton carried on twenty-four, or about 34 percent of the voyages from Darien, meaning this percentage of the total amount of cotton shipped from the town may have gone unrecorded. Under this assumption, as many as 2,900 or so bales of cotton shipped from Darien have gone unlisted, meaning that as many as 11,400 bales of mostly upland cotton may have been shipped from the town. This was approximately 3.65 million pounds of cotton, representing an 1820 monetary value of $554,800, or about $12.6 million in modern (2020) values.[63] This estimate relies on assumptions about missing information in the shipping lists, but

[61] Gray, *History of Agriculture*, 2:1027.

[62] MeasuringWorth.com, "real price" of a commodity calculator.

[63] Ibid.

these assumptions do not seem unreasonable. This close to thirteen million dollars' worth of cotton emphasizes the great importance of the cotton trade between Darien and Savannah.

Rice was identified as a cargo on only seven coasters sailing from Darien in 1820. On five of these voyages, the cargo is identified as "rice," presumably rough rice. Although, Darien is listed as the origin of these shipments of rough rice, it is suspected this grain was loaded at plantations along the Altamaha River near Darien. The other two rice cargoes consisted of forty-five casks and forty tierces of clean rice carried by the sloops *Traveller* and *Two Sisters*. This clean rice might have been loaded in Darien, which had rice milling facilities by 1817. In the following decade, with the perfection of tidal irrigation for growing rice, the lower Altamaha River would become one of the major rice-producing areas in Georgia.

Besides cotton and a modest amount of rice, few local products were shipped from Darien in 1820. Molasses was named as a cargo on two voyages, oranges on one voyage, and "sundries," "assorted cargo," and "merchandise" were each listed once. Passengers were aboard three vessels when they arrived in Savannah: the sloops *Morning Star* and *Ricebird* and the schooner *Harmony*. The *Harmony* reportedly took eight days to sail from Darien to Savannah, suggesting the single passenger on the schooner had a tedious and possibly cramped voyage, given the 270 bales of cotton aboard.

The lack of early newspapers from smaller ports makes it difficult to discern specifics of their involvement in the coasting trade. Darien, however, is unusual because it had a newspaper before 1830 that provided information on ship traffic. The *Darien Gazette* newspaper published shipping lists like those in Savannah and Charleston papers. From November 1820 to October 1821, the *Darien Gazette* listed 228 ship arrivals at Darien by 103 different vessels. Forty-nine of these ships were sloops; forty-three were schooners. The remaining eleven vessels consisted of five brigs, two ships, one United States cutter, one United States schooner, and two steamboats, the *Georgia* and the *Samuel Howard*. Most of the sloops and schooners were sailing from Savannah or Charleston, as well as other regional ports such as St. Marys and St. Simons Island in Georgia and St. Augustine and Amelia Island in Florida. A smaller number of vessels were sailing from more northerly locations, particularly New York and North Carolina, and a few larger brigs and ships were arriving from England, Cuba, or elsewhere overseas. Those sloops and schooners sailing from Savannah and Charleston carried sugar, salt, mackerel, groceries, dry goods, flour, stationery, coal, corn, whiskey, gin, bricks, nails, hats, bacon, lime, furniture, and iron. Many transported goods identified only as "assorted cargo" or "sundries." These cargoes were destined for merchants in or near Darien and various plantations. Many individuals and firms receiving the goods are identified in the shipping lists, and some advertised the wares they received in the *Darien Gazette*. Darien merchants such as Hall, Cooke & Company; C. G. Jones; and Barrington King & Company all received shipments from Savannah and Charleston aboard coasters consisting of corn, whiskey, groceries, molasses, salt, sundries, and "assorted cargo." These firms also received merchandise from distant ports, such as New York and Liverpool.

A few ships arriving in Darien in 1820 carried coastal agricultural products, most destined for transshipment to Savannah, Charleston, or more distant ports. These "products of the country" came from local ports such as St. Marys and St. Simons Island. Among those coasters arriving with local commodities were the sloop *Mathews* sailing into Darien in early February from Crooked River in Camden County with a cargo of cotton destined to John L. K. Holzendorf, a Darien merchant; the schooner *Fire Fly* from St. Marys with cotton for J. H. Giekie & Company; the sloop *Ann*, also from St. Marys, with oranges; and the sloop *Frederica*, under a Captain Grandison, from St. Simons Island with cotton and oranges.[64]

*Savannah's Trade with Other Local Ports in 1820*

Although Darien was Savannah's most important local trading partner in 1820, coasters came into the city from thirty-three reported locations (see Figure 7.5). The towns of St. Marys and Riceboro lagged far behind Darien in terms of coaster traffic into Savannah, with sixteen arrivals from each. The trade between St. Marys and Savannah was carried on by several vessels, and most of the captains involved were local men, residents of Camden County or elsewhere along the Georgia coast. These included John Chevalier, captain of the sloop *John Chevalier* and James Vincent, master of the sloop *James*, both residents of St. Marys; James Deverger of Darien and his sloop *Favorite*; and David Pidge of Savannah, captain of the sloop *Hermit*. Only a single person who might be a Northern captain, an individual named Burr, sailed from St. Marys in 1820. The dominance of local captains in this trade is in sharp contrast to the heavy participation of Northern captains and ships in the trade between Darien and Savannah.

Cotton was the principal cargo carried from St. Marys. The type of cotton is unidentified, but most would have been Sea Island cotton. Only three

[64] *Darien Gazette*, November 1820-October 1821.

vessels transported the variety of manufactured goods that had been common cargoes out of St. Marys in earlier years. Now that American ships could trade directly with Florida and other foreign ports, there was little need to transship these cargoes through St. Marys.

In 1820, eight different coasters made sixteen trips into Savannah from Riceboro. Unlike St. Marys, the trade with Riceboro was dominated by Northern captains from Sippican, including Captain Prince Snow with the sloop *Leopard*, James Blankenship with the sloop *Howard & James*, and captains named Luce and Bolles, both of which are Sippican family names. Only two local captains are identified among those sailing this route. These were James Deverger of Darien, with the sloop *Favorite*, and Nicholas Sallowich with his sloop *Union*.

Despite its name, cotton dominated the cargoes carried out of Riceboro in 1820. Cotton appears as a cargo on eleven arrivals at Savannah and rice on only six. Most of the rice shipped was clean rice in tierces, casks, or barrels.

In 1820, coasters arrived in Savannah from other small port towns in the region, including Beaufort in South Carolina and Sunbury and Brunswick in Georgia. Cotton was the most common cargo carried from Beaufort. Coasters in the Beaufort-Savannah trade included the schooner *Isabella*, under her captain, Isaac Pierce; the sloop *Lawrence*, under Captain John Turner; and the schooner *Little Ann*, commanded by a captain named Drayton, possibly Alvin Drayton. The forty-four-ton *Little Ann* was enrolled at Beaufort in 1815 with the notation that she was formerly a "U.S. Boat," one of the American gunboats used through the War of 1812. This same document also notes that the schooner's papers were surrendered at Beaufort in November 1822 because she was "broken up."[65]

By 1820, the town of Hardwick was no longer named as a port of call for coasters. However, Sunbury remained an important port, with twelve sailings to Savannah that year. Only two men sailed the Sunbury-Savannah route: Captain Nicholas Sallowich, a resident of Savannah and master of the sloop *Union*, and Sippican resident Ephraim Allen, captain of the sloop *Hesper*. Cotton was the principal cargo from Sunbury in 1820, along with small quantities of clean rice.

Five coaster arrivals in Savannah in 1820 originated at Brunswick. These were made by Captain Charles Porquet and his sloop *Maria* and the sloops

[65] Enrollment No. 2, "Master Abstracts," Port of Beaufort, schooner *Little Ann*, October 27, 1815.

*Governor Shelby* and *Harriet*. Cotton comprised the primary cargo carried from Brunswick.

By 1820, coasters were sailing into Savannah from almost every one of the Sea Islands and from plantations located on the region's major rivers. Captain John Fowler and the sloop *Francis* brought one hundred bales of cotton from St. Helena Island in South Carolina; a Captain Risely carried cotton from "Fuskay Island" (Daufuskie Island) aboard the sloop *Diana*; the sloop *Joseph* transported cotton from Skidaway Island. Four coasters sailed from Sapelo Island, three from St. Catherines Island, and three from St. Simons Island, most carrying Sea Island cotton. In January, the schooner *Cheves* sailed into Savannah from Blackbeard Island on the central Georgia coast with a cargo of wood. Commonly, when wood is listed as cargo, it appears to be firewood instead of lumber or timber, both of which are occasionally mentioned in shipping lists. However, Blackbeard Island was a United States government timber reserve, supplying live oak for naval vessels, meaning the wood carried by the *Cheves* was almost certainly live oak timber, not firewood.

By 1820, coasters were visiting plantations along the lower reaches of the region's major rivers, where planters cultivated Sea Island cotton and rice. Six arrivals were reported from the May River in South Carolina, most carrying cotton and only one with a cargo of rice. The Ogeechee River's potential to be one of the great rice-growing areas in Georgia was becoming evident in 1820 when rice was reported aboard five of the seven coaster arrivals from there. Four of these rice cargoes consisted of casks and tierces of clean rice while one is identified only as "rice."

In 1820, coasters also arrived from the Little Satilla and the Great Satilla rivers on the lower Georgia coast. The Little Satilla River was served by a single vessel, the sloop *Governor Shelby*, which made three voyages into Savannah carrying cotton on every trip. Three coasters sailed from the Satilla River; cotton was the only cargo carried.

No rice was reported to have been shipped into Savannah from either the Little Satilla or Great Satilla rivers in 1820. Little rice was carried into Savannah from any of the major rivers south of the Ogeechee. In 1820, these areas were entering a period of expansion in rice production. Within a few years, increasing quantities of rice began to be shipped from all the major rivers in Georgia, reflecting the full adoption of tidal flow agriculture. By 1825, eleven of the twelve coasters sailing from the Ogeechee River into Savannah carried rough rice.[66] For the Satilla River, shipment of large quantities of

[66] *Savannah Republican*, 1825.

rough rice did not begin until the 1840s. Many planters along the Georgia and lower South Carolina coasts maintained ties with Charleston merchants and shipped their crops there. This was especially true of the early rice planters in Georgia, many of whom were natives of South Carolina or members of South Carolina families with long-established relationships with Charleston merchants. Additionally, Savannah had little rice milling capacity in 1820, and it behooved Georgia planters to ship their rough rice to Charleston, where these facilities were available. Much of the rice transported by coasters into Savannah in 1820 is identified as clean rice, processed at the few rice mills then in existence along the Georgia coast.

*The Savannah-Florida Trade in 1820*

The sailing trade between Savannah and Florida was a modest activity in 1820, considerably lessened from the years after the embargo, when Florida provided an important outlet for Southern cotton. Florida was still a Spanish possession in 1820 but was acquired by the United States the following year. In 1820, there were sixteen coaster arrivals in Savannah from Florida. These ships arrived from Amelia Island, the town of Fernandina on Amelia Island, St. Augustine, the St. Johns River, and "Cape Florida" or the "coast of Florida." Several ships were involved in the Florida-Savannah trade, some of which sailed this route for many years. The most active of these coasters included the North Carolina-built schooner *Mary McKay*, sailing under captains Richard Marcellin and Edmund Richardson; the schooner *Choctaw*; the sloop *Caroline*, under a Captain Egery; and the sloop *James*, commanded by James Vincent. Most of these captains were residents of St. Marys and Camden County or elsewhere in Georgia. Northern captains had little involvement in the Florida-Savannah trade in 1820 compared to their heavy participation on that route ten years earlier.

Cotton was the most common cargo carried by the ships sailing from Florida although they occasionally carried more unusual items. The schooner *Choctaw* and sloop *Harriet* each sailed into Savannah with a cargo of mahogany from the "coast of Florida." The arrival of both vessels was reported on June 2, and it appears they sailed together from Florida. The mahogany represented material salvaged from a shipwreck or stranded vessel. The *Choctaw* seems to have specialized in salvage operations because a month earlier, the *Daily Georgian* had reported her arrival from a "wrecking voyage" off Florida. This voyage had been unsuccessful, forcing the *Choctaw* to return to Savannah with a cargo of turtles. In July, the schooner *Mary McKay* also carried turtles from Florida. This voyage received special attention in the newspaper, with the notation that the *Mary McKay* had been "attacked by pirates" off the Florida coast. The attackers made off with the ship's documents, and on his return to Savannah, Captain Marcellin had to secure a new enrollment because his papers had been "taken by a Pirate."[67]

*Savannah Merchants in the Coasting Trade in 1820*

In 1820, the *Daily Georgian* named consignees for 152 of the 213 coasters arriving in Savannah. In twenty-one instances, the terms "to master" or "to order" were used rather than a specifically named firm or individual. The term "to master" was often used when a ship arrived in ballast. In a few instances, it was used when miscellaneous cargoes such as wood, staves, oranges, or corn were carried. It appears that when a cargo was consigned to the master, he was responsible for it, from its initial purchase to its disposal in Savannah.

When the term "to order" is used in a commercial transaction, it typically means that a cargo is not assigned to a specific consignee because it has not been purchased in advance and has been assigned by the owner (shipper or planter) to the shipmaster to dispose of.[68] Most of the cargoes assigned "to order" in the 1820 *Daily Georgian* consisted of cotton or rice presumably shipped by a planter under the knowledge that the ship captain would assign the cargo to a specific merchant in Savannah.

In 1820, seventy-six named consignees received 249 individual shipments by coasters at Savannah. These consignees included several well-established city merchant houses, as well as newer businesses established in the previous decade to take advantage of the expanding mercantile opportunities in Savannah. Many of these merchants are included in an 1823 Savannah commercial directory where most are identified as "Factors," "Commission Merchants" or both.[69]

Forty-eight of the consignees are listed a single time in receipt of a single shipment. Table 7.7 lists the twenty-eight Savannah merchants receiving two or more shipments aboard coasters in 1820. Nine firms received ten or more shipments during the year. The factor and commission merchant John McNish received twenty-seven individual cargoes in 1820, more than any other consignee. Cotton made up most of these cargoes although the firm did take delivery of small quantities of rice and molasses. Most shipments to John

[67] Enrollment No. 24, "Master Abstracts," Port of Savannah, schooner *Mary McKay*, July 29, 1820.

[68] Collinsdictionary.com, "Order bill of lading."

[69] Kayser & Co, *Commercial Directory*, 44–45.

McNish originated at Darien, Brunswick, St. Simons Island, and the Satilla River area, suggesting the firm concentrated its business on the central and lower Georgia coast. McNish's association with these planters likely arose from his work as a clerk and bookkeeper for two prominent St. Simons Island planters, John Couper and James Hamilton.[70] R. & J. Habersham, one of the oldest merchant firms in the city, received twenty-two shipments, consisting of both cotton and rice. Most of these cargoes came from the Ogeechee River, Sunbury, Riceboro, and Beaufort. The company received a single shipment of cotton from Darien and none from the lower Georgia coast. The fact that the Habersham firm received little cotton from Darien, then the principal source of upland cotton for the sailing coasters, suggests they dealt mainly with Sea Island cotton.

These firms received cargoes by several vessels; none seem to have relied on one or two ships to the exclusion of others. For example, seventeen different ships delivered the twenty-seven shipments to John McNish while R. & J. Habersham obtained their twenty-two shipments aboard fourteen different vessels. Merchants received shipments aboard coasters commanded by local captains, as well as Northern men. Exceptions were shipments arriving from the lower Georgia coast, where locally owned vessels commanded by local men dominated the trade.

Sometimes, coasters arriving at Savannah carried cargo for a single merchant, but more often, it was destined for two or more consignees. When one merchant was named consignee, the cargo was likely derived from a single planter. However, planters who grew both cotton and rice, or operated more than one plantation, might use different factors for their two crops or their various plantations.[71] Typically, two or three consignees are named for each arriving coaster, but several vessels arrived with cargoes destined for as many as six or seven firms or individuals. When the sloop *Rosetta* arrived from Darien in early March, seven different Savannah merchants were named as consignees of the 192 bales of cotton she carried. Large cargoes did not necessarily mean numerous consignees. A. B. Fannin & Company, which principally handled Darien cotton, was the only consignee named for a shipment of 203 bales carried by the sloop *Morning Star*, as well as 200 bales aboard the sloop *Ann*. In later years, as rough rice became a more important cargo, it was common for multiple consignees to be named for cotton shipments while rough rice most often went to a single merchant.

[70] Vanstory, *Georgia's Land*, 153.

[71] House, *Planter Management*, 44.

At the start of the second decade of the nineteenth century, the Georgia coasting trade was entering into a period of significant growth. The 213 coaster arrivals in Savannah in 1820 were more than twice as many as had occurred ten years earlier. The Savannah newspapers were beginning to publish information on the types and amounts of cargo. According to the published shipping lists, cotton was the principal cargo of the Georgia coasters in 1820. However, the number of bales carried was given for only half of the shipments; upland cotton was rarely distinguished from Sea Island cotton. The 1820 *Daily Georgian* records that 12,053 bales of cotton were transported into Savannah aboard coasters, but the actual number of bales of cotton carried is given for only 75 of the 139 voyages where cotton was a cargo. Darien was the principal source of cotton in 1820, and the *Daily Georgian* reports that cotton was aboard seventy-one of the ninety-two arrivals from the town. The number of bales carried was given for only forty-six of these arrivals, constituting 8,572 bales. As discussed previously, relying on assumptions about the amounts of cotton carried aboard the 34 percent of vessels where the number of bales is not provided, it can be estimated that as many as 11,400 bales of cotton were shipped from the town. Most of the cotton from Darien was upland cotton, meaning the 11,400 bales would have been worth about $554,800, given cotton prices in 1820, equivalent to approximately $12.6 million in modern (2020) values.[72]

Unlike Dairen, the cotton carried from the Sea Islands and the small ports and plantations along the adjacent mainland was principally Sea Island cotton. In 1820, cotton from these areas was aboard sixty-eight of the coaster arrivals in Savannah. The *Daily Georgian* provides information on the number of bales carried for only twenty-nine of these arrivals. This total, 3,481 bales, encompasses about 43 percent of the cotton cargoes from Sea Island cotton-growing locations. These 3,481 bales represented 1,218,350 pounds of cotton with an 1820 value of approximately $399,618. However, less than half (twenty-nine out of sixty-eight shipments) of what are likely Sea Island cotton shipments are enumerated in the 1820 *Daily Georgian*. Thus, it may be that as much as eight hundred thousand to one million dollars' worth of Sea Island cotton was carried by the coasting fleet into Savannah in 1820.

It is impossible to estimate the quantities of rough rice carried into Savannah in 1820 because specific amounts are not published. However, it was not a great amount, considering that only seven cargoes identified as "rice" are

[72] Gray, *History of Agriculture*, 2:1030; MeasuringWorth.com, "real price" of a commodity calculator.

listed for the entire year. However, in 1820, coasters did carry 819 tierces and barrels of clean rice into Savannah. If these containers held about 600 pounds, the capacity of the average tierce, then this represented an estimated 491,400 pounds of rice. Gray notes that the average price for clean rice in 1820 was 2.8 cents per pound, meaning the clean rice transported into Savannah had a value of approximately $13,760. It is obvious that the quantity and value of rice carried by the coasters in 1820 was insignificant when compared to cotton. This would soon change as rice agriculture expanded along the coast and as rice processing facilities were established in and near Savannah.[73]

The shipment of Georgia coastal crops to Charleston continued in 1820. The specifics of this trade, in terms of the types and quantities of commodities shipped, are impossible to estimate due to a lack of records. However, an indication of the volume of coaster traffic sailing from Georgia locations to Charleston is seen in the shipping lists published in the *Charleston Courier*. In the five months from January through May 1820, the newspaper listed thirty-four arrivals of coasting vessels from Georgia locations. Eighteen of these were from Savannah, ten from Darien, three from St. Marys, two from St. Simons Island, and one from the "Satilla." The newspaper listed the types of cargoes carried by these coasters but not the quantities. Not all carried plantation products; two arrived in ballast, and a few transported commodities such as dry goods, tobacco, and rum. However, sixteen shipments of cotton, nine shipments of "rough rice," and seven shipments of "rice" were aboard twenty-eight of the coaster arrivals from Georgia during this period. Most of the rice and rough rice was shipped out of "Savannah," apparently referring to plantations along the lower Savannah River. Many of these rice plantations were located on the South Carolina side, so it is assumed that some of the eighteen voyages from Savannah were carrying products from South Carolina, not Georgia. Rice was also carried from Darien, while cotton was shipped from St. Marys, the Satilla River, St. Simons Island, Darien, and Savannah. The cotton from Darien and Savannah may have been the upland variety brought down the Savannah and Altamaha rivers; the remainder was most likely Sea Island cotton. On two arrivals from Savannah and one from St. Simons Island, "cotton seed" was listed as a cargo.

During these five months, there were ninety-nine arrivals of coasters in Savannah from Georgia locations. It appears that close to one-quarter of the coaster voyages carrying products from Georgia locations were sailing to Charleston rather than Savannah. This proportion might be high, as some

[73] Gray, *History of Agriculture*, 2:680.

voyages into Charleston from "Savannah" likely sailed from plantations on the South Carolina side of the river. Regardless, vessels sailing from these Savannah River plantations with rough rice were purposefully bypassing Savannah in favor of Charleston. The sixteen shipments of "rice" and rough rice carried to Charleston from Georgia were considerably more than the seven shipments carried into Savannah from Georgia locations. It is obvious that a considerable proportion of unmilled Georgia rice was being shipped to Charleston as late as 1820. However, the bulk of the cotton crop carried from the Georgia coast was taken into Savannah. The disparity between where the two commodities were shipped supports the contention that a lack of rice milling facilities in Savannah convinced Georgia planters to send their rough rice to Charleston, which had the necessary pounding mills to process the grain.

The coasters sailing to Charleston from Georgia locations in 1820 represented only a small proportion of the total coaster traffic into the city. Between January and May, the *Charleston Courier* listed 395 coaster arrivals from South Carolina locations. The principal cargo was rice although a considerable number also transported cotton, and a few had minor cargoes aboard, such as pitch, tar, and groundnuts. Around 95 percent of the local coaster trade into Charleston between January and May in 1820 depended on South Carolina–grown commodities. The cargoes from Georgia represented less than five percent of the local coaster traffic into the city during this period. Thus, relative to the total volume of its local rice and cotton commerce, Charleston was minimally dependent upon Georgia crops.

On the other hand, in terms of the volume of their business, Savannah merchants were losing a considerable portion of the Georgia and lower South Carolina rice crop to Charleston. This loss was recognized by Savannah businessmen and was remedied with the construction of several pounding mills in and near Savannah during the 1820s. With the erection of these mills, increasing numbers of Georgia planters began to ship their rough rice to merchants in Savannah, with many planters in nearby areas of South Carolina following suit.

It is apparent that in 1820, some captains were wide-ranging in their trading, arriving in Savannah from disparate locations along the South Carolina and Georgia coasts. Others tended to concentrate on very specific areas. This was particularly true for captains sailing on the Darien-Savannah route, where the tremendous amount of cotton being shipped could support many vessels. It is also evident that there was an insufficient number of local ships and captains to handle the quantities of goods, particularly cotton, which had to be shipped. This need attracted Northern captains to Georgia, where they sailed their ships over the six- or seven-month cotton and rice shipping season, before returning north.

## The Georgia Coasting Trade in 1830

The ten years between 1820 and 1830 saw unprecedented growth in the Georgia coasting trade, particularly as it related to the importance of Savannah as the principal port. This growth went hand in hand with the great expansion in rice cultivation along Georgia's major rivers and with the continued development of Savannah as a commercial center. Rice production expanded, and by 1830, rice plantations were in operation along the lower reaches of the major rivers in Georgia and South Carolina. Sea Island cotton remained an important crop, but it peaked well before 1830, with the crop that year similar to what it had been twenty-five years earlier. Few reliable statistics on the production or exports of these two principal crops exist for Georgia during the 1820s, but some scattered information is available. In 1825, Savannah reportedly exported 7,231 tierces of rice. This had risen to 11,455 by 1826 and to 14,707 by 1827. Sea Island cotton exports from Savannah in 1827–1828 are reported at 11,648 bales, only slightly more than the 10,888 bales exported in 1820–1821. Total cotton exports from Savannah increased greatly during the 1820s, entirely due to the expansion in cultivation of upland cotton. Sugar production on the Georgia coast also peaked around 1830 and then declined because of a decrease in import duties on sugar and falling foreign markets, leading planters to abandon sugar in favor of cotton and rice.[74]

The population of Savannah grew slightly during the 1820s, from 7,523 in 1820 to just 7,773 in 1830.[75] However, the city's commercial infrastructure expanded considerably during this decade. The city was badly damaged by fire in January 1820, after which the central commercial district was rebuilt, initiating a spirit of optimism. The 1820s saw a series of schemes for commercial development, including discussions about canals and efforts at harbor improvements. In addition, facilities for milling rice were constructed in Savannah, attracting planters to ship their rice there rather than Charleston for processing and sale. On August 28, 1828, the *Daily Georgian* reported that two new rice mills were being erected in the city. These were steam mills, the first

[74] "Commerce of Savannah," *DeBow's Review*, June 1851, 10:685; *Daily Georgian*, October 2, 1828; Gray, *History of Agriculture*, 2:744, 748.

[75] "The City of Savannah," *DeBow's Review*, July 1853, 106.

in the city, and the paper noted that only two tide mills were available for pounding rice.

These expansions in crop production and commercial infrastructure are directly reflected in an increase in the volume of local coaster traffic at Savannah. Newspaper shipping lists show that the number of coaster arrivals into Savannah grew from 213 in 1820, to 251 in 1825 and to 479 in 1830, more than doubling in ten years. The ports of origin for the coasters rose from 33 in 1820 to 46 in 1830 (Figure 7.6). This increase is associated principally with the transport of rough rice from areas where its cultivation had spread or increased during the 1820s and with the transport of clean rice from recently established pounding mills along the lower Savannah River. Darien retained the premier position as the port of origin for coasters arriving in Savannah, a rank it had achieved just before 1820. In 1830, 225 coaster arrivals originated at Darien, more than double the 91 voyages into Savannah ten years earlier. However, the relative importance of Darien as a port of origin did not change significantly; it continued to account for about 45 percent of coaster arrivals at Savannah.

The shipping lists of the 1830 *Daily Savannah Republican* are used to examine the structure and workings of the Georgia coasting trade as it existed that year. The *Daily Savannah Republican*, also known as the *Savannah Republican* and the *Savannah Daily Republican*, was published daily except Sundays from October to June and then every other day except Sunday during the summer. Other Savannah newspapers adopted this limited publication schedule during the summer because many residents departed the city during what was known as the "sickly season," partly to escape the heat but, more importantly, to avoid the dangerous diseases common to coastal areas during the summer. As a result, many businesses in the city closed or curtailed their activities between June and September. In addition, port business slowed considerably as these were the months when little cotton or rice was shipped to market.

The *Daily Savannah Republican* published shipping information in a column titled "Marine List Port of Savannah." By this time, shipping columns were regularly including information on the types and amounts of the principal cargoes carried by arriving coasters, in addition to the names of the vessel, master, port of origin, and consignee. The information on departing vessels consisted of the names of the vessel, master, and the port of destination. From 1830 on it becomes possible to reasonably estimate the volume and value of cargoes carried by individual vessels or transported from locations into Savannah, at least for rice and cotton.

The 479 arrivals of coasters into Savannah in 1830 represent the most recorded for any year between 1800 and 1861. By 1835, the number of arrivals dropped to 308, and while there were some fluctuations in arrivals over the following fifteen years, the general trend was downward to just 132 arrivals in 1861. The decrease in coaster arrivals directly reflects two factors: the increasing participation of steamboats in the coasting trade and the expansion of the network of railroads out of Savannah that transported cotton from the interior. After 1830, steamers quickly garnered most of the passenger traffic and began carrying more cargo, especially cotton, reducing the quantities available to sailing vessels. The steamboats' reliable scheduling was an important reason for their popularity. However, steamboats rarely carried rough rice because they lacked the deep, commodious hulls of sailing vessels into which the loose grain was loaded for transport. The great expansion of rice production along coastal Georgia and South Carolina after 1820 contributed significantly to the continued importance of sailing coasters long after steamboats were introduced.

The second factor contributing to a decline in the sailing trade after 1830 was the extension of railroads from Savannah into the interior cotton-growing regions of Georgia. By the late 1830s, railroads were transporting much of the upland crop that had formerly been carried down the Altamaha River to Darien. By the early 1840s, the elimination of this flow of cotton essentially ended Dairen's relevance as a port in the Georgia sailing trade.

### *Georgia Coasting Ships and Captains in 1830*

Eighty-three different sailing vessels, consisting of fifty-nine sloops and twenty-four schooners, made the 479 arrivals into Savannah from local ports in 1830. This is, coincidentally, the exact number of ships sailing in the coasting fleet ten years earlier, when only 213 arrivals at Savannah were recorded. Further, the proportion of sloops to schooners was the same as it had been in 1820, and this dominance of sloops had been evident in the local trade since the late eighteenth century. After 1830, however, the number of sloops sailing in the Georgia trade began to decrease relative to schooners, such that by 1861 schooners outnumbered sloops four to one.

The most active trader in 1830 was the Rochester-built sloop *George Washington*, which sailed into Savannah twenty times with coastal cargoes. Walter Smith, a resident of Savannah, was named the sloop's master on most of these voyages. Captain Smith seems to have purchased the fifty-three-foot *George Washington* in 1829 from a group of owners in the town of Rochester, and he sailed the sloop in the Georgia trade until his death sometime in

1840.[76] Thirteen of the *George Washington*'s voyages in 1830 originated at named rice mills or plantations, most located on the lower Savannah or the Combahee rivers. Clean and rough rice made up the sloop's cargoes on every trip. The places of origin included "Haywoods Mill," "McLarens Mill," "Roses Place," and Roseland Plantation.[77] The *George Washington* also made six voyages from Darien, transporting rice, cotton, and on one voyage, syrup. The sloop made a single journey from Hilton Head Island in South Carolina with a cargo of fifty-four bales of Sea Island cotton consigned to the merchant Robert Habersham.

Among the other active traders in 1830 were the sloops *Bolivar*, making nineteen voyages into Savannah; the *Othello* and *Albert*, each with fifteen voyages; the *Conductor*, with fourteen; the *Three Brothers*, *Excel*, and *Leader* with thirteen; and the *Mariner* and *Eliza* with twelve each. Local men commanded and owned several of these ships. In this group were long-time captain John Chevalier, master of the Connecticut-built sloop *Leader*, and Richard Hill, captain of the *Conductor*. Both men resided in St. Marys, where most of their voyages into Savannah originated. Another local captain, Claude Phillip Leset, a resident of Liberty County, made fourteen trips from the town of Riceboro with the sloop *Albert* in 1830. The only other ships sailing from Riceboro with coastal produce were the sloops *America* and *Portsmouth*. The master of both seems to have been Leonard Bolles, one of the many Sippican captains now participating in the Georgia trade.

The sloop *Eliza*, another of the active traders in 1830, was not only commanded and owned by local men but also locally built at St. Marys in 1828. Her master and owner in 1830, Edmund Richardson of St. Marys, sailed several vessels in the Georgia trade from 1809 to the early 1840s.[78] On seven of Captain Richardson's nine voyages in 1830, he sailed from the Turtle River in Glynn County. Sea Island cotton was his principal cargo, but he also carried miscellaneous items such as hides, skins, tallow, and beeswax.

John M. McColley brought the sloop *Othello* into Savannah fifteen times in 1830. All but one of these trips originated at locations in lower South Carolina and the Ogeechee River. Captain McColley was a resident of Derby, Connecticut, but he was a part owner of the *Othello* with Chatham County

[76] Enrollment No. 6, Port of Savannah, sloop *George Washington*, May 2, 1829; *Daily Georgian*, January 1, 1841.

[77] McKinnon, "Chart of Savannah River, 1825."

[78] Enrollment No. 5, Port of St. Marys, sloop *Eliza*, August 10, 1828; Enrollment No. 1, Port of St. Marys, sloop *Eliza*, June 7, 1832.

resident Daniel Blake, a prominent rice planter. At the time, Blake owned Vallambrosa Plantation on the Ogeechee River and had business interests in Savannah's recently established steam-powered rice mills.[79] All, or most, of the 14,400 bushels of rough rice the *Othello* carried from the Ogeechee likely came from his plantation.

In addition to Captain McColley of the *Othello*, several Northern captains sailed in the Georgia trade in 1830. Captains and ships from the Sippican area of Massachusetts were particularly numerous. Twenty-four, or approximately one-third of the seventy-eight names of masters listed in the 1830 *Daily Savannah Republican* shipping lists, can be identified as Sippican men or possess names of known maritime families of that area. This represents an increase over 1820 when Sippican men made up about 20 percent of the captains in the trade.

These Northern men were energetic traders, making 203 of the 479 arrivals in Savannah in 1830. Among the most active Sippican captains were Elisha Luce, master of the sloops *Angel* and *William*; James Blankenship of the sloop *Excel*; Leonard Bolles of the sloop *America*; Sylvester Bates Jr. of the sloop *Dirigo*; and Ezra Sturtevant of the sloop *Mariner*. As they were in earlier years, Sippican captains were very active in the Darien-Savannah trade and were principal traders with the lower Ogeechee River, which had become an important rice-producing area. Two Sippican sloops, the *Excel* and *Wave*, transported at least 49,250 bushels of rough rice from the Ogeechee in 1830. At least one Sippican master, Captain Nathan Briggs, with his sloop *Mary Howard*, ventured south of Darien and carried cotton from the Satilla River area on the lower Georgia coast, but not a single Sippican man seems to have sailed on the St. Marys-Savannah route. In 1830, that trade was almost totally in the control of local captains residing in St. Marys and Camden County, as it had been for at least three decades.

### *Ports in the Coasting Trade in 1830*

In 1830, Darien was the most important port of origin for the coasters sailing into Savannah, as it had been since 1820. There were 225 coaster arrivals in Savannah from Darien in 1830, representing almost half of the 479 arrivals in the city that year. The ship captains from Sippican concentrated their activities in the Darien-Savannah trade. While these men made about 42 percent of all coaster arrivals into Savannah in 1830, they made about 65

[79] Enrollment No. 4, Port of Savannah, sloop *Othello*, February 15, 1830; J. F. Smith, *Slavery and Rice*, 34–35.

percent of the 225 trips from Darien. The principal cargo for the Sippican captains and others sailing this route was the upland cotton Darien received in great quantities from the Georgia interior. Only occasionally were other commodities, such as rice, syrup, or hides shipped from Darien.

The town of St. Marys retained its secondary position relative to Darien in terms of coaster traffic to Savannah. In 1830, forty-two voyages into Savannah originated at this South Georgia port. The vessels most active in the St. Marys-Savannah trade were the sloops *Betsey Maria* (Captain Malatiah H. Hubbard), *Leader* (Captain John Chevalier), *Conductor* (Captain Richard A. Hill), and *George* (Captain William Laen), and a single schooner, the *Mary Adams*, sailed by Captain Manuel Riberon, a member of one of several Spanish families from Florida who settled in the St. Marys area.

Sea Island cotton was the most common cargo shipped from St. Marys. Oranges also were a common St. Marys export, such as the three thousand carried by Captain Dennis Pacetty aboard the appropriately named schooner *Orange* in February, and the thirty thousand carried by Captain Richard Hill on the sloop *Conductor* in January. Other minor cargo from St. Marys included live oak timber, lightwood, hides, and beeswax. Captain Malatiah Hubbard transported two shipments of otter skins aboard his sloop *Betsey Maria*. No rice was shipped out of St. Marys to Savannah in 1830.

Among the important ports of origin for the coasters in 1830 were those areas where rice cultivation had expanded in the late 1820s. These included the lower Savannah River, the Ogeechee River, the Altamaha River, and several rivers in lower South Carolina, including the Combahee and New rivers, as well as rice plantations around Beaufort. One of these rice-growing areas, the Ogeechee River, ranked third as a port of origin for coasters sailings into Savannah in 1830. Several ships made the thirty-two voyages from the Ogeechee that year, and rough rice was carried on thirty-one. Ships from the Ogeechee River also transported lesser quantities of clean rice packed in tierces and a modest amount of cotton. Between 1820 and 1830, the shipment of rough rice from the Ogeechee River had risen from almost none to at least sixty-six thousand bushels, an expression of the success and expansion of the tide flow technique, as well as the new milling facilities in Savannah, which attracted planters to ship their rough rice there.

A considerable amount of clean and rough rice was carried from mills and plantations along the lower Savannah River. Lesser amounts were shipped from the other major rivers in Georgia. Rice was listed as a cargo on thirty coaster voyages from Darien, much of it representing crops shipped from nearby rice plantations along the lower Altamaha. The rice shipped from

Darien consisted of 647 casks and tierces of clean rice and at least 30,000 bushels of rough rice.

In 1830, rough rice was a cargo on three of the fourteen coaster voyages from the Satilla and Little Satilla rivers. A single ship, the Connecticut-built sloop *George*, commanded by William Laen of St. Marys, carried these five thousand bushels. At the time, cotton was the principal commodity shipped from the two Satillas. Rice had been grown along the Satilla River since the eighteenth century, and why so little was shipped to Savannah in 1830 is unknown.[80] Oranges were carried on several voyages from the "Satilla," and hides, sugar, and molasses constituted minor cargoes from the two rivers.

Riceboro was the origin of twenty-four coaster voyages 1830, making the town the fourth most important port of origin in terms of number of sailings. Sea Island cotton, carried on twenty-one voyages, was the main cargo; rice was listed on four. Miscellaneous items, such as "cane," hides, and "live oak wood," were also shipped from the town. Only three ships sailed the Riceboro-Savannah route. These were the sloop *Albert*, commanded by local captain Claude Phillip Leset, and two Sippican vessels, the sloops *America* under Captain Bolles and the *Portsmouth* under Captain Bates.

The Turtle River in Glynn County was the origin of sixteen voyages into Savannah in 1830. Sea Island cotton, carried on twelve voyages, was the primary commodity. Considerable amounts of syrup, sugar, molasses, and hides were also shipped, as well as small quantities of beeswax, tallow, and oranges. Sailing the Turtle River-Savannah route were several sloops captained by residents of St. Marys: Edmund Richardson with the *Eliza*, Manuel Riberon with the *Mill Maid*, William Laen with the *George*, and John Chevalier with the *Leader*. Two Sippican captains sailed from the Turtle River. These were a Captain Burr, with the sloop *Merchant*, and a Captain Handy, possibly Elisha Handy, with the sloop *Rosetta*.

The town of Sunbury was a moderately important local port in 1830, with twelve coaster voyages from there to Savannah. Sea Island cotton was the principal cargo, but vessels also transported small quantities of hides, beeswax, and tallow. Three local captains handled the Sunbury-Savannah trade in 1830. These were George Rentz and Nicholas Sallowich, both serving as master of the sloop *Ann*, and Edmund Richardson, with his sloop *Eliza*.

Harris Neck, on the central Georgia coast in McIntosh County, had become more important in the trade since 1820. In that year, just one coaster carried cargo from there; this number increased to five in 1827–1828. Eleven

[80] J. F. Smith, *Slavery and Rice*, 21, 35.

voyages were made from Harris Neck in 1830, with Sea Island cotton listed as cargo on every trip. The three vessels sailing from Harris Neck were all sloops: the *Angelica* (Captain Burgo), *Two Friends* (Captain Cannon), and *Ann* (Captain George Rentz).

Coasters transported cargoes into Savannah from several of the Sea Islands in 1830. These were Ossabaw, St. Catherines, Sapelo, St. Simons, and Skidaway. Three ships, a surprisingly high number, made eight trips from Ossabaw Island. Sea Island cotton, the island's main crop, was a cargo on every voyage, and live oak or wood were carried on several. Two ships transporting 157 bales of cotton sailed from St. Catherines while three coasters from Sapelo Island carried 126 bales of cotton and 25 hogsheads of sugar. Similarly, three coasters sailed from Skidaway Island, all carrying Sea Island cotton. A single vessel sailed from St. Simons Island, the schooner *Enterprise*, under a captain named Eldridge. The *Enterprise* carried no cargo for Savannah, and the paper noted the vessel was "bound for Charleston."

The trade between Savannah and lower South Carolina in 1830 was similar to 1820 when twenty-four coasters arrived from that region. In 1830, the paper reported twelve arrivals from Beaufort, three from the Combahee River, two from "Euhaw" (referring to the Euhaw Creek area along the Broad River), as well as single arrivals from Hilton Head Island, the May River, and the New River.

Sea Island cotton was the principal cargo carried from Beaufort; a single shipment of rice is listed for 1830. Much of the rice grown in the Beaufort area was shipped to Charleston. Coasters sailing from other locations in lower South Carolina carried rough rice, clean rice, and Sea Island cotton.

The trade between Savannah and Florida proceeded modestly through the 1820s, at least for the sailing coasters. Sixteen arrivals were from Florida in 1820, nineteen in 1825, and thirteen in 1830. The 1830 arrivals comprised seven from St. Augustine, five from the St. Johns River, and one from Fernandina on Amelia Island. In addition to passengers, coasters from Florida carried Sea Island cotton, oranges, hides, and skins. In November, the sloop *Import* sailed from St. Augustine with a cargo of "tar and rosin." This represents one of earliest mentions of the transport of naval stores in the Georgia trade. Shipment of naval stores would increase over time, eventually becoming important cargoes during the summer when cotton and rice crops were unavailable for shipment.

### *The Coasting Cargoes in 1830*

In 1830, the *Daily Savannah Republican* provided cargo or passenger information for 422 of the 479 coaster arrivals. Table 7.8 provides information on the number of shipments of various commodities identified in the paper's shipping lists that year. Estimates of the minimum quantities of these items are given where those data are available.

The *Daily Savannah Republican* distinguished "Sea Island cotton" from "cotton" in its 1830 shipping lists. Most of the cargoes identified as "cotton" originated at Darien, indicating the term was used for upland cotton. In a few instances, based on where the cargo originated, the term "cotton" was used when Sea Island cotton was shipped. In these obvious cases, the "cotton" has been classified as Sea Island cotton. In 1830, Sea Island cotton was a cargo on 135 coaster arrivals in Savannah, and the number of bales or bags carried is given for 113 of these. The total amount listed on these arrivals was 6,022.25 bales or bags. The quantity of Sea Island cotton aboard individual vessels ranged from the two bales carried by the sloop *Wave* from the Ogeechee River in February to the 313 bales the sloop *Sapelo* transported from Darien that same month. The amount of Sea Island cotton aboard the *Sapelo* was considerably greater than the normal amount carried and might be incorrect or should be listed as upland cotton, considering it was shipped from Darien. More typically, individual shipments of Sea Island cotton were smaller, averaging around fifty-three bales and rarely above a hundred. This is in marked contrast to the average of 236 bales of upland cotton per shipment, almost five times that of the Sea Island variety, and few coasters carried fewer than one hundred bales of upland cotton on any given voyage. The main reason seems to be that Sea Island cotton shipments typically represented the crop of a single planter or a small number of planters, each of whose cotton production was small. On the other hand, most of the upland cotton carried by coasters represented the production of large numbers of inland planters collected in Darien for reshipment to Savannah. During the shipping season, there were large quantities of cotton on hand at Darien, and merchants and agents took every opportunity to ship as much as possible on an available vessel, regardless of the cotton's origin.

Upland cotton was identified as cargo on 167 coaster arrivals in 1830, with the number of bales given for 158 arrivals. A total of 37,361.5 bales of cotton are listed for these arrivals. This represents more than six times the number of bales of Sea Island cotton carried into Savannah that year and reflects the huge quantity of upland cotton arriving in Darien from the interior. The number of bales of upland cotton carried by individual vessels ranged from the twenty-seven carried by the sloop *Catherine & Elizabeth* to the 522 reported aboard the sloop *Advance* when she arrived in Savannah in late January.

Rough rice was listed as a cargo for seventy-four shipments, sixty-three of which included specific amounts in terms of number of bushels. These sixty-three arrivals carried 134,224 bushels of rough rice, representing about 6,040,080 pounds of grain. The amounts of rough rice carried ranged from the one hundred bushels brought from Darien by the sloop *Franklin* to the fifty-five hundred bushels shipped from the Ogeechee River aboard the Sippican sloop *Excel.* In addition to the large rough rice cargo, the *Excel* carried fifty tierces of clean rice and fourteen bales of Sea Island cotton.

Clean rice was listed as a cargo on fifty coaster arrivals, with specific amounts given in terms of tierces, barrels, or casks for every arrival.

In addition to rice and cotton, the coasters in 1830 carried a variety of other coastal commodities, including syrup (typically identified as "Georgia syrup"), sugar, molasses, hides, skins, otter skins, deer skins, oranges, "live oak wood," corn, potatoes, beeswax, and tallow. The quantities carried are not listed for many of these items, excepting syrup, sugar, and molasses. The shipping lists show that coasters carried into Savannah fifty-seven barrels and forty-six hogsheads of "Georgia syrup," twelve barrels and seventy-one hogsheads of sugar, and seventeen hogsheads of molasses. These sugarcane derivatives all came from locations south of Sunbury, such as Sapelo Island, Darien, the Turtle River, and the "Satillas."

The coasters transported enormous quantities of rice, Sea Island cotton, and upland cotton into Savannah in 1830. Relying on average weights of bales, tierces, barrels, and casks and on the average prices for rice and cotton in 1830, rough estimates of the quantities and values of these cargoes are possible. The rice cargoes included 134,224 bushels of rough rice with a value of about $99,728, at 74.3 cents per bushel, and 2,529,600 pounds of clean rice with a value of $75,888, assuming a price of three cents per pound. The 6,022.25 bales of Sea Island cotton, representing about 2,107,788 pounds of fiber (assuming 350-pound bags or bales), had a value of $526,947 at 25 cents per pound, the average price of Sea Island cotton that year. The 37,361.5 bales of upland cotton constituted about 11,955,680 pounds (assuming an average bale of about 320 pounds). This upland cotton was worth about $1,016,233, given the average price of 8.5 cents per pound for cotton that year.[81]

[81] Commodity prices from Gray, *History of Agriculture*, 2:1027, 1030–1031.

The monetary figures presented here are minimal values for these principal commodities because the specific quantities of rice and cotton are not provided for every coaster arrival into Savannah. However, estimating those voyages for which no quantities were published makes it possible to derive approximate figures on the total cargoes. For example, no information on the number of bales is provided for twenty-two of the 135 arrivals of coasters carrying Sea Island cotton. These twenty-two arrivals represent approximately 16 percent of all Sea Island cotton shipments, suggesting an additional 964 unrecorded bales were carried into the city. Thus, it might be argued that coasters carried approximately 6,986 bales of Sea Island cotton into Savannah in 1830. Upland cotton was a cargo on 167 coaster voyages from Darien in 1830, and the number of bales (37,361.5) is provided for 158 of these arrivals. The nine arrivals with no quantities represent about 5 percent of the total number of voyages. If these nine arrivals carried amounts similar to the 158 arrivals with known numbers of bales, then an additional 1,868 bales may have been carried, representing a total of 39,230 bales of upland cotton over the entire year.

In 1830, the *Daily Savannah Republican* names rough rice as a cargo on seventy-four coaster arrivals, sixty-three of which included specific numbers of bushels. The eleven arrivals with no quantities given represent about 15 percent of all shipments of rough rice. The sixty-three arrivals with known quantities carried 134,224 bushels, the additional 15 percent that might have been aboard the unrecorded voyages accounts for an estimated 20,134 bushels, bringing the total number of bushels possibly carried into the city to 154,358, representing about 6,946,110 pounds of grain.

These estimates on the total amounts of cotton and rice carried by the coasting fleet into Savannah are just that, estimates, and they cannot be fully verified with the available data. However, they likely provide a realistic picture of the approximate quantity and value of these two important cargoes carried in the coasting trade in 1830. Relying on these estimated quantities, the average weights of Sea Island and upland cotton bales in 1830, and the prices for these commodities, the coasting fleet transported about $114,688 worth of rough rice, $611,275 worth of Sea Island cotton, and $1,067,056 worth of upland cotton into Savannah in 1830. With the $75,888 worth of clean rice, the total value of these principal commodities was close to $1,900,000.

The estimated amount of Sea Island cotton carried by the coasters in 1830 represented a substantial percentage of that exported from Savannah. The *Merchants' Magazine and Commercial Review* reported that 9,579 bales of Sea Island cotton were exported from Savannah between October 1829 and

September 1830, which would have included most of the cotton carried by coasters into Savannah in 1830.[82] The estimated 6,986 bales of Sea Island cotton arriving aboard coasters in 1830 constitute approximately 73 percent of the 1829–1830 export figure and indicate how important the sailing coasters were in the Sea Island cotton trade.

Various reports indicate that Savannah exported about 240,000 bales of upland cotton in 1830. The estimated 39,230 bales of upland cotton carried into Savannah by coasters is about 16 percent of the total exported from the city that year. Most of the upland cotton exported arrived in some other manner; some was carried by steamers from Darien, but much more came into Savannah via the Savannah River aboard pole boats, bateaux, barges, and steamboats. During the cotton shipping season, Savannah newspapers listed the almost daily arrival of pole boats from Augusta, each carrying as much as five hundred or six hundred bales of cotton. By 1830, several steamboats operated on the Savannah River, towing barges laden with cotton from inland landings to Savannah. These steamers moved a considerable quantity of cotton from the interior. For example, in 1841, one Savannah River steamer, the *Mary Summers,* transported at least 9,664 bales of upland cotton down the river to Savannah.[83]

Lieber reports that 11,455 tierces of rice were exported from Savannah for the year ending September 30, 1830.[84] This figure omits October through December, but it seems to be the only export figure available. In 1830, the sailing coasters carried 4,017 tierces, 90 barrels, and 109 casks of clean rice into the city. As noted previously, it is unknown if the terms "barrel" and "cask" were consistently used to specify containers distinct from tierces, so these containers are lumped together as "tierces," for a total of 4,216. Although there was variation, approximately 59 percent of a bushel of rough rice was converted into clean rice during milling, meaning the estimated 154,358 bushels of rough rice carried by coasters into Savannah in 1830 represented approximately 6,830 tierces of clean rice weighing 600 pounds each.[85] Thus, in 1830 the rice carried by the coasting fleet into Savannah represented an estimated 11,046 tierces of clean rice, extremely close to the 11,455 reportedly exported from the city between October 1829 and September 1830. Admittedly, the figures developed here are estimates, but they show the sailing

[82] Hunt, "Commercial Cities," 504.

[83] Lieber, *Encyclopaedia Americana*, 439; Pearson et al. "Mary Summers," 170.

[84] Lieber, *Encyclopaedia Americana*, 439.

[85] Milling and weight figures from House, *Planter Management*, 65.

coasters transported a high proportion of the rice exported from Savannah. Further, the amount of rice carried by coasters in 1830 was a considerable increase over previous years, substantiating the expansion in rice production occurring in Georgia in the late 1820s, as well as the increasing tendency for Georgia planters to ship their rice to the city where factors could now handle all aspects involved in marketing the crop, including milling.

Despite the increasing amounts of Georgia rice shipped into Savannah, some Georgia planters continued to send their crops to Charleston in 1830, meaning the amounts carried into Savannah are an incomplete measure of the state's total production. Charleston newspapers show that shipments from the Georgia coast below Savannah into that city continued into the 1830s. However, by the late 1820s, these shipments were declining, and most coasters arriving in Charleston were sailing from South Carolina locations such as Georgetown, Santee, Combahee, Ashepoo, Euhaw, and Beaufort.

In 1830, the sailing coasters still carried passengers although steamboats had captured most of the passenger traffic along the coast. Passengers were reported aboard twenty of the coaster arrivals in Savannah in 1830. The number of passengers carried ranged from a single individual to the twenty-nine people traveling on the schooner *William* from St. Augustine in early November.

### *Seasonality of the Georgia Coasting Trade in 1830*

By 1830, the seasonal pattern in the Georgia sailing trade that had begun to emerge in 1820 was fully established. This seasonality was driven by the availability of the primary commodities, rice and cotton. The normal shipping season for these two crops extended from October or November into April, so the figures used here for 1830 encompass two shipping seasons, 1829–1830 and 1830–1831. However, the patterns of shipment seen over one calendar year do mirror the patterns over a shipping year. In 1830, as in earlier years, sailing coasters arrived in Savannah every month, but 387 arrivals, representing 81 percent of the total, occurred between November and April. The shipment of rice usually began in October; in 1830 the first rough rice shipped for the 1830–1831 season may have been the seventeen hundred bushels carried by the sloop *Albert* from Riceboro on October 2. Rice shipments peaked in December and January and then tapered off, with little or none shipped during the summer. In 1830, coasters carried thirty-one rice shipments into Savannah during December and the same in January. These represented the highest number of rice shipments reported for any month, and together, they constituted close to half of the rice shipments for the entire calendar year.

Only a few shipments of rice, usually clean rice, were made during the summer. In 1830, only three individual shipments of rice arrived at Savannah during the four months from June through September, and no shipments at all are reported for July and September. The last shipment of rice for the season seems to have been the 111 tierces of clean rice carried from McLarens Mill on the Savannah River by the sloop *George Washington* on June 14.

Shipments of Sea Island cotton into Savannah occurred every month in 1830, but, like rice, most cotton shipments were made between November and April. However, the peak of the Sea Island cotton shipping season occurred in February or March, a month or two after the peak of the rice-shipping season. There were twenty-four individual shipments of Sea Island cotton in February and twenty-seven shipments in March, the highest monthly numbers reported for 1830. Together, these two months encompassed nearly 40 percent of all the Sea Island cotton shipped that year. A small number of cotton shipments were made during summer 1830—three in both June and July, two in August, and one in September. In some years, no cotton was shipped during the summer.

As in previous years, there weren't enough local captains and ships to handle the agricultural cargoes during the shipping season, so Northern captains came South to participate in the trade. Notices of the arrival of these Northern captains usually appeared in Savannah newspapers in October. The first Northern coaster to arrive for the 1830–1831 season seems to have been the Sippican sloop *Excel*, James Blankenship, master. The *Excel*'s arrival from New York was reported in the *Daily Savannah Republican* on October 2. On October 9, the newspaper reported the arrival of another Sippican ship, the sloop *America*, with Leonard Bolles as master. The *America* was sailing from New Bedford, and the newspaper noted she was "bound for Riceboro" with merchandise and passengers. Over the next several weeks, other vessels arrived at Savannah from Northern ports, most worked in the coasting trade over the next seven or eight months before returning North.

Locally owned ships, with the advantage of being on hand at the start of the shipping season, typically began to arrive in Savannah with rice and cotton a couple of weeks before the Northern coasters. In 1830, local vessels, such as the sloops *Albert*, *Catherine & Elizabeth*, *Mill Maid*, *George Washington*, and *Flora*, all sailed into Savannah with rice and cotton during October, before the Northern sloop *America* arrived with her first local cargo late in the month.

In April and May, as the shipping season was winding down, the Northern captains began to sail home. Notices of these departures were reported in Savannah newspapers. In 1830, most of the Northern ships left the Georgia coast by the end of May, but a few lingered, as they did in most years. The last Northern vessel to work in the 1829–1830 season was the sloop *Good Intent*, whose arrival in Savannah with a load of cotton from Darien was reported on June 19. That same day, the *Daily Savannah Republican* published an advertisement noting, "The fast Sailing Packet Sloop GOOD INTENT, Capt. Allen" was departing for New York.

Through the summer, locally owned coasters continued sailing in and out of Savannah, but at a greatly reduced level. In 1830, the *Daily Savannah Republican* reported only twelve coaster arrivals in July, six in August, and six in September. These local ships carried small quantities of rice and cotton, apparently crops remaining from the previous season, but they more often arrived in ballast or with other cargoes, such as corn or live oak timber. These vessels continued to carry a variety of merchandise, hardware, foodstuffs, and luxury goods out from Savannah to coastal towns and plantations, but records of these outbound shipments are scarce. Those records that are available, mostly plantation accounts and journals, generally apply to later years, but they indicate that the shipment of even these items declined during the summer.[86]

### *Savannah Merchants in the Coasting Trade in 1830*

The mercantile and financial infrastructure of Savannah grew significantly during the 1820s. An expansion in banking, insurance, and business establishments was accompanied by a similar expansion in physical facilities directly related to the coasting trade, such as docks, wharves, warehouses, and steam-driven rice mills. Though not extensive, ship construction and repair facilities were available in the city. On January 2, 1830, the *Daily Savannah Republican* noted that the firm of "Willink & Wood" had dissolved, and future business would be conducted solely by Henry F. Willink. The Willink firm became the largest shipbuilding and ship repair facility in the city.

In 1830, the *Daily Savannah Republican* listed names of consignees for 379 of the 479 coaster arrivals into the city. Forty-five arrivals were destined only "to master," and two were "to order"; fifty-three arrivals included no

[86] The accounts that have been examined in detail apply to years after 1830 and include the estate records of James Hamilton and James Troup, Glynn County Wills and Appraisements, Books E and F; Charles C. Jones Plantation Books, Howard-Tilton Library, Tulane University; Roswell King's journals for 1838–1844 in "Roswell King Diary," Hill Memorial Library, Louisiana State University; Roswell King's journals for 1845–1854 in Sullivan, *All Under Bank*, and the Hugh Fraser Grant accounts in House, *Planter Management.*

information on who the shipments were destined for. However, only three of the arrivals with no listed consignee carried coastal cargo. Many arrived in ballast, or the newspaper noted the ship was bound for another port. As in earlier years, when the term "to master" was used as the consignee, vessels typically arrived in ballast or were transporting some of the minor coastal commodities, such as oranges, live oak timber, or hides.

Seventy-two mercantile firms or individuals are named as consignees for 679 individual shipments aboard the 379 arrivals. The number of named consignees is four fewer than in 1820 even though the number of shipments in 1830 was almost triple the number arriving ten years earlier. Twenty-four consignees are named a single time in receipt of a shipment. Some of those receiving a single shipment appear to be individuals rather than firms, possibly handling the shipment of their own crops while others are merchant firms that had little dealings with coastal produce.

Eighteen firms received ten or more shipments of coastal produce in 1830. Among these were several businesses that had long been involved in coastal commerce. Principal among these was the Robert Habersham firm, which was named as consignee on eighty-one individual shipments in 1830, more than any other concern. The firm was now preeminent among merchants in handling coastal products, a position they would retain until the Civil War. The firm was heavily involved in the rice trade, but it also handled considerable quantities of Sea Island and upland cotton. In 1830, coasters delivered cargoes to Robert Habersham from locations between Beaufort in lower South Carolina and the Satilla River on the lower Georgia coast. However, most shipments destined for Robert Habersham came from the rice-producing areas along the Savannah, Ogeechee, and Combahee rivers and from the town of Riceboro.

The consignments to Robert Habersham arrived aboard twenty-four different sloops and schooners. These included locally owned and Northern vessels, but the firm relied heavily on three Northern ships. These were the sloop *Excel*, James Blankenship master, which sailed into Savannah twelve times during 1830 with shipments for Robert Habersham; the sloop *Othello*, John McColley master, which arrived eleven times with cargo for the firm; and the sloop *Wave*, commanded by another captain named Blankenship, which arrived eight times. Another of the ships transporting cargoes to the firm in 1830 was the sixty-four-foot sloop named *Robert Habersham*, built in Derby, Connecticut, in 1829. Early records on the ownership of this vessel have not been located, but the merchant Robert Habersham was an owner in 1834, and it is likely he was an original owner.[87]

When Robert Habersham received a shipment consisting only of rice, the firm tended to be the sole consignee for the entire cargo. However, when the firm received a mixed cargo of rice and Sea Island cotton, or a cargo of more than ten or so bales of cotton, it tended to be one of several consignees receiving the cargo. This pattern of single consignees for rice shipments, and multiple consignees for Sea Island cotton or mixed cargoes is a pattern characteristic of all shipments into Savannah, not just those involving Robert Habersham. Rice shipments tended to come from a single plantation or mill, as did small shipments of Sea Island cotton. However, shipments of large quantities of Sea Island cotton and mixed shipments of cotton and rice, as well as other commodities such as hides, oranges, or sugar, likely represented the products of more than one plantation consigned to different Savannah merchants. When Captain Nicholas Sallowich sailed his sloop *Ann* into Savannah with 111 bales of Sea Island cotton from Sunbury in late January, the cargo was consigned to seven merchants. Earlier in the month, the sloop *Albert* arrived from Riceboro with a mixed cargo of 150 bales of Sea Island cotton and 50 tierces of rice. Six merchants and the master were named as consignees for these goods.

The pattern of multiple consignees seen for large shipments of Sea Island cotton in 1830 does not hold for upland cotton, most of which came from Darien. Of the 150 arrivals carrying only upland cotton, 93, or 62 percent, were shipped to a single consignee, regardless of the number of bales involved.

In addition to Robert Habersham, the Savannah merchants most heavily involved in handling coastal produce in 1830 included Thomas Butler & Company, which received fifty-five shipments; R. & W. King, fifty-two shipments; Elias Fort, fifty-one shipments; Silas & Fannin, forty-eight shipments; and Butts & Patterson, thirty-four shipments. Many merchants received both rice and cotton although some dealt primarily or solely in just one of these crops. Thomas Butler & Company, Elias Fort, Butts & Patterson, William Bowers & Company, J. Stone & Company, and Hall Shafer & Tupper dealt mainly or exclusively in cotton. Robert Habersham and R. & W. King specialized in rice, and Petit de Villars and Ketchum & Burroughs handled a considerable amount of the clean rice shipped from pounding mills on the lower Savannah River.

[87] Enrollment No. 34, port of New Haven, sloop *Robert Habersham*, October 12, 1829.

As in previous years, some Savannah merchants tended to receive goods from areas where they had ties with local planters or merchants. In 1830, Elias Fort received fifty-two shipments, mainly Sea Island cotton, from locations on the central and lower Georgia coast, such as St. Marys and the Satilla and Turtle rivers. Several firms—J. Stone & Company, J. W. Long, Andrew Low & Company, Elias Bliss, and W. Bowers & Company—dealt almost exclusively in the upland cotton shipped from Darien.

The sailing trade along the Georgia coast peaked in 1830, as evidenced by the number of coaster arrivals in Savannah over the year and the number of ports of origins of these vessels. Newspaper shipping lists report the coasters carried almost nine million pounds of rice and fourteen million pounds of cotton into Savannah, goods with a value at the time of almost two million dollars. In modern (2020) monetary value, this quantity was valued close to fifty-eight million dollars, a reflection of the importance of the coasting fleet in the local and national economies.[88] Even this amount, however, is a significant underestimate of the monetary worth of the coasters' cargoes, partly because newspapers do not provide quantities for all shipments of the principal commodities rice and cotton. Further, the values of miscellaneous items, such as sugar, molasses, syrup, hides, oranges, live oak wood, groundnuts (peanuts), and the like cannot be calculated because newspapers rarely published the amounts carried. This amount also doesn't include the value of the cargoes the coasters carried out of Savannah and Charleston to local ports and plantations.

After 1830, the importance of the Georgia sailing coasters in the movement of commodities declined as the more efficient steamboats became more involved in transporting agricultural products. The following chapter shows that the sailing fleet shrank, and its share of coastal commodities likewise decreased. However, sailing vessels remained particularly important in transporting rough rice, and expanding rice production along the Southern coast after 1830 assured the continued existence of the Georgia coasters.

[88] MeasuringWorth.com, "real price" of a commodity calculator.

# Chapter 8

## "RICE: The market is well supplied with this article": The Georgia Coasting Trade from 1831 to 1861

On January 31, 1839, the Savannah *Daily Georgian* reported in its "Ship News" column the arrival of five sloops from local ports carrying Sea Island cotton, upland cotton, and rice. These were the *Pearl*, *Two Friends*, *Angel*, *William*, and *William Wray*, and they were among the forty-three sailing ships then working in the Georgia coasting trade. According to the *Daily Georgian*, these twenty-three sloops and twenty schooners made 123 arrivals in Savannah in 1839, a considerable reduction from the 479 arrivals reported nine years earlier, and less than half the 308 arrivals reported in 1835. The decade of the 1830s saw a steady decline in the activities of the Georgia coasters. This decline was evident in all aspects of the local sailing trade: a decrease in the number of coaster arrivals into Savannah, a decrease in the number of sailing vessels participating in the trade, and a decrease in the number of local ports served. During the three decades from 1831 to 1861, the activities of the coasters fluctuated, but the trade showed a persistent downward trend and never again reached the heights seen in 1830. As shown in Figure 7-1, between 1835 and 1861, thirty to forty ships were sailing in the trade, making from as few as 110 to as many as 308 voyages into Savannah. This chapter examines the workings of the Georgia sailing trade over the three decades after 1830, including those factors contributing to its slow decline.

### The Georgia Coasting Trade in 1839

In 1830, the peak of the sailing trade in Georgia, eighty-three sloops and schooners made 479 arrivals into Savannah carrying as much as two million dollars' worth of rice and cotton. The sailing trade remained important through the first half of the 1830s, showing only a slight decline in activity. In 1835, fifty-four coasters made 308 arrivals into Savannah, carrying six million pounds of rough and clean rice and eleven million pounds of Sea Island

and upland cotton, commodities valued at least 2.2 million dollars.[1] However, the second half of the decade saw a decline in the sailing trade. It reached a low point in 1839, a year examined in detail to provide information on the Georgia trade at the decade's end. The shipping lists published by the Savannah newspaper, the *Daily Georgian*, in 1839 are used to examine the trade. That year, the *Daily Georgian* was published daily except Sundays from October through June, and as was typical for city papers, it was issued with less frequency during the summer. In this instance, the paper was published Tuesdays, Thursdays, and Saturdays. Shipping information was carried in a column titled "Ship News." The information presented for calendar year 1839 encompasses portions of the 1838–1839 and 1839–1840 seasons.

### *The Crisis of 1839*

With just 123 coaster arrivals in Savannah, 1839 was unique for the trade as it was an extreme low point for the coasters, and their activities were typically greater after 1840. Several factors explain the downturn in the sailing trade seen through the 1830s and for the very low numbers of coasters sailing in 1839 specifically. The most direct cause was the economic depression known as the Crisis of 1839. This depression arose from a nationwide economic difficulty known as the Panic of 1837. The panic's causes and the following depression were complex and included American as well as international factors. The Panic of 1837 was preceded by prosperity, speculation, and an enormous extension of credit from banks. At the time, many Americans were buying recently opened Western lands on credit in the form of bank notes, often at inflated prices, and state governments, as well as individuals, were extended credit to develop infrastructure such as canals, roads, and railroads.[2]

In 1836, President Andrew Jackson issued the "Specie Circular," requiring the use of gold and silver currency to purchase government lands to protect settlers from paying inflated land prices with devalued paper money. Earlier, Jackson had refused to renew the charter of the Second Bank of the United States, resulting in the withdrawal of government funds. At the same time, the British Central Bank, uncomfortable with the large flow of funds into the United States by British investors, increased its deposit rate. These factors made it more attractive for British investors to invest within Britain, thus pulling their funds out of the United States. As a result, there was a run on

[1] Sailing and commodity data from *Daily Savannah Republican*, 1835.
[2] Wallis, "Depression of 1839."

banks to exchange bank notes and paper money for specie while the specie available was declining, as was the credit formerly available to American banks, states, and individuals from British sources. The commercial failures began in New York in May 1837, when banks began to stop payments in specie. Soon, commercial businesses began to fail, and prices plummeted. Business and bank failures spread across the country. It is reported that out of the 850 banks in the United States, 343 closed entirely, and 62 failed partially.

The South was particularly hard hit. Along with a lack of specie, bank failures, and an inability to obtain credit, upland cotton prices fell dramatically to ten cents a pound, partly because cotton was commonly used as collateral on bank loans. With so many loans recalled, cotton lost its value. The crisis mainly affected banks, merchants, and possibly planters due to depressed prices. The average citizen felt few repercussions because the costs of imported goods also fell. Most banks in the United States suspended specie payments until summer 1838. After a short recovery in 1838 and 1839, banks in the South and West again suspended specie payments in October 1839, a reflection of the drop in cotton prices, and the economy slid into the Crisis of 1839, the most severe recession and depression the nation had seen.[3]

Although this nationwide financial crisis affected the American coasting trade, other strictly local factors damaged the Georgia sailing trade. One was the increasing presence of steamboats in the coastal trade. By the late 1830s, steamers had captured most passenger traffic along the coast and were transporting a considerable quantity of the cotton once shipped aboard sailing vessels. On January 31, 1839, the *Daily Georgian* reported, in addition to the *Pearl*, *Two Friends*, *Angel*, *William*, and *William Wray*, the arrival of two steamboats in Savannah. These were the *J. Stone* and *Ocmulgee*, together carrying 1,089 bales of upland cotton and eleven bales of Sea Island cotton from Darien. A newspaper advertisement noted the *J. Stone* was running between Savannah and Darien three times a week and could carry seven hundred bales of cotton.[4] Other steamers were operating along the coast by 1839, and many carried cotton into Savannah, cotton that ten years earlier was shipped aboard sailing coasters. Also damaging to the sailing trade was the expansion of railroads out of Savannah into interior Georgia, which began to transport cotton that had once been floated down the state's rivers to coastal locations where it would be carried by sailing coasters to Savannah and Charleston. Darien

[3] Faragher, et al. *Out of Many*, 263; Taylor, *Transportation Revolution*, 342–43; Wallis, "Depression of 1839," 1.

[4] *Daily Georgian*, January 15, 1839.

was especially hard hit by the capture of cotton shipments by railroads. In 1830, the town had been the origin of 225 sailing trips into Savannah, with more than thirty-seven thousand bales of cotton shipped. In 1835, the number of trips from Darien to Savannah fell to 132 and by 1839, was reduced to just thirty-nine. Darien never regained its preeminent position in the local trade. In the early 1840s, the town entered a period of economic depression from which it never recovered.

The decline in the sailing trade through the 1830s was somewhat arrested during the early 1840s, with the end of the 1839 economic depression and, more specifically, with the continued expansion of rice production in Georgia and South Carolina. As previously noted, rough rice was ill-suited for transport by steamboats as the grain was typically dumped loose into a vessel's hold. The shallow-hulled steamboats, lacking the spacious and easily accessible holds of sailing vessels, rarely carried rough rice. The increasing rice production after 1840 prolonged the participation of sailing vessels in the local trade. When the January 20, 1841, issue of the *Daily Georgian* reported that "the market is well supplied with this article," it referred to rice, the commodity that played a key role in sustaining the sailing trade from the 1830s to the Civil War.

Despite the economic crises of the late 1830s, the population of Savannah grew considerably during the third decade of the nineteenth century, from 7,773 in 1830 to 11,214 by 1840.[5] The commercial standing of Savannah also grew during this decade, reflected in increases in exports of rice and cotton. Rice exports from Savannah tripled during this period, from 8,663 tierces in 1827–1828 to 24,392 tierces in 1840 while shipments of upland cotton rose from 139,238 bales in 1830 to 199,176 in 1839–1840. Sea Island cotton agriculture had reached its geographic limits earlier, and its production decreased slightly during the 1830s. In 1829–1830, 9,579 bales of Sea Island cotton were shipped from Savannah, dropping to 8,642 bags/bales in 1839–1840.[6]

Although the decrease in coaster traffic in 1839 does reflect the economic depression of that year, the effects on Savannah merchants and coastal planters are difficult to gauge. Some planters certainly faced hardships, considering the depressed prices they received and their inability to obtain credit. Mary

[5] "The City of Savannah," *DeBow's Review*, July 1853, 106.

[6] Export figures from *Daily Georgian*, October 2, 1828; *DeBow's Review* 10 (June 1851): 685; Lee and Agnew, *Historical Record*, 137–39; Hunt, "Commercial Cities," 504–505.

Bullard notes that Phineas Nightingale, a planter on Cumberland Island, was financially ruined by the crises and left the island.[7]

Other planters, especially those growing rice, may not have fared as poorly as Phineas Nightingale. The *Daily Georgian*, on January 5, 1839, reported rice was selling for $3.25 to $3.50 per hundred pounds, down from $5.00 a year earlier. This price drop was certainly felt by coastal planters, but for some, it was offset by an increase in rice production brought about by the continued expansion of tide flow agriculture. Additionally, by September 1839, rice prices in Savannah rose by almost one dollar. Sea Island cotton prices were fairly high in 1839 despite the country's economic difficulties. According to the *Daily Georgian*, Sea Island cotton was selling for thirty-five to forty-five cents per pound in Savannah in January 1839, higher than it had been in several years and considerably more than the crop would sell for in subsequent years. Upland cotton prices seem to have been more adversely affected by the depression. It was selling for thirteen cents a pound in early 1839 but would drop to ten cents per pound by September of that year.[8]

Another factor that seems to have contributed to the lessened coaster traffic into Savannah in 1839 was the city's outbreak of yellow fever. No yellow fever was recorded in Savannah until 1820, when a minor epidemic occurred. Another minor outbreak hit in 1827, but from 1830 to 1839, no cases were reported. In summer 1839, one of the driest and hottest on record, fever outbreaks did occur. The epidemic in Savannah became severe, and ultimately, 1839 was considered one of the sickliest ever known. The deaths in Savannah were often attributed to "bilious fever" rather than yellow fever, making it impossible to say how many cases of yellow fever there were.[9] Captains were hesitant to bring their ships into ports where there was yellow fever. Coasters that refrained from entering Savannah in late summer and fall 1839 contributed to the decrease in arrivals that year.

### *Georgia Coasting Ships and Captains in 1839*

The forty-three coasters sailing into Savannah in 1839 consisted of twenty-three sloops and twenty schooners. Since the eighteenth century, sloops had outnumbered schooners in the trade, but the proportional number of sloops had been slowly decreasing in favor of two-masted schooners. Just

[7] Bullard, *Robert Stafford*, 203.

[8] Commodity prices for 1839 from the *Daily Georgian*, January 5 and September 28; and Gray, *History of Agriculture*, 2:1031.

[9] Lee and Agnew, *Historical Record*, 133.

three years later, in 1842, schooners would first outnumber sloops, twenty to eighteen, and through 1861, the number of sloops continued to decrease. In 1839, these forty-three ships made 123 arrivals in Savannah from twenty-eight named coastal locations. As noted for previous years, the information presented here relates to coasters sailing into Savannah; it does not include ships that traded between other local ports and plantations but never visited Savannah. The number of vessels that fell into this category is unknown, but a few can be identified. For example, on April 8, 1839, the "Memoranda" section of the *Daily Georgian* "Ship News" column reported that the schooner *Betsey Maria* had arrived in Brunswick from Burnt Fort, the sawmilling center on the Satilla River. The *Betsey Maria* had sailed in the Georgia trade since launching in 1826, including many voyages into Savannah. The schooner was working in the local trade in 1839 but apparently never sailed into Savannah, possibly because of the yellow fever outbreak in the city.

While the number of coaster arrivals at Savannah in 1839 was 75 percent lower than in 1830, the number of ships making these voyages was reduced by only half. Fourteen of the coasters sailing in 1839, or about one-third of the fleet, had sailed in the Georgia trade in 1830. Most ships that sailed in the trade for nine years or more were commanded or owned by captains residing in Georgia and South Carolina. This pattern of older vessels belonging to residents and newer ones to Northern men existed from 1800 to 1861. A principal reason for this was that many of the Northern masters were involved in shipbuilding and ship ownership. These Northern captains, most from the Sippican area of Massachusetts, often brought their newly built ships South to trade. Rather than build their own vessels, it was common for Southern captains to purchase ships from these Northern men after some years of use.

Before 1839, it was typical for several coasters to make ten or more voyages into Savannah annually. In 1839, no coaster made even ten arrivals. Two sloops, the *Eagle* and the *Virginia*, were the most active coasters, each sailing into Savannah nine times. The *Eagle* was a Northern vessel commanded by Sippican men Leonard Bolles and a captain named Briggs (Table 8.1). The *Virginia* was owned and sailed by Georgia resident Francis Chevalier of St. Marys (see Table 8.1).[10]

Among the other active coasters sailing in 1839 was the Virginia-built schooner *Hamilton*, which sailed into Savannah eight times. The master of the *Hamilton* was James Ridley, who seems to have been a native of Maine

[10] Enrollment No. 124, Port of New Bedford, sloop *Eagle*, October 4, 1839; *Ship Registers of New Bedford 1796–1850*, sloop *Virginia*.

but who may have lived in Savannah by 1839 (see Table 8.1).[11] All but one of Captain Ridley's voyages with the *Hamilton* originated at the Ogeechee River. His cargoes consisted of rice and Sea Island cotton. In 1839, the sloop *Macon* made eight trips into Savannah. The sloop *Georgia* made seven voyages, the sloop *William* made six voyages, and the sloops *America* and *Sapelo* each made five trips (see Table 8.1).

There appear to be forty-nine different coaster captains named in the shipping lists published in 1839. Out of these, eleven are identified as men from the Sippican region, but there may have been a few more. Thus, approximately 22 percent of the captains sailing in the Georgia trade in 1839 were from Sippican. This is lower than the participation by Sippican men in 1830, when they represented about one-third of the shipmasters in the trade. The lessened involvement of Sippican captains may relate to the unfavorable economic conditions of 1839. As in previous years, the Sippican captains were quite active, making at least forty-two, or a little over one-third of all the coaster arrivals into Savannah in 1839.

### *Coasting Ports in 1839*

Despite the slower pace of activity in the sailing trade in 1839, Darien retained its position as the leading port of origin for coasters sailing into Savannah (Figure 8.1). The *Daily Georgian* reported thirty-nine coaster arrivals from Darien this year, down considerably from the 225 arrivals in 1830. As in previous years, the Sippican captains were heavily involved in the Darien-Savannah trade, accounting for twenty-six of the thirty-nine voyages from Darien that year.

The Darien of 1839 was not the Darien of 1830. In 1837, the population of the town reportedly reached five hundred, but the prosperity and growth Darien had experienced for thirty years was interrupted by the Panic of 1837 and the Crisis of 1839. Additionally, the newly constructed railroads into interior Georgia from Savannah were diverting cotton shipments that had formerly been carried down the Altamaha River to Darien. The economic crises resulted in the Bank of Darien's failure in 1842. Even with the economic slowdown, cotton exports out of Darien continued through the early 1840s. On February 22, 1838, the *Brunswick Advocate* reported that 56,100 "bags" of cotton were exported from Darien between June 1837 and February 1838. The newspaper does not state how much was upland cotton coming down the

[11] Enrollment No. [illegible], Port of Savannah, schooner *Hamilton*, August 30, 1838.

Altamaha River, but most of it was. Sailing coasters and steamboats transported most of this cotton, 33,496 bags, to Savannah. The rest was carried to Charleston (9,694 bags), New York (8,060 bags), and Providence, Rhode Island (500 bags). Some of the larger ships that called at Darien carried 3,350 bags directly to Liverpool. Direct cotton exports out of Darien to Northern and overseas ports continued at a similar level over the next several years; 10,537 bales were shipped to New York in 1840, and 13,656 bales were shipped directly to New York and Providence in 1843. After that, however, direct exports of cotton from Darien dropped precipitously. Only 1,411 bales were shipped in 1844, and shortly afterward, direct exports to overseas ports essentially ended, though coasters still carried small amounts to Savannah and Charleston. By the mid-1840s, Darien's economic boom was over.[12]

In 1839, cotton was the principal commodity carried by the sailing coasters out of Darien, as it had been in earlier years. Upland cotton was listed as a cargo on twenty-seven of the thirty-nine coasters sailing from Darien. The shipping lists in the *Daily Georgian* provide the number of bales carried on each of the twenty-seven voyages, for a total of 6,839 bales. Sea Island cotton was listed on five voyages, consisting of 472 bales and two "packets." These 7,311 bales represented only the cotton carried by sailing coasters; by this time, steamers transported a considerable amount. Rough rice was carried on only seven coasters sailing out of Darien, and the number of bushels was given for only three, totaling 5,480 bushels. In addition, two of the Sippican coasters, the sloops *Georgia* and *William Wray*, carried ninety-two casks of clean rice. Unlike previous years, when small quantities of hides, syrup, and wood were often shipped aboard sailing coasters from Darien, only cotton and rice were reported in 1839.

According to the published shipping lists, no vessels sailed into Savannah from any of the rice plantations along the Altamaha River, suggesting that any rice shipped from these plantations was listed under Darien. However, the quantity of rice shipped from Darien to Savannah in 1839 seems surprisingly small. Some of the clean rice may have been carried aboard steamers, but little, if any, of the rough rice. Some Altamaha River planters were still shipping their rice to Charleston, not Savannah. One who did was Hugh Fraser Grant of Elizafield Plantation. In 1839, Grant dealt with the Charleston factor Robertson & Thurston, shipping 2,903 bushels of rice to them. The following year, however, he began to ship some of his rice crop to Robert Habersham in Savannah, and by 1845, he had ended his dealings with

[12] Donnell, *Chronological and Statistical*, 252, 300, 347.

Charleston factors entirely. Other area planters also began to shift their business to Savannah factors during this period.[13]

Nineteen coaster voyages originated at the central coast town of Riceboro in 1839, making it the second most important port of origin for coasters (see Figure 8.1). The town continued to be important for sailing coasters because steamers rarely visited Riceboro. After all, it was far up the narrow and winding North Newport River. By 1839, steamers regularly visited most of the Sea Islands and the principal port towns along the coast, like Beaufort, Brunswick, and St. Marys, easily reached from the inland passage. As a result, fewer and fewer sailing coasters were transporting cargoes from these locations. Driven by economic necessity, the sailing ships concentrated their business on individual rice-growing plantations and smaller ports, like Riceboro and Sunbury, not consistently served by steamboats. The one exception was the town of Darien, from which such large amounts of cotton were shipped through the early 1840s that the services of both steamboats and sailing coasters were required to transport it to Savannah and other ports.

Two men dominated the Riceboro-Savannah trade in 1839. These were Sippican captain Leonard Bolles with his sloop *Eagle* and local captain John L. Grovenstein, who sailed the sloop *Macon*. Each of these men made eight trips into Savannah from Riceboro. As in earlier years, Sea Island cotton was the principal commodity shipped, and it was specifically listed as cargo on ten trips; "cotton" was named on three others, together totaling 1,862 bales. Only two shipments of rice from Riceboro were reported, both carried by the sloop *Macon* and totaling 4,100 bushels. The only other commodities from Riceboro in 1839 were eight barrels of "Georgia syrup" and one shipment of hides.

The Ogeechee River was the third most import place of origin for the sailing coasters in 1839, with eleven voyages originating there. This trade was dominated by the Habersham-owned schooner *Hamilton*, which made seven trips. Rough and clean rice were the principal commodities carried by the *Hamilton* and the four other coasters sailing this route in 1839. These other ships each made a single voyage from the Ogeechee, and all appear to be locally owned vessels. The cargoes from the Ogeechee River consisted of 15,170 bushels of rough rice, 213 casks of clean rice, and 162 bales of Sea Island cotton.

Eleven coaster voyages originated at the "Satillas" in 1839, referring to both rivers of this name, and a single vessel sailed from the town of "Jefferson" (Jeffersonton) located on the Satilla River. Three local coasters sailed in this

[13] House, *Planter Management*, 58, 171–79.

trade. The sloop *Virginia* and her captain Francis Chevalier, making nine voyages, dominated the Satilla-Savannah trade. The sloop *Argo*, under a master named Taylor, made two trips while St. Marys resident George Ratcliff, with the schooner *Margaret*, made the single trip from the town of Jefferson laden with thirty-three hundred bushels of rough rice. Sea Island cotton and rice were the only commodities shipped from the Satillas.

Other Georgia ports served by sailing coasters in 1839 included the town of Sunbury, origin of six voyages, and the Turtle River, origin of seven. Sea Island cotton was the main commodity shipped from both locations. Two vessels served the Turtle River, the local sloop *Argo*, under Captain Taylor, and the Sippican sloop *America*, commanded by Captain Leonard Burr. Five different coasters made the six voyages from Sunbury, and all appear to be local ships with local masters. These were the sloops *Angelica* (Smith, master); *Pearl* (Ketchum, possibly James Ketchum of Savannah, master), *Swallow* (Jeremiah Boor, or Boar, of Savannah, master), and *Two Friends* (Smith, master), and the schooner *Orange* (James Lane of Savannah, master).

In 1839, two coaster voyages were made from St. Marys, which had been such an important port in the local trade in earlier years. The sloop *Bolivar* made one of these sailings, with Edmund Richardson of St. Marys as master, carrying thirty-five bales of Sea Island cotton. A schooner named *Yankee* made the only other voyage. The *Yankee* was not a true Georgia coaster, and on this voyage, she was sailing in ballast (with no cargo).

Unlike previous years, very few coasters sailed into Savannah from the Sea Islands in 1839. Sea Island cotton was the principal crop on most of the islands, and much of it was now shipped aboard steamers. In 1839, two local coasters, the sloops *George Washington* and *Swallow*, carried Sea Island cotton and corn from Sapelo Island; the schooner *Orange* made one trip from Ossabaw Island with thirty bales of Sea Island cotton; and the sloop *Eagle* carried eleven bales of Sea Island cotton from Wilmington Island.

In 1839, several coasters were transporting rice into Savannah from locations along the lower Savannah River. This consisted of rough rice and clean rice that had been processed at the rice mills operating along the river. Several of these mills were identified, including Drakies, Hamilton's, Manigault's, Mercer's, Rutledge's, and Manor's. Most of the coasters carrying crops from the lower Savannah River were local ships, including the sloops *Hamilton*, *Jackson*, and *Science* and schooners *Albemarle* and *Vesta*.

The sailing trade between Savannah and lower South Carolina was minimal in 1839. Only four voyages originated there; the schooners *Idonius* and *Young Eagle* sailed from the May River with rice and cotton. The sloop

*Georgia* arrived from "Dawfusky" Island with fifty-two bales of Sea Island cotton, and the schooner *Aurora* came from Beaufort in ballast. Regular steamboat service to Beaufort and the South Carolina Sea Islands had taken much of the trade with these locations.

By 1839, the coaster trade between Savannah and Florida was essentially nonexistent. Seven arrivals from Florida into Savannah were reported, one from Jacksonville and six from St. Augustine. These trips were made by seven different vessels, and all but one arrived at Savannah in ballast. The single exception was the sloop *Science*, sailing from St. Augustine with an unidentified cargo destined for the firm of Cohen & Miller. As with other locations, steamers ran between the principal ports on the Northeastern Florida coast and Savannah and carried most locally produced agricultural commodities.

### *Seasonality of the Trade in 1839*

Despite the small number of coaster sailings in 1839, the strong pattern of seasonality of the Georgia trade that had developed in the 1820s was evident. No coaster arrivals in Savannah are reported for June and July; one arrival occurred in August, two in September, and one in October. Typically, coasters began to sail into Savannah with coastal crops by late October. The small number of coasters sailing in 1839 might account for the single arrival in October, but yellow fever in the city that fall possibly kept coasters away until November. Most of the 1839 voyages occurred between November and May, peaking between January and March when the bulk of the cotton and rice crops were available for shipping. The first captain to carry new rice into Savannah in 1839 seems to have been John Grovenstein with his sloop *Macon*, arriving in Savannah on October 23 with twelve hundred bushels of rough rice from Riceboro.

As was typical, only locally owned ships sailed into Savannah during the off-season. The only coasters sailing into Savannah between the end of May and the end of October were the sloops *Hamilton* and *Macon*, the former owned by Robert Habersham with James Ridley as master, while the latter was owned and sailed by local captain John Grovenstein. Northern vessels began to arrive in Savannah at the start of the shipping season. The first to come South for the 1839–1840 season was the sloop *Eagle*, whose arrival from New Bedford was reported in the *Daily Georgia* on October 12, 1839. The following week, the *Daily Georgian* ran an advertisement noting the *Eagle* was "receiving freight" in preparation for a voyage to Riceboro.[14] Captain Leonard

[14] *Daily Georgian*, October 22, 1839.

Burr's arrival in Savannah from Newport, Rhode Island, with his sloop *America* was reported over a month later, on November 16. Sloops *Angel* and *Stranger* sailed into Savannah from Hartford, Connecticut, two days later. These ships all sailed in the Georgia trade for the next five or six months.

The first of the Northern vessels to head North after the 1838–1839 shipping season was the schooner *Florida,* under Captain Timothy Savory, whose departure for New Bedford was advertised in the May 1, 1839, issue of the *Daily Georgian.* Notices for other coasters sailing North appeared over the following weeks.

### *The Coasting Cargoes in 1839*

For the 123 coaster arrivals in Savannah in 1839, the *Daily Georgian* lists cargo information for all but one. This was the sloop *Eagle,* whose arrival from Riceboro was reported on May 29. This was the sloop's final voyage of the season, and she may have carried no cargo at all, or at least none destined for Savannah because no consignees are named. Ten vessels are reported to have arrived in ballast over the year. Two coasters arrived carrying "merchandise" from Darien rather than the typical agricultural commodities of the trade. The remaining arrivals were vessels transporting coastal products of various sorts.

Table 8.2 presents information on the number of shipments of various commodities carried into Savannah by coasters in 1839 and the minimum estimated amounts for those commodities where this information is available. The 1839 shipping lists in the *Daily Georgian* identify twenty coaster arrivals in Savannah carrying specified amounts of rough rice, totaling 41,700 bushels, or 1,876,500 pounds. There were eleven shipments of clean rice in 1839, consisting of 964 casks or barrels. Assuming each container carried about 600 pounds of rice, the total amount in 1839 was about 578,400 pounds. Over the year, coasters carried fifty-four shipments of Sea Island cotton into Savannah, totaling 3,820 bales, and thirty-one shipments of upland cotton, totaling 7,387 bales. Assuming an average weight of 350 pounds per bale, this represents an estimated 1,337,000 pounds of Sea Island cotton and 2,585,450 pounds of upland cotton. The amounts shown in Table 8.2 are considerably less than those obtained for 1830, as shown in Table 7-8, reflecting the decreased coaster traffic in 1839.

The amount of rough rice carried into Savannah in 1839 was certainly more than shown in Table 8.2. The newspapers show that "rice," presumably rough rice, was carried on five arrivals, with no quantities given. These five coasters represent 20 percent of the twenty-five arrivals carrying rice, and, assuming they carried amounts similar to the twenty coasters laden with known amounts of rice, they may have transported an additional 8,340 bushels of rough rice. Thus, a projected 50,040 bushels of rough rice, representing 2,251,800 pounds of grain, may have been transported by coasters into Savannah in 1839. Even this amount, though, is significantly less than the more than six million pounds of rough rice the coasters carried into Savannah in 1830 and the more than four million pounds transported in 1835.

Other than rice and cotton, the coasters carried few other local goods in 1839. Two shipments of corn were reported, brought into Savannah by the *Swallow*, a forty-four-foot sloop built in Charleston in 1828. One of these corn shipments came from Sapelo Island, the other from Sunbury. Other miscellaneous cargoes in 1839 consisted of six shipments of hides, five of wood, one of beeswax, and one of eight barrels of Georgia syrup carried from Riceboro aboard the sloop *Eagle* (see Table 8.2).

The decline in cargo carried by the coasting fleet in 1839 is expressed in lowered values. The value of the 1.8 million pounds of rough rice and the 578,400 pounds of clean rice reportedly transported into Savannah was approximately $111,000, given the average price of rough and clean rice in 1839. The actual value of the rice was slightly higher because newspapers did not report the quantity carried on five coaster entries in 1839. However, the value of rice carried in 1839 was less than the estimated $190,576 worth of rice transported into the city by coasters in 1830. The Sea Island cotton carried by the coasters in 1839 had an estimated value of $534,800. This was somewhat less than the $611,275 worth of Sea Island cotton carried by the sailing fleet in 1830, but it was the most valuable commodity the coasters carried. The upland cotton carried aboard sailing coasters in 1839 had an estimated value of $310,254, a significant drop relative to the over one million dollars' worth of upland cotton transported in 1830. The great decrease in the value of upland cotton cargoes in 1839 reflects the effects of the nationwide economic depression that year, the decline in the amount of upland cotton shipped out of Darien, and the loss of coastal cotton cargoes to steamers.

Together, these principal commodities had a minimum value of approximately $956,000 dollars in 1839. Relying on the computations provided by Measuringworth.com, these commodities' present-day (2020) value is approximately $27.4 million. Despite the effects of the Crisis of 1839, the coasting fleet retained its importance in the movement of commodities of great value.

### *Savannah Merchants in the Coasting Trade in 1839*

In 1839, the *Daily Georgian* named consignees for 109 of the 123 coaster arrivals in Savannah. Most of those arrivals where no consignee was identified were vessels arriving in ballast. Three arrivals list cargoes destined "to master." As expected, in light of the fewer sailings in 1839, far fewer consignees received cargo than in earlier years. In 1830, there were seventy-two named consignees; in 1839, just thirty-four received 226 individual shipments. Three merchant firms dominated the coasting trade in 1839 in terms of shipments received: Robert Habersham, R. & W. King, and George W. Anderson & Brother. Together, these firms were named as consignees on 110 individual shipments, almost half of all the shipments received in 1839. Robert Habersham was a consignee on forty-three shipments, more than any other merchant. Before October 1839, this company was known as "Robert Habersham"; however, when Robert brought his son William Neyle into the business that month, the firm became known as "Robert Habersham & Son."[15] The firm retained this name until 1857, when it appears as "Robert Habersham & Sons," indicating other sons had joined the company.

The Habersham firm principally received rice and Sea Island cotton. It was named as consignee on thirty-two rough and clean rice shipments in 1839, more than any other firm. Robert Habersham received most shipments from the town of Riceboro, the Ogeechee River, and the Turtle River, with lesser numbers coming from Darien, Sunbury, the "Satillas," and from mills and plantations along the lower Savannah River. The company was well positioned to handle rough rice, as Robert Habersham owned the Upper Steam Rice Mill, where he could pound and polish the rice cargoes he received. By 1839, the Habersham firm had become Savannah's leading rough rice dealer.

Robert Habersham received cargoes aboard sixteen different coasters in 1839, and, as in earlier years, these included locally owned and Northern vessels. The schooner *Hamilton* arrived in Savannah seven times that year with rice and Sea Island cotton for the firm, every voyage originating at the Ogeechee River. Robert Habersham owned the *Hamilton*, so there is no surprise that he used the schooner in his business. However, the firm was not the only consignee for cargo carried by the *Hamilton*, indicating that Habersham made money transporting cargo for others.

The firm of R. & W. King received thirty-six shipments aboard sailing coaster in 1839, second only to Robert Habersham & Son. This business had been prominent in the local coastal trade since before 1830. In 1830, they

[15] Ibid., October 1, 1839.

mainly received rice; however, in 1839, Sea Island cotton was the main cargo, with lesser quantities of rice, syrup, hides, and upland cotton. Most of the firm's cargo came from Riceboro, the Satilla River area, the Turtle River, and the Ogeechee River. R. & W. King received shipments aboard thirteen sailing vessels, including Northern and locally owned ships.

George W. Anderson & Brother was named a consignee on thirty-one shipments of coastal commodities in 1839. These cargoes consisted mostly of Sea Island cotton from Riceboro and the "Satillas."

In addition to these "big three" firms, other merchants receiving coastal products in 1839 included Elias Reed, with thirteen shipments; Hardee & Harden, also with thirteen shipments; W. Patterson & Company, with nine; J. Cummings & Company, with eight shipments; E. P. Butts & Company, with eight shipments; and Washburn, Lewis & Company, with seven. Elias Reed and Hardee & Harden received primarily Sea Island cotton, much of it from locations on the lower Georgia coast, such as the Turtle and Satilla rivers. W. Patterson & Company dealt heavily in clean rice, mainly derived from mills and plantations on the lower Savannah River. E. P. Butts & Company and Washburn, Lewis & Company dealt in upland cotton, and their shipments in 1839 came from Darien.

## The Georgia Coasting Trade at Mid-Century

After the disastrous year of 1839, the economy of the Southeastern coast began to recover. There was continued expansion in rice cultivation, and prices for rice and Sea Island cotton were stable. The 21,332 tierces of rice exported from Savannah in 1839 rose to 29,217 tierces by 1844–1845, and 30,136 tierces in 1848. Sea Island cotton production remained steady over this period; 8,009 bales were exported from Savannah in 1844–1845 and 11,261 bales in 1851–1852.[16]

The 1840s and 1850s were prosperous for the coastal planters of South Carolina and Georgia. In 1850, census returns list seventy-one planters in Georgia whose annual rice yields were over a hundred thousand pounds. A few growers produced over a million pounds. The importance of Savannah as the regional mercantile center grew considerably during this period. In 1850, there were 246 merchants and ninety-eight commission houses in the city, and five new banks opened between 1848 and the mid-1850s. Twenty-seven steamboats were operating out of Savannah in the mid-1850s, and seven

[16] White, *Statistics*, 159; Gray, *History of Agriculture*, 2:1030–1031; Hunt "Commercial Statistics," 619.

packet lines ran regular schedules between Savannah, New Orleans, and Northern ports such as New York, Boston, and Philadelphia.[17]

The sailing trade along the Georgia coast experienced a slight revival after 1839 in terms of the number of vessels working and the quantities of cargo. The number of coaster arrivals into Savannah between 1840 and 1861 fluctuated, as seen in Figure 7-1, and, except in 1841, when the *Daily Georgian* listed 253 arrivals by forty-seven different vessels, the number of arrivals were never as high as they had been during the peak years 1827–1835. Generally, after 1840, between thirty and forty coasters were sailing in the Georgia trade in any given year. The lowest number of coaster arrivals during these two decades occurred in 1860, when thirty-three different sloops and schooners sailed into Savannah 113 times.

The shipping lists from the *Daily Georgian* (also called the *Savannah Georgian*) are used to examine the Georgia coasting trade as it existed at the middle of the nineteenth century. Additionally, newspaper advertisements for arriving and departing sailing coasters were examined; however, these had declined over the 1840s. In their stead, there were more advertisements and notices for the steamers now regularly traveling between Savannah and local ports.

### *Georgia Coasting Ships and Captains in 1850*

In 1850, the *Daily Georgian* lists 225 arrivals at Savannah made by thirty-one individual sailing coasters. These thirty-one ships consisted of twelve sloops and nineteen schooners. Schooners had first outnumbered sloops sailing in the trade in 1842, and they slowly increased that proportion over time. The composition of the coastal fleet was changing in terms of sloops versus schooners—and aging. In 1850, the average age of the sloops was fifteen years. Schooners were younger, with an average age of 10.7 years. Twenty years earlier, in 1830, at the high point of the sailing trade, the average age of sloops was just 7.3 years, while that of schooners was similar, 7.9 years. In the two decades after 1830, fewer and fewer new sailing ships entered the Georgia trade; they were no longer economically feasible, given the growing competition from steamboats. Instead, old ships were kept working for as long as possible.

Related to the falling numbers of new ships entering the trade was a decline in Northern captains and Northern-owned vessels. In 1850, only four men can be positively identified as Northern captains. These were James Burr,

[17] J. F. Smith, *Slavery and Rice*, 36; Fraser, *Savannah in the Old South*, 253.

master of the schooner *Company*; Luther Cudworth, master of the schooner *Harriet Lewis*; and a Captain J. Dexter of the schooner *David Belknap*, all from the Sippican region, and John Godfrey, from Sandwich, Massachusetts, master of the schooner *Savannah*. These four men made up 12 percent of the estimated thirty-four individual masters sailing in 1850. Together, these Northern captains made eleven voyages into Savannah in 1850, accounting for 5 percent of the 225 arrivals sailing coasters. The participation of Northern captains was declining and had been since 1830 when Northerners made up more than 30 percent of the coasting captains and made more than 40 percent of the sailings into Savannah.

With the withdrawal of Northern captains from the Georgia trade, the number of local captains increased. Even though these men resided locally, most were natives of Northern states or foreign countries (Table 8.3). Census records for 1850 and 1860 show that less than one percent of the coasting captains living in Georgia were native to the state. In 1850, immigrants from Denmark, England, Norway, France, Ireland, and Italy composed 80 percent of the coasting captains residing in Georgia (Table 8.3). The avoidance of a maritime life by native Southern men reflected cultural norms that valued other endeavors, particularly land ownership, agriculture, and, to a lesser extent, mercantile interests.

The most active traders in 1850 were the schooner *Cotton Plant*, making sixteen voyages into Savannah, and the sloops *Eagle* and *Splendid*, each sailing into the city fifteen times. The master of the *Cotton Plant* was John Arnaud, a native of France, who sailed as captain of the *Cotton Plant* from 1848 to 1861. Captain Arnaud principally sailed from the Ogeechee River carrying rough rice and occasionally Sea Island cotton. The *Cotton Plant* was owned by Savannah merchant Robert Habersham, and most voyages made by Captain Arnaud included cargo for the firm of Robert Habersham & Son(s). No records have been found to indicate that John Arnaud ever held ownership in a coaster, and he may have been employed by the Habersham firm during his entire career.[18]

The sloops *Eagle* and *Splendid*, each making fifteen voyages into Savannah, were also Southern-owned ships. When the *Eagle* was enrolled in Savannah in September 1850, Henry J. Dickerson was named as owner. The sloop had sailed in the Georgia trade since 1827, the year she was built. First owned and sailed by Rochester men, the *Eagle* was in the possession of two

[18] Enrollment No. 6, Port of Savannah, schooner *Cotton Plant*, February 23, 1846; Enrollment No. 3, Port of Savannah, schooner *Cotton Plant*, August 31, 1860.

Georgians by 1843, coasting captain John L. Grovenstein and Savannah merchant Henry Brigham. In 1850, Captain Grovenstein was master on eight of the *Eagle*'s voyages. The other men serving as her master that year were Peter Maris (sometimes Merris or Morris), who made five voyages; Peter Collins, who made one; and a captain named Thompson, who sailed the sloop on a single trip from Sapelo Island. Captains Maris and Collins resided in Savannah, but both were foreign-born, the former from Italy and the latter from Norway.[19]

In 1850, under John Grovenstein's command, the *Eagle* sailed exclusively from Riceboro, transporting rice and Sea Island cotton. Captain Grovenstein had sailed in the Riceboro-Savannah trade since the late 1830s. However, beginning in September 1850, under the new ownership of Henry Dickerson, the *Eagle*'s six arrivals all originated at the Ogeechee River, and all carried rough rice. The only consignee was Robert Habersham & Son.

The owner of the sloop *Splendid*, which made fifteen voyages into Savannah in 1850, was Danish-born Charles Stevens. Captain Stevens purchased the *Splendid* in 1843 from Henry Dickerson of Savannah.[20] The *Splendid* sailed into Savannah from numerous locations along the coast, unlike many coasters that served a single location or small area. Five of the *Splendid*'s voyages originated at rice plantations and mills on the Back River near Savannah, but she also sailed from Darien, the Satilla and Little Satilla rivers, and Burnt Fort on the lower coast, as well as from the New River in lower South Carolina. The *Splendid*'s principal cargo was rough rice, much of it destined for Robert Habersham & Son. However, three of the sloop's voyages were made in July and September, the off-season for the coasters. The cargo on one of these voyages consisted of 249 barrels of rosin and forty-seven barrels of spirits of turpentine, shipped from Burnt Fort. Wood was the only cargo mentioned on another of these summer voyages, and no cargo was identified on the third. By the mid-1840s, naval stores and other miscellaneous items became important cargo for local coasters operating during the summer.

In 1850, Charles Stevens also owned the sloop *America*, which made eleven voyages into Savannah. Stevens served as master on these trips, and he sailed from two locations, Darien and the Satilla River. The principal cargo carried from Darien was rough rice, while Sea Island cotton was the *America*'s

[19] Enrollment No. 6, "Master Abstracts," Port of Savannah, sloop *Eagle*, September 14, 1850; Enrollment No. 16, Port of Savannah, sloop *Eagle*, December 16, 1843; Otto, *1850 Census Chatham County*, 66; Georgia Historical Society, *1860 Census*, 304.

[20] Otto, "Naturalization Oaths of Allegiance," 15; Pearson, *Charles Stevens*, 34–36.

main cargo when sailing from the Satilla River. Most of the rough rice carried from Darien came from rice plantations along the lower Altamaha River. Hugh Fraser Grant, who owned rice plantations on the south side of the Altamaha River opposite Darien, specifically recorded one shipment of rice on the *America* in 1850. This consisted of 2,689 bushels of rough rice, for which Captain Stevens was paid $134.45, or five cents per bushel in freight charges on December 6. This rice was carried to Hugh Grant's factor in Savannah, Robert Habersham & Son. On April 23, Grant recorded a payment of $128.56 to "Capt Stevens," but it is unclear if this payment was for transporting rice to Savannah or for freight on supplies delivered to Grant.[21]

Other active coasters in 1850 included the schooner *American Coin* and sloop *Science*, each of which sailed into Savannah twelve times over the year; the schooner *Levant* and sloop *Virginia*, each with eleven trips; and the schooner *Fort George Packet* and sloop *Catherine Chard*, each making ten voyages. In June 1850, Savannah resident and Irish immigrant Patrick Doyle became captain of the *American Coin* and sailed from locations on the central Georgia coast, mainly carrying naval stores. The sloop *Science* was commanded by Captain Charles Thompson, known as "Captain Charley," on her twelve voyages into Savannah in 1850. The Ogeechee River was the origin of eleven of these trips while one originated at "Heyward's Plantation," probably referring to one of the several rice plantations owned by Daniel Heyward on the South Carolina side of the Savannah River.[22] Rough rice was the principal cargo of the *Science*, but small amounts of clean rice and Sea Island cotton were carried on some trips from the Ogeechee River. Robert Habersham & Son was named as a consignee on every cargo during the year; the firm was the only consignee on eleven of the sloop's twelve arrivals. Robert Habersham and his son William purchased the *Science* from Gazaway Bugg Lamar in 1840. The Habersham's ownership of the sloop into the 1850s is implied in an 1856 newspaper advertisement stating that the *Science* was departing "For Little Ogeechee and Big Ogeechee," and anyone interested in shipping freight should "apply to R. Habersham & Son or to the captain at Habersham's Wharf."[23]

The schooner *Levant*, built as a sloop in Rochester, Massachusetts, in 1833 and re-rigged as a schooner in 1844, sailed into Savannah eleven times under five different masters in 1850. The *Levant*'s sailings were from the

[21] House, *Planter Management*, 204–205.

[22] Rowland et al., *History of Beaufort*, 315.

[23] *Savannah Daily Republican*, May 26, 1856.

Ogeechee River and the Back River, mainly carrying rough rice consigned to Robert Habersham & Son. The owner of the *Levant* in 1850 was Henry J. Dickerson.[24]

All but one of the sloop *Virginia*'s eleven voyages into Savannah in 1850 originated at the Satilla River. Before July 18, England-born William Laen of St. Marys served as master of the *Virginia*, which was apparently owned by his wife, Catherine, when enrolled in Darien in 1849.[25] The *Daily Georgian* shipping lists reveal that rough rice was a cargo on the *Virginia*'s arrivals in 1850, most consigned to Robert Habersham & Son. In addition to rice, the *Virginia* occasionally carried Sea Island cotton. Only on these occasions are consignees other than the Habersham firm named for the sloop's cargoes.

The aging schooner *Fort George Packet* made ten arrivals into Savannah in 1850 under two masters. One was Thomas Roome, and the other was named Thompson, possibly Charley Thompson. All of the *Fort George Packet*'s voyages originated at Riceboro, and her cargoes included rough rice, Sea Island cotton, upland cotton, and corn. The sloop *Catherine Chard* made ten sailings into Savannah. Her captain was Lewis Peter Wiggins, a native of Russia who had sailed in the Georgia trade since the mid-1840s. On January 4, 1850, the *Daily Georgian* reported the arrival of the *Catherine Chard* from the town of Jeffersonton, located at the head of navigation on the Big Satilla River. This seems to be the first voyage of the sloop in the Georgia trade, and she carried ninety-two bales of Sea Island cotton, forty barrels of spirits of turpentine, and three hundred barrels of rosin. The entire cargo was consigned to Noble A. Hardee, one of the sloop's owners. During 1850, the sloop's voyages originated at locations in Camden County, and on each voyage, one or both of her Savannah owners, N. A. Hardee and the firm of Boston & Gumby, were listed as consignees.[26]

Every one of the ten coasters discussed here, those making ten or more voyages into Savannah in 1850, were owned by Southerners, and their masters lived locally. There had been Southern captains and owners in the Georgia trade since 1800, but the shift toward almost total dominance by Southern captains and owners began only after 1830. As the Northern captains

[24] Enrollment No. 7, Port of Savannah, schooner *Levant*, April 28, 1845; Enrollment No. 4, Port of Savannah, schooner *Levant*, September 27, 1857.

[25] Enrollment No. 2, Port of Darien, sloop *Virginia*, January 1, 1849.

[26] "Conveyances of Enrolled Vessels, Savannah," sloop *Catherine Chard*, June 1, 1852; Enrollment No. 21 (?), Port of Savannah, sloop *Catherine Chard*, June 21, 1853.

withdrew from the Georgia trade, they often left their ships behind, selling them to Southerners.

Among the few Northern captains still involved in the Georgia trade in 1850, the most active was James Burr of Freetown, Massachusetts. Captain Burr sailed his schooner *Company* into Savannah six times over the year. Five voyages originated at Darien and one at the Ogeechee River. The *Company* carried cotton on only two voyages from Darien, reflecting the tremendous decline in upland cotton shipments from the town since the early 1840s. On three of her voyages from Darien, the main cargo of the *Company* was naval stores consisting of spirits of turpentine and rosin.

The only other Northern captain to make more than a single voyage into Savannah in 1850 was Luther Cudworth, from Freetown, Massachusetts, and captain of the schooner *Harriet Lewis*. The *Harriet Lewis* made three trips into Savannah from Darien carrying rough rice on every voyage, for a total of 11,500 bushels. As in earlier years, the few Northern captains still sailing along the Georgia coast concentrated their activities in the Darien-Savannah trade. However, the modest amounts of goods they carried in 1850 was a far cry from the huge volume of cotton and rice Northern ships had transported into Savannah in earlier years.

### *Ports in the Coasting Trade in 1850*

In 1850, sailing coasters arrived at Savannah from twenty-eight named locations (Figure 8.2). Many of the landings that had been important to sailing coasters in the past—Beaufort, Darien, Brunswick, St. Marys, and the Sea Islands—no longer were in 1850. By 1850, steamers had captured most of the trade with these locations. That year, the *Daily Georgian* ran numerous advertisements for the main steamboat lines operating out of Savannah. These included the "Semi Weekly U.S. Mail Steam Packet Line," which operated the steamers *Ocmulgee*, *Wm. Gaston*, and *St. Mathews* between Savannah and Palatka, Florida, on the upper St. Johns River. One of these boats departed Savannah every Tuesday and Saturday and stopped at Darien, Brunswick, St. Marys, and Jacksonville on their route. The independent steamboat *Ivanhoe* traveled weekly between Savannah and St. Marys, stopping at Darien, St. Simons, Brunswick, Bethel (on the Turtle River), and Jeffersonton on the Big Satilla River. In the summer, the *Ivanhoe* regularly visited the sawmilling town of Burnt Fort located almost fifty miles up the Big Satilla.[27] When picking up or delivering cargo or passengers, some steamers called at those

[27] *Daily Georgian*, January 1 and June 1, 1850.

plantation landings along their route that were convenient to reach. By 1850, much of the trade left to the sailing coasters consisted of those landings inaccessible or impractical to reach by steamboats.

In 1850, the *Daily Georgian* published ports or places of origin for all but three of the 225 coaster arrivals in Savannah. The Ogeechee River was listed as a place of origin for fifty-six arrivals, more than any other location (Figure 8.2). The Ogeechee River had become the most common place of origin for the sailing coasters in the early 1840s, replacing Darien for the top position. The importance of the Ogeechee River and other rice-producing areas had grown such that many of them became increasingly dependent on the coasters for the transport of rough rice.

Eleven different ships made the fifty-six trips from the Ogeechee River. These vessels sailed from the several rice plantations located along the lower river. However, the Savannah newspapers rarely name a specific Ogeechee River plantation as a point of origin for a coaster, in contrast to the Savannah River, where individual plantation names were often published. Two ships dominated the Ogeechee River-Savannah trade, both locally owned and commanded. These were the schooner *Cotton Plant*, John Arnaud master, and the sloop *Science* under Captain Charley Thompson. Most of the other coasters sailing in this trade were also local vessels, such as the sloop *Visitor*, under her captain and owner Domingo Galleo; the sloop *Eagle*, under Captains Peter Maris and Peter Collins; the sloop *Washington*, sailed by Captain Alexander Wilson; and the schooner *Albemarle*, commanded by a master named Thompson, possibly Peter Thompson. These men were residents of Savannah, as were John Arnaud and Charley Thompson.

Carried aboard all but one of the fifty-six voyages, rough rice was by far the most important cargo transported from the Ogeechee River in 1850. The *Daily Georgian* shipping lists report that sailing coasters transported 174,820 bushels of rough rice from the Ogeechee during the year. More than 90 percent of this rice, 160,507 bushels, was consigned to a single Savannah merchant, Robert Habersham & Son. Twenty years earlier, in 1830, coasters had transported an estimated 80,000 bushels of rough rice from the Ogeechee, but the amounts carried dropped over the next decade to 55,000 bushels in 1835, and a low of just 15,170 in 1839. However, 1839 was unusual in that the economic crises of that year and, possibly, the Savannah yellow fever epidemic depressed the shipment of coastal commodities of all types. Subsequently, during the 1840s, the amount of rough rice transported from the Ogeechee River increased steadily. In 1841, just two years after the Crisis of 1839, sailing coasters carried 60,000 bushels of rough rice into Savannah from the

Ogeechee. By 1845, this number had risen to 95,800 bushels, then to 118,850 bushels in 1848 before reaching almost 175,000 bushels transported in 1850.[28]

In addition to rough rice, small amounts of clean rice and Sea Island cotton were shipped from the Ogeechee River. In 1850, just four shipments of Sea Island cotton, consisting of 104 bales, and two shipments of clean rice, consisting of 206 casks, were shipped from the Ogeechee. Several rice planters on the Ogeechee River, such as Richard Arnold, who owned White Hall and Cherry Hill plantations in Bryan County, had rice mills. Arnold's journals for the years 1847 to 1849 show that he commonly shipped rice, Sea Island cotton, and small amounts of syrup aboard Captain Thompson's sloop *Science* to Robert Habersham & Son and Andrew Low & Company in Savannah. In addition, Richard Arnold's accounts list a variety of foodstuffs and commodities received from Savannah aboard coasters. Among the items delivered to Arnold by the *Science* and *Cotton Plant* were lumber, lime, rice flour, flour, butter, tar, and bricks, the last to be used to construct a sugar mill.[29]

In 1850, Riceboro was the origin of thirty-two of the coaster arrivals in Savannah, second to the Ogeechee River. Five vessels made these voyages: the schooners *Fort George Packet* and *Northern Belle* and sloops *B. S. Newcomb*, *Eagle*, and *Julia Ann*. Making eleven voyages from Riceboro that year, John Grovenstein, master and owner of the *Fort George Packet*, *Eagle*, and *B. S. Newcomb*, dominated the Riceboro-Savannah trade in 1850,.[30]

The chief commodities shipped from Riceboro were Sea Island cotton and rough rice. The *Daily Georgian* shipping lists report that coaters transported 27,785 bushels of rough rice and 1,281 bales of Sea Island cotton from Riceboro. A single shipment of twenty-five bales of cotton, identified specifically as "upland cotton," was carried from the town by the *Fort George Packet* in November. Small amounts of syrup, corn, and rosin were also shipped during 1850.

[28] Amounts of rough rice shipped from the Ogeechee River for various years are from the *Daily Savannah Republican*, 1830 and 1835, and the *Daily Georgian*, 1839, 1841, 1845, 1848, and 1850.

[29] Sullivan, *Beautiful Zion*, 121; Hoffman and Hoffman, *North by South*, 72, 126–46, 194.

[30] John Grovenstein's ownership of the *B. S. Newcomb*, *Eagle*, and *Fort George Packet* is found in Enrollment No. 24, Port of Savannah, sloop *B. S. Newcomb*, December 23, 1848; Enrollment No. 24, Port of Savannah, sloop *Eagle*, November 24, 1841; Enrollment No. 16, Port of Savannah, sloop *Eagle*, December 13, 1842; and Enrollment No. 5, Port of Savannah, schooner *Fort George Packet*, February 3, 1847.

The town of Darien was the third most common port of origin for coasters sailing into Savannah in 1850, with twenty-nine arrivals from there. Carried on nineteen of the voyages into Savannah, rough rice was the most common cargo. This rice came from the several large plantations along the lower Altamaha River near Darien, and most, if not all, was loaded and carried from plantation landings, not the town of Darien itself.

Upland cotton, which had been such an important cargo in earlier years, was carried on only two voyages from Darien in 1850. The two shipments, totaling 282 bales, were carried by the schooner *Company*, one of the few Sippican vessels remaining in the Georgia trade. The transport of upland cotton by coasters from Darien had fallen after steamboats entered the trade and had plummeted when the railroad reached the inland branches of the Altamaha River. Sailing coasters carried approximately 38,000 bales of upland cotton from Darien to Savannah in 1830. This dropped to around 32,000 bales in 1835. In 1841, however, sailing coasters carried only 2,096 bales of upland cotton from Darien into Savannah; in 1845, that number fell to a paltry 423 bales. The transport of interior cotton by rail had all but eliminated the sailing coasters from this specific trade.

Four shipments of Sea Island cotton were reported from Darien in 1850. Charles Stevens and his sloop *America* carried two bales on each of two voyages; the schooner *American Coin*, under Captain Patrick Doyle, carried thirty-four bales on one voyage; and the schooner *Company*, James Burr master, transported three bales. A modest quantity of naval stores was shipped from Darien: 1,749 barrels of rosin, turpentine, and spirits of turpentine. Most of this, 1,174 barrels, was carried by the Sippican schooner *Company*.

Darien was active in the shipment of lumber, but most was loaded on seagoing vessels for transport to distant American and overseas ports. In 1850, a single vessel, the sloop *Visitor*, owned and commanded by Savannah resident Domingo Galleo, carried thirty-three thousand feet of lumber from Darien to Savannah. This cargo was shipped in August, when local captains often transported such miscellaneous goods.

In 1850, a number of coasters sailed into Savannah from rice plantations and rice mills located along the lower Savannah River near the city. Twenty arrivals originated at the Back River, the branch of the Savannah that flows on the South Carolina side of Hutchinson Island opposite the city of Savannah. Individual mills and plantations named in the shipping lists include Hamilton's Mill, Taylor's Mill, Hayward's Plantation, Potter's Plantation, Dr. Pritchard's Plantation, and Williamson's Plantation. Rough and clean rice were the commodities shipped from these locations. Charles Stevens's sloop

*Splendid* was one of the most active coasters in this lower river trade, making seven voyages from Back River and Abercorn with almost thirteen thousand bushels of rough rice, most consigned to Robert Habersham & Son.

Nine coaster arrivals into Savannah originated at the New River, a major rice-producing area on the lower South Carolina coast. Other than ten bales of Sea Island cotton and an unstated amount of wood transported by the schooner *Young Eagle*, rough rice was the only cargo.

In 1850, several coasters sailed into Savannah from ports on the central Georgia coast. These locations included the South Newport River, the Sapelo River, "Sapelo," which might refer to the island or river, Sunbury, Brunswick, and the Turtle River. The locally built sloop *Liberty* made all five voyages originating at the South Newport River, carrying Sea Island cotton and naval stores. Several coasters made the nine voyages reported from the Sapelo River and "Sapelo," carrying mostly Sea Island cotton and naval stores.

Thomas Snow and his small, twenty-eight-ton schooner *Sarah* dominated the trade between Savannah and the town of Sunbury in 1850. Captain Snow, a native of the island of St. Bartholomew who resided in Liberty County, made eight of the twelve voyages from Sunbury carrying Sea Island cotton, some rough rice, and, on one voyage, seven hundred bushels of corn. Thomas Snow had been sailing in the Sunbury-Savannah trade since 1841, and in most years, he was the captain traveling this route most frequently.[31]

The schooner *Elias Reed* and her captain, Richard Owens, a native of Wales and resident of Brunswick, made four voyages from the Turtle River and four from the town of Brunswick into Savannah in 1850. The *Elias Reed* was the only coaster to sail from these locations that year. She hauled some Sea Island cotton and rice as well as naval stores and hides, sugar, and beeswax.

In 1850, most of the coaster traffic from the lower Georgia coast into Savannah came from the Satilla River, or, as it was still occasionally called, the "Great Satilla River" or the "Big Satilla River." Twenty arrivals are reported from there, with two from the Little Satilla River. In addition, there was one arrival from Burnt Fort and one from Jeffersonton, the small towns located inland on the Satilla River. There were only two voyages from the town of St. Marys although seven originated at the town of Centervillage, located forty miles inland on the St. Marys River. Three ships, all locally owned coasters, dominated the trade between the lower coast and Savannah. These were the sloops *Virginia*, *America*, and *Catherine Chard*. The *Virginia*

[31] Otto, *1850 Census Liberty County*, 17.

made eleven voyages into Savannah, ten originating at the Satilla River. The sloop's captain on six of these voyages was William Laen of St. Marys. The *Catherine Chard*, with Lewis Wiggins as master, made the voyages from the towns of St. Marys and Centervillage. The sloop *America*, under her owner and captain Charles Stevens, made six trips into Savannah from the Satilla River.

The cargoes carried from the Satilla River consisted of rough rice, Sea Island cotton, and naval stores. The *Daily Georgian* for 1850 reported that fourteen shipments of rough rice totaling 37,160 bushels were made from the Satilla River. The coasters had carried relatively little rough rice into Savannah from the Satilla River in the 1830s, and even during most of the 1840s, annual shipments were on the order of only 8,000 to 10,000 bushels. In 1849, however, there was a significant jump in rough rice shipments from the Satilla River when coasters transported a little over 27,000 bushels to Savannah, three times more than in previous years. Through the 1850s, rice production along the Satilla River expanded, and shipments to Savannah increased. These reached 70,550 bushels by 1855, peaking at 109,000 bushels in 1859. There is no record that clean rice was shipped from the Satilla River aboard sailing vessels, suggesting either a lack of pounding mills or that any clean rice was carried by steamboats.[32]

No sailing coaster came into Savannah from Florida in 1850. The Florida portion of the trade largely ended before 1840, and coasters rarely sailed from Florida locations in later years. Those few that did typically transported green turtles or salvaged cargo recovered during a "wrecking voyage." A fair number of coasters were sailing into Savannah from South Carolina locations in 1850. Most came from rice-growing areas along the New River and the Carolina side of the Savannah River, such as the Back River. One vessel, the schooner *Morning Star*, sailed from Okatee Landing with rough rice and Sea Island cotton for Robert Habersham & Son. No coasters came from Beaufort, and few had sailed from there in recent years since that trade was now mainly carried by steamboats.

No coasters sailed into Savannah from any of the Sea Islands in 1850 unless the five arrivals from "Sapelo" refer to Sapelo Island rather than Sapelo River. After steamers began to serve many of the Sea Islands in the late 1820s, they were visited by fewer and fewer sailing coasters. In most years after 1840,

[32] Rough rice shipment figures are from the shipping lists in the *Daily Georgian*, 1841–1854, *Daily Morning News*, 1855, 1858–1860, and *Savannah Daily Republican*, 1856–1857.

there were one or two arrivals at Savannah from one or more of the Sea Islands.

### *The Coasting Cargoes in 1850*

Table 8.4 provides information on the shipments of various commodities transported into Savannah by sailing coasters derived from the shipping lists published in the 1850 *Daily Georgian*. Information on the minimum amounts of these goods is provided where available. Rough rice was listed as a cargo on 157 of the 225 arrivals of coasters. The number of bushels, given for all but one arrival, totaled 394,512. Clean rice was listed as a cargo on nine arrivals, totaling 664.5 casks. Eighty-one of the coaster arrivals carried 4,265 bales of Sea Island cotton. Upland cotton, totaling just 351 bales, was identified on five arrivals. By 1850, the sailing coasters transported modest naval stores into Savannah. That year, these commodities were aboard twenty-seven arrivals and consisted of more than five thousand barrels of turpentine, spirits of turpentine, and rosin.

The commercial production of sugar along the Georgia coast had peaked in the 1830s, but small amounts continued to be shipped in later years. In 1850, just two shipments of sugar were reported. One shipment of twenty barrels came from the Turtle River in Glynn County, and the other, consisting of three casks, was shipped from Centervillage on the St. Marys River. Three shipments of syrup, consisting of seventy-nine barrels, all from Riceboro, were reported for 1850. Other miscellaneous items carried by coasters included more than forty-four hundred bushels of corn, "wood," staves, lumber, "cedar posts," hides, deer skins, potatoes, beeswax, groundnuts (peanuts), and two bales of wool.

The single arrival of rough rice, for which no quantity was given, represents 0.6 percent of the 157 shipments of rough rice into Savannah. Assuming this one vessel carried amounts similar to the other arrivals, it would have been laden with 2,367 bushels of rice. If so, it can be argued the coasting fleet transported an estimated 396,879 bushels of rough rice into Savannah in 1850, representing 17,859,555 pounds of rice. Prices for rough rice were rarely quoted in the commercial section of the *Daily Georgian* in 1850, but the average price for clean rice over the year was 3.4 cents per pound. Generally, the price relationship between clean rice and rough rice was computed at one to four, meaning that if one hundred pounds of clean rice sold at $3.40 (the average price in 1850), then a bushel of rough rice (forty-five pounds) would bring a quarter of this amount or eighty-five cents.[33] Assuming an average price of eighty-five cents per bushel, the rough rice carried into Savannah by coasters in 1850 was worth approximately $337,347.

Rice production in Georgia and South Carolina, and the amounts of rough rice carried by the coasting fleet, had grown considerably since 1830. In 1830, at the high point of the sailing trade, the coasters transported an estimated 164,353 bushels of rough rice into Savannah. This dropped to 110,885 bushels in 1835, but the amounts rose again in later years, increasing to 234,625 bushels in 1845, before reaching the almost 397,000 bushels carried in 1850. On the other hand, the amount of clean rice carried by sailing coasters declined after 1830. In 1830, coasters carried 4,216 casks, barrels, and tierces of clean rice into Savannah. This number dropped to 2,487 containers in 1835, then to 1,444.5 containers in 1845 before falling to the 664.5 tierces and casks reported in 1850. This significant decline in clean rice cargoes directly reflects the effect of steamboats on the coasting trade. Clean rice, unlike rough rice, was suitable to transport by steamers, and over the years, they carried an increasingly larger share of this commodity.[34]

As in earlier years, it is difficult to estimate precisely how many pounds of clean rice were carried by the coasters in 1850 because it is unknown if the recorder inconsistently used the terms tierces and casks, the only two containers mentioned. However, assuming each of the 664.5 containers carried about 600 pounds of rice, the capacity of an average tierce, the total amount of clean rice carried was 398,700 pounds. With clean rice bringing an average of 3.4 cents per pound, the estimated value of this rice was $13,556. Thus, the approximate value of all rice carried by sailing coasters into Savannah in 1850 was $350,903. This was twice the value of rice carried at the peak of the sailing trade in 1830 and 30 percent greater than that of the rice carried in 1845. The total worth of the rice cargoes rose over the years despite the considerable decline in the amount of clean rice transported. This increase in value was due to the expansion of rice production along the coast after 1830, the growth of rice-milling facilities near Savannah that attracted planters to ship their crops there, and the continued and expanding participation of sailing coasters in the transport of rough rice.

Sea Island cotton represented the most valuable commodity carried by sailing coasters in 1850. The 4,265 bales transported into Savannah

[33] Gray, *History of Agriculture*, 2:1030; House, *Planter Management*, 67–68.

[34] Amounts of rice carried into Savannah are derived from the shipping lists in the *Daily Savannah Republican*, 1830 and 1835 and the *Daily Georgian*, 1845 and 1850.

represented an estimated 1,492,750 pounds of fiber, assuming an average weight of 350 pounds per bale or bag. Gray notes that the average price for Sea Island cotton sold in Charleston in 1850 was 27.8 cents per pound, meaning the value of the Sea Island cotton carried by the Georgia coasters can be estimated at $414,985. The approximate value of the small quantity of upland cotton carried, just 351 bales, is $14,373, assuming an average price of 11.7 cents per pound.[35]

The 4,265 bales of Sea Island cotton carried into Savannah in 1850 was a little over half of the 7,188 bales estimated to have been carried by the sailing fleet in 1830. Some of this decline might be related to lower production in 1850, given that Savannah exported about 2,000 fewer bales of Sea Island cotton in 1850 than in 1830. However, most of this reduction is due to the increasing amounts of cotton carried by steamers after 1830. During the 1850 shipping season, steamers like the *William Gaston*, *Ocmulgee*, and *St. Mathews* regularly arrived in Savannah from locations as far south as the St. Johns River with cotton and other goods. In January 1850, one of these steamers was arriving in Savannah every three or four days, and over the month they carried several hundred bales of Sea Island cotton into the city. In addition these steamers transported other cargoes, including wool, moss, sugar, oranges, hides, and merchandise. Another steamer, the *Ivanhoe*, provided weekly service between Savannah and St. Marys with stops at "Darien, St. Simons Island, Brunswick, Bethel, Jeffersonton &c." In January 1850, in the middle of the rice and cotton shipping season, the *Ivanhoe* arrived in Savannah three times, transporting 225 bales of Sea Island cotton and 72 bales of upland cotton. Other steamers operated out of Savannah and carried similar commodities into the city.[36] If this single month of January can be considered a guide, it appears that by 1850, the steamboats serving the coast were transporting about the same amount or more of Sea Island cotton as the sailing coasters.

It was not uncommon for coasters to arrive in Savannah with quite small quantities of Sea Island cotton. Several ships carried just two or three bales on individual voyages. The average carried per voyage over the year was 53 bales. The sloop *Catherine Chard* carried the three largest shipments of Sea Island cotton. Two of these, 303 and 287 bales, were transported from the inland town of Centervillage while the third, 200 bales, was shipped from the town of St. Marys. These large shipments may represent the production of several growers accumulated by a merchant or agent for shipping.

[35] Gray, *History of Agriculture*, 2:1027, 1031.

[36] Steamboat data from the *Daily Georgian*, January 1850.

Sailing coasters carried twenty-seven shipments of naval stores into Savannah in 1850 (see Table 8.4). Specific amounts are given for twenty-five shipments. These consisted of 970 barrels of turpentine, 588 barrels of spirits of turpentine, 2,240 barrels of resin, 807 barrels identified as "turpentine and resin," and 500 barrels described as "spirits of turpentine and rosin." Presumably, the barrels identified specifically as "turpentine" represent non-distilled tree sap, or "soft" or "crude" turpentine as it was often called. Tar, another class of naval store produced from the controlled burning of pine wood, was not transported in 1850.

Transporting naval stores in any quantity by sailing coasters was a recent phenomenon in 1850. As early as 1830, the sloop *Import* reportedly carried "tar and rosin" into Savannah from Florida. However, mentions of these commodities were rare until the late 1840s when newspapers began to report naval stores as regular cargo, such as the hundred barrels of turpentine carried by the sloop *Liberty* from Harris Neck in November 1847.[37] Naval stores continued to be carried in modest quantities by coasters into Savannah to 1861. The twenty-seven shipments made in 1850 represent the largest number of deliveries recorded in the Savannah newspapers between 1800 and 1861. By mid-century, several small turpentine distilleries were established along the coast. A notice of sale of one facility appeared in the *Daily Georgian* on February 5, 1850. The notice described the distillery as a "fifteen barrel still" just two miles from the Sapelo River "where vessels of 100 tons can receive their cargoes." Six thousand acres of pineland were included in the sale, and interested parties were to contact George W. Anderson & Brother, the prominent Savannah factor acting as the seller's agent. George W. Anderson & Brother received several shipments of naval stores in 1850, including some from the South Newport River area, possibly from the distillery mentioned in the sale notice.

After 1850, the number of naval store shipments into Savannah ranged from a high of twenty in 1851 to three in 1861. In 1855, coasters carried 5,682 barrels of naval stores into Savannah, the largest quantity reported for any year. Unlike the shipment of rice and cotton, the transport of naval stores was not seasonal; in 1850, these commodities were shipped every month except December.

Some idea of the value of naval stories carried by the coasters can be obtained using the amounts shipped in 1850 and the prices these commodities brought in Savannah that year. The *Daily Georgian* regularly published a

[37] *Daily Georgian*, November 9, 1847.

"Savannah Market" section in its "Commercial" column that provided information on the prices of various local commodities imported into the city. In addition, the newspaper published a "Savannah Wholesale Prices Current" table that gave wholesale prices for a wide range of goods. These sources reveal that spirits of turpentine shipped into the city brought about 32 cents per gallon on average in 1850, while "crude turpentine" and rosin brought about $2.20 per barrel. Thus, the estimated value of the 970 barrels of turpentine carried by coasters in 1850 was $2,134, and that of the 3,047 barrels of rosin and "turpentine and rosin" was $6,703. The estimated value of the 588 barrels of spirits of turpentine, assuming forty gallons per barrel, was about $7,526. The value of the 500 barrels of "spirits of turpentine and rosin" is more difficult to determine, but assuming 250 barrels of each, this cargo had a value of $3,750. The total value of the naval stores carried into Savannah in 1850 is estimated to be $20,113.

The coasters carried a great quantity of coastal produce into Savannah in 1850 that had a considerable value, but what proportion of these goods did they carry relative to the amounts exported from the city that year? Clearly, the share of some cargoes, particularly cotton, carried by coasters had declined over time as steamboats and railroads became more important in the movement of commodities into Savannah. In 1850–1851, Savannah reportedly exported 35,602 tierces of clean rice, 11,642 bales of Sea Island cotton, and 305,792 bales of upland cotton.[38] As noted above, 396,879 bushels of rough rice were reportedly carried into Savannah aboard coasters in 1850. This represents 29,700 tierces of clean rice weighing 600 pounds apiece. When the 664.5 containers of clean rice carried by coasters are added, the total amount of clean rice transported in 1850 is estimated to have been 30,430 tierces. This represents 85 percent of the total amount of rice exported from the city in 1850–1851 and is somewhat lower than for 1830, when coasters carried about 96 percent of the rice exported from the city.

Two factors account for this decrease. One is steamboats, which carried a substantial amount of clean rice into Savannah by 1850. In addition, clean rice, rough rice, and Sea Island cotton were carried into Savannah aboard plantation boats and flats, not sailing coasters. Most small boats came from plantations and mills along the lower Savannah River that were within easy rowing or poling distance of the city. Savannah newspapers published information on the arrivals of these plantation boats, but these entries were inconsistent in earlier years. In 1850, the *Daily Georgian* did include these types of

[38] Hunt, "Commercial Statistics," 619.

boats in its shipping lists, and these entries reveal the significant quantities of produce carried by these small boats. For example, on February 19, 1850, the *Daily Georgian* reported the arrival of "Dr. Potter's boat from Plantation" with sixty casks of clean rice. The previous day, the newspaper noted the arrival of three boats, owned by Mr. Middleton, Mr. B. Hinse, and Mr. Holiday, all arriving "by canal" carrying a total of twenty-two hundred bushels of rough rice. These three boats came through the Savannah and Ogeechee Canal, which connected the city with the Little Ogeechee and Ogeechee rivers, providing rice planters there convenient access to the Savannah market. The canal had been completed in 1830 but had experienced numerous problems and little success during its early years of operation. However, in the 1840s, it became important in the transport of timber from the Ogeechee River to sawmills at Savannah, and Ogeechee River planters began to use the canal to ship their rice to market. The three boats passing through the canal in February 1850 were among a large number that transported more than twelve thousand bushels of rough rice in the early months of that year.[39]

The small plantation boats traveling into Savannah also carried cotton, such as "Dr. Waring's Boat from Plantation," whose arrival with ten bales of Sea Island cotton was reported in the *Daily Georgian* on February 13, 1850. Over the year, these small boats certainly carried a considerable amount of rice and cotton into Savannah. However, an accurate assessment of their role in transporting coastal products into the city is difficult to obtain, given how little information is available. The newspaper shipping lists could be used to track the arrival of plantation boats in Savannah for some years, but this information was published intermittingly and is not as complete or as comprehensive as the record for sailing coasters.

### *Seasonality of the Trade in 1850*

Sailing coasters arrived in Savannah every month in 1850, but as was typical, arrivals were at their lowest in July, August, and September. Seventy-six percent of the coaster voyages in 1850 occurred during the six months between November and April, the primary shipping season for rice and cotton. The first rice shipment of the 1850–1851 shipping season was apparently the thirty-two hundred bushels of rough rice carried by the schooner *Cotton Plant* on October 16. No place of origin was published for the *Cotton Plant* on this voyage, but she normally sailed in the Ogeechee River-Savannah trade. Rough rice shipments peaked between November and January, with the *Daily*

[39] Sullivan, *Beautiful Zion*, 152–54.

*Georgian* reporting thirty-two coaster arrivals with rice in November, the highest for any month of the year. Sea Island cotton shipments were highest in February (eleven shipments), March (eighteen shipments), and April (twelve shipments).

The last rough rice shipment into Savannah in 1850 was the fifteen hundred bushels carried from Riceboro by John Grovenstein and his sloop *Eagle* on July 18. The rice aboard the *Eagle* was a small amount for the sloop that typically carried thirty-five hundred bushels per trip during the height of the shipping season. Two other small rice shipments, five hundred and one thousand bushels, arrived in Savannah in July. In 1850, there were no rice shipments reported for August and September and no Sea Island cotton shipments in either July or August.

### *Savannah Merchants in the Coasting Trade in 1850*

In 1850, the *Daily Georgian* named consignees for 219 of the 225 coaster arrivals at Savannah. These 219 arrivals carried 370 identified shipments of coastal produce to thirty-five named consignees. A single commodity is identified as cargo aboard 129 arrivals, representing slightly over half of all arrivals that year. Most single cargo arrivals, 113, consisted of shipments of rough rice. When Sea Island cotton was carried aboard a coaster, it was generally loaded with other items, such as rice, naval stores, hides, or the like. Sea Island cotton was listed as the sole cargo on sixteen arrivals. Only five Savannah merchants received more than ten individual shipments over the year; sixteen received just a single shipment. Most of those receiving one shipment were factors, commission merchants, or other mercantile firms, but a few apparently were private individuals, such as the "Mrs. McIntosh-to order" listed as one of several consignees on a shipment of mixed cargo carried from Riceboro by the sloop *Liberty*. The *Liberty*'s cargo included Sea Island cotton, corn, rosin, spirits of turpentine, potatoes, and groundnuts (peanuts). The other consignees listed for this arrival, Andrew Low & Company, Way & King, and Yonge & Gammell, were established Savannah merchants who typically dealt in Sea Island cotton or naval stores. When multiple consignees are named for multiple shipments, as in this instance, it is difficult to determine what was being shipped to whom. However, in this case, the potatoes and groundnuts likely were being shipped *by*, rather than *to*, Mrs. McIntosh since the term "to order" typically meant that a cargo was not assigned to a specific consignee and had yet to be disposed of by the owner.

In 1850, Robert Habersham & Son was named consignee on 127 shipments, twice as many as any other merchant. The other firms receiving more than ten shipments were George W. Anderson & Brother with sixty-four shipments, Way & King with thirty-nine, Noble A. Hardee & Company with thirty-six, and Elias Reed with twenty-seven. Robert Habersham & Son dominated the Georgia coasting trade in 1850, as the firm had since the 1820s. They were especially prominent in the rice trade. The firm was listed as the sole consignee on ninety-six of the 225 arrivals for 1850, and these arrivals carried ninety-two shipments of rough rice, four of clean rice, and three of Sea Island cotton. Robert Habersham & Son was named as one of multiple consignees on twenty-six other arrivals. Most of these twenty-six arrivals included rough rice as a cargo, and the Habersham firm probably received some or all of it. Where listed as the sole consignee, Robert Habersham & Son received 282,094 bushels of rough rice (representing 13.1 million pounds of grain), 299 casks of clean rice, and 58 bales of Sea Island cotton. When the company was the only consignee, it received 71 percent of the rough rice and 34 percent of the clean rice carried aboard coasters into Savannah in 1850. These commodities were valued over $250,000 in 1850, equivalent to about $8.5 million today (2020), reflecting the Habersham company's dominance in the Georgia rice trade.[40]

Most of the rough rice consigned to the Habersham firm in 1850 came from the Ogeechee River, with fifty-five total shipments representing about 160,507 bushels. Smaller amounts came from the Satilla River (20,500 bushels) and the New River, South Carolina (18,100 bushels).

In 1850, Robert Habersham & Son received shipments of coastal produce aboard twenty-seven sailing vessels. However, they relied heavily on three ships: the schooner *Cotton Plant*, which carried sixteen shipments to the firm; the sloop *Science*, which delivered twelve shipments; and the sloop *Eagle*, which carried eleven shipments. The company's heavy use of the *Cotton Plant* and *Science* makes sense because they owned both in 1850. There is no evidence that the firm owned the *Eagle*, but over the years, they relied heavily on the sloop's master, John Grovenstein, because of his prominence in the Riceboro-Savannah trade.

The Savannah firm George W. Anderson & Brother was named as a consignee on sixty-four shipments carried by sailing coasters in 1850. They received shipments from Riceboro, Sunbury, the Satilla River, and the South Newport River. Like the Habersham firm, Anderson & Brother dealt heavily in rice but also received numerous shipments of Sea Island cotton, some naval

[40] Modern equivalence derived from "relative value" calculator at MeasuringWorth.com.

stores, and miscellaneous cargoes, such as syrup. The firm had handled coastal products since at least 1823, when they were known as George Anderson & Son.[41]

The mercantile business Way & King was named a consignee on thirty-nine coaster arrivals in 1850. The principals of this company were William Way and William King of Savannah. They received shipments of rice, Sea Island cotton, and, possibly, naval stores. Most of these shipments originated at Riceboro while others came from the Ogeechee River, the town of Centervillage on the St. Marys River, and from other locations between Sunbury and St. Marys.

Noble A. Hardee & Company was named as a consignee for thirty-six shipments of rice, Sea Island cotton, and naval stores. The principal of the firm, Noble Hardee, was a native of Camden County, and his business was heavily involved with the South Georgia coast, receiving shipments from Centervillage, St. Marys, the Satilla River, and the town of Jeffersonton. The company was often listed as a consignee when naval stores were shipped. Noble Hardee was a part owner of the *Catherine Chard* in 1850, and his firm received ten shipments aboard the sloop that year. The other owners of the sloop, merchants John Boston, John Gumby, and Elias Reed, were also listed as consignees on most of her arrivals in Savannah.

One of the owners of the *Catherine Chard*, Elias Reed, was named as a consignee for twenty-seven shipments of rice, Sea Island cotton, and naval stores. Reed, identified as a merchant in the 1850 Chatham County census, mainly received cargoes from locations along the central and southern Georgia coast, such as Riceboro, the Sapelo River, Brunswick, the Turtle River, the Satilla River, the Little Satilla River, and Centervillage. The company relied heavily on the schooner *Elias Reed* and sloop *Catherine Chard.*

Some Savannah merchants who received a small number of shipments by coasters in 1850 seem to have dealt with very specific commodities. The merchants J. Gammell and Yonge & Gammell, possibly the same firm, dealt exclusively with naval stores, receiving shipments from Darien, Sunbury, and the Satilla and Sapelo rivers. The principals in these firms seem to have been William P. Yonge, a Savannah commission merchant and native of the

[41] Kayser & Co., *Commercial Directory*, 38–39.

Bahamas, and John Gammell, a native of Massachusetts, identified in 1860 as a "corn merchant & factor."[42]

In 1850, T. M. Turner & Company received two shipments of spirits of turpentine and rosin from the South Newport River. Thomas M. Turner was a Savannah druggist, and he may have used these goods in his business. In the nineteenth century, pine sap products, including spirits of turpentine and rosin, were taken internally as a remedy for respiratory complaints and used externally as liniment plasters and inhalants.[43]

By the middle of the nineteenth century, the plantation system along the lower Southeastern coast was at its zenith. The cultivation of Sea Island cotton and rice was profitable, and it brought great wealth to many coastal planters and to merchants in Charleston and Savannah who marketed these crops. The fleet of coasting vessels operating along the coast was still vital in transporting coastal commodities to market and moving merchandise from the region's two commercial centers to smaller towns and plantations. However, the role of the sailing ships in this fleet had begun to change in the 1830s. The number of sailing vessels in the trade had declined, falling from eighty-three vessels in 1830 to forty-seven in 1841 and to just thirty-two in 1850. More importantly, the number of voyages made by the sailing vessels into Savannah had dropped considerably over this period, from a high of 479 in 1830 to 225 in 1850.

Most of the Northern captains were gone from the Georgia trade by 1850, and the business was left mostly in the hands of local masters and shipowners, many of whom were foreign immigrants. Further, the sailing fleet had aged over the years as Northerners abandoned the trade. Many Northern captains were involved in shipbuilding and traditionally had brought their newly built ships South to trade. Because relatively few coasters were built in the South, Southern captains and shipowners often purchased their ships from Northern owners, usually after several years of use.

Compared to earlier years, the Georgia sailing fleet in 1850 was relegated to a peripheral position in the coastal economy. It was now heavily involved in transporting rough rice and other miscellaneous commodities, like naval stores. Despite this, the fleet remained viable and important to the region's economy. In 1850, the thirty-two ships working in the trade transported more than eighteen million pounds of rough and clean rice and about 1.5 million pounds of Sea Island and upland cotton into Savannah. The value of the rice

[42] Georgia Historical Society, *1860 Census*, 141.

[43] Otto, *1850 Census Chatham County*, 99; Grieve, *Modern Herbal*, 635.

was $350,903, almost twice the estimated value of rice ($190,576) carried into the city by coasters twenty years earlier, reflecting the dramatic increase in rough rice shipped aboard sailing vessels. The Sea Island cotton carried by the coasters had an estimated value of $414,985. This was about 65 percent of the value of Sea Island cotton carried by the sailing fleet in 1830 ($611,275), an expression of the loss of this cargo to steamers. The small quantity of upland cotton carried by the sailing coasters in 1850 had an estimated value of only $14,373, a significant contrast to the over one million dollars' worth of upland cotton transported in 1830. The value of the naval stores carried is estimated at $20,113.

Together these principal commodities had a value of approximately $800,000 dollars in 1850. Relying on the computations provided by Measuringworth.com, it is estimated the present-day (2020) value of the $350,903 worth of rice shipped in 1850 is about $12 million, the $414,985 worth of Sea Island cotton has a contemporary value of approximately $14.2 million, the $14,373 worth of upland cotton a contemporary value of about $491,000 and the $20,113 worth of naval stores a modern value of approximately $688,000. Together, in terms of their present-day worth, these commodities have a modern value of $27.3 million.[44]

## The Final Decade of the Georgia Sailing Coasters

The 1850s were an era of prosperity for the Southeastern coast and the South as a whole. The coastal rice and Sea Island cotton planters prospered as a group, and many Savannah merchants grew wealthy. Sea Island cotton prices rose from about thirty cents per pound in 1850 to an average price of about forty-seven cents in 1860, with the highest quality cotton sometimes bringing as much as a dollar per pound at the start of the Civil War. Rice prices remained relatively stable over the decade. The average price for clean rice exported from the United States in 1850 was 3.4 cents per pound and rose to 4.3 cents per pound in 1854 before dropping to 3.2 cents in 1860. Although rice prices remained level, rice production in Georgia expanded significantly during the 1850s. In 1849, Georgia produced 38,950,691 pounds of rice. Ten years later, in 1859, the state's production was 52,507,652 pounds. Rice production in South Carolina was considerably greater than that of Georgia;

[44] Computations of contemporary worth and value are from "relative value" calculator at MeasuringWorth.com.

however, it dropped over the decade, from 159,930,613 pounds in 1849 to 119,100,528 pounds in 1859.[45]

Despite the significant increase in rice production in Georgia over the 1850s, rice exports through Savannah remained relatively stable. In 1850–1851, the city exported 35,602 casks of rice and 39,929 casks the following year. The year 1854–1855 was a poor one for rice planters, when only 8,220 casks were reportedly shipped through the city. A hurricane in September 1854 destroyed much of the Georgia rice crop, resulting in a drop in production and exports that year. A serious outbreak of yellow fever in Savannah in late summer and fall 1854 contributed to a decline in rice shipments into the city. The scarcity of rice on the market drove the price up to four cents per pound. However, rice exports recovered quickly, and in 1855–1856, Savannah shipped 29,907 casks, representing over nineteen million pounds of rice with a value of $780,000. By 1859–1860, rice exports had risen to 35,588 casks, close to the amount reported at the start of the decade.[46] Why rice exports through Savannah did not increase in conjunction with the reported increase in rice production in the state is unknown.

Sea Island cotton exports through Savannah also increased during the 1850s. In 1850–1851, the city exported 11,642 bales. In 1855, this figure was 15,484 bales, and by 1859–1860, Sea Island cotton exports had risen to 24,941 bales. This increase in exports presumably represents an increase in production following the rise in Sea Island cotton prices, particularly toward the end of the decade. Upland cotton exports through Savannah grew considerably between 1850 and the start of the Civil War. The city exported 305,792 bales of upland cotton in 1850–1851; by 1854–1855, this figure had risen to 373,908 bales and to 522,096 bales in 1859–1860. The year 1860 was particularly prosperous for Savannah, with the city exporting significant amounts of cotton and rice. This occurred on the eve of the start of the Civil War and the ultimate collapse of the Georgia coasting trade.[47]

Despite the period's prosperity, the population of the coastal counties of Georgia (except the city of Savannah) changed little in the 1850s. In the five

[45] Gray, *History of Agriculture*, 2:723, 738–739, 1030–1031; J. F. Smith, *Slavery and Rice*, 87.

[46] Savannah rice export figures from Hunt, "Commercial Statistics," 617; Lee and Agnew, *Historical Record*, 134, 138; and *DeBow's Review* 22 (February 1857): 202, and 29 (November 1860): 669–670.

[47] Savannah cotton export figures from Hunt "Commercial Statistics," 617, 619; Lee and Agnew, *Historical Record*, 138; and *DeBow's Review*, 22 (February 1857): 202, and 29 (November 1860): 669–670.

coastal counties of Bryan, Camden, Glynn, Liberty, and McIntosh, the White population grew slightly between 1850 and 1860, from 7,257 to 7,673. The enslaved population of these counties decreased from 21,260 in 1850 to 19,507 in 1860. However, throughout the decade, the enslaved population of these counties was always large, making up 70 to 75 percent of the total population and reflecting the labor requirements of rice and Sea Island cotton agriculture.[48]

In contrast to the other Georgia coastal counties, Chatham County, which included the city of Savannah, experienced considerable population growth during the decade. The county's population grew from 23,901 in 1850 to 31,043 in 1860. The enslaved Black population of the county remained at around 14,000 persons, while the White population grew by almost 70 percent, from 9,152 in 1850 to 15,511 in 1860. This growth reflects the expansion of the urban population of Savannah and is a measure of the city's expanding commercial and economic importance.[49]

Although the 1850s in general represented a decade of commercial expansion, there was a nationwide economic slowdown in 1857 known as the Crisis of 1857. This crisis resulted partly from a drop in foreign investments, particularly British investments going to railroads then being built at a ferocious rate in the United States. Large banks, especially those in New York, began to recall loans as they weakened from a lack of foreign capital and, particularly, when Western branches of these banks began to fail. As these Western banks attempted to withdraw funds from their New York branches, panic ensued. In September 1857, the New York–bound steamer *Central America*, laden with $1.5 million in gold bullion from California, sank off the North Carolina coast, depriving Northern banks of desperately needed specie. By mid-October, massive runs on New York banks forced them to suspend payments. The bank runs led to suspensions by banks in other states. An additional complication was a drop in prices for exports, particularly Western grains, as the European demand for these commodities declined with the end of the Crimean War in 1856.

The Crisis of 1857 was followed by a depression, resulting in massive unemployment, particularly in the Northeast. The crisis did influence shipping in the United States, with many ships and men left idle between 1857 and 1859 and a significant drop in the construction of new vessels during the

[48] J. F. Smith, *Slavery and Rice*, 217.
[49] Ibid.

same period. However, the depression was short-lived mainly because of an influx of gold into New York banks from the goldfields of California.[50]

The Crisis of 1857 was most damaging in the Northeast and the West and had a smaller effect on the Southern economy. The specific effects of the crisis and subsequent depression on the Georgia coasting trade are difficult to measure. The European demand for cotton remained steady, so this aspect of the coastal economy was little affected. However, there was a reduced demand in Europe for food grains, which might have affected rice exports. An examination of the arrivals and the cargoes carried by the coasters sailing into Savannah between 1857 and 1859 seems to indicate the economic downturn did have a minor influence on the trade. For one, the number of coaster arrivals in Savannah declined in 1857. That year, the *Savannah Daily Republican* listed 158 sailing coaster arrivals at Savannah, thirty fewer than the previous year. The number of arrivals increased in 1858, to 164, and then to 208 in 1859. If the decline in arrivals in 1857 was related to the crisis, then the recovery was swift. Possibly a more telling indicator of the economic difficulties was a sizeable decline in the amount of rice shipped aboard the sailing coasters. In 1856, coasters transported about 473,000 bushels of rough rice into Savannah. This number fell to about 362,000 bushels in 1857, a decline of almost 25 percent. In 1858, the coasters carried about 423,000 bushels of rough rice into Savannah, an increase from 1857 but still 50,000 bushels below what had been carried in 1856, the year before the economic crisis began. The drop in rice shipments presumably reflects a decline in production by planters in response to a declining market for these crops in Europe and, possibly, in the Northeast due to the depression.[51]

Sea Island cotton shipped aboard coasters also declined during the crisis. In 1856, approximately three thousand bales of Sea Island cotton were shipped aboard coasters into Savannah; twenty-three hundred bales were shipped in 1857, and fifteen hundred in 1858.

Whatever the effects the Crisis of 1857 might have had on the Georgia coasting trade, they were not lasting, and by 1859, the effects had disappeared. That year, the number of sailing coaster arrivals in Savannah rose to 208, the amount of rough rice carried by the coasters increased to about 642,000 bushels (77 percent more than 1857), and the quantity of Sea Island cotton shipped rose to 2,097 bales, an increase of about 600 bales over the previous year.

[50] Taylor, *Transportation Revolution*, 345–50; Huston, *Panic of 1857*, 14–30.

[51] Shipping information from the *Savannah Daily Republican*, 1856 and 1857, and the *Daily Morning News*, 1858 and 1859.

The work of the sailing coasters was certainly sustained by the period of prosperity during the 1850s. Still, there was no dramatic expansion of the sailing trade in terms of the number of vessels working or the number of arrivals in Savannah. Essentially, the sailing trade remained stable during this decade; it did not grow in concert with the regional economy because it was already marginalized well before 1850, displaced by the efficiency, speed, and dependability of steamboats and railroads. The one factor that did the most to sustain the sailing trade during the 1850s was the continued increase in rice production and the shipment of rough rice, a commodity suited to transport by the deep-hulled sailing ships. The sailing vessels also could visit remote landings and spend time loading rice at individual plantations, something that steamboats traveling on a set schedule were reluctant to do. During the 1850s, however, at least one steamboat was constructed at Savannah with a hull suitable for transporting rough rice. This was the 111-foot *Robert Habersham*, launched at the Henry F. Willink Ship Yard in September 1859. Owned by former coasting captain Henry J. Dickerson, the *Robert Habersham* was designed to serve two purposes: to carry rice and to serve as a lighter and tow boat on the Savannah River.[52] The *Robert Habersham* came into the city several times in 1859 carrying as much as fifteen thousand bushels of rough rice on a single trip, four to five times the capacity of the typical coaster. There is no doubt that the *Robert Habersham* represented damaging competition to sailing vessels.

While the transport of rough rice by sailing coasters expanded during the 1850s, ships continued to transport most of the miscellaneous items carried in earlier years though often in small quantities. The shipment of sugarcane products had almost ended by 1850, and from 1850 to 1861, only a few shipments of syrup were reported. Steamers carried some sugar products, but very little, because the commercial cultivation of sugarcane along the Georgia coast was near its end.[53] Wood of various types is mentioned as a cargo on a few coasters every year of the decade. In January 1860, the sloop *Science* carried 1,700 "hoop poles" into Savannah from the Ogeechee River. These were shipped to Robert Habersham & Sons and would have been used as hoops for barrels, possibly rice casks. Hides were carried by a few coasters every year. Typically, neither the quantity nor type of hide is given, but most were cowhides. Occasionally, hides were specifically identified, such as the six bundles of deerskins and two bundles of "otter and coon skins" carried by the *Elias*

[52] *Daily Morning News*, September 5, 1859.

[53] Gray, *History of Agriculture*, 2:748 .

*Reed* from the Turtle River in December 1854. By this time, steamboats normally transported fruit, such as oranges and limes, and only rarely are these items listed as cargo aboard sailing ships. Most fruit shipments originated in Florida; however, in 1851, the *Harriet Lewis* carried 10,000 oranges from Darien. In 1856, the sloop *H. Gorgas* transported 1,023 watermelons from Burnside Island, one of the more unusual food items listed in shipping records. Sailing coasters infrequently carried live turtles from Florida, as they had done in earlier years.[54]

Occasionally, the coasters transported more unusual cargoes. In 1855, the schooner *Company* transported "120 pieces of stone" from the Ogeechee River, although what these might have been used for is unknown. In 1854, four shipments of "bars of railroad iron" were carried from Brunswick to Savannah aboard the schooners *W. D. Jenkins* and *Elias Reed* and the sloop *Virginia*. Ironically, these three ships were transporting cargo that would threaten their own existence.[55]

## The Civil War and the Georgia Coasting Trade in 1861

The year 1861 was to be pivotal for the Georgia coasting trade, as it was for most Southern economic and social institutions. The Civil War disrupted the Georgia trade and, essentially, marked the end of any meaningful participation by sailing vessels. In summer 1861, federal naval forces arrived off the mouth of the Savannah River and, by late fall, had gained almost complete control of the coast, blockading the major inlets along Georgia, South Carolina, and Northeastern Florida. With the establishment of the blockade, it became unsafe for coasting vessels to operate, and the movement of goods by ship ended.

The events of 1861 were the culmination of a long period of sectional disagreement between the North and the South, primarily over the issue of slavery and states' rights. The election of Abraham Lincoln in fall 1860 brought these differences to a climax, leading to the secession of Southern states and, ultimately, to war. At a convention in Charleston on December 20, 1860, delegates voted unanimously for South Carolina to secede from the Union. Other Southern states soon followed: Florida on January 10, 1861, and Georgia on January 19. The feeling was that an armed conflict of some

[54] *Daily Morning News*, January 16, 1860; *Daily Georgian*, December 15, 1851, and December 16, 1854; *Savannah Daily Republican*, July 19, 1856.

[55] *Daily Morning News*, October 29, 1855; *Daily Georgian*, August 16 and September 4, 1854.

sort was inevitable, and Southern states began to strengthen their militias before the official formation of the Confederacy. Even before his state seceded, Governor Joseph Brown of Georgia ordered militia units to the coast to guard key entrances and harbors. On January 3, 1861, a small force from Savannah captured the nearly deserted Fort Pulaski on Cockspur Island near the entrance to the Savannah River.[56]

In early 1861, the Savannah newspapers were full of news of secession and the increasing tensions between North and South. Announcements and notices about the organization and training of local military units became common items in the papers. In January, the *Daily Morning News* published a notice that "all persons capable to bear arms in White Bluff District and Islands" were to meet on Isle of Hope to organize a unit to "repel any hostile aggression" and enable "Coast defense." That same month, the newspaper contained notices of meetings of a variety of military groups such as the "Chatham Artillery," the "Pulaski Guards," the "Mounted Guards," the "Republican Blues," the "Phoenix Riflemen," and the "Dekalb Rifleman," whose members were to depart for Fort Pulaski.[57]

Despite the grave nature of the times, the shipping lists published in the Savannah *Daily Morning News* during most of 1861 reveal little of the tumultuous events. These lists show the coasting trade out of Savannah experienced no serious disruptions by the events of secession and the establishment of a wartime footing in Georgia, South Carolina, and Florida. The year began with heavy coaster traffic driven by the bumper crops of 1860. In just the first two months of 1861, twenty-six different sailing coasters made forty-five arrivals in Savannah laden with 130,694 bushels of rough rice, 212 casks of clean rice, and 1,070 bales of Sea Island cotton. These two months' shipments were nearly half of the rough rice and Sea Island cotton carried the previous year. Coasters were arriving from the same locations as in the past, such as the Ogeechee, Satilla, Altamaha, Savannah, and Turtle rivers and Riceboro and Skidaway Island in Georgia; the Combahee and Santee rivers in South Carolina; and the Indian River in Florida.

Nor did the seriousness of the times immediately drive Northern captains and ships from the Southern trade. Captains Leonard Bolles of the schooner *Benjamin English*, Burr of the schooner *Mary A. Rowland*, Freeman C. Keene of the schooner *Paugassett*, and Franklin Hathaway of the schooner *Alexander Blue* were all Northern men, and their ships were Northern-owned.

[56] Candler, *Confederate Records*, 29; Heard, "St. Simons Island," 250.
[57] *Daily Morning News*, January 1861.

Each of them sailed into Savannah with coastal cargoes in 1861. Except for Freeman Keene, these men departed for the North by May, the typical pattern for the Northern captains. Captain Freeman Keene, a native of Mattapoisett in the Sippican area of Massachusetts, sailed the *Paugassett* into Savannah on November 5, 1861, with a cargo of rice from Darien for Robert Habersham & Sons. This voyage occurred long after the Union blockade was in effect and when it was illegal for Northern vessels to work in the Southern coasting trade. Why Captain Keene remained in Georgia is unknown, but he was still there in September 1863, when men from the United States gunboat *Seneca* encountered Keene and his wife near Darien. The couple was taken aboard the gunboat, whereupon Freeman Keene offered his services to the United States government.[58]

On April 12, 1861, Southern troops in Charleston fired on Fort Sumter, commencing hostilities. A few days later, President Lincoln issued a proclamation calling seventy-five thousand militiamen into active service, intending to use them to force the seceded states back into the Union. On April 19, 1861, President Lincoln issued the order to blockade the coastline of the Confederacy. Immediately, Georgia's Governor Brown rushed all available state troops to Savannah and Brunswick.[59] Despite Lincoln's order, the United States had neither the ships nor men to implement a successful blockade. As a result, coasters continued to sail in and out of Savannah long after the blockade order was issued. The monthly arrivals of coasters into Savannah during most of 1861 followed the pattern seen in earlier years. Arrivals, highest in January, February, and March, when the bulk of the rice and Sea Island cotton crops were shipped, then declined. Arrivals were at their lowest during the summer; only one was reported in July and two in August.

On March 22, 1861, the Savannah *Daily Morning News* published a notice by the "Georgia Naval Coast Guard" asking for volunteers, and in May, the paper published an advertisement stating that fifty seamen were wanted for privateer service. It is unknown if these requests took crewmen away from the coasting trade; however, by summer 1861, there were expressions of concern over the blockade even though it had not been fully implemented. In July 1861, the *Daily Morning News* published a "Notice to Planters" signed by twenty-one merchants in Savannah, recommending that planters refrain from shipping their cotton into the city and instead hold it on their plantations "until the blockade is fully and entirely abandoned." This notice partly

[58] *ORN*, ser. 1, 1:669.

[59] Heard "St. Simons Island," 250.

reflected a fear that coasting vessels carrying cotton into the city might be captured by blockading forces but also was partly related to the difficulties and dangers of shipping the cotton safely out of Savannah to foreign ports. Further, Southern merchants, planters, and politicians had decided to withhold cotton from shipment to European countries, hoping that they, especially Britain, would try to break the blockade as they became starved for cotton. Ultimately, this cotton embargo was ineffective, and many Southern merchants, despite their continued public support for the idea of an embargo, were making every effort possible to ship cotton out aboard blockade-runners. This activity could reap enormous profits. In August 1861, in another sign of the fear of the blockade, General Alexander Lawton, commanding the Military District of Georgia at Savannah, issued an order requiring all "steamers and sailing vessels passing along the inland navigation between Savannah and St. Marys" to obtain instructions on how to proceed.[60]

After President Lincoln issued his proclamation for a blockade of Southern states on April 19, Flag Officer S. H. Stringham was put in command of the Atlantic Blockading Squadron. Blockading forces were ordered to "seize and capture all privateers or armed vessels acting under authority or pretended authority of the insurrectionary States, [and] all vessels with arms, munitions, or articles contraband of war." Ships attempting to carry other types of critical supplies into Southern ports were to be ordered off and seized if they persisted in their efforts. Neutral vessels were given time to leave Southern ports with no interference.[61]

There was an immediate desire to blockade Charleston and Savannah, and two ships were ordered to proceed there: the steam frigate USS *Niagara* to Charleston and the two-gun steamer USS *Dawn* to the Savannah River. The *Niagara* was off Charleston by May 12; however, the *Dawn* was considered too small to serve off the Georgia coast and went to the York River in Virginia instead. In her place, Flag Officer Stringham ordered the steam propeller USS *Union* to the Savannah River and sent the sailing brig USS *Perry* to implement a "strict blockade" off the mouth of the St. Marys River and the port of Fernandina.[62] Over the summer, other vessels joined the blockade of ports along the Southeastern coast, and by October, approximately a dozen steamers and sailing vessels were operating off the principal ports and entrances between Charleston and Fernandina. On October 29, 1861, the

[60] Wise, *Lifeline*, 28; *Daily Morning News*, August 29, 1861.
[61] *ORN*, ser. 1, 4:367.
[62] Ibid., 5:629, 640.

Atlantic Blockading Squadron was split into the North and South Atlantic Blockading Squadrons. The South Blockading Squadron, under the command of Flag Officer Samuel Francis du Pont, operated between Cape Henry, North Carolina, and Key West in Florida.[63]

The blockading ships did inhibit ship traffic in and out of Southern ports, and they captured several prizes in 1861. However, Flag Officer Stringham considered the blockading force too small to be effective, noting that it required additional vessels, particularly steamers that could carry an adequate supply of coal to remain at sea for long periods.[64] Most scholars agree with Silas Stringham that the Union blockade was ineffective during its first two years of implementation. During this period, blockade-runners passed through the blockade into ports and inlets along the coast south of Charleston with relative ease. Over time, as ships became available, the blockade was strengthened. Still, through most of 1861, it had little effect on the vessels sailing in the Georgia coasting trade because vessels continued to travel between Savannah and Northern ports through much of May. As late as the middle of May, five steamship lines connecting to Northern ports were advertising in the *Daily Morning News* though this traffic stopped by the end of the month, and traffic with overseas ports declined significantly before ending in the second week of June. However, advertisements for steamers and sailing vessels departing for and arriving from locations all along the coast between Charleston and Jacksonville continued to be published through September 1861. In that month, the sloop *Mary Elizabeth* advertised it would depart for Bluffton and Beaufort, South Carolina, and the steamer *Dixie* that it would leave for Brunswick and Fernandina.[65]

Efforts were made to protect coaster traffic. In July 1861, Confederate Secretary of the Navy Stephen R. Mallory reported to President Jefferson Davis that four steamers had been equipped to protect the coasts of South Carolina and Georgia, and other gunboats were under construction. He stated, optimistically, that this force could protect the "entire inland navigation between Charleston and Savannah."[66]

On November 8, 1861, combined Union naval and army forces captured Port Royal Sound in lower South Carolina, and it became a main supply and repair depot for the South Atlantic Blockading Squadron. The occupation of

[63] Ibid., 6:282.

[64] Ibid., 5:664–665.

[65] *Daily Morning News*, May and September 1861.

[66] *ORN*, ser. 2, 2:76.

Port Royal Sound and blockading forces' subsequent armed reconnaissance expeditions against coastal locations made it almost impossible for sailing coasters to operate safely along the inland passage. These actions resulted in the evacuation of most of the inhabitants of the Sea Islands. A few steamboats dared to operate along the coast after this; as late as January 1862, advertisements appeared for steamers traveling between Brunswick and Jacksonville, but this activity did not last.[67]

In February 1862, General Robert E. Lee, then in command at Savannah of the forces of the Southern region, concentrated all available forces at Savannah, considered the most important position on the Georgia coast. The withdrawal of Confederate troops from most of the coast was achieved by March 1862, leaving United States forces in complete control of the coastal islands and most of the inland passage.[68]

### *Georgia Coasting Ships and Captains in 1861*

Mostly, the Georgia coasting trade operated normally during much of 1861 despite the beginnings of war. The *Daily Morning News* reported 133 coaster arrivals in the city that year, made by thirty sailing vessels (twenty-four schooners and six sloops). These vessels arrived from nineteen named coastal locations, a decrease from the twenty-eight points of origin named in 1850. An estimated thirty different masters sailed in the trade in 1861 (Table 8.5).

The coasting fleet had continued to age. The average age of the entire fleet in 1861 was 16.5 years, and the oldest vessel sailing was the thirty-six-year-old schooner *Fort George Packet.* As a group, however, sloops were older than schooners with an average age of 23.6 years versus 14.8 years for the schooners.

Over time, larger vessels that could carry more cargo were brought into the Georgia trade. In the early 1850s, coasters rarely carried more than four thousand bushels of rough rice on any one voyage. By 1855, three coasters, all schooners, carried more than six thousand bushels of rough rice on individual voyages into Savannah. These were the *Leopold O'Donnell*, the *L. N. Godfrey*, and the *Marietta Hand.* In 1861, the largest single cargo of rough rice reported for a sailing coaster was the seventy-two hundred bushels carried by the schooner *Alexander Blue.* Although a few large schooners entered the Georgia

[67] *Daily Morning News*, January 1862.

[68] Candler, *Confederate Records*, 51–153.

trade after 1850, none had even half the fifteen thousand–bushel capacity of the steamer *Robert Habersham*, built as a rice carrier in 1859.

The most active vessel in the Georgia trade in 1861 was the schooner *Cotton Plant*, making thirteen arrivals in Savannah during the year under her master John Arnaud. The *Cotton Plant* sailed mostly in the Ogeechee River-Savannah trade. The sloop *Swallow* sailed into Savannah eleven times in 1861 with Charleston resident David Little listed as master on ten arrivals. Swedish native and Savannah resident Andrew Backman was master on one. The *Swallow* was constructed in Charleston in 1854. The vessel's owner in 1861 seems to have been prominent Chatham County planter Daniel Blake, who possessed extensive rice lands along the Ogeechee River.[69] Other active coasters in 1861 were the schooner *William D. Jenkins*, making nine arrivals in Savannah; the schooner *William Totten*, with eight arrivals; and the sloop *Science* and schooner *Elias Reed*, each of which sailed into Savannah seven times.

As was true in 1850, most of the Georgia coasters were commanded and owned by Southern men, and participation by Northern men in the trade was minimal. In 1855, only four of the approximately thirty-three masters sailing in the trade are positively identified as Northern men, and in 1860, only three of the thirty-two or thirty-three coasting captains were from the North. In 1861, seven of the thirty or so masters can be specifically identified as Northern captains, a slightly higher number than in the immediately preceding years. Most were from the Sippican area of Massachusetts.

Traditionally, the Northern captains had been most heavily involved in the Darien-Savannah trade. Some did sail this route in 1861, such as Captain Franklin Hathaway, who carried seventy-two hundred bushels of rough rice from Darien aboard the schooner *Alexander Blue*, and Freeman Keene, master of the schooner *Paugassett*, who transported seven thousand bushels of rough rice from Darien on his two voyages from there. However, in 1861, several of the Northern captains sailed into Savannah from the Satilla River, a trade that had grown with the expansion of rice cultivation along that river after 1840. These Northern ships primarily carried rough rice. The *Daily Morning News* reported that coasters carried 81,985 bushels of rough rice into Savannah from the Satilla River in 1861, and more than 40 percent, consisting of 37,200 bushels, was carried in Northern ships commanded by Northern captains. Northern captains rarely sailed in the Satilla River-Savannah trade before

[69] Enrollment No. 18, Port of Charleston, sloop *Swallow*, April 8, 1854; Register No. 1, Port of Savannah sloop *Swallow*, April 8, 1856; "Vessel Papers," File S-188.

1852, but after that year, they assumed an increasing role in that trade, even though their overall participation in the Georgia trade declined.

*Ports in the Coasting Trade in 1861*

The Ogeechee River was the principal port of origin for coasters in 1861, with ten different ships making thirty-five voyages into Savannah from there (Figure 8.3). The Ogeechee had been the primary port of origin for Georgia coasters for almost twenty years, an expression of the quantities of rice grown there and the increasing reliance of the sailing coasters on the transport of rough rice. Rough rice (totaling 110,617 bushels) was the only cargo carried by the coasters sailing from the Ogeechee in 1861. In March, Swedish-born Captain Andrew Backman brought the sloop *Science* into Savannah from "Ossabaw Island and the Ogeechee" with thirteen bales of Sea Island cotton and a thousand bushels of rough rice. The cotton was probably shipped from Ossabaw, while the rice came from the Ogeechee.

Savannah resident John Arnaud was the most active captain in the Ogeechee River-Savannah trade, making ten trips in 1861 with the *Cotton Plant*. Other captains and ships prominent in this trade were David Little and the sloop *Swallow*, Andrew Backman and the sloop *Science*, Domingo Galleo and his schooner *Eliza Ann*, Captain Worthington (probably Richard Worthington) with the schooner *William D. Jenkins*, and Andrew Hanson with his aging schooner *Fort George Packet*. It appears that all the ships sailing in the Ogeechee River-Savannah trade were Southern-owned vessels, and the masters were all residents of Georgia or South Carolina. Southern captains had always dominated the trade with the Ogeechee River, and it is unknown why Northerners rarely participated in it.

The Satilla River was the origin of twenty coaster arrivals in Savannah in 1861, making it the second most important port of origin in the Georgia trade that year (see Figure 8.3). The Satilla had become increasingly more important for the sailing coasters during the decade of the 1850s as rice cultivation expanded along the lower river. In 1850, the Satilla occupied fifth place as a port of origin for sailing coasters, and by 1855, it was second, a position it retained through 1861. Most of the large rice plantations along the Satilla were on the Camden County side of the river. In 1849, Camden County produced 6,400,940 pounds of rice, and by 1859, production increased to 10,330,068 pounds, putting Camden County second to Chatham County in terms of rice production in Georgia. Among the larger rice planters along the

Satilla River in 1860 were Stephen King, F. M. Adams, Duncan L. Clinch, J. B. Guerrard, L. W. Hazlehurst, and George Lang.[70]

Eight different coasters carried cargoes from the Satilla River in 1861. Three were Northern vessels with Northern masters, and the remainder were Southern-owned and Southern-manned. The coasters sailing from the Satilla carried rice, Sea Island cotton, and naval stores. The *Daily Morning News* reported that 81,985 bushels of rough rice were aboard seventeen voyages while 442 bales of Sea Island cotton were on six. On one of his two voyages from the Satilla with his schooner *Northern Belle*, Charles Stevens carried fifty-four bales of Sea Island cotton, in addition to 470 barrels of rosin and twenty-six barrels of spirits of turpentine, the only shipment of naval stores reported from the Satilla River in 1861. Since 1847, the Satilla had been the source of a modest quantity of naval stores, with several shipments typically carried from there each year, so the single shipment made in 1861 was unusual.

The Little Satilla River was not named as a place of origin for coasters sailing into Savannah in 1861. In previous years, the Little Satilla was occasionally listed as a port of origin, and the cargoes shipped from there were Sea Island cotton, naval stores, and wood—never rice. Similarly, not a single coaster sailed into Savannah from St. Marys in 1861. The town, once the most important local port in the trade, was last listed as a port of origin for a coaster in 1859 when Charles Stevens sailed his schooner *Northern Belle* from there with a cargo of Sea Island cotton and merchandise.

In 1861, one coaster did visit Camp Pickney, the river landing about forty miles up the St. Marys River. This was the schooner *William D. Jenkins*, commanded by Captain Richard Worthington, who sailed from there with five hundred barrels of turpentine.

Fourteen coaster voyages were made from the Altamaha River to Savannah in 1861, making it the third most common port of origin. Rough rice was listed as a cargo on every arrival, Sea Island cotton on two, and hides on one. Five vessels made these trips, and they carried 46,450 bushels of rough rice, twenty-eight bales of Sea Island cotton, and forty hides. These coasters were transporting cargoes from unnamed plantations located along the lower river. The newspaper did report on the arrival of the schooner *Fort George Packet* from "Butler Island," located in the Altamaha estuary, with a cargo of rough rice.

Darien was named as the origin of fourteen coaster voyages in 1861. Except for one voyage, rough rice was the only cargo carried from Darien. The

[70] J. F. Smith, *Slavery and Rice*, 226.

single exception was a cargo of "merchandise" transported aboard the sloop *Swallow*. Other than the *Swallow*, eight different coasters carried 56,400 bushels of rough rice from Darien. This rice certainly derived from rice plantations along the Altamaha River, and Darien likely was named as the port of origin simply because it was nearby. Unlike the coasters sailing from the "Altamaha River," which were all Southern ships, five or, possibly, six of those sailing from "Darien" appear to be Northern vessels.

The Back River, the Northern branch of the Savannah River located opposite the city of Savannah, was the origin of thirteen coasters voyages (see Figure 8.3). Six ships, all local vessels, made these trips, carrying rough rice on every voyage. The only other commodity carried consisted of seventy-three tierces of clean rice transported aboard the schooner *Cotton Plant*, along with a load of rough rice.

Ships serving the Back River often also carried cargo from locations along the Savannah River. In 1861, three different coasters made five voyages from the Savannah River. These were the schooner *Levant*, the sloop *Virginia*, and the schooner *Emma Julia*, which typically sailed in the Ogeechee River-Savannah trade. Rough rice was the only cargo carried from the Savannah River.

The coasters did not carry all the rice from the Savannah and Back rivers in 1861. The plantations and mills along these rivers were near Savannah, making it convenient to ship rice by plantation boats and flats, and the arrivals of many of these boats are listed in the *Daily Morning News*. Nor was all the rice grown along the lower Savannah River shipped to Savannah; some planters here, particularly those with holdings on the South Carolina side of the river, continued to send their crops to Charleston.

Riceboro was the origin of eight coaster voyages into Savannah in 1861. The town had been one of the most important ports of origin for the coasters since about 1820, retaining its importance because it was not typically visited by steamboats. However, the number of coasters sailing to and from Riceboro declined during the 1850s, as they concentrated their efforts on the major rice-producing regions. Only two ships sailed from Riceboro to Savannah in 1861, each making four trips. These were the schooners *Fort George Packet*, owned and commanded by Andrew Hanson, and *William Totten*, owned and commanded by Charley Thompson.[71] As was true in the past, shipments from

[71] Enrollment No. 1, Port of Savannah, schooner *Fort George Packet*, November 13, 1858; Enrollment No. 1, Port of Savannah, schooner *Fort George Packet*, July 15, 1859;

Riceboro were mixed, consisting of 574 bales of Sea Island cotton, 4,000 bushels of rough rice, 1,480 bushels of corn, and 100 bags of cotton seeds.

The Sapelo River (referring to the area around Harris Neck on the central Georgia coast) was listed as a port of origin for four coaster voyages. A single ship, the North Carolina-built schooner *E. B. Hackburn,* under her master Patrick Doyle, made these voyages. The *Hackburn* transported Sea Island cotton on every trip for a total of 280 bales. Corn, wood, and "merchandise" were also shipped from the Sapelo River.

Only two voyages originated in Sunbury, both made by the sloop *Splendid*, Stephen Williams, master and part owner.[72] Captain Williams carried forty-seven bales of Sea Island cotton on his two trips.

Ossabaw Island is the only one of the Sea Islands mentioned in the *Daily Morning News* shipping lists in 1861. Steamboats had captured most of the trade from the islands before 1840, and few coasters sailed from them after that date. Three different coasters made the three trips from Ossabaw, all Southern-owned and commanded. These were the schooners *Northern Belle*, Charles Stevens, master; *E. B. Hackburn*, Patrick Doyle, master; and the sloop *Science*, Andrew Backman, master. Sea Island cotton (a total of ninety-nine bales) was the only cargo.

Several coasters sailed into Savannah from South Carolina. Three ships, the sloop *Swallow* and schooners *Zaidee* and *John W. Anderson* sailed from the Combahee River, together carrying 18,150 bushels of rough rice and 212 casks of clean rice. The sloop *Family* made one voyage from Beaufort with seven bales of Sea Island cotton. The schooner *Zaidee* made one voyage from the Santee River and one from the "South Santee River" carrying rough rice. The Santee River is north of Charleston, and it was unusual for a coaster to bypass Charleston to sail into Savannah. Interestingly, the *Zaidee* carried rice from the Santee River to Savannah several times in the late 1850s.

No coasters sailed from Florida into Savannah in 1861. During the entire decade of the 1850s, there were normally only two or three arrivals from Florida annually, and no arrivals at all in some years. Those few coasters sailing from Florida in the 1850s departed from locations such as the Indian River, New Smyrna, and Fernandina. Some sailed into Savannah in ballast, while

Charley Thompson's ownership of the *William Totten* is found in "Conveyances of Enrolled Vessels, Savannah," schooner *William Totten*, June 21, 1855; and "Vessel Papers", File B-24–2.

[72] Stephen Williams' ownership of the *Splendid* is found in "Vessel Papers," File B-24–2.

others carried more unusual commodities, such as lemons and turtles. Among these was the sloop *Convert*, which arrived in Savannah in September 1853 from New Smyrna, Florida, with a cargo of "100 Green turtles."[73]

### *The Coasting Cargoes in 1861*

The cargoes carried by the sailing coasters in 1861 were the same as those carried during the previous two decades, with one exception: There wasn't a single shipment of identified upland cotton. The last report of a sailing coaster carrying a specifically identified cargo of upland cotton into Savannah appeared in the February 4, 1860, shipping list of the *Daily Morning News*. On that day, the newspaper reported the arrival of the sloop *Splendid*, Antonio Lawrence, master, from "Turbridge, South Carolina" with a cargo of ninety-six bales of upland cotton, four bales of Sea Island cotton, and three hundred bushels of corn. These ninety-six bales were the only shipment of upland cotton reported for the entire year of 1860.

During 1861, the Savannah *Daily Morning News* listed a cargo for all 133 coaster arrivals into the city. A single commodity was identified as cargo aboard 105 arrivals, and 89 single cargo arrivals consisted of shipments of rough rice. Sea Island cotton was carried aboard twenty-five arrivals, and as in previous years, this cotton was normally loaded with other items, such as rice, naval stores, or corn.

Rough rice was the principal cargo carried by the coasters in 1861, and it was aboard 113, or 85 percent of all arrivals in Savannah (Table 8.6). According to the shipping lists of the *Daily Morning News*, the amount of rice carried on these 113 arrivals was 398,106 bushels, representing 17,914,770 pounds of grain.

The schooner *Cotton Plant*, commanded by John Arnaud and apparently owned by Robert Habersham, carried more rough rice into Savannah than any other coaster. The *Cotton Plant* sailed into the city thirteen times during 1861 with 40,767 bushels or 1.8 million pounds of rough rice. The most the *Cotton Plant* carried on any voyage was 3,900 bushels, close to her maximum capacity. Other ships transported considerable quantities of rice over the year. The sloop *Swallow* carried 29,770 bushels on eleven voyages, and the schooner *William D. Jenkins* transported 26,900 bushels on eight sailings. These coasters were all Southern-owned and commanded; no Northern ship carried this much rice during the year.

[73] *Daily Georgian*, September 6, 1853.

Clean rice was listed as a cargo on two arrivals in 1861, making up seventy-three tierces and 212 casks (see Table 8.6). The 212 casks were shipped from the Combahee River aboard the schooner *Zaidee*, while the 73 tierces were carried by the *Cotton Plant* from the Back River.

Sea Island cotton was listed as a cargo aboard twenty-five coaster arrivals (see Table 8.6). Ten different ships made these trips, carrying a total of 1,477 bales of cotton. Most of this cotton was shipped from Riceboro, the Satilla River, the Sapelo River, and Ossabaw Island. Every ship carrying Sea Island cotton was a Southern coaster; no cotton was shipped aboard Northern vessels. The average number of bales carried by individual coasters was about sixty, but most carried fewer, and there were only five shipments of a hundred bales or more. The largest single shipment of Sea Island cotton was the 225 bales brought from Riceboro by the schooner *William Totten*. This unusually large shipment was consigned to a single merchant, John W. Anderson.

The last shipment of Sea Island cotton carried by coaster into Savannah consisted of thirty-two bales from Ossabaw Island aboard the schooner *E. B. Hackburn* on September 6. This was very early for the crop of 1861, meaning this cotton could have been from the previous year's planting. Typically, Sea Island cotton shipments began in November or early December, and a fair amount was shipped into the city by the end of the year. However, in 1861 the federal naval blockade disrupted this pattern. Cotton was a valuable prize to the United States Navy, and its shipment was curtailed early in the blockade as planters feared it would be captured. Further, the slow strengthening of the blockade over summer and fall 1861 inhibited the export of cotton out of Savannah, except on the occasional blockade-runner. As noted previously, as early as July, Savannah merchants requested planters not to ship their cotton unto the city. The thirty-two bales shipped from Ossabaw Island in early September were the only cargo of cotton carried into Savannah by a coaster after May. This cotton was consigned to John W. Anderson & Company, one of Savannah's most prominent commission merchants. The fact that the cotton was shipped might indicate the firm believed they could sell it, possibly placing it aboard a blockade-runner, such as the fast steamer *Bermuda*, which departed the Savannah River on October 29, 1861. The *Bermuda* safely made it to England, and the two thousand bales of cotton she carried reaped a huge profit.[74]

However, in general, there was no market for cotton in Savannah after June 1861. Before that date, the *Daily Morning News* published a weekly

[74] Wise, *Lifeline*, 52.

"Commercial Record." It provided information on the sales of various commodities, but the "Commercial Record" was discontinued in June because few, if any, sales took place. The newspaper did continue to print information on the receipt of commodities after June, including small amounts of upland cotton brought by railroad. Although there was essentially no market for cotton in Savannah after June or July, a few factors published notices that they would continue business "on their own account," meaning they would rely on their own financial resources to buy cotton. Only a small number of Savannah merchants could afford to do this, and one suspects they extended this service only to their most valued planter customers. The inability to sell their cotton and the increasing danger presented by the ever-tightening blockade drove some coastal planters to take drastic measures. In late November, after the fall of Port Royal, the *Daily Morning News* reported that planters near Charleston were burning their cotton to prevent it from falling into the hands of Union forces, and the newspaper was confident that all Sea Island planters would follow this example.[75]

Although coasters carried no Sea Island cotton into Savannah after early September 1861, they continued to transport a considerable amount of rice. Between September 6, the day the *Daily Morning News* reported the last shipment of Sea Island cotton, and the end of the year, the newspaper listed thirty-six coaster arrivals laden with 120,297 bushels of rough rice. Rice shipments apparently continued long after cotton shipments because some could be sold to residents as well as to state and Confederate authorities for use as military rations. Rice was issued to Confederate troops; it was part of the official ration allowance at the start of the war. It may have been a particularly common issue to troops assigned to the coastal areas of South Carolina and Georgia.[76]

Robert Habersham & Sons received a considerable quantity of rough rice in late 1861, suggesting the firm was able to sell some of it, but in some cases they may have been providing payments to planters out of "their own account." The Habersham firm continued to process the rough rice it received at its mill in Savannah long after the imposition of the blockade. In late November 1861, the company published a notice stating that they were still accepting rice at the Upper Rice Mills, but "in the present emergency," planters were advised to send their own hands to discharge the rice.[77] Even though rice continued to be processed and sold in Savannah into late 1861, the prices

[75] *Daily Morning News*, November 27, 1861.
[76] United Daughters of the Confederacy, *Patriot Ancestor*, 152.
[77] *Daily Morning News*, November 27, 1861.

paid are unknown because this information was not published in the *Daily Morning News* as it had been during the first half of the year.

Only two sailing coasters transported naval stores into Savannah in 1861. In March, Captain Charles Stevens carried 470 barrels of rosin and twenty-six barrels of spirits of turpentine from the Satilla River on his schooner *Northern Belle*. Also aboard the *Belle* on this voyage were fifty-four bales of Sea Island cotton. Three months later, in June, the schooner *William D. Jenkins*, under Captain Richard Worthington, brought five hundred barrels of turpentine into the city from Camp Pickney on the St. Marys River. These two shipments of naval stores was a historically low number, considering that the previous five years had seen nine to twelve shipments of naval stores each year. Some of the naval stores produced along the coast may have been diverted to local military uses, accounting for the small amounts shipped to Savannah. In addition, military call-ups in Georgia had begun before March, reducing the workforce formerly available for the production of naval stores.

Minor cargoes carried in 1861 included six shipments of corn, a single shipment of "wood," one shipment of one hundred bags of cottonseed, one shipment of forty hides, and one shipment of "21 bundles of hay" (see Table 8.6).

The 1861 *Daily Morning News* reported coasters carried 398,106 bushels of rough rice into Savannah, representing 17,914,770 pounds. This was almost the same amount carried by the coastal fleet ten years earlier in 1851 and considerably more than the 273,730 bushels shipped in 1860. The amount of rough rice carried in 1861 is particularly high considering the Civil War curtailed the shipment of all coastal cargoes, especially after the strengthening of the federal blockade in fall 1861. The average price brought by rough rice in Savannah during 1861 is difficult to determine because after June local newspapers stopped publishing information on the sales of rice and other commodities. However, between January and May, the "Commercial Record" in the *Daily Morning News* reported that rough rice was bringing from seventy-five cents to one dollar per bushel in Savannah. Assuming an average of about eighty-five cents per bushel, the 398,106 bushels of rice carried by coasters into the city had a value of approximately $338,390.

Only 285 tierces and casks of clean rice were shipped aboard sailing coasters in 1861, less than half the 664.5 tierces and casks shipped aboard coasters in 1850, a reflection of the continually increasing role of steamboats in the transport of clean rice. The 285 containers of clean rice shipped by coaster represent 171,000 pounds of rice, assuming these containers held 600 pounds each. The average price of clean rice sold in Savannah in 1861 was about 3.2 cents per pound, meaning the clean rice carried by coasters over the entire year had a value of only $5,472.[78]

The 1,477 bales of Sea Island cotton carried into Savannah in 1861 represented 516,950 pounds of cotton. Most of this cotton was shipped early in the year, between January and March, as was typical for this cotton crop. Commodity prices in the Savannah *Daily Morning News* reveal that most Sea Island cotton was bringing between twenty-one and twenty-six cents per pound in the first several months of 1861. This was a considerable drop from the approximately forty-seven cents per pound the previous year, a drop certainly brought on by the early effects of secession. In light of these low prices, the 1,477 bales of Sea Island cotton carried into Savannah had a value of $119,000, considerably less than the approximately $339,000 worth of Sea Island cotton carried into the city by coasters one year earlier.[79] There is the question of how much of the cotton shipped into Savannah in 1861 was successfully sold, given the effects of secession and the establishment of the blockade. Except for the single shipment of Sea Island cotton from Ossabaw Island in September, all Sea Island cotton carried by coasters into Savannah in 1861 arrived before May 10. Cotton sales in Savannah continued for the first five months of 1861, and ships laden with cotton and other commodities departed the city for Northern ports until as late as April. Vessels bound for foreign ports continued to leave Savannah for an additional six weeks, but their numbers were reduced. The last ship to depart Savannah carrying cotton for an overseas port seems to have been the British ship *Eliza Bonsall*, which cleared for Liverpool on June 8 carrying 2,794 bales of upland cotton, 1,075 bales of Sea Island cotton, and 559 barrels of rosin. Three days later, the Swedish bark *Westerbotton* left Savannah for Liverpool carrying lumber, apparently the last foreign vessel sailing from the city before the imposition of the blockade. It appears that most, if not all, of the Sea Island cotton carried into Savannah by sailing coasters in 1861 was successfully marketed and shipped before events prevented it.[80]

The estimated value of all rice and Sea Island cotton carried into Savannah by coasters in 1861 was $462,862. The small amounts of naval stores and other lesser commodities added little to this value. The value of the 1861 cargoes is about $100,000 less than the estimated $560,475 worth of

[78] *Daily Morning News*, 1861.

[79] 1861 Sea Island cotton prices from the *Daily Morning News*, January through May 1861; 1860 prices from Gray, *History of Agriculture*, 2:1031.

[80] Shipping information from the *Daily Morning News*, May and June 1861.

commodities carried by coasters into Savannah in 1860 and about 60 percent of the value in 1850. This drop in value over the decade reflects the steady loss of cargo by sailing vessels to steamboats, especially in terms of Sea Island cotton. Of course, 1861 was an unusual year, with the shipment and sale of coastal commodities impeded by the blockade and other events related to the outbreak of the Civil War. The estimated present-day (2020) value of the $343,862 worth of rice shipped on coasters in 1861 is approximately $10.4 million; the $119,000 worth of Sea Island cotton has a modern (2020) value of approximately $3.6 million. In terms of their present-day worth, these principal cargoes carried by the coasters in 1861 had an estimated value of $14 million.[81]

No reliable 1861 export statistics exist for Savannah because the blockade disrupted normal shipping out of Savannah and other Southern ports. Thus, it is impossible to state what proportion of the city's exports derived from the cargo carried by the sailing coasters.

*Seasonality of the Trade in 1861*

The strong seasonality of the coasting trade that developed by the 1820s was evident in 1861, despite the imposition of the naval blockade by the United States. Coasters sailed into Savannah every month, but arrivals were at their lowest during the summer, with one arrival in July, two in August, and three in September. As in previous years, most coaster arrivals occurred between October and March, the principal shipping season for rice and cotton. February saw twenty-five arrivals, more than any other month, followed by March with twenty-two arrivals, and January with twenty. The one exception to the pattern of previous years was the two arrivals during December, a month when coaster sailings into Savannah were typically high. These December arrivals were the schooner *William D. Jenkins* from the Back River with 4,000 bushels of rough rice and the schooner *Fort George Packet* with 2,030 bushels of rough rice from Butler Island. The Union blockade, which had been in place since the summer, seems to have had little effect on the coasting trade at Savannah until December, considering that fourteen local coasters sailed into the city in October and eighteen in November. The coasters arriving before December sailed from locations scattered all along the Georgia coast between Savannah and the Satilla River. Before November, the Union blockading forces concentrated their efforts on stopping blockade-

[81] Computations of contemporary worth and value are from MeasuringWorth.com, purchasing power calculator for commodities.

runners leaving Southern states and vessels trying to reach Southern ports. They made no robust attempts to intercept coasting ships traveling the inland passage until after the capture of Port Royal in early November and the institution of armed reconnaissance expeditions against coastal locations that same month.

On September 9, 1861, the *Daily Morning News* reported the arrival of the schooner *Cotton Plant*, John Arnaud, master, from the Ogeechee River with 2,667 bushels of rough rice. This shipment might represent the first new rice of the year, but it was early in the growing season and could have been rice from the previous year's harvest. Two weeks later, on September 23, the newspaper reported the arrival of the *William Totten* from Darien with 1,800 bushels of rough rice. This shipment may be a better candidate for the first rice from the crop of 1861.

In 1861, there were several unusual shipments of rice in June. These probably represent late shipments from the crop of 1860, held back because of the threat of capture posed by the blockade. It is also true that by May, the shipment of rice out of Savannah to overseas ports was ending, and Savannah merchants may have encouraged their planter clients to withhold rice shipments until it was safe to sail or until they were able to find a buyer for the crop.

Typically, Sea Island cotton shipments began in November. However, in 1861, no cotton was shipped aboard sailing coasters after the first week of September, a factor attributed directly to the blockade and the lack of a market in Savannah.

### *Savannah Merchants in the Coasting Trade in 1861*

In 1861, the *Daily Morning News* named specific consignees for 130 of the 133 coaster arrivals, two arrivals were shipped "to order," and no consignee was listed for one arrival. These vessels carried 171 individual shipments of coastal produce to 23 named consignees. Table 8.7 lists the consignees, the number of shipments they received, and the principal goods received. The twenty-three firms and individuals named as consignees represented about one-quarter of the ninety-eight firms identified as "Commission and Forwarding" houses in the 1860 Savannah city directory.[82] Many of the other commission merchants in the city dealt with upland cotton, huge quantities of which were shipped through Savannah each year. As seen in Table 8.7,

[82] Haunton, "Savannah in the 1850s," 99.

only four Savannah merchants received ten or more individual shipments aboard coasters over the year while ten received a single shipment.

As had been true for almost forty years, the Habersham family firm, Robert Habersham & Sons, was preeminent in the coasting trade out of Savannah. The business was named as consignee on eighty-seven arriving shipments in 1861, three times more than any other merchant. The other merchant houses receiving ten or more shipments by coaster were John W. Anderson & Company with twenty-five shipments, Tison & Gordon with eleven shipments, and Noble A. Hardee & Company with ten.

In 1861, Robert Habersham was seventy-seven years old, and he had headed the family firm since succeeding his father, Joseph, in 1810. By this time, it is suspected that his sons, William, Frederick, Alexander, and possibly John, were in charge of the company's daily operations. As in the past, Habersham & Sons was particularly prominent in the rice trade. The firm was listed as the sole consignee on seventy-three of the 133 coaster arrivals, carrying seventy-one shipments of rough rice, one of clean rice, and three of Sea Island cotton. Robert Habersham & Sons was named as one of the multiple consignees on eleven other coaster arrivals. Each included rough rice as a cargo, and it is likely the Habersham firm received some or all of this rice. Where listed as the sole consignee, Robert Habersham & Sons received 260,917 bushels of rough rice, 212 casks of clean rice, and 31 bales of Sea Island cotton. The firm certainly received more rough rice than this aboard sailing coasters although it is impossible to determine how much more. Even so, when named as the only consignee, Robert Habersham & Sons received 65 percent of all the rough rice and 75 percent of the clean rice carried by coasters in 1861.

Most of the rice shipped to Habersham & Sons in 1861 came from the Ogeechee River. They also received rice from other locations, such as the Altamaha, Back, Combahee, and Savannah rivers, and from the town of Darien. In 1861, Robert Habersham & Sons was a consignee on three coaster arrivals from Riceboro, whereas in the past the firm had typically received numerous shipments from there. All shipments from Riceboro in 1861 were consigned to several firms, making it impossible to determine what commodities were destined for Robert Habersham & Sons. However, rough rice was aboard every arrival when they were listed as a consignee.

In 1861, Robert Habersham & Sons was named as the sole consignee for three shipments of Sea Island cotton, totaling thirty-one bales. This cotton was carried from the Altamaha River, Beaufort, and Ossabaw Island. The company certainly received more Sea Island cotton because they are named as one of two or more consignees for other shipments of cotton.

In 1861, Robert Habersham & Sons received shipments of coastal produce aboard twenty-one different coasters. As in past years, they relied heavily on a small number of these ships. These included the schooners *Cotton Plant*, *William D. Jenkins*, and *Blooming Youth* and the sloops *Science* and *Virginia*. It does appear the firm owned the *Cotton Plant* and *Science* in 1861, accounting for their heavy use of these two coasters.

The 260,917 bushels of rough rice received by Robert Habersham & Sons, when they were the sole consignee, had an 1861 value of about $221,780, assuming an average price of 85 cents per bushel. The price may have risen after June when Habersham & Sons received a considerable amount of rough rice. However, what this rice might have sold for is unknown as the *Daily Morning News* did not publish commodity price information during this period. The value of the known amounts of rough rice, clean rice, and Sea Island cotton shipped aboard coasters to the Habersham firm in 1861 was about $358,000. This represents a minimum value of the goods the firm received because it includes only those instances when the company was named as the sole consignee for an arriving cargo. Even so, the $358,000 is equivalent to almost eleven million dollars today (2020), reflecting the Habersham company's continued heavy participation in the Georgia coasting trade.

This trade, as well as other business and planting ventures, had made the Habersham family among the wealthiest in Savannah. The 1860 federal census noted that Robert Habersham owned $100,000 worth of real estate and had a personal estate valued at $150,000, among the highest in the city. The real estate included his personal residence on Orleans Square, other city property, the Union Rice Mill, and a considerable amount of agricultural land, including several rice plantations on the lower Savannah River. Much of Robert's personal estate was tied up in the approximately 275 people he held in bondage in 1860. That year, his eldest son, William Neyle Habersham, owned $15,000 worth of real estate and had a personal estate valued at $50,000. Another son, John R. Habersham, who still resided in his father's household, had a personal estate valued at $20,000.[83]

The John W. Anderson firm, which received the second largest number of shipments aboard coasters in 1861, was the successor to the mercantile

[83] Modern equivalence derived from CPI index at MeasuringWorth.com; wealth of the Habershams from US Census Bureau, "US Census of Population, Chatham County, 1860"; and Shelnutt, "Robert Habersham," 32.

house that began as George Anderson & Son in the 1820s. By 1850, George W. Anderson was president of the Planters Bank of Savannah and had either withdrawn from the family's factorage business or left its day-to-day operation in the hands of his brother John because by 1855, the firm was known only as John W. Anderson & Company.[84]

John W. Anderson & Company was named a consignee on twenty-five shipments carried by coasters in 1861. These included ten shipments from the Satilla River, five from Riceboro, and lesser numbers from the Ogeechee, Savannah, Altamaha, and Sapelo rivers, as well as from Ossabaw Island and the towns of Darien and Sunbury. The firm dealt heavily in rice, but they also received a modest quantity of Sea Island cotton. In the twelve instances where Anderson & Company is listed as sole consignee, the firm received 48,250 bushels of rough rice and 275 bales of Sea Island cotton. The company was particularly active in the Satilla River rice trade, receiving at least 30,650 bushels of rough rice from there, more than any other Savannah merchant.

The mercantile company Tison & Gordon received eleven shipments by sailing coaster in 1861 (Figure 8.4). The principals of the firm were William Hayes Tison, a native of Glynn County, and William Washington Gordon, a member of an old and distinguished Savannah merchant family.[85] In 1861, Tison & Gordon were listed as sole consignees on only three coaster arrivals, all from the Back River with clean and rough rice. Tison & Gordon also were named, with other merchants, as consignees of several shipments of Sea Island cotton from locations such as the Satilla River, Sunbury, and Riceboro. In earlier years, the firm seems to have handled more Sea Island cotton than rice.

Noble A. Hardee & Company received ten shipments of coastal commodities in 1861. As in the past, the firm dealt principally with the Southern coast, with five of their ten shipments coming from the Satilla River. The firm received rough rice, Sea Island cotton, and a few miscellaneous commodities, such as wood.

When 1861 drew to a close, the coasting trade in Georgia, as it had existed for six decades, ended. The Union blockade tightened during the fall, making it unsafe for the coasters to operate. The blockade and the voluntary withholding of cotton by merchants ended the regular markets in a principal cargo of the coasters. The outward shipment of coastal commodities did not stop entirely, as blockade-runners successfully carried some cargoes out.

[84] Otto, *1850 Census Chatham County*, 2.

[85] Georgia Historical Society, *1860 Census*, 372; "Biographical Note," Gordon Family Papers, Southern Historical Collection, University of North Carolina.

During the war years after 1861, however, there is no evidence that sailing coasters engaged in any meaningful trade with Savannah. The arrival of the aging schooner *Fort George Packet* in Savannah in the second week of December 1861 seems to have been the last sailing coaster to carry plantation crops into the city. In later years, the coasting trade never rebounded to its pre-war levels, and the meaningful participation by sailing vessels ended with the *Fort George Packet*'s voyage.

# Chapter 9

## The End of an Era

On December 13, 1861, the *Daily Morning News* reported the arrival of the schooner *Fort George Packet* at Savannah with a cargo of 2,030 bushels of rough rice. The schooner's captain, Danish-born Andrew Hanson, brought his forty-one-foot ship from Butler Island, located at the mouth of the Altamaha River and site of one of the great tideland rice plantations. The arrival of the *Fort George Packet* marked a defining moment in the Georgia sailing trade. Savannah newspapers record her as the last sailing vessel to transport a coastal cargo into the city before the trade was entirely shut down by the events of the Civil War.

The Civil War devastated the sailing trade in Georgia. The war destroyed the agricultural infrastructure that had supported the trade for more than sixty years, brought financial ruin to many coastal planters, and, most importantly, ended slavery, the structure that supported the tideland rice and Sea Island cotton economy. Rice cultivation, which had been so important in sustaining the sailing trade after the 1830s, was particularly hard hit. During the war, dikes, canals, trunks, and other features associated with growing rice were destroyed, or they deteriorated through four years of neglect. The end of slavery was especially hard on rice agriculture, which depended heavily on enslaved labor. After the war, the freed people who remained in the coastal area were reluctant to return to the rice fields, and it was difficult to find paid workers willing to undertake the onerous, backbreaking manual labor involved in rice agriculture.[1]

The prosperous Sea Island cotton plantations located on the coastal islands were mostly abandoned after the capture of Port Royal Sound in November 1861, when island residents fled their homes and moved to the mainland. Throughout the war, cotton fields were neglected, and farming equipment was lost, damaged, or destroyed. Shortly after capturing Savannah

[1] Sullivan, *Beautiful Zion*, 136.

in December 1864, General William T. Sherman issued "Special Field Order No. 15," stipulating that all the coastal Sea Islands between Charleston, South Carolina, and Jacksonville, Florida, and an area extending thirty miles inland, were to be "reserved and set apart for the settlement of the negroes now made free by the acts of war and the Proclamation of the President of the United States." This order was partly an effort to punish Southern planters along the rice coast, who had been among the leaders of secession, and partly to serve the practical purpose of providing for the newly freed people who were a burden on the United States military. President Andrew Johnson rescinded Field Order No. 15 in summer 1865, and by the end of 1867, the disputed land had been returned to its prewar owners. The formerly enslaved people did not benefit from Sherman's order. However, for a short time, the order led to confusion over land ownership in the entire region, and the affair may have set back efforts to reestablish agricultural production in the coastal region.[2]

The devastation of the agricultural economy that supported the Georgia coasting trade was accompanied by the decimation of the sailing fleet itself, with many vessels captured, lost, or purposefully sunk during the war. In April 1862, fifteen ships were seized by Confederate authorities and scuttled as obstructions in the lower Savannah River to block federal incursions into the channel. Nine were Georgia coasters that worked in the local trade in the prewar years, and three were Savannah pilot boats. That July, Henry F. Willink, owner of the largest shipyard in Savannah, appraised the fifteen vessels so the owners could be reimbursed. The coasters and pilot boats in Willink's list, the values he assigned, and the ships' owners are presented in Table 9.1.

The first nine vessels listed in Table 9.1 were coasters that sailed in the Georgia trade, some for many years. Eight had sailed into Savannah in 1861, and they represented almost one-third of the entire Georgia coasting fleet that year. These scuttled ships were not the only Georgia coasters lost during the Civil War. The sloop *Science* and the schooner *Fort George Packet* were both sunk as obstructions in the Savannah River in 1863, and several coasters were captured by Union forces. Among these was Charles Stevens's schooner *Northern Belle*. Captain Stevens had renamed his ship *Southern Belle* and hidden her in a creek along the Altamaha River above Darien. In June 1862, the *Southern Belle* was found, seized by men from the federal steamer *Madgie*, and taken to St. Simons Island with a load of rice to feed the formerly enslaved people ("contrabands") then encamped on the island. Apparently, while the *Northern Belle* was docked at or near Captain Stevens's old home at Frederica,

[2] Ibid., 199–200; B. Myers, "Sherman's Field Order No. 15."

her cables parted in a wind, and she was driven ashore. She broke up in the Frederica River.[3]

With their livelihood shut down by the war, some of the coasting captains used their ships in the Confederate cause, such as the schooner *Elias Reed*, which was paid fifty dollars to transport the soldiers of the "St. Marys Volunteers" from the town of St. Marys to Brunswick in April 1862. In early 1862, Charles Stevens used his schooner *Northern Belle* to transport materials to obstruct the inland passage between Brunswick and Darien. Other coasters were used in similar service, and a few became blockade-runners. Among these latter was the schooner *John W. Anderson*, captured by the blockading vessel USS *Dale* off St. Simons Sound on November 18, 1861. The *John W. Anderson*, one of the first blockade-runners to be captured, had been renamed *Mabel* and was sailing under the British flag, purportedly from Havana. However, it was learned that the schooner was sailing out of Nassau bound for St. Catherines Sound. Found aboard the schooner was a cargo consisting of a variety of foodstuffs, as well as blankets, cloth, saddles, bridles, and "1 case pistols (revolvers) and 2 cases cavalry swords." The *John W. Anderson* was taken to Philadelphia as a prize of war. On September 28, 1861, less than a month before her capture, the *John W. Anderson* had received the first Confederate States registration document issued at Savannah. At the time, the schooner was apparently owned by Henry J. Dickerson of Savannah, and her captain may have been William Watson, a former owner who had served as the schooner's master on several coasting voyages made earlier in the year.[4]

Another former Georgia coaster owned by Henry Dickerson, the schooner *Emma Julia*, was taken while attempting to run the blockade. Renamed *Chance*, the *Emma Julia* was captured by blockading forces off Wassaw Sound in June 1862 while trying to make a run to Nassau with a cargo of salt. In November 1863, the schooner *Elias Reed* was captured by the United States Navy ship *Octorara* in the Bahamas with a cargo of Sea Island cotton, turpentine, and rosin. The *Elias Reed* had sailed out of St. Marys bound for Nassau. In March 1864, the blockade vessel *Tioga* captured the sloop *Swallow* attempting to make a run out of the Combahee River for Nassau. This was apparently the same vessel that had sailed in the coasting trade under Captain

[3] "Vessel Papers," File F-22, schooner *Fort George Packet;* and File S-245, sloop *Science*; Pearson, *Charles Stevens*, 55–56.

[4] "Vessel Papers," File E-159, schooner *Elias Reed*; Pearson, *Charles Stevens*, 52; *ORN* ser. 1, 12:345–46; Register No. 1, Port of Savannah, schooner *John W. Anderson*, September 28, 1861.

David Little. On board the *Swallow* were found 180 bales of cotton, 80 barrels of rosin, 25 boxes of tobacco, and "an old set of Confederate colors." Another Georgia coaster captured by the United States Navy was the sloop *Julia Ann*, identified as the *Julia* at the time of her capture off Sapelo Sound in 1864. The *Julia Ann* did not sail into Savannah in 1861, but she worked in the local trade in earlier years.[5]

By the end of the Civil War, at least fifteen of the ships sailing in the Georgia trade in 1861 had been scuttled, lost, or captured. There were possibly others for which no record has been found. Even these fifteen, however, represented half the total number of vessels sailing in 1861 and included the most active ships trading in the years just before the war. All were Southern-owned ships, and their loss, for all intents and purposes, eliminated the coasting fleet sailing out of Savannah.

A number of the Georgia coasting captains became involved with military activities during the Civil War. Their maritime experience and their knowledge of coastal waters made them valuable as pilots and ship captains, for both the Confederacy and the Union. As noted, several of them also became involved in blockade running, and some were casualties of the war. William W. Austin of Savannah, who served as master of the schooners *Albemarle*, *Blooming Youth*, and *Company* before the war, joined the Confederate Navy in July 1861. In June 1863, while serving as pilot on the ironclad ram *Atlanta*, Captain Austin was wounded and taken prisoner by the gunboat *Weehawken* during action in Wassaw Sound. Austin died in September 1866, apparently from the wounds he received during this encounter.[6]

Among the men captured with William Austin on the *Atlanta* was Lewis Wiggins, who served as master of several coasters before the war. Lewis Wiggins's activities during the Civil War were more adventurous than most. He apparently was released soon after his capture aboard the *Atlanta* because in December 1863, he and Savannah dry goods merchant Henry Lathrop purchased a half interest in a sloop named *Buffalo*, reportedly "lately built and of the burthen of about forty tons." Wiggins and Lathrop paid seven thousand dollars for their half interest in the *Buffalo*, historically a very high price for such a small vessel and a reflection of the inflationary prices seen during the

[5] Capture of the *Emma Julia/Chance* in *ORN*, ser. 1, 13:151; capture of the *Elias Reed* in *ORN*, ser. 1, 1:533–34; capture of *Swallow* in *ORN*, ser. 1, 17:669–670; and capture of the *Julia Ann* in *ORN*, ser. 1, 15:543–44.

[6] "Affidavit of Jane Austin, widow of W. W. Austin," Georgia Archives, Georgia's Virtual Vault.

Civil War. The two men apparently purchased the *Buffalo* as a blockade-runner because on February 1, 1864, the sloop was captured by Union forces at Cabbage Bluff near Brunswick with a load of seventy-two bales of cotton. Lewis Wiggins, one of several men captured with the sloop, was carried north to Philadelphia as a prisoner of war. He supposedly escaped from prison in September 1864 and made his way north to Canada and freedom. From there, he took ship to England, where in November he obtained a position as the signal quarter master on board the recently commissioned Confederate raider *Shenandoah*. Wiggins served on the *Shenandoah* until the end of the war, eventually making his way back to Savannah in spring 1866.[7]

John Q. A. Butler, resident of Glynn County, who in 1860 was part owner of the schooner *Florida* with Charles Stevens, was master of the schooner *Lida* when she was captured by the USS *Saint Lawrence* at Port Royal while trying to run the blockade in December 1861. Butler was imprisoned in New York but soon released at the request of F. W. Seward, assistant secretary of state.[8]

Henry J. Dickerson, although he gave up working as a coasting captain long before the Civil War, was an owner of several coasters when the war began. During the war, Dickerson became quite active in Savannah business and military affairs. He was an agent for the Savannah Manufacturing Company, which provided materials such as salt, oil, and turpentine to the Confederate government, and he served on the committee to organize arms-bearing citizens in the city. In 1863, Dickerson purchased the steamer *Jeff Davis* to use as a privateer. The federal Army captured the *Jeff Davis* when Savannah fell in December 1864.[9]

The tragedy wrought by the Civil War on Denmark native Charles Stevens has been noted. In 1862, his schooner *Northern Belle* was captured by a Union raiding party and lost in the Frederica River. In December 1864, Stevens and several others serving picket duty on the Altamaha River were captured by Union forces, carried North, and imprisoned at Fort Delaware on the Delaware River. Captain Stevens died in the prison hospital on February 1, 1865.[10]

[7] Chatham County, Georgia, deed of sale, Joseph Dangaix to Henry Lathrop and Louis Wiggins, December 28, 1863, Deed Book 3W, 94; Culver, "Louis Peter Wiggins," 5–7; *ORN*, ser. 1, 15:260–61.

[8] *OR*, ser. 2, 2:182, 327.

[9] Walker, "Henry James Dickerson."

[10] Pearson, *Charles Stevens*, 55–62.

Charles Stevens had sailed in the Georgia trade since the early 1840s and became one of the most successful and prosperous of the Southern coasting captains, deriving his income from his shipping business and the small cotton plantation he developed at the old colonial town of Frederica on St. Simons Island. His family returned to Frederica after the war, one of a few families to return and reestablish life on the island. They had property and land, but Stevens's shipping business was destroyed, his vessel captured and lost by the federal Navy. His two sons, John and George, were too young to operate and manage the family lands. His widow, Sarah, attempted to profit from her land by leasing it out, but these ventures were generally unsuccessful. After the war, Sarah Stevens derived a small income as postmistress at Frederica, and in the 1890s she was allotted a Confederate widow's pension by the state of Georgia.[11]

It's unsurprising that a few of the Southern coasting captains offered their services to the United States, not the Confederacy. Many of them were foreign born, had relatively little wealth, and owned few if any slaves. These men lacked the intensity of feelings about slavery, states' rights, and secession held by most native-born Southerners. In addition, their livelihood was destroyed early in the war, and many likely blamed their losses on secession. One of the Southern captains who sided with the Union was Savannah resident Peter Collins, a native of Norway, who served as master of the schooner *Mary Ann* before the war. Collins, who had intimate knowledge of the coastal waterways, became a pilot on the USS *Braziliera* for the United States Navy. In October 1864, Peter Collins was killed by Confederate snipers while guiding the steamer *Mary Sanford* up the Satilla River.[12]

Several of the Northern captains who traded along the Southeastern coast also served during the Civil War. Like their Southern counterparts, these men were valuable because of their intimate knowledge of the sailing conditions in the area. John W. Godfrey, a resident of Massachusetts who sailed in the Georgia trade in the 1850s as master of the schooners *Savannah* and *L. N. Godfrey*, became a pilot for the United States Navy. In January 1863, Godfrey piloted the Union fleet up the Ogeechee River in the attack on Fort McAllister. Before the war, John Godfrey had occasionally carried rice from the Ogeechee River, and he would have been familiar with its navigation. Freeman C. Keene, master of the schooner *Paugassett*, inexplicably remained in Georgia after the start of the war, even though he was a native and resident

[11] Ibid, 62–63.
[12] *ORN*, ser. 1, 16:15.

of Rochester, Massachusetts. In September 1863, the United States gunboat *Seneca* encountered Captain Keene and his wife near Darien, where Keene offered his services to the Union Navy.[13]

After the war, few former coastal planters had the capital necessary to finance the reestablishment of rice agriculture, and the number of rice-planters declined significantly. The total rice production in Georgia in 1869 was about 21.5 million pounds, less than half of the 51.7 million pounds produced in 1859. Some planters did continue to cultivate rice, but most faced severe labor problems in an agricultural endeavor that had relied and depended upon enslaved labor. The number of rice planters in Georgia dropped from ninety-six in 1859 to seventy in 1869. Those planters who reestablished rice cultivation tended to grow smaller amounts than in the prewar years. In 1859, thirty-six planters in South Carolina and Georgia each produced a million or more pounds of rice; however, in 1869, only two did. These were Daniel Heyward of Beaufort County, South Carolina, and Thomas C. Arnold, whose rice lands were located along the Ogeechee River in Bryan County, Georgia. Bryan County was the only county in Georgia to show an increase in rice production after the Civil War.[14]

The 1870s saw a gradual increase in rice production in coastal Georgia and South Carolina, but it never reached its pre-war levels. In the 1880s, rice agriculture was in decline in the tidewater region as its cultivation expanded along the lower Mississippi River in Louisiana. By 1889 Louisiana became the leading grower of rice in the United States, producing about 70 percent of the total national crop. In addition, several severe hurricanes in the 1890s dealt devastating blows to Georgia and South Carolina planters, hastening the end of rice cultivation. Rice cultivation along the Atlantic coast essentially ended in the first decade of the twentieth century.[15]

The sailing trade along the Georgia coast was decimated by the Civil War; the agricultural basis for the trade was badly damaged, many of the ships that worked in the trade were sunk or captured, and many of the Georgia captains died in the war or never returned to their old occupation. Despite these severe blows, the local coasting trade out of Savannah did revive after the war. However, the sailing trade never reached the volume of the pre-war years because neither Sea Island cotton nor rice cultivation revived to their

[13] *ORN*, ser. 1, 13:195, 544, and ser. 1, 14:669.

[14] U.S. Secretary of the Interior, *Agriculture of the United States in 1860*, 22–27; Sullivan, *Beautiful Zion*, 135–36.

[15] Sullivan, *Beautiful Zion*, 136.

earlier positions. After the war, steamers dominated the trade as they had begun to do many years earlier. Only a small number of sailing vessels remained in the trade after the Civil War.

A few of the Georgia captains returned to their old trade after the war. Danish-born Stephen Williams, former owner and master of the sloops *B. S. Newcomb* and *B. F. Sherwood* and the schooner *Fort George Packet*, was living near or in the old port town of Sunbury in Liberty County in 1870 working as a "sea captain." Residing in Savannah in 1870 were former captains Lewis Wiggins, David Little, and Andrew Backman, all employed as ship captains or seamen. Captain Wiggins was serving as master of the steamboat *E. D. Morgan*, and David Little was captain of the steamer *Mary Ellen*. In 1885, Lewis Wiggins became port warden of Savannah, giving up his long career on the water aboard both sailing and steamships. In his later years, he moved to live with his daughter in Columbus, Georgia, where he died on August 15, 1902, at the age of eighty-two.[16]

Captain Charley Thompson was sailing as master of the tiny, twenty-one-ton sloop *Fleet* in 1865, the year the newspaper the *National Republican* described him as "one of the oldest coasters in the trade, and we can recommend him to all concerned." Charley Thompson died in Savannah on February 4, 1878, and Andrew Hanson, a friend and fellow coasting captain who sailed with Thompson in the Riceboro-Savannah trade before the war, was appointed administrator of his estate. The entire value of the Thompson estate was only $650.00. Peter Marris, captain of the sloops *Eagle*, *Visitor*, and *Virginia* before the war, was living in Brunswick in 1870, working as a "river pilot." John Louis Grovenstein, who began his career as a coasting captain in 1831 as the twenty-year-old master of the schooner *Flora* and had held ownership in at least seven sloops and schooners, lived in St. Marys after the Civil War. In 1880, forty-nine years after taking command of the *Flora*, John Grovenstein was still involved in the coasting trade, working as a "steamboat captain" just two years before his death.[17]

In 1870, Andrew Hanson, owner and master of the *Fort George Packet* in the years immediately before the Civil War, was residing with his wife and

[16] US Census Bureau, "US Census of Population Chatham County and Liberty County, 1870"; Culver, "Lewis Peter Wiggins," 9.

[17] Enrollment No. 11, Port of Savannah, sloop *Fleet*, November 10, 1865; *National Republican*, December 6, 1865, in R. M. Myers, *Children of Pride*, 1701; Chatham County, Georgia, estate of Charles Thompson, Appraisement, Estate Accounts, File T-173; US Census Bureau, "US Census of Population Glynn County, 1870," and "US Census of Population Camden County, 1880."

two children in Effingham County, just west of Savannah. The census for that year gives his occupation as "pilot-(sea)," revealing he continued his life aboard ships. By 1882, Captain Hanson was living in Savannah, where he is named in several city directories through 1890. These directories indicate Hanson worked as a "pilot," though he is not named as one of the authorized pilots for the port of Savannah during this period. Andrew Hanson apparently died in 1890 or 1891.[18]

Some of the captains left their old occupation after the war. Danish-born William Watson, close friend and countryman of Charles Stevens, moved inland with his wife, Rosetta, and by 1870, was living in Aiken County, South Carolina, near Augusta, Georgia, working as a farmer. John Arnaud, who with the schooner *Cotton Plant* had been one of the most active of all the coasting captains in the years before the Civil War, was living in Savannah in 1870 and working as a clerk at the Upper Rice Mills. He continued working in that position at least through 1879. This was the rice mill operated by the Habersham family, who had employed Captain Arnaud to command their ships for many years.[19]

When Andrew Hanson sailed the schooner *Fort George Packet* into Savannah in the second week of December 1861 with a cargo of rough rice, he had no idea that his voyage marked the end of an era. It does seem appropriate that such a venerable vessel as the *Fort George Packet* should be the last of the coasters to sail into Savannah. Built at Fort George Island, Florida, in 1825, the schooner sailed in the Georgia trade for her entire thirty-six-year career, transporting great quantities of rough rice, Sea Island cotton, and other commodities into Savannah from the plantations and farms along the Southeastern coast.

The events of the Civil War ended the coasting trade as it had existed for over six decades. Many coasting ships were destroyed during the war, and some captains lost their lives while others took up different professions. Only a few returned to their old trade, now dominated by steam vessels. Newspaper shipping lists, official vessel documents, and a small number of journals, account books, and letters left by merchants, planters, and ships' captains are all that remain to tell the story of the Georgia coasting trade. The men who sailed

[18] US Census Bureau, "US Census of Population Effingham County, Georgia, 1870"; Sholes, *Savannah Directories 1882–1890.*

[19] US Census Bureau, "US Census of Population Chatham County, Georgia, and Aiken County, South Carolina, 1870"; Estill, *Savannah Directory*, 13; Sholes, *Savannah Directory 1879*, 128.

the coasters are long gone. None of the vessels remain, except for a few archaeological examples. The commodities most important to the trade, Sea Island cotton and rice, are no longer grown in the region, and the economic system that supported the trade has disappeared. Even the memory of these coasting ships and their activities has faded almost entirely. Today, few residents of Savannah or small towns such as Beaufort, Riceboro, or St. Marys, have any idea of the great importance that the coasting trade once had to their communities. The story told here of these men, their ships, and the cargoes they carried is an effort to bring to life this forgotten piece of Southern maritime history.

# Tables

**Table 4.1. Agricultural Production of Counties Involved in the Georgia Coasting Trade, 1840-1860**

| COUNTY | RICE (pounds) | COTTON[a] (400-lb bales) | SUGAR (1000-lb hogsheads) | MOLASSES (Gallons) |
|---|---|---|---|---|
| **1840** | | | | |
| Colleton S.C. | 5,483,533 | 420,340 | 0 | |
| Beaufort S.C. | 5,629,482 | 1,514,850 | 0 | |
| Chatham Ga. | 6,158,516 | 1,157,106 | 0 | |
| Bryan Ga. | no returns | no returns | no returns | |
| Liberty Ga. | 223,297 | 1,347,424 | 8.5 | |
| McIntosh Ga. | 2,826,203 | 512,877 | 7.3 | |
| Glynn Ga. | 1,937,200 | 2,322,000 | 22 | |
| Camden Ga. | 1,006,440 | 2,639,740 | 20.5 | |
| Nassau Fla. | 315,000 | 66,425 | 0 | |
| St. Johns Fla. | 0 | 0 | 25 | |
| **1850** | | | | |
| Colleton S.C. | 45,308,600 | 6,592 | 33 | 8,520 |
| Beaufort S.C. | 47,230,082 | 12,672 | 20 | 6,621 |
| Chatham Ga. | 19,453,750 | 580 | 0 | 246 |
| Bryan Ga. | 2,409,387 | 536 | 4 | 2,996 |
| Liberty Ga. | 1,892,462 | 1,883 | 24 | 11,640 |
| McIntosh Ga. | 3,122,919 | 520 | 3 | 7,217 |
| Glynn Ga. | 3,820,875 | 1,036 | 71 | 5,766 |
| Camden Ga. | 6,400,949 | 858 | 45 | 7,502 |
| Nassau Fla. | 404,305 | 279 | 44 | 4,964 |
| St. Johns Fla. | 426 | 0 | 290 | 6,325 |
| **1860** | | | | |
| Colleton S.C. | 22,808,981 | 9,731 | 166 | 0 |
| Beaufort S.C. | 18,790,918 | 19,121 | 24 | 6,767 |
| Chatham Ga. | 25,934,160 | 933 | 127 | 4,516 |
| Bryan Ga. | 1,609,676 | 402 | 6 | 4,507 |
| Liberty Ga. | 2,548,382 | 2,405 | 36 | 3,790 |
| McIntosh Ga. | 6,421,100 | 752 | 2 | 4,464 |
| Glynn Ga. | 4,842,755 | 688 | 13 | 3,600 |
| Camden Ga. | 10,330,068 | 630 | 7 | 5,749 |
| Nassau Fla. | 2,400 | 154 | 41 | 5,730 |
| *St. Johns Fla.* | 8,900 | 1 | 30 | 4,572 |

Sources: Data from U.S. Department of State, *Compendium of the Sixth Census*, 204-206; DeBow, "Agricultural Statistics," 378-380; U.S. Secretary of the Interior, *Agriculture of the United States in 1860*, 22-28.

a. The 1840 census presents cotton production in "pounds Gathered" rather than in 400-lb-bales of ginned cotton.

**Table 5.1. Information on Several Georgia Coasting Schooners Contained in 1857-1861 New York Marine Survey of Vessels.**

*Alexander Blue* - Built: Port Jefferson, N.Y./1856; Wood: white oak and chestnut; Fasteners: iron; Hull: full modeled; Other: centerboard, half poop

*Blooming Youth* - Built: Sussex Co., Del./1848; Wood: white oak; Fasteners: iron; Hull: full modeled

*Challenge* - Built: New London, Conn./1854; Wood: white oak and chestnut; Fasteners: iron and copper; Hull: full bodied; Other: centerboard, trunk cabin

*Champion* - Built: Charleston, S.C./1848; Wood: white oak; Fasteners: iron and copper; Hull: full bodied; Other: trunk cabin

*Dusky Sally* - Built: Duxbury, Mass./1834; Wood: white oak; Fasteners: iron; Hull: full bodied; Other: break to deck

*Florida* - Built: Freetown, Mass./1834; Wood: white oak; Fasteners: iron; Hull: medium bodied

*Harriet Lewis* - Built: Newport, R.I./1850; Wood: white oak; Fasteners: iron and copper; Hull: full modeled; Other: centerboard

*John Fraser* - Built: Rochester, Mass./1849; Wood: white oak; Fasteners: iron and copper; Hull: full modeled; Other: break to deck

*L. N. Godfrey* - Built: Brookhaven, N.Y./1854; Wood: white oak and chestnut; Fasteners: iron and copper; Hull: full modeled; Other: centerboard, trunk cabin

*Roswell King* - Built: Rochester, Mass./1837; Wood: white oak; Fasteners: iron and copper; Hull: full modeled

*Satilla* - Built: Brunswick, Ga./1851; Wood: live oak; Fasteners: iron and copper; Hull: full modeled; Other: trunk cabin

*Specie* - Built: Tappen, N.Y./1827; Wood: white oak; Fasteners: iron and copper; Hull: full modeled; Other: centerboard

*Zaidee* - Built: Perth Amboy, N.J./1853; Wood: white oak; Fasteners: iron; Hull: full modeled; Other: centerboard, trunk cabin

Source: Data from Board of Underwriters, *New York Marine Register*, 1857-1858; Board of American Lloyd's, American Lloyds', 1859-1861.

**Table 5.2. State of Build of Vessels Working in the Georgia Coasting Trade, 1800-1861.**

| NORTHERN STATES | NUMBER OF VESSELS |
|---|---|
| MAINE | 2 |
| NEW HAMPSHIRE | 1 |
| MASSACHUSETTS | 87 |
| CONNECTICUT | 45 |
| RHODE ISLAND | 11 |
| TOTAL | 146 |

| MIDDLE ATLANTIC STATES | NUMBER OF VESSELS |
|---|---|
| NEW YORK | 22 |
| PENNSYLVANIA | 3 |
| NEW JERSEY | 9 |
| DELAWARE | 5 |
| MARYLAND | 10 |
| VIRGINIA | 10 |
| NORTH CAROLINA | 13 |
| TOTAL | 72 |

| SOUTHERN STATES | NUMBER OF VESSELS |
|---|---|
| SOUTH CAROLINA | 16 |
| GEORGIA | 23 |
| FLORIDA | 2 |
| TOTAL | 41 |

**Table 6.1. Numbers of Individuals in Coastal Counties Involved in Navigation in 1840.**

| STATE/COUNTY | NAVIGATION OF THE OCEAN | NAVIGATION OF CANALS,LAKES AND RIVERS | TOTAL POPULATION | SLAVE POPULATION |
|---|---|---|---|---|
| South Carolina | | | | |
| Charleston | 292 | 30 | 82661 | 58539 |
| Colleton | 0 | 3 | 25548 | 19246 |
| Beaufort | 6 | 7 | 35794 | 29685 |
| Georgia | | | | |
| Chatham | 201 | 40 | 18808 | 9542 |
| Bryan | 0 | 0 | 3182 | 2407 |
| Liberty | 0 | 2 | 7241 | 5583 |
| McIntosh | 11 | 97 | 5360 | 3923 |
| Glynn | 6 | 8 | 5302 | 4644 |
| Camden | 6 | 45 | 6075 | 4065 |
| Florida | | | | |
| Nassau | 0 | 0 | 1892 | 908 |
| Duval | 6 | 0 | 4156 | 1801 |
| St. Johns | 6 | 0 | 2694 | 888 |

Source: U.S. Department of State, *Compendium of the Sixth Census.*

**Table 6.2. Chatham County Coasting Captains Identified in the 1840 Census.**

| NAME | OCCUPATION | BIRTHPLACE |
|---|---|---|
| William F. Baker | Navigation of the ocean | Unknown |
| Jeremiah Boar | Navigation of canals, etc. | Unknown |
| William Cannett (Canuet) | Navigation of the ocean | South Carolina |
| Sylvanus Church | Navigation of the ocean | Pennsylvania |
| James P. Dent | Navigation of the ocean | Georgia |
| Henry Dickerson | Navigation of the ocean | Maryland |
| Alexander Johnston | Navigation of the ocean | Unknown |
| James Lain (Lane) | Navigation of canals, etc. | Germany |
| John H. Lightbourn | Navigation of canals, etc. | Georgia |
| John Moore | Navigation of canals, etc. | Unknown |
| Joseph Rifile (Rafile) | Navigation of canals, etc. | Italy |
| John Robbins | Navigation of the ocean | Unknown |
| John Russell | Navigation of the ocean | Bahamas ? |
| Nicholas Solewitch (Sallowich) | Navigation of canals, etc. | Italy |
| Noah B. Sisson | Navigation of the ocean | Rhode Island |
| William Smith | Navigation of the ocean | Unknown |
| Charles Thompson | Navigation of the ocean | Germany |
| William Thompson | Navigation of the ocean | Unknown |
| ? Wendell | Navigation of the ocean | Unknown |
| William White | Navigation of the ocean | Connecticut |
| John Williams | Navigation of the ocean | Germany |
| Alexander Wilson | Navigation of canals, etc. | Unknown |
| Joseph Wood | Navigation of canals, etc. | Unknown |

*Source*: U.S. Census Bureau, "U.S. Census of Population, Chatham County, Georgia, 1840."

**Table 6.3. Coasting Captains Residing in Georgia Listed in the 1850 and 1860 Censuses.**

| COUNTY | NAME | OCCUPATION | AGE | PLACE OF BIRTH |
|---|---|---|---|---|
| *Chatham* | | | | |
| 1850 | John Arnowe | coaster | 40 | France |
| | William Crowell | seaman | 45 | Massachusetts |
| | Patrick Doyle | mariner | 30 | Ireland |
| | John L. Grovenstein | coaster | 40 | Georgia |
| | Francis Hernandez | ship carpenter | 44 | Florida |
| | Alexander Johnston | pilot | 27 | Scotland |
| | John H. Lightburn | seaman | 36 | Georgia |
| | Peter Morris | coaster | 30 | Italy |
| | W. Ross Postell | seaman | 34 | Georgia |
| | Joseph Rafile | mariner | 48 | Italy |
| | Daniel Reddick | coaster | 38 | Georgia |
| | John Russell | ship carpenter | 37 | Bahamas |
| | Patrick Ryan | coaster | 25 | Ireland |
| | Peter Thompson | seaman | 30 | Norway |
| | William Watson | coaster | 32 | Germany |
| | John Williams | coaster | 35 | Germany |
| | Richard Worthington | coaster | 25 | England |
| 1860 | John Arnold (Arnow) | captain of vessel | 40 | France |
| | William W. Austin | capt. of schooner | 37 | Dumfries, Scotland |
| | Andrew Backman | capt. coasting vessel | 30 | Sweden |
| | Peter Collins | mariner | 30 | Norway |
| | Patrick Doyle | mariner | 35 | Mayo, Ireland |
| | William Frazier | mariner sloop | 27 | Providence, R.I. |
| | Andrew Hanson (Hansen) | capt. coasting vessel | 35 | Denmark (census states Baden, Germany) |
| | Barron W. Hadley | captain of a sloop | 30 | Rochester, Mass. |
| | Francis Hernandez | pilot | 54 | St. Augustine, Fla. |
| | Alexander Johnston | pilot | 39 | Scotland |
| | Antonio Lawrence | mariner | 30 | Fiat, Portugal |
| | William R. Postell | mariner | 45 | Beaufort Dist., S.C. |
| | James Smith | mariner sloop | 36 | New York City |
| | Charles Thompson | capt. river sloop | 30 | Hanover, Germany |
| | William Watson | mariner | 41 | Denmark |
| | Stephen Williams | mariner | 32 | Denmark |
| | Richard Wurthington | capt. coasting vessel | 40 | London, England |
| *Liberty* | | | | |
| 1850 | Thomas Snow | sea farer | 35 | Sweden |
| *McIntosh* | | | | |
| 1850 | George Russell | pilot | 38 | Virginia |
| *Glynn* | | | | |
| 1850 | John D. Denson | seaman | 34 | Massachusetts |
| | Richard Owens | coasting captain | 35 | Wales |
| | Charles Stevens | coasting captain | 36 | Denmark |
| 1860 | John D. Denson | seaman | 50 | Massachusetts |
| | John Q. A. Butler | mariner | 30 | Massachusetts |
| | Charles S. Stevens | mariner | 45 | Denmark |

| | | | | |
|---|---|---|---|---|
| *Camden* | | | | |
| 1850 | Samuel Brockington | farmer | 51 | South Carolina |
| | Sylvanus Church | pilot | 46 | Pennsylvania |
| | William Laen | capt. sloop | 43 | England |
| | Dennis Pacetty | capt. sloop | 40 | Florida |
| | Edmund Richardson | seaman | 72 | Massachusetts |
| 1860 | John L. Grovenstein | sea captain | 48 | St. Marys, Ga. |

*Source*: Data from U.S. Census Bureau, "U.S. Censuses of Population, Georgia, 1850 and 1860."

**Table 6.4. Mariners on Coasting Vessels Listed in the 1860 Chatham County Census.**

| NAME | AGE | RACE[a] | OCCUPATION | PLACE OF ORIGIN |
|---|---|---|---|---|
| George Armstrong | 22 | B | Mariner coasting vessel | Chatham Co., Ga. |
| Antonio Brassos | 27 | W | Mariner sloop | St. Augustine, Fla. |
| Elijah Cooper | 33 | M | Mariner coasting vessel | State of N.Y. |
| Edward Frazier | 22 | M | Mariner coasting vessel | New York City |
| William Frazier | 27 | W | Mariner sloop | Providence, R.I. |
| Thomas Murphy | 27 | W | Mariner coasting vessel | Baltimore, Md. |
| William Pollard | 21 | W | Mariner coasting vessel | Lancashire, Eng. |
| George Robertson | 30 | W | Mariner coasting vessel | Derbyshire, Eng. |
| Joseph Robertson | 28 | W | Mariner coasting vessel | Derbyshire, Eng. |
| James Robinson | 31 | B | Mariner coasting vessel | Philadelphia, Penn. |
| James Smith | 36 | W | Mariner sloop | New York City |
| Edward Sullivan | 30 | W | Mariner coasting vessel | Liverpool, Eng. |
| George Toombs | 26 | W | Mariner coasting vessel | New York City |

*Source*: Information from Georgia Historical Society, 1860 *Census*.
a. B = Black, M = Mulatto, W = White

**Table 6.5. Wealth Information on Georgia Coasting Captains and Savannah Merchants from the 1860 Census.**

| RESIDENCE/NAME | VALUE OF REAL ESTATE | VALUE OF PERSONAL ESTATE |
|---|---|---|
| SOUTHERN CAPTAINS | | |
| Georgia | | |
| John Arnold (Arnow) | | $300.00 |
| William W. Austin | | $500.00 |
| Andrew Backman | | $500.00 |
| Antonio Brassos | | $100.00 |
| John Q.A. Butler | $650.00 | $100.00 |
| Peter Collins | | |
| John D. Denson | | |
| William Dixon | | $2,400.00 |
| Patrick Doyle | | $200.00 |
| William Frazier | | |
| John L. Grovenstine | $1,500.00 | $1,130.00 |
| Barron W. Hadley | | $1,000.00 |
| Andrew Hanson | | $200.00 |
| Christy Holverson | $1,000.00 | $1,000.00 |
| Antonio Lawrence | | $150.00 |
| William Phillips | | $200.00 |
| William R. Postell | | $3,000.00 |
| Charles Stevens | $2,500.00 | $11,250.00 |
| Charles Thompson | | $250.00 |
| William Watson | $3,000.00 | $2,000.00 |
| Stephen Williams | | $50.00 |
| Richard Wurthington | | $1,000.00 |
| Florida | | |
| Dennis Pacetty | $1,600.00 | $500.00 |
| NORTHERN CAPTAINS | | |
| Massachusetts | | |
| Pelag Blankenship | $2,500.00 | $10,000.00 |
| Leonard Bolles | $1,000.00 | |
| Clark Delano | $1,000.00 | $2,000.00 |
| Henry D. Delano | $5,000.00 | $14,000.00 |
| Franklin Hathaway | $1,600.00 | $2,700.00 |
| Hiram Look | $600.00 | $200.00 |
| William Mendall | $1,200.00 | $2,500.00 |
| James Pitcher | $4,000.00 | $4,000.00 |
| Stephen Hadley | $6,000.00 | $15,000.00 |
| Luther Cudworth | $600.00 | $2,400.00 |
| Jabez Delano | $1,500.00 | $50.00 |
| Henry D. Chase | $1,000.00 | |
| Luther Cudworth | $600.00 | $2,400.00 |
| Connecticut | | |
| Philander P. Coe | $4,000.00 | $150.00 |
| SAVANNAH SHIP OWNERS AND MERCHANTS | | |
| J. W. Anderson | $34,000.00 | $70,000.00 |
| Henry Brigham | $32,000.00 | $35,000.00 |

| | | |
|---|---|---|
| W. H. Burroughs | $10,000.00 | $30,000.00 |
| Henry J. Dickerson | $75,000.00 | $26,500.00 |
| John C. Frazer | | $1,500.00 |
| John Gammell | | $9,000.00 |
| Robert Habersham | $100,000.00 | $150,000.00 |
| William N. Habersham | $15,000.00 | $50,000.00 |
| Noble A. Hardee | $50,000.00 | $100,000.00 |
| Andrew Low | $54,000.00 | $40,000.00 |
| William H. Tison | $75,000.00 | $52,000.00 |
| John L. Villalonga | $4,000.00 | $100,000.00 |
| Henry F. Willink | $30,000.00 | $29,000.00 |

*Source*: Data from Data from U.S. Census Bureau, "U.S. Censuses of Population, Connecticut, Georgia, Florida, and Massachusetts, 1860."

**Table 7.1. Arrivals of Coasters into Savannah in 1800 from Local Ports, Excluding Charleston.**

| DATE | VESSEL | MASTER | ORIGIN |
|---|---|---|---|
| Jan 2 | Slp. *Patty* | Landon | Brunswick |
| Jan 16 | Slp. *Eliza* | Dexter | Brunswick |
| Apr 10 | Sch. *Lady Washington* | Burrell | St. Augustine |
| Apr 24 | Sch. *Alective* | Rudolph | St. Marys |
| May 8 | Sch. *Alective* | Chevalier | St. Marys |
| May 22 | Sch. *Alective* | Rudolph | St. Marys |
| May 22 | Sch. *Florida* | Segar | St. Augustine |
| Jun 5 | Slp. *Harriet* | Lindgren | Beaufort |
| Jun 12 | Sch. *Alective* | Chevalier | St. Marys |
| Jun 19 | Sch. *Experience* | Burrell | St. Augustine |
| Jun 19 | Slp. *Susannah* | Waterman | St. Marys |
| Jul 4 | Sch. *St. Aclare* | Aurnoue | St. Augustine |
| Jul 29 | Sch. *Alective* | Rudolph | St. Marys |
| Aug 12 | Sch. *John* | Burr | St. Augustine |
| Sep 12 | Sch. *Fox* | Wilson | Beaufort |
| Sep 26 | Sch. *Fox* | Pierce | St. Marys |
| Sep 30 | Sch. *Jolly Rover* | Records | St. Marys |
| Sep 30 | Sch. *Alective* | Rudolph | St. Marys |
| Jul 3 | Sch. *Friendship* | Brewster | St. Marys |
| Dec 11 | Sch. *Kitty* | Bennett | St. Augustine |

*Source*: Data from the Savannah newspapers *Georgia Gazette* and *Columbian Museum & Savannah Advertiser* for the year 1800.

**Table 7.2. Departures of Coasters from Savannah to Local Ports in 1800, Excluding Charleston.**

| DATE | VESSEL | MASTER | DESTINATION |
|---|---|---|---|
| Jan 9 | Sch. *Sally* | Chalker | Sunbury |
| Jan 16 | Sch. *Alective* | Rudolph | St. Marys |
| Jan 23 | Sch. *William and Sally* | Ellis | Brunswick |
| Feb 6 | Sch. *Abigail* | Chadwick | St. Augustine |
| Feb 6 | Sch. *Alective* | Rudolph | St. Marys |
| Feb 20 | Slp. *Three Sisters* | Reynolds | Brunswick |
| Feb 27 | Sch. *Alective* | Chevalier | St. Marys |
| Mar 6 | Slp. *William and Sally* | Bragg | Brunswick |
| Mar 6 | Slp. *Feronia* | Latham | St. Marys |
| Mar 6 | Slp. *Three Sisters* | O'Bryan | Brunswick |
| Mar 6 | Sch. *Regulator* | Williams | St. Marys |
| Apr 3 | Slp. *Harriet* | Lindgren | Beaufort |
| Apr 10 | Sch. *Eliza* | Whitfield | St. Marys |
| Apr 17 | Sch. *Nabby* | Cobb | St. Marys |
| Apr 24 | Sch. *Friendship* | Brewster | St. Marys |
| May 1 | Slp. *Friendship* | Guillenet | St. Marys |
| May 1 | Sch. *Sally* | Hatch | Brunswick |
| May 1 | Sch. *Alective* | Rudolph | St. Marys |
| May 15 | Sch. *Alective* | Chevalier | St. Marys |
| May 15 | Sch. *Lady Washington* | Pearson | St. Augustine |
| May 22 | Slp. *Harriet* | Lindgren | Beaufort |
| May 22 | Sch. *Alective* | Rudolph | St. Marys |
| May 29 | Slp. *Cinderilla* | Phipps | Beaufort |
| Jun 5 | Sch. *Florida* | Segar | St. Augustine |
| Jun 12 | Slp. *Three Sisters* | O'Bryan | St. Simons |
| Jul 10 | Sch. *Alective* | Rudolph | St. Marys |
| Jul 17 | Slp. *Friendship* | Guillenet | St. Marys |
| Jul 17 | Sch. *Alective* | Rudolph | St. Marys |
| Jul 31 | Sch. *Alective* | Rudolph | St. Marys |
| Aug 7 | Sch. *Fox* | Wilson | Beaufort |
| Aug 21 | Sch. *Alective* | Rudolph | St. Marys |
| Sep 4 | Sch. *Alective* | Rudolph | St. Marys |
| Sep 18 | Sch. *Kitty* | Harding | St. Augustine |
| Sep 25 | Slp. *Friendship* | Hillary | Sunbury |
| Oct 2 | Sch. *Alective* | Chevalier | St. Marys |
| Oct 2 | Slp. *Mary* | Joad | St. Simons |
| Oct 30 | Sch. *Alective* | Rudolph | St. Marys |
| Nov 6 | Slp. *William* | Lightbourn | Hardwick |
| Nov 13 | Slp. *Two Friends* | Bailey | Beaufort |
| Nov 13 | Slp. *Betsey* | Hoadley | St. Marys |
| Nov 20 | Sch. *Alective* | Rudolph | St. Marys |
| Dec 4 | Slp. *Patty* | Chalkery | St. Marys |
| Dec 4 | Sch. *Alective* | Rudolph | St. Marys |
| Dec 18 | Sch. *Kitty* | Wood | St. Augustine |
| Dec 25 | Slp. *Friendship* | Guillenet | St. Marys |
| *Dec 25* | *Sch. Alective* | *Rudolph* | *St. Marys* |

*Source*: Data from the Savannah newspapers *Georgia Gazette* and *Columbian Museum & Savannah Advertiser* for the year 1800.

**Table 7.3. Georgia Coaster Arrivals at and Departures from Savannah in 1805 (Charleston is Excluded).**

| DATE | VESSEL | MASTER | ORIGIN/DESTINATION |
|---|---|---|---|
| *Arrivals* | | | |
| Jan 10 | Slp. *Eagle* | Suarez | St. Augustine |
| Jan 13 | Slp. *Ranger* | Davis | St. Augustine |
| Feb 18 | Sch. *Patsey* | Bacon | St. Augustine |
| Apr 1 | Slp. *Eliza* | Allen | Sunbury |
| May 28 | Sch. *Two Friends* | Davidson | Beaufort |
| Jun 14 | Slp. *Cinderilla* | Genett | Beaufort |
| Jun 14 | Slp. *Conception* | Gomez | St. Augustine |
| Jul 9 | Sch. *Rainbow* | Smith | Beaufort |
| Oct 29 | Slp. *Polly* | Chevalier | St. Marys |
| Nov 5 | Slp. *Polly* | Lightbourn | Hardwick |
| *Departures* | | | |
| Jan 7 | Slp. *Eliza* | Allen | Sunbury |
| Jan 10 | Slp. *Cinderilla* | Genett | Beaufort |
| Feb 23 | Slp. *Eliza* | Denning | St. Marys |
| Apr 1 | Slp. *Experiment* | Rackins | Darien |
| Apr 14 | Slp. *Franklin* | Grants | Beaufort |
| Jun 25 | Slp. *Polly* | Lightbourn | Hardwick |
| Oct 1 | Sch. *George* | Fowler | St. Marys |
| Oct 11 | Slp. *Polly* | Chevalier | St. Marys |
| Oct 29 | Sch. *Three Sisters* | Beard | St. Marys |
| Nov 6 | Slp. *Polly* | Chevalier | St. Marys |
| Nov 8 | Slp. *Lucy* | Stanton | Sapelo |
| Dec 20 | Slp. *Fairhaven* | Toy | St. Marys |
| Dec 20 | Slp. *Ruth* | Philips | St. Marys |
| Dec 20 | Sch. *Susan* | Allen | St. Augustine |

*Source*: Data from the Savannah newspaper *Georgia Republican & State Intelligencer* for the year 1805.

**Table 7.4. Cargoes Carried by Coasters into Savannah from Charleston, January 1810 to April 1810.**

| DATE | VESSEL | MASTER | CONSIGNEES | CARGO |
|---|---|---|---|---|
| Jan 11 | Slp. *Delight* | Cooper | Dr. Harra;<br>Wm. Defaure | glassware and flour |
| Jan 16 | Sch. *Three Friends* | Burrows | master | flour, coal,<br>tobacco, and gin |
| Jan 20 | Sch. *Hope* | Kehr | (none listed) | ballast and passengers |
| Jan 23 | Slp. *Halcyon* | Leslie | Wm Woodbridge;<br>Dufore & Maher | flour, gin, and rum |
| Jan 23 | Slp. *Republican* | Brown<br>P | Marquand<br>alding & Co. | molasses and<br>rum |
| Feb 8 | Slp. *Columbia* | Beecher | J. Meigs; master | salt and tobacco |
| Feb 13 | Slp. *Alpha* | Ward | master | wine |
| Mar 6 | Sch. *Enterprise* | Williams | master | ballast |
| Mar 6 | Slp. *Republican* | Myers | (none listed) | bricks |
| Mar 6 | Slp. *Columbia* | Beecher | J. Reid | bricks |
| Mar 6 | Slp. *Halcyon* | Leslie | R. Richardson;<br>T. V. Gray;<br>Williford & Stevens | cotton bagging,<br>coffee, and bacon |
| Mar 17 | Slp. *Delight* | Cooper | John Henry | sugar, iron, and wine |
| Mar 17 | Slp. *Republican* | Brown | master | (none listed) |
| Apr 3 | Slp. *Delight* | Cooper | Minis & Henry;<br>Steinert & Co;<br>Knox & Pope | rope, sugar, lemons,<br>and merchandise |
| Apr 5 | Slp. *Republican* | Brown | Williford & Stevens;<br>master | bricks, sugar,<br>and coffee |

*Source*: Data from the Savannah newspaper *Republican & Savannah Evening Ledger*, January to April, 1810.

**Table 7.5. Cargoes Carried by Coasters into Savannah from Local Ports, January 1810 to April 1810 (Charleston is Excluded).**

| DATE | VESSEL | MASTER | ORIGIN | CONSIGNEES | CARGO |
|---|---|---|---|---|---|
| Jan 9 | Sch. Delancey | Blythewood | Beaufort | A. Low & Co.; Low & Wallace; S. Barrett | cotton |
| Jan 16 | Sch. *Traveller* | Saunders | Amelia | (none listed) | ballast |
| Jan 16 | Slp. *Concord* | Walter | Amelia | (none listed) | ballast |
| Jan 16 | Sch. Maria | Sola | St. Augustine | Small & M'Nish | oranges and lime juice |
| Jan 16 | Slp. *Happy Return* | Baker | St. Marys | master | nails and cotton bagging |
| Jan 16 | Slp. *Walcott* | Kirby | St. Marys | master | salt |
| Jan 16 | Slp. *Alpha* | Ward | St. Marys | (none listed) | ballast |
| Jan 18 | Slp. *Eliza* | Baker | St. Marys | Small & M'Nish; J & J. Carruthers | cotton bagging, salt, and negroes |
| Jan 18 | Sch. *Renown* | Allen | Amelia | master | ballast |
| Jan 18 | Slp. *Concord* | Dennison | Amelia | (none listed) | (none listed) |
| Jan 20 | Sch. *Thomas* | Irelan | Amelia | master | ballast |
| Jan 23 | Slp. *Polly* | Fowler | St. Marys | C & H Fistner; master | salt, cheese, and bagging |
| Jan 23 | Sch. *Kitty Ann* | Slater | Amelia | master | 3000 bushels salt |
| Jan 27 | Sch. Young Sea-Horse | Hoadley | Amelia | master | ballast |
| Jan 27 | Slp. *Diana* | Ceasar | Amelia | Peter Mitchell | ballast |
| Jan 27 | Slp. *George* | Wood | Amelia | master | ballast |
| Jan 30 | Sch. *Fanny & Maria* | Scail | Amelia | (none listed) | return cargo |
| Jan 30 | Slp. *Rising Sun* | Percy | Amelia | master | ballast |
| Jan 30 | Slp. *Roxana* | Homer | Amelia | master | ballast |
| Jan 30 | Slp. *President* | *Swain* | *Amelia* | *master* | *ballast* |
| Jan 30 | Slp. *Diana* | Ceasar | Amelia | master | ballast |
| Feb 8 | Slp. *Thomas* | White | Sunbury | (none listed) | rice and cotton |
| Feb 13 | Sch. *Renown* | Allen | Amelia | master | ballast and passengers |
| Feb 13 | Sch. Carlton | Johnson | Amelia | master | ballast and passengers |

| DATE | VESSEL | MASTER | ORIGIN | CONSIGNEES | CARGO |
|---|---|---|---|---|---|
| Feb 20 | Sch. *Maria* | Sola | Amelia Is. | Small & M'Nish | oranges |
| Feb 20 | Sch. *Commerce* | Foster | Amelia Is. | master | ballast |
| Feb 20 | Slp. Thomas | White | Hardwick | Joseph Habersham | rice |
| Feb 24 | Sch. *Nancy White* | Newall | Amelia | R. Richardson & Co. | ballast |
| Mar 15 | Slp. *Concord* | Walter | Amelia | master | ballast |
| Mar 20 | Sch. *Rambler* | Sayer | Amelia | master | ballast |
| Mar 20 | Slp. *Patty* | Brooks | Amelia | master | ballast |
| Mar 20 | Slp. *Liberty* | Wing | Darien | Benjamin Manrice; George Scott | rice and cotton |
| Mar 27 | Slp. *Reformation* | Delano | Amelia | master | ballast |
| Mar 27 | Slp. *Packet Handy* | Amelia | Ogden & Bates | lime juice | |
| Apr 3 | Slp. *Joy* | Ray | Amelia | Sturges, Burroughs & Butler | salt and merchandise |

Source: Data from the Savannah newspaper *Republican & Savannah Evening Ledger*, January to April, 1810.

**Table 7.6. Coasting Masters Making Two or More Arrivals into Savannah in 1820.**

| MASTER | VESSEL | ARRIVALS |
|---|---|---|
| Ephraim Allen | Slp. *Hesper* | 4 |
| Joseph Allen | Slp. *Good Intent* | 3 |
| John Anderson | Slp. *Governor Shelby* | 6 |
| James Blankenship | Slp. *Howard & James* | 2 |
| Leonard or Savory Bolles | Slps. *Harriet, Mercy* | 7 |
| Allen Chase | Slp. *Rosetta* | 7 |
| John Chevalier | Slp. *John Chevalier* | 6 |
| Church | Sch. *Harmony* | 2 |
| Thomas Clark | Slp. *Joseph* | 3 |
| Clark | Slp. *Excellent* | 2 |
| Crawford | Slp. *Lion* | 3 |
| Alvin Drayton | Sch. *Little Ann* | 3 |
| Curtis | Sch. *Three Sisters* | 3 |
| Delphry | Slp. *Suffolk* | 2 |
| James Deverger | Slp. *Favorite* | 6 |
| R. (?) Drinkwater | Slp. *Neptune* | 3 |
| William Egery | Slps. *Caroline, Ricebird* | 4 |
| John Fowler | Slp. *Frances* | 2 |
| Garrett | Slp. *Sally* | 2 |
| Robert Gibbs | Slp. *Spartan* | 4 |
| Israel P. Hand | Slp. *Traveller* | 2 |
| Samuel (?) Harrison | Slps. *Harriet, Mariner* | 2 |
| Joseph Howland | Slp. *Rosetta* | 6 |
| Isaacs | Slps. *Lydia, Mary* | 2 |
| Francis Jason | Sch. *Cheves* | 2 |
| Jones | Slp. *Washington* | 4 |
| Lee | Sch. *Choctaw* | 2 |
| George A. Luce | Slp. *Ann* | 3 |
| Stephen (?) Luce | Slps. *Suffolk, Support* | 2 |
| Richard Marcellin | Sch. *Mary McKay* | 3 |
| Mitchell | Slp. *Mary* | 2 |
| Barnabas Nye | Slp. *Morning Star* | 3 |
| David Pidge | Slp. *Hermit* | 2 |
| Isaac Pierce | Sch. *Isabella* | 3 |
| Charles Porquet | Slp. *Maria* | 7 |
| Redding | Sch. *Mary* | 2 |
| Edmund Richardson | Slps. *Dosoris, Providence*; Sch. *Mary McKay* | 10 |
| Risely | Slp. *Diana* | 2 |
| Nicholas Sallowich | Slp. *Union* | 11 |
| Prince Snow | Slps. *Leopard, Support* | 3 |
| John W. Turner | Slp. *Lawrence* | 3 |
| James Vincent | Slp. *James* | 3 |
| Joseph Woodward | Slp. *Bridgeport* | 3 |
| Woodward | Slp. *Atlanta* | 3 |
| James Worth | Slps. Favorite, Marion | 3 |

*Source*: Data from the Savannah newspaper *Daily Georgian*, 1820.

**Table 7.7. Savannah Merchants Receiving Two or More Shipments of Coastal Commodities in 1820.**

| CONSIGNEE | NUMBER OF SHIPMENTS | COMMODITIES RECEIVED |
|---|---|---|
| J. McNish | 27 | cotton, some rice and molasses |
| R. & J. Habersham | 22 | cotton and rice |
| A. B. Fannin & Co. | 13 | cotton |
| John Lewis (J. Lewis) | 13 | cotton and rice |
| J. S. Pelot | 12 | cotton, some rice |
| Th. Butler | 12 | cotton |
| Bullock & Sadler (Saddler) | 10 | cotton and rice, some molasses |
| Miller & Fort (Fort & Miller) | 10 | cotton |
| R. Richardson & Co. | 10 | cotton, some rice and molasses |
| Andrew Low & Co. | 9 | cotton, some rice |
| J. P. Williamson | 9 | cotton, some rice and molasses |
| Scott and Fannin | 7 | cotton |
| Baker & Welman | 6 | cotton |
| William Jenner | 5 | cotton, some rice |
| Carnochan & Mitchell | 4 | cotton |
| J. S. Bulloch | 4 | cotton |
| Johnston & Hills | 4 | cotton, some rice |
| Shaffer & Davenport | 4 | cotton |
| Francis H. Welman | 2 | cotton |
| Gaudry & Dufaure | 2 | cotton and oranges |
| Greene & Lippitt | 2 | cotton |
| Guerard & Polhill | 2 | cotton |
| Hall & Hoyt | 2 | cotton |
| Lawrence & Thompson | 2 | cotton |
| Major Napier/T. Napier | 2 | cotton |
| Meigs & Reed | 2 | cotton |
| Sc. Slaughter | 2 | cotton |
| Th. Stubbs | 2 | cotton |

*Source*: Data from the Savannah newspaper *Daily Georgian*, 1820.

**Table 7.8. Shipments of Various Commodities Carried by Sailing Coasters into Savannah in 1830.**

| COMMODITY | NUMBER OF SHIPMENTS | MINIMUM AMOUNTS |
|---|---|---|
| Rough rice | 74 | 6,040,080 pounds |
| Clean rice | 50 | 2,529,600 pounds |
| Sea Island cotton | 135 | 2,107,788 pounds |
| Upland cotton | 167 | 11,955,680 pounds |
| Oranges | 29 | 33,000 |
| Orange trees | 1 | |
| Lemons | 2 | |
| Hides and skins | 41 | |
| Otter skins | 2 | |
| Deer skins | 2 | |
| Syrup | 11 | 57 barrels, 46 hogsheads |
| Sugar | 7 | 12 barrels, 71 hogsheads |
| Molasses | 2 | 17 hogsheads |
| Live Oak | 10 | |
| Beeswax | 2 | |
| Tallow | 2 | |
| Wood | 3 | |
| Corn | 2 | |
| Cane | 1 | |
| Tar and rosin | 1 | |
| Potatoes | 1 | |
| Groundnuts | 1 | |

*Source*: Data from the Savannah newspaper *Daily Savannah Republican*, 1830.

**Table 8.1. Coasting Masters Making Two or More Arrivals into Savannah in 1839.**

| MASTER | VESSEL | ARRIVALS |
|---|---|---|
| William (?) Baker | Slps. *Science, Jackson* | 2 |
| Leonard Bolles | Slp. *Eagle* | 8 |
| Paul Briggs | Slp. *Angel* | 2 |
| Nathan (?) Briggs | Slp. *Eagle*; Sch. *Florida* | 3 |
| Daniel C. Brown | Slp. *William Wray* | 4 |
| Leonard Burr | Slp. *America* | 5 |
| Caswell | Slp. *William* | 3 |
| Francis Chevalier | Slp. *Virginia* | 9 |
| John L. Grovenstein | Slp. *Macon* | 7 |
| Leonard (?) Hammond | Slps. *Georgia, Stranger* | 2 |
| W. Hathaway | Slp. *Sapelo* | 5 |
| James (?) Ketchum | Slp. *Pearl* | 2 |
| James D. Lane | Sch. *Orange* | 2 |
| Lawrence | Sch. *Savannah* | 2 |
| James Lee | Slp. *Science* | 2 |
| John Lightbourn | Sch. *Vesta* | 3 |
| Elisha Luce | Slp. *William* | 3 |
| Stephen or Rowland Luce | Slp. *Georgia* | 2 |
| Stephen or Rowland Luce | Slp. *Stranger* | 2 |
| Allen (?) Payne | Sch. *Canton* | 2 |
| Edmund Richardson | Slps. *Bolivar, Macon* | 2 |
| James Ridley | Sch. *Hamilton* | 8 |
| Rodgers | Sch. *Cornelia* | 3 |
| Timothy Savory | Sch. *Florida* | 3 |
| Sherman/Shearman | Slp. *Georgia* | 4 |
| Walter Smith | Slp. *George Washington* | 4 |
| William (?) Taylor | Slp. *Argo* | 4 |

*Source*: Data from the Savannah newspaper *Daily Georgian*, 1839.

**Table 8.2. Shipments of Various Commodities Carried by Sailing Coasters into Savannah in 1839.**

| COMMODITY | NUMBER OF SHIPMENTS INTO SAVANNAH | MINIMUM AMOUNTS |
|---|---|---|
| Rough rice | 25 | 1,876,500 pounds |
| Clean rice | 11 | 578,400 pounds |
| Sea Island cotton | 54 | 1,337,000 pounds |
| Upland cotton | 31 | 2,585,450 pounds |
| Hides | 40 | |
| Syrup | 1 | 8 barrels |
| Beeswax | 1 | |
| Wood | 5 | |
| Corn | 2 | |

*Source*: Data from the Savannah newspaper *Daily Georgian*, 1839.

**Table 8.3. Ship Masters Sailing in the Georgia Coasting Trade in 1850.**

| NAME | VESSELS | RESIDENCE | PLACE OF ORIGIN |
|---|---|---|---|
| John Arnaud/Arnow | Sch. *Cotton Plant* | Georgia | France |
| Barillo | Slp. *Eutaw* | unknown | unknown |
| James Burr | Sch. *Company* | Massachusetts | Massachusetts |
| Peter Collins | Sch *Levant*, Slp. *Eagle* | Georgia | Norway |
| Luther Cudworth | Sch. *Harriett Lewis* | Massachusetts | Massachusetts |
| J. Dexter | Sch. *David Belknap* | unknown | Massachusetts (?) |
| Patrick Doyle | Sch. *American Coin* | Georgia | Ireland |
| Domingo Galleo | Slp. *Visitor* | Georgia | Italy |
| John W. Godfrey | Sch. *Savannah* | Massachusetts | Massachusetts |
| John L. Grovenstein | Slp. *Eagle*, Slp. *B.S. Newcomb* | Georgia | Georgia |
| Heald (J. S. Heald ?) | Slp. *Liberty* | Georgia (?) | unknown |
| Holmes | Sch. *Elias Reed* | unknown | unknown |
| Johnson (two persons) | Sch. *Edmund & Francis*, Sch. *Maria* | unknown | unknown |
| William Laen/Lane | Slp. *Virginia* | Georgia | England |
| Loose | Sch. *Albemarle* | Georgia (?) | unknown |
| Millen/Miller | Sch. Albemarle | Georgia (?) | unknown |
| Peter Maris/Morris | Sch. *Levant*, Slp. *Eagle* | Georgia | Italy |
| Richard Owens | Sch. *Elias Reed* | Georgia | Wales |
| Reed | Sch. *Florida* | unknown | unknown |
| Ret (?) | Sch. Morning Star | unknown | unknown |
| Thomas Room/ Roome | Sch. *Fort George Packet*, Slp. *B.S. Newcomb* | Georgia (?) | unknown |
| George Russell | Sch. *Levant* | Georgia | Virginia |
| Simon Santina/ Santini | Sch. *Levant*, Slp. *Washington* | Georgia | France |
| Thomas Snow | Sch. *Sarah* | Georgia | Bartholomew Island, W.I. |
| Charles Stevens | Slps. *America*, *Splendid* | Georgia | Denmark |
| Charles Thompson | Sch. *Northern Belle*, Slp. *Science* | Georgia | Germany |
| Peter Thompson (?) | Slp. *Young Eagle* | Georgia | Norway |
| Thompson (possibly 2 persons serving as master of Schs. *Albemarle*, *Levant*, *Ft. George Packet* and Slps. *Eagle*, *Eutaw* and *Virginia*) | | | |
| William Watson | Sch. *William D. Jenkins* | Georgia | Denmark |
| Lewis Wiggins | Slp. *Catherine Chard* | Georgia | Russia |
| Stephen Williams | Slp. *Julia Ann* | Georgia | Denmark |
| Williams | Sch. *Young Eagle* | unknown | unknown |
| Alexander Wilson (?) | Slp. *Washington* | Georgia | unknown |
| Richard Worthington | Slp. *Splendid* | Georgia | England |

*Source*: Data from the Savannah newspaper *Daily Georgian*, 1850.

**Table 8.4. Shipments of Various Commodities Carried by Sailing Coasters into Savannah in 1850.**

| COMMODITY | NUMBER OF SHIPMENTS | MINIMUM AMOUNTS |
|---|---|---|
| Rough rice | 157 | 17,866,845 pounds |
| Clean rice | 9 | 398,700 pounds |
| Sea Island cotton | 81 | 1,492,750 pounds |
| Upland cotton | 5 | 122,850 pounds |
| Turpentine | 10 | 970 barrels |
| Spirits of turpentine | 17 | 588 barrels |
| Rosin | 22 | 2,240 barrels |
| Sugar | 2 | 20 barrels, 3 casks |
| Syrup | 3 | 79 barrels |
| Hides | 6 | 407 hides |
| Deer skins | 1 | |
| Beeswax | 2 | 3 tierces, 2 casks |
| Wood | 6 | |
| Staves | 2 | 100 staves |
| Lumber | 1 | 33,000 feet |
| Cedar posts | 1 | 424 pieces |
| Corn | 10 | 4,410 bushels |
| Peas | 1 | |
| Wool | 1 | 2 bales |
| Groundnuts (peanuts) | 2 | 200 bushels |
| Potatoes | 1 | |

Source: Data from the Savannah newspaper *Daily Georgian*, 1850.

**Table 8.5. Ship Masters Sailing in the Georgia Coasting Trade in 1861.**

| NAME | VESSEL | RESIDENCE | PLACE OF ORIGIN |
|---|---|---|---|
| John Arnaud/Arnow | Sch. *Cotton Plant* | Georgia | France |
| William Aston | Sch. *John W. Anderson* | Georgia | England |
| William H. Austin | Sch. *Blooming Youth* | Georgia | Scotland |
| Andrew Backman | Slps. *Science, Swallow* | Georgia | Sweden |
| Barett | Slp. *Virginia* | unknown | unknown |
| Leonard Bolles | Sch. *Benjamin English* | Massachusetts | Massachusetts |
| Burr | Sch. *Benjamin English,* Sch. *Mary A. Rowland* | Massachusetts | Massachusetts |
| Butler | Sch. *Emma Julia* | unknown | unknown |
| H. Chase | Schs. *Emma Julia, Levant* | unknown | unknown |
| Patrick Doyle | Sch. *E. B. Hackburn* | Georgia | Ireland |
| Evers | Sch. *Morning Star* | Unknown | Unknown |
| Domingo Galleo | Sch. *Eliza Ann* | Georgia | Italy |
| John L. Grovenstein | Sch. *Elias Reed* | Georgia | Georgia |
| Barron W. Hadley | Sch. *Zaidee* | Georgia | Massachusetts |
| Andrew Hanson | Sch. *Fort George Packet* | Georgia | Germany |
| Franklin L. Hathaway | Sch. *Alexander Blue* | Massachusetts | Massachusetts |
| Johan | Sch. *Adelle* | Unknown | Unknown |
| Freeman C. Keene | Sch. *Paugassett* | Massachusetts | Massachusetts |
| David S. Little | Slp. *Swallow* | Georgia/ South Carolina | New York |
| Luce | Sch. *Island Queen* | Massachusetts (?) | Massachusetts (?) |
| Morrell | Slp. *Virginia* | unknown | unknown |
| Nye | Sch. *Home* | Massachusetts (?) | Massachusetts (?) |
| Benjamin F. Pickens | Sch. *Challenge* | Massachusetts | Massachusetts |
| L. H. (?) Reed/Read | Sch. *Charles W. Bentley* | unknown | unknown |
| James (?) Smith | Sch. *Emma Julia*, Slp. *Virginia* | Georgia | New York |
| Charles Stevens | Sch. *Northern Belle* | Georgia | Denmark |
| Charles Thompson | Schs. *Levant, William Totten* | Georgia | Germany |
| William Watson | Sch. *John W. Anderson* | Georgia | Denmark |
| Lewis Wiggins | Slp. *Catherine Chard* | Georgia | Russia |
| Stephen Williams | Slp. *Splendid* | Georgia | Denmark |
| Richard Worthington | Sch. *William D. Jenkins* | Georgia | England |

Source: Data from the Savannah newspaper *Daily Morning News*, 1861.

**Table 8.6. Shipments of Various Commodities Carried by Sailing Coasters into Savannah in 1861.**

| COMMODITY | NUMBER OF SHIPMENTS | MINIMUM AMOUNTS |
|---|---|---|
| Rough rice | 113 | 17,914,770 pounds |
| Clean rice | 2 | 171,000 pounds |
| Sea Island cotton | 25 | 516,950 pounds |
| Turpentine | 1 | 500 barrels |
| Spirits of turpentine | 1 | 26 barrels |
| Rosin | 1 | 470 barrels |
| Corn | 6 | 2,700 bushels |
| Hides | 1 | 40 hides |
| Cottonseed | 1 | 100 bags |
| Wood | 1 | |
| Hay | 1 | 21 bundles |

Source: Data from the Savannah newspaper *Daily Morning News*, 1861.

**Table 8.7. Savannah Merchants Receiving Shipments of Coastal Commodities in 1861.**

| MERCHANT | NUMBER OF SHIPMENTS | COMMODITIES |
|---|---|---|
| Robert Habersham & Sons | 87 | rice, Sea Island cotton |
| J. W. Anderson & Co. | 25 | rice, Sea Island cotton |
| Tison & Gordon | 11 | rice, Sea Island cotton |
| Noble A. Hardee & Co. | 10 | rice, Sea Island cotton, naval stores, hides |
| J. C. Fraser | 9 | rice, possibly Sea Island cotton |
| William H. Burroughs & Co. | 3 | rice, possibly Sea Island cotton |
| Andrew Low & Co. | 3 | rice |
| Behn & Foster | 3 | rice |
| King & Baker | 3 | Sea Island cotton |
| Hunter & Gammel | 2 | naval stores, possibly Sea Island cotton |
| Erwin & Hardee | 2 | sea island cotton, rice |
| Jones & Cassels | 2 | rough rice and Sea Island cotton |
| Davant & Lawton | 2 | rice, corn |
| E. Molyneux | 1 | merchandise |
| J. Waldburg | 1 | Sea Island cotton |
| Glen & Herley | 1 | merchandise |
| J. Williamson | 1 | rough rice |
| E. C. Wade | 1 | corn (?) |
| Jho. F. Kesty | 1 | corn (?) |
| Laroche & Bell | 1 | ballast |
| T. M. Forman | 1 | hay |
| J. B. Fairchild | 1 | cotton seed |
| Way & Jones | 1 | rough rice |

Source: Data from the Savannah newspaper *Daily Morning News*, 1861.

**Table 9-1. List of Ships Seized and Sunk as Obstructions in the Savannah River in April 1862.**

| VESSEL | APPRAISED VALUE | OWNER |
|---|---|---|
| Schooner *Blooming Youth* | $6,000 | Henry J. Dickerson |
| Schooner *Catherine Chard* | $4,000 | Louis Wiggins |
| Schooner *Cotton Plant* | $2,500 | Robert Habersham |
| Schooner *Eliza Ann* | $1,200 | Domingo Galleo |
| Schooner *Levant* | $3,500 | Henry J. Dickerson |
| Sloop *Splendid* | $1,500 | Stephen Williams (1/2) |
| Schooner *William Totten* | $3,500 | Charley Thompson |
| Schooner *William D. Jenkins* | $4,000 | Henry J. Dickerson |
| Sloop *Liberty* | $ 500 | Christy Holverson |
| Pilot Boat *J. R. Wilder* | $6,100 | |
| Pilot Boat *H. F. Willink* | $6,200 | |
| Pilot Boat *G. B. Cumming* | $3,500 | |

# APPENDIX A: SAILING VESSELS IN THE GEORGIA COASTING TRADE, 1800-1861.

## SLOOPS (SLP)

| NAME OF VESSEL | RIG | YEARS IN TRADE |
|---|---|---|
| ADVANCE | SLP | 1830 |
| ALAMODE | SLP | 1842-1847 |
| ALBERT | SLP | 1820-1831 |
| ALMYRA | SLP | 1815 |
| ALONZO | SLP | 1820 |
| ALPHA | SLP | 1809-1810 |
| ALPHA | SLP | 1836-1837 |
| ALPINE | SLP | 1815 |
| AMAZON | SLP | 1830 |
| AMBITION | SLP | 1857-1864 |
| AMELIA | SLP | 1811-1827 |
| AMERICA | SLP | 1826-1856 |
| ANGEL | SLP | 1830-1843 |
| ANGELICA | SLP | 1830-1839 |
| ANN | SLP | 1821-1837 |
| ANN ELIZA | SLP | 1818-1819 |
| ANN MARIA | SLP | 1818-1831 |

| NAME OF VESSEL | RIG | YEARS IN TRADE |
|---|---|---|
| ANNA MATILDA | SLP | 1810 |
| ANUBIS | SLP | 1808-1809 |
| ARGO | SLP | 1807-1808 |
| ARGO | SLP | 1830-1843 |
| ARK | SLP | 1814-1815 |
| ATLANTIC | SLP | 1819-1820 |
| ATLAS | SLP | 1830 |
| AUGUSTA | SLP | 1830 |
| B.F. SHERWOOD | SLP | 1846-1852 |
| B.S. NEWCOMB | SLP | 1844-1853 |
| BENJAMIN | SLP | 1805 |
| BERRIEN | SLP | 1830 |
| BETSEY | SLP | 1799-1800 |
| BETSEY | SLP | 1820 |
| BOLIVAR | SLP | 1830-1843 |
| BRANDT | SLP | 1830 |
| BRIDGEPORT | SLP | 1818-1820 |
| BRIGHT PHOEBUS | SLP | 1818 |

| NAME OF VESSEL | RIG | YEARS IN TRADE |
|---|---|---|
| CAMDEN PACKET | SLP | 1796-1797 |
| CAROLINA | SLP | 1805 |
| CAROLINE | SLP | 1808-1821 |
| CARRIER DOVE | SLP | 1854-1859 |
| CASHIER | SLP | 1835 |
| CATHERINE | SLP | 1820-1821 |
| CATHERINE & ELIZABETH | SLP | 1827-1834 |
| CATHERINE CHARD | SLP | 1843-1861 |
| CELIA | SLP | 1808-1813 |
| CHARITY | SLP | 1821 |
| CHAUNCEY | SLP | 1820-1821 |
| CINDIRILLA | SLP | 1800-1809 |
| CITIZEN | SLP | 1818-1820 |
| COLLECTOR | SLP | 1808-1820 |
| CONCORD | SLP | 1809-1812 |
| CONDOMA | SLP | 1821 |
| CONDUCTOR | SLP | 1828-1842 |

| NAME OF VESSEL | RIG | YEARS IN TRADE |
|---|---|---|
| CONVERT or CONVENT | SLP | 1853-1855 |
| CORDELIA | SLP | 1805 |
| CRAWFORD | SLP | 1828 |
| CUMBERLAND | SLP | 1814-1815 |
| CYNTHIA | SLP | 1814-1817 |
| CYNTHIA | SLP | 1820-1831 |
| DART | SLP | 1813 |
| DELIGHT | SLP | 1805-1810 |
| DELIGHT | SLP | 1813-1831 |
| DIAMOND | SLP | 1805 |
| DIANA | SLP | 1809-1820 |
| DIRIGO | SLP | 1827-1832 |
| DOLPHIN | SLP or SCH | 1820-1828 |
| DOSORIS | SLP | 1814-1821 |
| DOVE | SLP | 18,001,821 |
| EAGLE | SLP | 1803-1805 |
| EAGLE | SLP | 1827-1851 |
| EARL | SLP | 1807-1808 |
| ECLIPSE | SLP | 1830 |
| EDWARD | SLP | 1813 |

| NAME OF VESSEL | RIG | YEARS IN TRADE |
|---|---|---|
| EDWARD H. FITLER | SLP | 1852 |
| ELEANOR | SLP | 1808 |
| ELE ANOR | SLP | 1825-1830 |
| ELIZA | SLP | 1807-1818 |
| EL IZA | SLP | 1830-1840 |
| ELIZA | SLP | 1828-1835 |
| ELIZA ANN | SLP | 1825 |
| ELIZA NICOLL | SLP | 1825-1828 |
| ELIZABETH | SLP | 1812-1818 |
| ELIZABETH | SLP | 1857 |
| ELLEN GALLAGHER | SLP | 1859 |
| EMMA | SLP | 1841 |
| EMMA & ELIZA | SLP | 1835-1843 |
| ENTERPRISE | SLP | 1820-1821 |
| EPHRAM | SLP | 1828 |
| EUTAW | SLP | 1841-1850 |
| EVENING STAR | SLP | 1861-1861 |
| EXCEL | SLP | 1828-1830 |
| EXCELLENT | SLP | 1820 |
| EXPERIMENT | SLP | 1805 |

| NAME OF VESSEL | RIG | YEARS IN TRADE |
|---|---|---|
| EXPERIMENT | SLP | 1819-1820 |
| EXPRESS | SLP | 1827-1841 |
| FACTOR | SLP | 1800-1814 |
| FAIR TRADER | SLP | 1828-1836 |
| FAIRHAVEN | SLP | 1805 |
| FALCON | SLP | 1824-1827 |
| FAME | SLP | 1828 |
| FAME | SLP | 1809 |
| FAMILY | SLP | 1861 |
| FASHION | SLP | 1856 |
| FAVORITE | SLP | 1808 |
| FAVORITE | SLP | 1819-1828 |
| FELLOWSHIP | SLP | 1808-1814 |
| FISH | SLP | 1808-1814 |
| FLORA | SLP | 1815-1830 |
| FLORENCE | SLP | 1827-1828 |
| FLORIDA | SLP | 1808-1810, 1847 |
| FRANCES or FRANCIS | SLP | 1820 |
| FRANKLIN | SLP | 1805-1807, 1830 |
| FREDERICA | SLP | 1813-1820 |
| FRIENDSHIP | SLP | 1800-1810 |

| NAME OF VESSEL | RIG | YEARS IN TRADE |
|---|---|---|
| FROLIC | SLP | 1820 |
| GENERAL ELMER | SLP | 1828 |
| GENERAL WASHINGTON | SLP | 1820 |
| GEORGE | SLP | 1805-1810 |
| GEORGE | SLP | 1830-1836 |
| GEORGE WASHINGTON | SLP | 1826-1842 |
| GEORGIA | SLP | 1838-1844 |
| GEORGIA | SLP | 1833 |
| GEORGIA | SLP | 1841-1843 |
| GEORGIA | SLP | 1830-1841 |
| GIPSEY | SLP | 1828 |
| GOLD HUNTER | SLP | 1818 |
| GOOD HOPE | SLP | 1816-1834 |
| GOOD INTENT | SLP | 1809-1816 |
| GOOD INTENT | SLP | 1819-1830 |
| GOOD RETURN | SLP | 1824-1825 |
| GOVERNOR | SLP | 1819 |
| GOVERNOR SHELBY | SLP | 1817-1823 |
| GUIDE | SLP | 1843 |
| H. GORGAS or N. GORGAS | SLP | 1853-1856 |
| HALCYON | SLP | 1810 |
| HAPPY RETURN | SLP | 1809-1810 |
| HARDWARE | SLP | 1819-1821 |
| HARMONY | SLP | 1817-1818 |
| HARRIET | SLP | 1799-1800 |
| HARRIET | SLP | 1819-1820 |
| HARRIET | SLP | 1819-1828 |
| HELEN | SLP | 1820 |
| HENRIETTA | SLP | 1801 |
| HENRY | SLP | 1813-1825 |
| HERALD | SLP | 1825 |
| HERMIT | SLP | 1816-1821 |
| HERO | SLP | 1820-1821 |
| HESPER | SLP | 1821 |
| HESTER | SLP | 1813-1814 |
| HOPE | SLP | 1830, 1857 |
| HOWARD & JAMES | SLP | 1819-1825 |
| HUDSON | SLP | 1835 |
| HUMBIRD | SLP | 1803 |
| HUNTRESS | SLP | 1825-1827 |
| IMPORT | SLP | 1830 |
| INDEPENDENCE | SLP | 1821-1847 |
| INDUSTRY | SLP | 1809 |
| JACK-O-LANTERN | SLP | 1809-1814 |
| JACKSON | SLP | 1830-1841 |
| JAMES | SLP | 1818-1827 |
| JAMES & LUCY | SLP | 1844 |
| JANE | SLP | 1806-1821 |
| JASPER | SLP | 1842 |
| JAY | SLP | 1810 |
| JOHN CHEVALIER | SLP | 1819-1832 |
| JOHN HENRY | SLP | 1852-1860 |
| JOHN M. PARKER | SLP | 1844 |
| JOHN MILLER | SLP | 1848 |
| JOHN SLEIGH | SLP | 1812-1813 |
| JOHN WELLS | SLP | 1827 |
| JONES HALL | SLP | 1821 |

| NAME OF VESSEL | RIG | YEARS IN TRADE |
|---|---|---|
| JOSEPH | SLP | 1819-1821 |
| JULIA | SLP | 1805 |
| JULIA ANN | SLP | 1844-1855 |
| JUPITER | SLP | 1808-1809 |
| LADY WASHINGTON | SLP | 1819 |
| LAWRENCE | SLP | 1820-1821 |
| LEADER | SLP | 1827-1835 |
| LEOPARD | SLP | 1824-1825 |
| LEVANT (converted to schooner, 1844) | SLP | 1833-1844 |
| LIBERTY | SLP | 1808-1816 |
| LIBERTY | SLP | 1846-1861 |
| LINDA | SLP | 1800 |
| LINNET | SLP | 1818-1823 |
| LION | SLP | 1820 |
| LITTLE BETTE | SLP | 1816-1825 |
| LITTLE GEORGE | SLP | 1817-1920 |
| LITTLE POLL | SLP | 1810 |
| LIVE OAK | SLP | 1813-1821 |
| LIVELY | SLP | 1809-1821 |
| LUCY | SLP | 1805 |
| LYDIA | SLP | 1835 |
| MACON | SLP | 1830-1847 |
| MAGNOLIA | SLP | 1851-1854 |
| MAJOR STEWART | SLP | 1835-1835 |
| MARGARET | SLP | 1835 |
| MARIA | SLP | 1801 |
| MARIA | SLP | 1813-1825 |
| MARIAN or MARION | SLP | 1818-1822 |
| MARINER | SLP | 1819-1820 |
| MARINER | SLP | 1827-1836 |
| MARLOW | SLP | 1827-1828 |
| MARSHALL | SLP | 1830 |
| MARY | SLP | 1800-1808 |
| MARY | SLP | 1798-1808 |
| MARY | SLP | 1820-1830 |
| MARY ANN | SLP | 1820-1821 |
| MARY ANN | SLP | 1849-1852 |
| MARY BAKER | SLP | 1855 |
| MARY CUMMING | SLP | 1835-1841 |
| MARY ELIZABETH | SLP | 1858 |
| MARY HOWARD | SLP | 1830 |
| MARY JANE | SLP | 1823-1825 |
| MATHEWS | SLP | 1820-1830 |
| MENTAR or MENTOR | SLP | 1827 |
| MERCHANT | SLP | 1830-1839 |
| MERCY | SLP | 1821-1830 |
| MILL MAID | SLP | 1828-1834 |
| MINK | SLP | 1808-1809 |
| MORNING STAR | SLP | 1820-1821 |
| MORNING STAR | SLP | 1821-1824 |
| NANCY | SLP | 1800-1807, 1825 |
| NEPTUNE | SLP | 1819-1830 |
| NIAGARA | SLP | 1821 |
| NIMROD | SLP | 1830 |
| NORTHERN LIBERTY | SLP | 1825 |
| OLIVE | SLP | 1806 |
| OLIVE BRANCH | SLP | 1815 |
| ORION | SLP | 1808 |

| NAME OF VESSEL | RIG | YEARS IN TRADE |
|---|---|---|
| OTHELLO | SLP | 1827-1835 |
| PACKET | SLP | 1808-1810 |
| PARAGON | SLP | 1818 |
| PATSEY | SLP | 1800-1807 |
| PATTY | SLP | 1800-1810 |
| PAULINA JULIA | SLP | 1820 |
| PEARL | SLP | 1839 |
| PEGGY | SLP | 1800 |
| PHILIP WALTER | SLP | 1821 |
| PILOT | SLP | 1830 |
| PLANTER (converted to schooner, 1815) | SLP | 1815 |
| POLLY | SLP | 1805-1807 |
| POLLY | SLP | 1808-1815 |
| POLLY & BETSY | SLP | 1805-1814 |
| PORTSMOUTH | SLP | 1830 |
| PRESIDENT | SLP | 1810-1814 |
| PROSPECT | SLP | 1808-1810 |
| PROVIDENCE | SLP | 1821 |
| PRUDENCE | SLP | 1820 |
| RACCOON | SLP | 1808-1809 |
| RACHEL | SLP | 1800 |

| NAME OF VESSEL | RIG | YEARS IN TRADE |
|---|---|---|
| RAMBLER | SLP | 1820-1824 |
| RANGER | SLP | 1802-1808 |
| REFORMATION | SLP | 1810-1815 |
| REGULATOR | SLP | 1808-1809 |
| REGULATOR | SLP | 1825 |
| REGULATOR | SLP | 1824-1827 |
| REPORT | SLP | 1841 |
| REPUBLICAN | SLP | 1810 |
| RESOLUTION | SLP | 1808-1809 |
| RETURN | SLP | 1819-1920 |
| RICE BIRD | SLP | 1820 |
| RISING PLANET | SLP | 1810-1821 |
| RISING SUN | SLP | 1805-1825 |
| RISING SUN | SLP | 1827-1830 |
| ROBERT HABERSHAM (converted to schooner, 1831) | SLP | 1830-1831 |
| ROSETTA | SLP | 1810-1816 |
| ROSETTA | SLP | 1818-1830 |
| RUTH | SLP | 1805-1807 |
| SAILOR'S RIGHTS | SLP | 1819-1821 |
| SALLY | SLP | 1800-1802 |

| NAME OF VESSEL | RIG | YEARS IN TRADE |
|---|---|---|
| SALLY | SLP | 1819-1822 |
| SALLY | SLP | 1807-1810, 1821 |
| SAMUEL MARTIN | SLP | 1849 |
| SAPELO | SLP | 1827-1835 |
| SARAH | SLP | 1820 |
| SAUCY JACK | SLP | 1835 |
| SCIENCE | SLP | 1835-1861 |
| SEA SERPENT | SLP | 1825 |
| SIGNAL | SLP | 1830 |
| SOPHIA | SLP | 1808-1809 |
| SPARTAN | SLP | 1820-1825 |
| SPLENDID | SLP | 1830 |
| SPLENDID | SLP | 1840-1861 |
| SQUAMSCOTT | SLP | 1800-1813 |
| ST. MARYS | SLP | 1825-1827 |
| STARR (STAR) | SLP | 1859 |
| STRANGER | SLP | 1835-1841 |
| SUFFOLK | SLP | 1820 |
| SUPERB | SLP | 1820 |
| SUPPORT | SLP | 1820-1827 |
| SUSANNAH | SLP | 1800 |

| NAME OF VESSEL | RIG | YEARS IN TRADE |
|---|---|---|
| SWALLOW | SLP | 1828-1844 |
| SWALLOW (listed as sloop with 2 masts) | SLP | 1854-1861 |
| SWAN | SLP | 1838 |
| SWIFT | SLP | 1808 |
| SWIFT PACKET | SLP | 1808 |
| TEAZER | SLP | 1818-1819 |
| TELEGRAPH | SLP | 1818-1825 |
| THOMAS | SLP | 1808-1810 |
| THOMAS BUTLER KING | SLP | 1837-1838 |
| THREE BROTHERS | SLP | 1819-1830 |
| THREE FRIENDS | SLP | 1806 |
| THREE SISTERS | SLP | 1800-1801 |
| TORMENTOR | SLP | 1808 |
| TRADER | SLP | 1825-1830 |
| TRAVELLER | SLP | 1821 |
| TWO FRIENDS | SLP | 1800-1839 |
| TWO SISTERS | SLP | 1805-1825 |
| TYGER | SLP | 1799-1800 |
| UNION | SLP | 1814-1827 |
| VICTORY | SLP | 1810 |
| VIRGINIA | SLP | 1831-1861 |
| VISITOR | SLP | 1839-1858 |
| VOLANT | SLP | 1815-1816 |
| VOLANT | SLP | 1819-1821 |
| VOLUCIA | SLP | 1819-1825 |
| WARREN | SLP | 1821 |
| WASHINGTON | SLP | 1819-1823 |
| WASHINGTON | SLP | 1830 |
| WASHINGTON | SLP | 1832-1854 |
| WAVE | SLP | 1830 |
| WILLIAM | SLP | 1800-1808 |
| WILLIAM | SLP | 1825-1828 |
| WILLIAM | SLP | 1820-1851 |
| WILLIAM & ANN | SLP | 1827-1828 |
| WILLIAM AND SALLY | SLP | 1800-1801 |
| WILLIAM C. BEE | SLP | 1859 |
| WILLIAM WRAY | SLP | 1835-1843 |
| WILLING MAID | SLP | 1802-1810 |

## SCHOONERS (SCH)

| NAME OF VESSEL | RIG | YEARS IN TRADE |
|---|---|---|
| ABIGAIL | SCH | 1800 |
| ADELLE | SCH | 1861 |
| ADMIRAL BLAKE | SCH | 1843-1852 |
| ADVANCE | SCH | 1830 |
| AFFY | SCH | 1813-1814 |
| AGNES | SCH | 1835 |
| ALBATR0SS | SCH | 1843-1844 |
| ALBEMARLE | SCH | 1837-1853 |
| ALECTIVE | SCH | 1800 |
| ALEXANDER BLUE | SCH | 1853-1861 |
| ALICE | SCH | 1813-1814 |
| ALLIGATOR | SCH | 1818-1820 |
| ALTAMAHA or ALATAMAHA | SCH | 1841-1853 |
| AMAZON | SCH | 1841-1843 |
| AMELIA | SCH | 1835 |
| AMERICAN COIN | SCH | 1846-1852 |

| NAME OF VESSEL | RIG | YEARS IN TRADE |
|---|---|---|
| ANDREW LLOYD | SCH | 1821 |
| ANGEL | SCH | 1843-1845 |
| ANN | SCH | 1821-1841 |
| ANN M. STILL | SCH | 1857 |
| ANNA MARIA | SCH | 1810 |
| ANNA MARIA | SCH | 1820-1826 |
| ASA ELDRIDGE | SCH | 1858 |
| ATLANTIC | SCH | 1843 |
| AURORA | SCH | 1839 |
| BENJAMIN | SCH | 1820 |
| BENJAMIN ENGLISH | SCH | 1860-1861 |
| BERRIEN | SCH | 1861 |
| BETSEY | SCH | 1800-1807 |
| BETSEY MARIA | SCH | 1826-1839 |
| BLAKE | SCH | 1820 |
| BLOOMING YOUTH | SCH | 1857-1858 |
| BOLD COMMANDER | SCH | 1841 |
| BOLIVAR | SCH | 1830 |
| BRANDYWINE | SCH | 1828 |
| BRUTUS | SCH | 1816-1817 |
| CANTON | SCH | 1841-1844 |
| CAROLINE | SCH | 1844 |
| CAROLINE HERSCHEL | SCH | 1837-1838 |
| CARRIER | SCH | 1850 |
| CATHERINE & CORNELIA | SCH | 1814-1816 |
| CHALLENGE | SCH | 1858-1861 |
| CHAMPION | SCH | 1858 |
| CHARLES | SCH | 1838 |
| CHARLES EDMUNDSON | SCH | 1851 |
| CHARLES W. BENTLEY | SCH | 1856-1861 |
| CHARLOTTE | SCH | 1814 |
| CHEVES | SCH | 1820-1821 |
| CHOCTAW | SCH | 1820-1821 |
| CLARION | SCH | 1827-1828 |
| CLARISSA & MARY | SCH | 1820 |
| CLUB | SCH | 1821 |
| COL. W.F. WINKOOP or WYNKOOP | SCH | 1855-1857 |
| COLLECTOR | SCH | 1805, 1813 |
| COLUMBIA | SCH | 1826-1831 |
| COLUMBIA | SCH | 1837-1845 |
| COMET | SCH | 1817 |
| COMMERCE | SCH | 1810, 1821 |
| COMPANY | SCH | 1841-1859 |
| CONCORD | SCH | 1805 |
| CONSTITUTION | SCH | 1821 |
| CORNELIA | SCH | 1839 |
| COTTON PLANT | SCH | 1842-1861 |
| COTTON PLANTER | SCH | 1800 |
| CYGNET | SCH | 1841 |
| DARIEN | SCH | 1861-1862 |
| DAVID BELKNAP | SCH | 1846-1850 |
| DECATUR | SCH | 1827 |
| DEFIANCE | SCH | 1830 |
| DELANCEY | SCH | 1805-1810 |

| NAME OF VESSEL | RIG | YEARS IN TRADE |
|---|---|---|
| DELIGHT | SCH | 1853-1854 |
| DIOMEDE | SCH | 1821 |
| DISPATCH | SCH | 1805-1808 |
| DOLPHIN | SCH | 1821-1822 |
| DRUNKARD'S RETURN | SCH | 1813-1815 |
| DUE BILL | SCH | 1815-1816 |
| DUSKY SALLY | SCH | 1839 |
| E.B. HACKBURN | SCH | 1859-1861 |
| E.J. MUNSELL | SCH | 1857 |
| EAGLE | SCH or SLP | 1822-1828 |
| ECLIPSE | SCH | 1823 |
| EDMUND & FRANCIS | SCH | 1850-1851 |
| EDWARD | SCH | 1842 |
| EDWARD FRANKLIN | SCH | 1835-1844 |
| ELIAS REED | SCH | 1849-1861 |
| ELIZA | SCH | 1800 |
| ELIZA & EMMA | SCH | 1838 |
| ELIZA ANN | SCH | 1847-1861 |
| ELIZABETH | SCH | 1818-1835 |

| NAME OF VESSEL | RIG | YEARS IN TRADE |
|---|---|---|
| ELIZABETH HELEN | SCH | 1853 |
| ELLEN | SCH | 1834-1853 |
| EMBLEM | SCH | 1841 |
| EMELINE | SCH | 1825 |
| EMILY | SCH | 1827-1828 |
| EMMA JULIA | SCH | 1855-1861 |
| ENERGY | SCH | 1843 |
| ENTERPRISE | SCH | 1830 |
| ERIE | SCH | 1841 |
| ESTHER | SCH | 1808 |
| EXPERIENCE | SCH | 1800 |
| EXPERIMENT | SCH | 1806 |
| EXPERIMENT | SCH | 1808-1828 |
| EXPERIMENT | SCH | 1851 |
| FACTOR | SCH | 1841 |
| FAINE | SCH | 1810 |
| FAIR AMERICAN | SCH | 1808, 1816 |
| FAIR PLAY | SCH | 1796-1808 |
| FANNY & MARIA | SCH | 1810 |
| FAVORITE | SCH | 1821 |

| NAME OF VESSEL | RIG | YEARS IN TRADE |
|---|---|---|
| FIDELIA | SCH | 1814 |
| FIDELITY | SCH | 1816-1817 |
| FIRE FLY | SCH | 1818-1821 |
| FLORA | SCH | 1822-1835 |
| FLORIDA | SCH | 1800, 1825 |
| FLORIDA | SCH | 1835-1860 |
| FORT GEORGE PACKET | SCH | 1825-1861 |
| FOUR BROTHERS | SCH | 1810 |
| FOX | SCH | 1800, 1825 |
| FREETOWN | SCH | 1854 |
| FRIENDSHIP | SCH | 1817 |
| FRIENDSHIP | SCH | 1800 |
| GENERAL | SCH | 1828 |
| GENERAL JACKSON | SCH | 1818-1819 |
| GENERAL MARION | SCH | 1820 |
| GENERAL WASHINGTON | SCH | 1848 |
| GEORGIA | SCH | 1830-1844 |

| NAME OF VESSEL | RIG | YEARS IN TRADE |
|---|---|---|
| GOOD INTENT | SCH | 1814-1816 |
| GOSSAMER | SCH | 1833-1836 |
| GOVERNOR WHITE | SCH | 1802-1803 |
| GRAMPUS | SCH | 1825 |
| H.F. WILLINK, Jr. | SCH | 1857 |
| HAMILTON | SCH | 1838-1845 |
| HARMONY | SCH | 1820 |
| HARRIET | SCH | 1800 |
| HARRIET LEWIS | SCH | 1850-1854 |
| HARVEST | SCH | 1839 |
| HENRY A. BREED | SCH | 1839 |
| HEYWOOD or HEYWARD | SCH | 1851 |
| HOME | SCH | 1842-1861 |
| HOPE AND HANNAH | SCH | 1830 |
| HOPETON | SCH | 1848-1853 |
| HORIZON | SCH | 1830 |
| HOTSPUR | SCH | 1827 |

| NAME OF VESSEL | RIG | YEARS IN TRADE |
|---|---|---|
| HUGH E. VINCENT | SCH | 1854 |
| IDONIUS or IDONEOUS or IDONIOUS | SCH | 1834-1835 |
| IMPERIAL | SCH | 1844 |
| INCOME | SCH | 1830 |
| INDUSTRY | SCH | 1800-1802 |
| ISABEL | SCH | 1848 |
| ISABELLA | SCH | 1820-1839 |
| ISLAND QUEEN or ISLAND BELLE | SCH | 1861 |
| J.G. KING | SCH | 1852 |
| J. HERMAN | SCH | 1857 |
| J.M.B. LOVELL | SCH | 1860 |
| J. TRUMAN or J. TRUEMAN | SCH | 1856-1859 |
| J.P. or J.R. COLLINS | SCH | 1852-1856 |
| JAMES & AUGUSTUS | SCH | 1850-1854 |
| JAMES A. LARTON | SCH | 1859 |

| NAME OF VESSEL | RIG | YEARS IN TRADE |
|---|---|---|
| JAMES FRANCIS | SCH | 1839 |
| JAMES MADISON | SCH | 1808-1809 |
| JANE | SCH | 1820-1824 |
| JANE P. GLOVER | SCH | 1852-1856 |
| JOHN | SCH | 1799-1800 |
| JOHN & WILLIAM | SCH | 1808 |
| JOHN FRASER or JOHN FRAZIER | SCH | 1851-1860 |
| JOHN W. ANDERSON | SCH | 1851-1861 |
| JOLLY ROVER or JOLLY ROGER | SCH | 1800 |
| JOLLY SAILOR or JOLLY SAYLOR | SCH | 1800 |
| JOSEPH | SCH | 1841-1845 |
| JULIA ANN | SCH | 1805-1808 |
| KITTY | SCH | 1800-1801 |
| KITTY ANN | SCH | 1808-1814 |
| KORET | SCH | 1828 |
| L.N. GODFREY | SCH | 1855-1859 |

| NAME OF VESSEL | RIG | YEARS IN TRADE |
|---|---|---|
| LADY WASHINGTON | SCH | 1800-1805, 1838-1839 |
| LEOPOLD O'DONNELL | SCH | 1851-1859 |
| LEVANT (converted to schooner 1844) | SCH | 1845-1861 |
| LISSI | SCH | 1852 |
| LITTLE ANN | SCH | 1821 |
| LLEWELLYN | SCH | 1842-1843 |
| LOVELY KEZIAH | SCH | 1810 |
| LUCRETIA | SCH | 1817-1821 |
| LYDIA | SCH | 1800-1821 |
| MAID OF SENSE | SCH | 1808-1809 |
| MALAPORTE | SCH | 1813 |
| MARGARET | SCH | 1813, 1839 |
| MARGARET ANN HOWARD | SCH | 1859 |
| MARIA | SCH | 1810-1825 |
| MARIA | SCH | 1846, 1849 |
| MARIETTA HAND | SCH | 1855-1856 |
| MARIO | SCH | 1820 |
| MARION | SCH | 1833-1841 |
| MARION | SCH | 1843-1846 |
| MARMION | SCH | 1830 |
| MARS | SCH | 1819-1820 |
| MARTHA | SCH | 1855 |
| MARTHA MOORE | SCH | 1858 |
| MARY | SCH | 1805-1825, 1857 |
| MARY & ADELAINE | SCH | 1861 |
| MARY & ELIZA | SCH | 1813 |
| MARY A. ROWLAND | SCH | 1859-1861 |
| MARY ADAMS | SCH | 1827-1830 |
| MARY AND SALLY | SCH | 1803 |
| MARY ANN | SCH | 1851-1853 |
| MARY ANN | SCH | 1847 |
| MARY ANN | SCH | 1825 |
| MARY LONISS | SCH | 1860 |
| MARY MCKAY or MARY MCKOY | SCH | 1820-1825 |
| MARY PERRY | SCH | 1830 |
| MARY SHIELDS | SCH | 1843 |
| MEDORA | SCH | 1838 |
| MINERVA | SCH | 1806 |
| MOONLIGHT | SCH | 1859 |
| MORNING STAR | SCH | 1850, 1859 |
| MORRIS | SCH | 1848 |
| N. HOWARD | SCH | 1859 |
| NABBY | SCH | 1800 |
| NEW YORK | SCH | 1830 |
| NILE | SCH | 1837-1838 |
| NORTHERN BELLE | SCH | 1850-1861 |
| OCEAN QUEEN | SCH | 1849 |
| ODIN | SCH | 1827 |
| OLIVE | SCH | 1808-1810 |
| OLIVE BRANCH | SCH | 1831-1832 |
| ONLY SON | SCH | 1820 |
| ORANGE | SCH | 1827-1843 |

| NAME OF VESSEL | RIG | YEARS IN TRADE |
|---|---|---|
| ORANGE | SCH | 1799-1800 |
| OSCAR | SCH | 1847 |
| PATRICK | SCH | 1853 |
| PATSEY | SCH | 1805, 1819-1820 |
| PAUGASSETT | SCH | 1861 |
| PAULINE | SCH | 1845 |
| PET | SCH | 1847-1849 |
| PHOEBE | SCH | 1805 |
| PLANET | SCH | 1844 |
| PLANTER | SCH | 1809 |
| PLANTER (converted to schooner, 1815) | SCH | 1816 |
| PLATUS | SCH | 1838 |
| POLLY | SCH | 1800-1809 |
| PORT REPUBLIC | SCH | 1838-1843 |
| PRESIDENT | SCH | 1835-1844 |
| PRUDENCE | SCH | 1814 |
| RAINBOW | SCH | 1805 |
| RAINBOW | SCH | 1851-1860 |
| REAPER | SCH | 1828 |
| RECOVERY | SCH | 1825 |
| RED ROVER | SCH | 1841 |

| NAME OF VESSEL | RIG | YEARS IN TRADE |
|---|---|---|
| REGULATOR | SCH | 1800 |
| RELIANCE | SCH | 1827 |
| RELIEF | SCH | 1808 |
| RENOWN | SCH | 1810 |
| REPORTER | SCH | 1825 |
| RICHARD | SCH | 1821 |
| RICHARD | SCH | 1830-1844 |
| RICHARD BRADLEY | SCH | 1826 |
| RIO | SCH | 1851 |
| RISING STATES | SCH | 1821-1835 |
| RISING SUN | SCH | 1802-1805 |
| ROBERT A. (P.?) WARING | SCH | 1839 |
| ROBERT HABERSHAM (converted to schooner, 1831) | SCH | 1831-1846 |
| ROCHESTER | SCH | 1843 |
| ROSWELL KING | SCH | 1838-1856 |
| S.C. KING | SCH | 1852 |
| SALLY | SCH | 1801-1828 |
| SALLY JASPER | SCH | 1816-1821 |
| SAMUEL & JANE | SCH | 1805 |

| NAME OF VESSEL | RIG | YEARS IN TRADE |
|---|---|---|
| SARAH or SARAH ANN | SCH | 1841-1859 |
| SARAH JAYNE or SARAH JANE | SCH | 1825-1830 |
| SARAH POTTER | SCH | 1819 |
| SARDINE | SCH | 1852 |
| SATILLA | SCH | 1851-1856 |
| SAVANNAH | SCH | 1815 |
| SAVANNAH | SCH | 1823-1835 |
| SAVANNAH | SCH | 1841-1850 |
| SAVANNAH PACKET | SCH | 1800-1804 |
| SCIENCE | SCH | 1825 |
| SEA FLOWER | SCH | 1805, 1820 |
| SOPHRONIA | SCH | 1818-1827 |
| SOUTHERNER | SCH | 1843 |
| SPARTAN | SCH | 1830 |
| SPECIE | SCH | 1859-1860 |
| ST. ACLARE or ST. ACLARO | SCH | 1800 |
| ST. LEON | SCH | 1841 |

| NAME OF VESSEL | RIG | YEARS IN TRADE |
|---|---|---|
| STEPHEN AND FRANCIS | SCH | 1841-1842 |
| SUPERB | SCH | 1827 |
| SUSAN | SCH | 1805 |
| SYLVIA E. | SCH | 1854 |
| THETIS | SCH | 1808 |
| THOMAS | SCH | 1827 |
| THOMAS SPALDING | SCH | 1856 |
| THORN | SCH | 1821-1824 |
| THREE FRIENDS | SCH | 1810 |
| THREE SISTERS | SCH | 1805-1825 |
| TIGER | SCH | 1837-1845 |
| TOM BULL | SCH | 1818 |
| TRADER | SCH | 1835 |
| TRAVELLER | SCH | 1810, 1827-1830 |
| TRIMMER | SCH | 1813 |
| TRIUMPH | SCH | 1816-1817 |
| TWO BROTHERS | SCH (?) | 1800-1804 |
| TWO FRIENDS | SCH | 1805-1815 |
| TWO MARYS | SCH | 1852-1856 |
| UNION | SCH | 1830 |
| UNITY | SCH | 1805 |
| VESTA | SCH | 1835-1847 |
| VEXATION | SCH | 1823-1825 |
| VICTOR | SCH | 1821 |
| VIRGINIA | SCH | 1828 |
| VIRGINIA | SCH | 1837-1842 |
| WARREN | SCH | 1819 |
| WATER WITCH | SCH | 1816-1835 |
| WATERBORO | SCH | 1825 |
| WAVE | SCH | 1835 |
| WEST FALMOUTH | SCH | 1859 |
| WILLIAM | SCH | 1800 |
| WILLIAM | SCH | 1830 |
| WILLIAM AND HENRY or WILLIAM HENRY | SCH | 1835 |
| WILLIAM D. BORDEN | SCH | 1830 |
| WILLIAM D. JENKINS | SCH | 1845-1861 |
| WILLIAM TOTTEN | SCH | 1855-1856 |
| WILLIAM WHITE | SCH | 1840 |
| YANKEE | SCH | 1839 |
| YANTICK | SCH | 1847 |
| YORK | SCH | 1809-1816 |
| YOUNG EAGLE | SCH | 1835-1856 |
| YOUNG SEA-HORSE | SCH | 1805-1810 |
| ZAIDEE | SCH | 1854-1856 |

# APPENDIX B: MASTERS IN THE GEORGIA COASTING TRADE, 1800-1861.

| SURNAME | FIRST NAME/INITIALS | YEARS IN TRADE |
|---|---|---|
| Aldert | Sikke | 1845 |
| Allen | Ephraim | 1815, 1821 |
| Allen | J.D. | 1856-1860 |
| Allen | Joseph | 1821-1832 |
| Allen | Joshua | 1828 |
| Anderson | F.C. | 1845 |
| Anderson | John | 1820-1821 |
| Arnaud (Arnow or Arno) | John | 1845-1861 |
| Aston | William | 1860 |
| Austin | William H. | 1852-1860 |
| Backman | Andrew | 1853-1861 |
| Baisden | McGregor | 1814, 1820s |
| Baker | Ensign | 1825-1828 |
| Baker | W.H. | 1838 |
| Barden | Frederick | 1830 |
| Barstow | Zach., Sr. | 1823 |
| Bates | Obed. | 1819 |
| Bates | Paddock | 1826-1836 |
| Bates | Sylvester | 1820, 1827-1845 |
| Bates | Wilm. | 1827-1828 |
| Bee | William C. or William H. | 1847-1851 |
| Beetle (?) | J. | 1821 |
| Bell | William | 1821-1822 |
| Bennett | George H. | 1818-1819 |
| Bennett | Sam. R. | 1830 |
| Berry | Jabez | 1821 |
| Berry | Leonard | 1826, 1836 |
| Bessey (Besse) | Sam. | 1841 |
| Bickford | William | 1859-1860 |
| Bird | Edmund | 1846 |

| | | |
|---|---|---|
| Blanc | Martin (?) | 1814 |
| Blankenship | Andrew | 1826-1829 |
| Blankenship | James | 1819-1832 |
| Blankenship | Job | 1820, 1827-1836 |
| Blankenship | Pelag | 1830-1846 |
| Blythwood | Daniel H. | 1805, 1810; 1813 |
| Boar (Boore) | Jeremiah | 1839-1842 |
| Bolith (?) | James | 1859 |
| Bolles | William | 1823, 1827 |
| Bolles (Boles) | Frederick Bartlet | 1838-1841 |
| Bolles (Boles) | Leonard | 1819-1858 |
| Bolles (Boles) | Savory | 1809, 1815 |
| Bolles (Boles) | Savory A. | 1842-1845 |
| Bolles (Boles or Bowles) | Ebenezer | 1808-1823 |
| Booth | Otis | 1830-1835 |
| Bowen | Jeremiah | 1826-1828 |
| Bradley | Thomas | 1826 |
| Bradley | William | 1819 |
| Briggs | Franklin | 1830 |
| Briggs | Nathan | 1820; 1828-1843 |
| Briggs | Paul | 1826-1830 |
| Briggs | Silas | 1815-1830 |
| Brightman | Jacob | 1820 |
| Brockington | Samuel A. | 1835-1856 |
| Brooks | Hosea | 1809-1810 |
| Broughton | Elijah | 1838, 1847 |
| Brown | Daniel C. | 1839; 1842-1843 |
| Brown | John | 1827, 1830 |
| Brown | William | 1805-1814 |
| Brown | William E. | 1825 |
| Brownell (Brownhill) | Jonathan | 1820-1821 |
| Bruce | William G. | 1825 |
| Buckley | John | 1835; 1846 |
| Bulkley | Walter | 1820-21; 1838 |
| Burns | John G. | 1841-1844 |
| Burr | James W. | 1828-1849 |
| Burr | Leonard | 1828-1852 |

| | | |
|---|---|---|
| Burroughs (Burrows) | William | 1820 |
| Burrows | Jeremiah or Ambrose H. | 1800; 1810 |
| Butler | John | 1815 |
| Butler | John Q.A. | 1858-60 |
| Butler | William | 1825 |
| Caeser (Cesar) | Peter (?) | 1809-1813 |
| Campbell | Colin | 1854 |
| Canuet (Cannet) | William | 1834-1841 |
| Capo | John | 1830, 1832 |
| Carmutel | Vincent, Sr. | 1826 |
| Chadwick | Samuel | 1815, 1831, |
| Chalker | Charles | 1800-1808 |
| Chase | Allen | 1820-1825 |
| Chase | Freeman | 1827 |
| Chase | H. | 1852-1859 |
| Chevalier | Francis | 1838-1844 |
| Chevalier | John | 1799-1831 |
| Chittenden (Chiddenden) | Joel | 1820-1821 |
| Christia | Peter | 1859 |
| Christie (Christy) | Luke | 1833-1860 |
| Church | Sylvanus | 1796 |
| Church | Sylvanus | 1847 |
| Clark | Thomas | 1820 |
| Clive | Alex. | 1844-1845 |
| Coe | Philander Parmalee | 1854 |
| Coleman | Stephen | 1831 |
| Collins | Peter | 1852-1855 |
| Conyers | William (?) | 1800-1801 |
| Cooper | Benj. | 1815 |
| Cooper (Couper) | Jonathan | 1805-1830 |
| Corbitt | Robert | 1841-1843 |
| Courter | Edward D. | 1813-1820 |
| Cox | Gustavus or John | 1830-1832 |
| Coxetter | Louis Mitchell | 1841-1861 |
| Crapo | Ira | 1828 |
| Crapo (Grapo) | Peter (?) | 1827-1828 |

| | | |
|---|---|---|
| Crayton (Graton or Graydon) | Alvin | 1820-1821 |
| Crowell | Obediah | 1825 |
| Crowell | William | 1838-1839 |
| Cruger (Croker) | Nicholas | 1848-1849 |
| Cudworth | Luther | 1842-1852 |
| Cumming | James | 1821-1822 |
| Cunningham | James | 1820, 1825 |
| Daniels | David | 1838-1840 |
| Davis | Charles G | 1841-1852 |
| Dean | George | 1825-1830 |
| Delano | Clark | 1841 |
| Delano | Harper, Jr. | 1818 |
| Delano | Henry D. | 1842-1845 |
| Delano | Jabez | 1849 |
| Delano | James | 1810; 1816 |
| Delano | Job | 1826-1827 |
| Delano | John | 1814 |
| Delano | Stephen | 1827-1845 |
| Dennison | James (?) | 1809-1816 |
| Denson (Dennison) | John D. | 1844-1850 |
| Detyns | H.J. | 1851-1852 |
| Deverger | James | 1819-1833 |
| Dexter | J. | 1841-1850 |
| Dexter | Leonard | 1834 |
| Dexter | Rueben | 1810 |
| Dexter | S. | 1800 |
| Dickerson | Henry J. | 1838-1860 |
| Dixon | Wm. | 1857-1858 |
| Doyle | Patrick | 1850-1860 |
| Drinkwater | R. | 1819-1841 |
| Driscoll | Timothy | 1814 |
| Dubignon | Joseph (?) | 1808-1809 |
| Durant | Wm. | 1816 |
| Durfee | Madison | 1837 |
| Duval | J.F. | 1820 |
| Eaton | William P. | 1833-1842 |
| Edy | Benjamin | 1842 |

| | | |
|---|---|---|
| Egery | William | 1816; 1819-1821 |
| Eldred | Ward | 1830, 1832 |
| Eldridge | Sylvester | 1839 |
| Emerus (Emurus) | John | 1815 |
| Evans | Cranston | 1830 |
| Fithian (Fethieu?) | Isaac | 1827-1828 |
| Fitspatrick | Philip | 1820 |
| Flanders | Charles M. | 1844-1845 |
| Flood | Samuel F. | 1820-1841 |
| Foster | William | 1825-1826 |
| Fowler | John | 1820 |
| Fowler | Russell | 1805-1808 |
| Fowler | William | 1855 |
| Franklin | Henry or Joseph | 1841 |
| Frazier (Fraser) | William | 1859-1860 |
| Freeman | William (?) | 1821; 1830; 1858 |
| Frewin | James | 1818-1823 |
| Frost | George | 1820 |
| Gaines | Jesse | 1830 |
| Gale | Worthington | 1814 |
| Galleo (Gallio) | Domingo | 1843-1860 |
| Gardner | Nicholas | 1819 |
| Gatchell | William | 1846 |
| Gattis | D.H. | 1849-1852 |
| Gee | Samuel | 1851-1852 |
| Geer | Erastus | 1821, 1831 |
| Gibbs | Robert | 1819-1828 |
| Gifford | Isaac | 1818 |
| Gilbert | Geo. | 1820 |
| Godfrey | John W. | 1850-1859 |
| Gorham | William | 1805 |
| Grandison | Charles F. | 1820-1821 |
| Grant | William | 1805-1806 |
| Grovenstein | John Louis | 1835-1860 |
| Gurney | Barnabas H. | 1827-1848 |
| Gurney | Richard | 1827-1832 |
| Hadley | Barron W. | 1852-1860 |
| Hadley (Hedley) | Stephen D. | 1847, 1857 |
| Hall | Larnet | 1819 |

| | | |
|---|---|---|
| Hall | Samuel D. | 1830 |
| Hall | William | 1822-1835 |
| Ham | Jessie (?) | 1827 |
| Hamilton | Daniel | 1818 |
| Hammett | Charles E. | 1821; 1827-1828 |
| Hammond | Leonard | 1825-1841 |
| Hand | Israel P. | 1819-1820 |
| Handy | Jabez | 1828-1830 |
| Handy | Luther | 1829 |
| Handy | Elishu (Elisha) | 1815 |
| Hanson (Hansen) | Andrew | 1857-1861 |
| Harden | C.H. | 1841; 1857-1860 |
| Harrison | Sam. | 1819-1820 |
| Hart | George P. | 1827 |
| Hart | John | 1841–1842 |
| Harvey | Samuel | 1805, 1809 |
| Hatch | Robert (?) | 1800; 1805; 1808 |
| Hathaway | Franklin L. | 1850-1861 |
| Hathaway | Jason | 1820-1823 |
| Hathaway | Malcolm | 1827 |
| Hathaway | Welcome | 1821-1838 |
| Hathaway | William | 1825, 1828 |
| Headley (Hadley) | Amos | 1819 |
| Heald | J.S.F (?) | 1849-1850 |
| Heath | John | 1825 |
| Hernandez | Francis or Emanuel | 1829-1845 |
| Hill | Richard A. | 1825-1835 |
| Hillary | George (?) | 1800 |
| Hilliard | Nathaniel | 1805-1806; 1856 |
| Hogan | John | 1857-58 |
| Holbrook | S.H. | 1851 |
| Holcomb | G.H. | 1850 |
| Holverson (Helverson, or Halverson) | Christy | 1858-1860 |
| Honiker | Benj. (?) | 1830 |
| Hopkins | Elisha | 1818, 1828 |
| Howard | Nathaniel | 1815, 1843 |
| Howland | Joseph | 1820-1827, 1845 |
| Hubbard | John | 1820-1827 |

| | | |
|---|---|---|
| Hubbard (Hibbard or Hebbard) | Elihu | 1808 |
| Hubbard (Hibbard or Hebbard) | Malathia Hala | 1826-1841 |
| Hulse | Ulysses | 1823 |
| Ives | Jeremiah | 1813 |
| Jason | Francis (E?) | 1813, 1815, 1820 |
| Jerrett | J.E. or J.W. | 1856 |
| Johnson | Alexander | 1839-1845 |
| Johnson | Benjamin | 1830 |
| Johnson (Johnston) | James | 1818-1820 |
| Jones | Ex. (?) | 1821 |
| Kean (Kain) | Alex (?) | 1841-45 |
| Keen | Thomas | 1800-1802 |
| Keene | Freeman C. | 1861 |
| Keeney | Lockwood | 1834 |
| Keller | P.M. | 1820 |
| Ketchum | Nathaniel B. | 1818-1819 |
| Keys | James | 1827-1833 |
| King | Roswell, Jr. | 1823 |
| King | W. | 1814 |
| King | William J. | 1848-1849 |
| Kingsbury | Nathan | 1857 |
| Landon | Jonathan | 1800-1805 |
| Lane (Laen ) | William | 1830-1850 |
| Lane (Lain) | James D. | 1839-1844 |
| Lard (Law?) | John | 1821 |
| Lawrence (Laurence) | Antonio | 1856-1860 |
| Lawrence (Laurence) | James A.D. | 1827-1838 |
| Learned | E.B. | 1840 |
| Lee | James V. (W.?) | 1824-1842 |
| Leset (Lissett, Lisete, or Lucette) | Claud Phillip | 1825-1836 |
| Lester | James F. | 1823 |
| Lewis | David | 1833-1852 |
| Lewis | John L. (?) | 1827, 1830; 1835 |
| Lewis | John N. | 1842-1843 |

| | | |
|---|---|---|
| Lightbourn | John H. | 1837-1843 |
| Lightbourn | Samuel Stiles | 1800-1821 |
| Lindgreen (Lindegreen) | Charles | 1800 |
| Little | David S. | 1853-1860 |
| Look | Hiram | 1830, 1835-1859 |
| Lopa (?) | Cornelias | 1833 |
| Loveland | Luther | 1826-1828 |
| Lovell | William (?) | 1860 |
| Lubbock | Richard | 1817-1821 |
| Luce | Barnabas | 1829 |
| Luce | David | 1830 |
| Luce | Elisha (Elishu?) | 1820-1854 |
| Luce | Geo. A. | 1819-21, 1835 |
| Luce | Rowland, Jr. | 1820-1841 |
| Luce | Sam. W. | 1848-1856 |
| Luce | Stephen C. | 1820-1853 |
| Magee | John | 1850-1855 |
| Makin | Jno. | 1854-1856 |
| Marble | Joseph (?) | 1827, 1830 |
| Marcellin | Richard | 1820-21, 1825 |
| Maris (Morris, or Marias) | Peter | 1849-1855 |
| Martin | Jessie | 1806-1808 |
| Mason | Ichabod | 1818-1821 |
| Mauran (Maupan) | Nathaniel S. | 1831 |
| Maxwell | David | 1811 |
| Maxwell | William | 1816 |
| McColley (M'Auley or McAuley) | John M. | 1825-1830 |
| McCorison | Asa | 1840 |
| McDonough | Andrew1 | 819 |
| McGee | John (?) | 1835; 1837 |
| McLean | Allen | 1806; 1813-1816 |
| McMellan (McMillan) | Thomas | 1821-1828 |
| Meehan | Edward | 1856-1857 |
| Meigs | Joseph | 1811-1828 |

| | | |
|---|---|---|
| Mendall | William C. | 1838-1844; 1855 |
| Milne | John | 1818 |
| Miserol | Abram | 1838 |
| Mitchell | John | 1828 |
| Moon | Sylvester | 1842 |
| Moore | Charles C. | 1855 |
| Moore | James | 1840 |
| Moore | John | 1839-1841 |
| Morgan | John J. | 1835, 1837 |
| Morrill (Morel or Morrell) | William | 1840-1850 |
| Morton | Bartlett | 1822 |
| Murray | John | 1835-1848 |
| Nye | Barnabas Bates | 1820, 1827-28 |
| Nye | Stephen, Sr. | 1825-1826 |
| O'Bryant (O'Brien) | Jessie | 1846-1848 |
| Osborne | Sylvanus | 1830-1831 |
| Owens | Richard | 1845-1850 |
| Pacetty (Pacetti) | Dennis | 1830-1850 |
| Papot | Samuel (?) | 1855 |
| Parker | Andrew F. | 1858 |
| Parker | Daniel | 1815-1838 |
| Parkinson | A. | 1853 |
| Patch | Nathan (?) | 1816 |
| Payne | Allen | 1830 |
| Payne | William | 1827 |
| Peck | Fenn | 1826-1860 |
| Peck | Michael | 1824 |
| Perry | Abraham | 1845, 1854 |
| Perry | Kimbal | 1818 |
| Peterson | James | 1847 |
| Peterson | John | 1859 |
| Phipps | Daniel Goffe | 1800-1805; 1816 |
| Pickens | Benjamin F. | 1859-1860 |
| Pidge | David | 1816-1828 |
| Pierce (Pearce) | Isaac | 1816; 1820-1830 |
| Pierce (Pearce) | John | 1800-1801 |
| Pilsberry | Levi (Louis?) | 1823 |
| Pindar | Wm. Jr. | 1801 |

| | | |
|---|---|---|
| Pitcher | James H. | 1827-1841 |
| Pitcher | Theophilus, Jr. | 1834 |
| Pittman | Nicholas P. | 1839 |
| Poinsett | Josiah | 1830; 1837-1841 |
| Pollard | John | 1821 |
| Porquet (Parquet) | Charles | 1812-1828 |
| Postell | William Ross | 1845-1860 |
| Potter | Dan. L. | 1819 |
| Pratt | Horace S. | 1820-1825 |
| Prebble (Preble) | John R. | 1824-1825 |
| Prentice | Thomas | 1806 |
| Prim | Stephen | 1840 |
| Prindle | Joseph H. | 1827 |
| Procter | Daniel | 1825-1833 |
| Purvis/Pervis | George W. | 1850-1856 |
| Rackliff (Ratcliff) | George W. | 1834-1843 |
| Rafille (Raphael or Rafile) | Joseph | 1836-1850 |
| Ramsbottom | Patrick | 1834 |
| Ray | Peter | 1810 |
| Reddock (Reddick or Readick) | Daniel | 1838-1852 |
| Reed | John | 1819 |
| Reed (Read) | L.H. | 1850-1856 |
| Rentz (Rents) | George | 1825-1835 |
| Riberon (Ribron) | Manuel | 1830-1834 |
| Richardson | Edmund | 1808; 1813; 1820-1842 |
| Ridley (Redley) | James | 1835-43 |
| Robach (Roebach) | George H. | 1830 |
| Robbins (Robins) | Jno./John | 1830-1839 |
| Room (Roome) | Thomas | 1841-1858 |
| Rorback | G. H. | 1830 |
| Ross | Thomas | 1800-1801 |
| Rudolph | John T. | 1800 (?)-1835 |
| Rudolph | Robert (?) | 1800-1801 |
| Russell | George | 1841-1853 |

| | | |
|---|---|---|
| Russell | John | 1839-1857 |
| Sabatty | Clement | 1820 |
| Saffer | James | 1849 |
| Sallowich (Selowich) | Nicholas | 1813-1835 |
| Santina (Santini) | Simon (?) | 1850-1853 |
| Savage | Giles | 1820 |
| Savory (Savery) | Timothy, Jr. | 1827-1839 |
| Sayers | Fred. | 1819-1825 |
| Seymans | Joseph | 1815 |
| Shaddock | Anthony | 1808-1821 |
| Shaw | William | 1816 |
| Shearman | John | 1799 |
| Sherman | Humphrey L. | 1853 |
| Sherwood | Robert | 1798 |
| Sibley | Ira | 1819 |
| Simpson | William | 1820, 1824-1825 |
| Sinclair | Thos. | 1820 |
| Sisson | Noah B. | 1813-1833 |
| Sly | Robert | 1818 |
| Smith | Benjamin | 1841 |
| Smith | Edward | 1842 |
| Smith | Geo. | 1808 |
| Smith | James | 1860 |
| Smith | John | 1830; 1841-1848 |
| Smith | Shadrick | 1823 |
| Smith | Thomas | 1845-1847 |
| Smith | Walter William | 1829-1840 |
| Snow | Prince | 1808-1827 |
| Snow | Thomas | 1841-1853 |
| Sorley | William | 1818 |
| Southwick | Lyman | 1830-1831 |
| Spencer | Joseph (?) | 1803 |
| Spriggs (Spreggo) | George (?) | 1820 |
| Stanton | Jabez | 1801; 1805 |
| Staples | John | 1853 |
| Stephens | Arnold | 1828-1829 |
| Stevens (Stephens) | Charles | 1837-1861 |
| Stirk | F.M. | 1841 |

| | | |
|---|---|---|
| Stockwell | George (?) | 1856-1857 |
| Stoddard | Sanford | 1820-1821 |
| Stone | William | 1821 |
| Stotesbury | Peter | 1840 |
| Stotsbury (Stotsburg) | John | 1813; 1816; 1834 |
| Stubbs | Henry | 1841 |
| Studley | Allen | 1832 |
| Sturtevant | Ezra | 1824-1830 |
| Sturtevant | Lemuel | 1822-1830 |
| Sturtevant | Samuel | 1818, 1828 |
| Swain | Richard | 1816 |
| Swan (Swain) | Labun (Labon) | 1817 |
| Swasey | A. P. | 1835 |
| Taber | George | 1801 |
| Taber | John R. | 1814 |
| Taber | Joseph B. | 1823-1824 |
| Taylor | William C. | 1849-1860 |
| Thewes (Thewer) | Clement (?) | 1821 |
| Thomas | Nathan | 1816 |
| Thomas | William | 1838, 1845 |
| Thompson | Peter | 1841-1852 |
| Thompson | Thomas A. | 1820 |
| Thompson | William | 1838-40 |
| Thompson (Thomson) | Charles (Charley) | 1841-1860 |
| Tomasson (Tomerson) | William V. | 1825-1830 |
| Tompkins | Donald | 1820-1821 |
| Trefeather (Trefather) | Daniel | 1841 |
| Trump (Troup) | Dan. | 1847-1848 |
| Turner | John | 1839 |
| Turner | John W. | 1820-1835 |
| Veal | Thomas (?) | 1821 |
| Ventress | James | 1818 |
| Vernard | Edward or Wm. V. or Jacob | 1822-1825 |
| Vincent | James, Jr. | 1817-1825 |

| | | |
|---|---|---|
| Waddington | Stephen | 1809-1810 |
| Walford | Thomas | 1818 |
| Walker | John | 1818 |
| Ward | H. | 1809-1810 |
| Watson | William | 1843-1861 |
| Whebell | R.H.D. | 1835 |
| White | A.C. | 1830 |
| White | Paul J. | 1846-1858 |
| White/Wight | Wright | 1825-1828 |
| Whiteside | W.D. | 1840 |
| Wiggins | Lewis P. | 1846-1860 |
| Wightman (Wrightman) | Timothy, Jr. | 1827-1828 |
| Wilkinson | George W. | 1841-1851 |
| William | Thomas | 1824; 1848 |
| Williams | John | 1850 |
| Williams | Stephen | 1847-1860 |
| Williams | William | 1834-1847 |
| Willington | R. | 1841 |
| Willink | Henry F. | 1813-1860 |
| Wilson | Alexander | 1837-1851 |
| Wilson | J.W. | 1851 |
| Wilson | James | 1842-1845 |
| Windsor | Thomas | 1817 |
| Wing | Alden | 1808-1815 |
| Wing | Benj. | 1841 |
| Wing | P. | 1835 |
| Wing | Paul | 1816-1820 |
| Wing | R. (?) | 1825-1841 |
| Winn | Thomas | 1848-1849 |
| Wood | Joseph | 1825-1830 |
| Wood | William | 1818 |
| Woodland | J. | 1819, 1821 |
| Woodward | Joseph | 1819-1821 |
| Woodward | Thomas | 1820 |
| Woodworth | Darius | 1805-1808; 1819-1821 |
| Worth | James | 1816-1820 |
| Worthington | Peter | 1835; 1843 |

| | | |
|---|---|---|
| Worthington | Richard | 1844-1859 |
| Worthington | William | 1851-1858 |
| Wray | William D. | 1836-1841 |
| Young | Alvan/Alvin | 1822-1831 |

# Bibliography

*Public Records and Manuscripts*

Baker Family Papers. *Sloop Book.* Baker Family Papers, Manuscript 11/539/2. South Carolina Historical Society, Charleston, South Carolina.

Bureau of Customs. *Georgia Customs Districts.* Unpublished finding aid, pp. 159–64. RG 36. Records of the US Customs Service. Records of Customshouses and Collection Districts. NARA, National Archives Building, Washington, DC.

Bureau of Customs. *Outward Foreign Manifests, Savannah, GA, April 1816–January 1817, February–May 1817.* Box 10. RG 36. Records of the US Customs Service. NA identifier: 2133240. NARA, National Archives Building, Washington, DC.

Bureau of Customs. *Registers, Enrollments, and Licenses Used in the Settlement of French Spoliation Claims, Collection District of Savannah, , 1793–1800.* RG 36. Textual Records. NA Identifier: 4482976. NARA, National Archives Building, Washington, DC.

Bureau of Marine Inspection and Navigation [BMIN]. *Bonds for Enrollments, Port of Mobile, May 16, 1853–November 14, 1856.* RG 41. Field Records of the Bureau of Navigation 1774–1973. Textual records. NARA, National Archives Building, Washington, DC.

Bureau of Marine Inspection and Navigation [BMIN]. *Conveyances of Enrolled Vessels, Savannah 1850–1856.* RG 41. Customs House Records, vol. 4991. NARA, National Archives Building, Washington, DC.

Bureau of Marine Inspection and Navigation [BMIN]. *Master Abstracts of Registers, Enrollments and Licenses, 1815–1911.* Ports of Beaufort, Savannah, Sunbury, Darien, Brunswick, and St. Marys. RG 41. Records relating to vessel documentation. Enrollments: NA Identifier 2765707, Entry UD-127, Registers: NA Identifier: 817226, Entry UD-138. Microfilm 2151. NARA, National Archives Building, Washington, DC.

Bureau of Marine Inspection and Navigation [BMIN]. *Registers and Certificates of Enrollment for the Ports of Charleston, South Carolina, and Brunswick, Darien, Savannah, St. Marys, and Sunbury, Georgia, 1815–1866.* RG 41. Records relating to vessel documentation. NA Identifier: 604564. NARA, National Archives Building, Washington, DC.

Camden County, Georgia. Court of Ordinary Records. *Will of Elihu Hubbard, May 21, 1806.* Camden County Court of Ordinary, Will Book A, 1795–1829, 109. Camden County, Georgia, Court Records, 1745–1989, Record Group 1-03-83. Bryan-Lang Archives Repository, Woodbine, Georgia. Image at:

https://www.familysearch.org/records/images/image-details?page=1&place=393770&rmsId=TH-1942–30352–8246–32&imageIndex=129&singleView=true. Accessed October 2024.

Camden County, Georgia. Superior Court Records. Deed Books G and M. Camden County, Georgia, Court Records 1777–1976, Record Group 1-01-03. Bryan-Lang Archives Repository, Woodbine, Georgia.

Chatham County, Georgia. Court of Ordinary Records. Estate Accounts. Chatham County Court of Ordinary, Chatham County Courthouse, Savannah.

Chatham County, Georgia. Court of Ordinary Records. Will of David Pidge, November 11, 1828. Court of Ordinary Will Book G, 69. Chatham County Court of Ordinary, Chatham County Courthouse, Savannah. Image at: https://www.familysearch.org/records/images/image-details?page=5&place=393781&rmsId=TH-1971-30372-20861-59&imageIndex=65&singleView=true. Accessed October 2024.

Chatham County, Georgia. Superior Court Records, 1784–1910. Deed Books 2W, 2Z, 3H, 3T, 3W. Chatham County Courthouse, Savannah.

Connecticut Ship Database. "Connecticut Ship Database, 1789–1939," Mystic Seaport Museum, Collections Research Center, Mystic, Connecticut. Images at: https://research.mysticseaport.org/databases/ct-ships/, accessed 2018–2022.

Georgia Archives. *Affidavit of Mrs. Jane Austin, widow of W. W. Austin*. Georgia's Virtual Vault, Confederate Pension Applications, Georgia Confederate Pension Office, RG 58-1-1, Georgia Archive, University System of Georgia, Atlanta. Georgia Archives. Image at: https://vault.georgiaarchives.org/digital/collection/TestApps/id/40250/rec/22. Accessed November 15, 2022.

Georgia Archives. *General Tax Return of Chatham County, for the year 1799*. Georgia's Virtual Vault, County Records on Microfilm, Georgia Archives, University System of Georgia, Atlanta. Image at: https://vault.georgiaarchives.org/digital/collection/tax/id/6341/rec/2. Accessed November 15, 2022.

Georgia Historical Society. Frewin, Stevens, Taylor Family Papers, 1802–1999. Manuscript No. 1909. Georgia Historical Society, Savannah.

Georgia Historical Society. *Savannah Steam Rice Mill*. Mary Lane Morrison Papers, No. 1320, Folio 10, 42. Georgia Historical Society, Savannah.

Georgia Historical Society. Unidentified Cash Book, Accounts of Sloops *Mill Maid* and *Conductor* and Schooner *Fox* 1834–1836. Manuscript No. 1048. Georgia Historical Society, Savannah.

Glynn County, Georgia. Court of Ordinary Records. Wills and Appraisements, 1810–1916, Books E, F. Glynn County Courthouse, Brunswick.

Glynn County, Georgia. Superior Court Records. Sale of one-half of Schooner *Florida*, John Q.A. Butler to Charles Stevens, April 5, 1860, Deed Book N, Part 1, 1859–1861, 177–78. Glynn County Courthouse, Brunswick. Image at: https://www.familysearch.org/records/images/image-details?page=4&place=395650&rmsId=TH-909-85325-4862-85&imageIndex=126&singleView=true. Accessed October 2024.

Gordon Family Papers. Accounts for the Schooner Elias Reed. Manuscript No. 2235, Southern Historical Collection, Wilson Library, University of North Carolina, Chapel Hill.

Jones, Charles C. "Carlawater and Maybank Plantations Accounts." Plantation Books, 1834–1849. Liberty County Georgia, Collection No. 154. In "Southern Plantation Records," Microfilm No. 1705, Series H, Reel 28. Fredericksburg, MD: University Publications of America, 1988.

King, Roswell, Jr. *Roswell King Diary.* Joseph Jones Papers. Manuscript 19, document 1351, folder 64, box 8. Louisiana and Lower Mississippi Valley Collection, Hill Memorial Library, Louisiana State University, Baton Rouge.

King, Thomas Butler. Papers. Manuscript 1252. Southern Historical Collection, Wilson Library, University of North Carolina, Chapel Hill.

Mariners' Museum. "List of Shipping, Owned in the District of New-Bedford: Comprising the Towns of New-Bedford, Fairhaven, Dartmouth, Rochester, Wareham, and Westport, May 1835." New Bedford, MA: Benjamin T. Congdon. Mariners' Museum, Newport News, Virginia.

National Archives and Records Administration. *Vessel Documents.* Discussion of records relating to the documentation of vessels held by NARA (2007), found at: https://www.archives.gov/research/maritime/vessel-documents.html#:~:text=Certificates%20of%20registration%2C%20enrollment%2C%20and%20license%20are,Marine%20Inspection%20and%20Navigation%2C%20Record%20Group%2041. Accessed October 15, 2024.

Port of New Bedford Customs House Papers, Licenses Oaths. Port of New Bedford Customs House Records. New Bedford Free Public Library, Archives Room. New Bedford, Massachusetts.

Sanders, Thomas L. *Thomas L. Sanders Accounts, 1839–1881.* Accounts: Sailmaking and Rigging, Manuscript Collection. Peabody Essex Library, Salem, Massachusetts.

Southern Claims Commission. *Petition by Mrs. Sarah D. Stevens for compensation for loss of the Schooner Northern Belle for $4,000.* Disallowed Claims, case file 3261, M1407. Records of the U.S. House of Representatives. RG 233. NARA, National Archives Building, Washington, DC. Image at: https://www.ancestry.com/search/collections/1218/?name=sarah_stevens&count=50&name_x=1_1&residence=_glynn-georgia-usa_1154&residence_x=_1-0. Accessed June 10, 2023.

Vessel Papers. War Department Collection of Captured Confederate Records. Microfilm M909 (32 rolls). RG 109. NARA, National Archives Building, Washington, DC.

Wing, Paul. *Paul Wing Accounts, 1805–1824.* Manuscript No. 215. Special Collections & Archives. W. E. B. Du Bois Library, University of Massachusetts, Amherst.

*Newspapers*

Microfilm and digitized newspaper records were examined for the important ports involved in the Georgia coasting trade. The ports and the newspapers examined are listed below. Sources for these newspaper records are: Georgia Historic Newspapers, Georgia Archives; microfilm and online collections, Georgia Newspaper Project, University of Georgia Libraries, Athens (https://gahistoricnewspapers.galileo.usg.edu); South Carolina Historical Society, Charleston; South Caroliniana Library, Newspapers, University of South Carolina Libraries, Columbia, South Carolina.

Charleston

*Charleston City Gazette*, 1800, 1805

*Charleston City Gazette & Commercial Daily Advertiser*, 1810, 1816–1817

*Charleston Courier*, 1805, 1810, 1820

*Charleston Mercury*, 1822, 1827

Savannah

*Columbian Museum & Savannah Advertiser*, 1799–1804, 1810, 1813–1814

*Columbian Museum & Savannah Daily Gazette*, 1818

*Daily Georgian*, 1819–1821, 1827–1828, 1838–1839, 1841–1854

*Daily Morning News*, 1853–1855, 1857–1862

*Daily Savannah Republican*, 1830, 1835

*Georgia Gazette*, 1770, 1792, 1799–1801

*Georgia Republican*, 1806–1807

*Georgia Republican & State Intelligencer*, 1805–1806

*Public Intelligencer*, 1807–1809

*Republican & Savannah Evening Ledger*, 1807–1813

*Savannah Daily Republican*, 1856–1857, 1860

*Savannah Republican*, 1825

Darien

*Darien Gazette*, 1819–1825

Brunswick

*Brunswick Advocate*, 1837–1839

Florida

*Florida News* (Jacksonville), 1851–1855

*Florida Republican* (Jacksonville), 1849–1855

*Compiled Ship Enrollments and Registers*

Several publications have been produced that compile enrollment and register documents for ports along the Atlantic coast of the United States. These have been examined to supplement information obtained from the original vessel documents found at the National Archive and Record Administration. These data have been used to obtain information on the physical characteristics of vessels, place of build, ownership, the names of ship masters and homeports as discussed in the text. These publications are listed below.

Work Projects Administration

1938–1941 *Ship Registers and Enrollments of Newport, Rhode Island, 1790–1939*, vol. 1. Prepared by the Survey of Federal Archives, Division of Professional and Services Projects, Work Projects Administration. State of Rhode Island, Department of State, Division of Archives, Sponsor. National Archives, cooperating sponsor. Providence: National Archives Project, 1938–1941. Mimeographed. Collections of the Mariners' Museum Library.

1939 *Ship Registers of Dighton-Fall River, Massachusetts, 1789–1938*. Compiled by the Survey of Federal Archives, Division of Women's and Professional Projects, Works Progress Administration. National Archives, cooperating sponsor. Boston: National Archives Project, 1939. Mimeographed. Collections of the Mariners' Museum Library.

1940 *Enrollments of New Bedford, Massachusetts, 1808–1840.* Supplementary to Ship Registers of New Bedford, Massachusetts. Compiled by the Survey of Federal Archives, Work Projects Administration. Boston: National Archives Project, 1940. Collections of the New Bedford Free Public Library.

1941 *Ship Registers of New Bedford, Massachusetts, 1796–1850,* vol. 1. Prepared by the Survey of Federal Archives, Division of Community Service Programs, Work Projects Administration. Providence: National Archives Project, 1941. Mimeographed. Collections of the Mariners' Museum Library.

1941 *Ship Registers of New Bedford, Massachusetts, 1851–1865*, vol. 2. Prepared by the Survey of Federal Archives, Division of Community Service Programs, Work Projects Administration. Providence: National Archives Project, 1941. Mimeographed. Collections of the Mariners' Museum Library.

1984 *Enrollments of New Bedford, Massachusetts, 1841–1939*. Supplementary to Ship Registers of New Bedford, Massachusetts. Compiled by the Survey of Federal Archives, Work Projects Administration. Boston: National Archives Project, 1940. Original transcript copy, 1984. Collections of the New Bedford Free Public Library.

*Books and Articles*

Albion, Robert G. "Early Nineteenth Century Shipowning—A Chapter in Business Enterprise." *Journal of Economic History* 1 (1941):1–11.

———. *Rise of New York Port, 1815–1860.* New York: Scribner, 1939.

Amer, Christopher F. "The Malcolm Boat: A Preliminary Report." *Underwater Archaeology Proceedings from the Society for Historical Archaeology Conference* (1993): 76–82.

Amer, Christopher F., and Frederick M. Hocker. "A Comparative Analysis of Three Sailing Merchant Vessels from the Carolina Coast." In *Tidecraft: The Boats of South Carolina, Georgia and Northeastern Florida, 1550–1950*, by William C. Fleetwood Jr., 295–303. Tybee Island, GA: WGB Marine Press, 1995.

Ancestors Unlimited, comp. "Land District 13—Original Drawers 1820 Lottery." https://gaclayton.genealogyvillage.com/Land/1820lotterydist13.htm. Accessed November 15, 2022.

Baker, William A. *Sloops & Shallops*. Columbia: University of South Carolina Press, 1966.

Bancroft, Joseph. *Census of the City of Savannah: Together with Statistics, Relating to the Trade, Commerce, Mechanical Arts & Health of the Same, with Historical Notices, and a List of the Incorporated Companies, and of the Charitable Societies; to which is Added a Commercial Directory of the Principal Mercantile Houses, Manufacturers, Mechanics, Professions; Together with Particulars Respecting the Rail roads, Steamers, Packets, &c. Connected with the City*. Savannah: Edward J. Purse, Printer, 1848.

Barber, Henry E., and Allen R. Gann. *A History of the Savannah District, U.S. Army Corps of Engineers*. Savannah: US Army Corps of Engineers, 1989.

Beckwith, Adam L., and Daniel Lewis, comps. *Papers of the American Slave Trade, Series D: Records of the U.S. Customhouses, Part 1: Port of Savannah Slave Manifests, 1790–1860*. Bethesda: LexisNexis, 2005.

Beers, J. H. & Company. *Representative Men and Old Families of Southeastern Massachusetts*. Vol. 1. Chicago: J. H. Beers, 1912.

Bell, Malcolm, Jr. *Major Butler's Legacy: Five Generations of a Slaveholding Family*. Athens: University of Georgia Press, 1987.

Blunt, Edmund M. "Savannah River, 1816." [map]. Scale 1:62,500. In *The American Coast Pilot*, by Edmund M. Blunt. Newburyport, MA: E. Blunt, 1816. Map S3. Hargrett Rare Book & Manuscript Library, University of Georgia Libraries, Athens.

Board of Underwriters. *New York Marine Register: A Standard Classification of American Vessels and of Such Other Vessels As Visit American Ports, 1857–1858*. New York: Anthony and Company, Stationers and Printers, 1857–1858.

Board of American Lloyds. *American Lloyds' Registry of American and Foreign Shipping, 1859–1861*. New York: Clayton & Ferris Printers, 1859–1861.

Bolles, John A. *Genealogy of the Bolles Family in America*. Boston: Henry A. Dutton & Son, 1865.

Bonner, James C. *A History of Georgia Agriculture, 1732–1860*. Athens: University of Georgia Press, 1964.

Bullard, Mary R. *Robert Stafford of Cumberland Island: Growth of a Planter*. Athens: University of Georgia Press, 1995.

Burr, David H. "Map of Georgia & Alabama exhibiting the post offices, post roads, canals, rail roads & c.; by David H. Burr Late topographer to the Post Office, Geographer to the House of Representatives of the U.S." [London, 1839] Map. https://www.loc.gov/item/98688462/.

———. "Map of North and South Carolina exhibiting the post offices post roads, canals, rail roads &c. By David H. Burr; Late topographer to the Post Office.

Geographer to the House of Representatives of the U.S." [London, 1839] Map. https://www.loc.gov/item/98688532/.

Candler, Allen D., ed. *Confederate Records of the State of Georgia.* Vol. 3. Atlanta: Charles P. Byrd, State Printers, 1910.

Chapelle, Howard I. *The History of American Sailing Ships.* New York: W. W. Norton, 1935.

———. *The Search for Speed Under Sail, 1700–1855.* New York: W. W. Norton, 1967.

Clark, Malcolm. "The Coastwise and Caribbean Trade of the Chesapeake Bay, 1696–1776." Master's thesis, Georgetown University, 1970.

Clifton, James M. *Life and Labor on Argyle Island.* Savannah: Beehive Press, 1978.

Clowse, Converse D. "The Charleston Export Trade, 1717–1737." PhD diss., Northwestern University, 1963.

Coker, P. C. *Charleston's Maritime Heritage, 1670–1865: An Illustrated History.* Charleston, SC: CokerCraft Press, 1987.

Coleman, Kenneth, ed. *A History of Georgia.* Athens: University of Georgia Press, 1977.

Coleman, Kenneth, and Milton Ready, eds. *Colonial Records of the State of Georgia.* Vol. 28, pt. 2. Athens: University of Georgia Press, 1979.

"Collections of the Georgia Historical Society." *Southern Quarterly Review* 3 (January 1843): 40–93.

Collinsdictionary.com. "Order bill of lading." *Collins Online English Dictionary*, https://www.collinsdictionary.com/us/dictionary/english/order-bill-of-lading. Accessed December 14, 2022.

Cooper, John M., & Co., pub. *Map of the City of Savannah.* [map]. Scale 1:6000. Savannah: John M. Cooper, 1856. Waring Maps, no. 1018, vol. 2, plate 26. Georgia Historical Society, Savannah.

Coulter, E. Merton. *Georgia: A Short History.* Chapel Hill: University of North Carolina Press, 1960.

———. "The Great Savannah Fire of 1820." *Georgia Historical Quarterly* 23/1 (1939): 1–27.

———. *Thomas Spalding of Sapelo.* Baton Rouge: Louisiana State University Press, 1940.

Crisp, Edw, Thomas Nairne, John Harris, Maurice Mathews, and John Love. *A compleat description of the province of Carolina in 3 parts: 1st, the improved part from the surveys of Maurice Mathews & Mr. John Love: 2ly, the west part by Capt. Tho. Nairn: 3ly, a chart of the coast from Virginia to Cape Florida.* London: Edw. Crisp, ?, 1711. Map. https://www.loc.gov/item/2004626926/.

Crook, Morgan C., Jr., and Patricia D. O'Grady. "Spalding's Sugar Works Site, Sapelo Island, Georgia." In *Sapelo Papers: Researches in the History and Prehistory of Sapelo Island, Georgia*: *West Georgia College Studies in Social Sciences* 19 (1980): 9–13.

Culver, Robert Meldrim, Sr. "A Biography of Lewis Peter Wiggins, 1972." Ancestry.com, https://www.ancestry.com/mediaui-viewer/collection/1030/tree/37999377/person/19218385492/media/ed80f6eb-5dfa-4d1e-9f38–42108bca2b27?_phsrc=yPe295&_phstart=successSource. Accessed December 15, 2022.

Daggett, Kendrick Price. *Fifty Years of Fortitude: The Maritime Career of Captain Jotham Blaisdell of Kennebunk, Maine, 1810–1860.* Mystic, CT: Mystic Seaport Museum, 1988.

Davis, Charles G. *American Sailing Ships: Their Plans and History.* New York: Dover Publications, 1984. Reprint of *Ships of the Past.* Salem, MA: Marine Research Society, 1929.

*DeBow's Review: Agricultural, Commercial, Industrial Progress and Resources.* 29 vols. New Orleans: J. D. B. DeBow, 1846–1860. https://catalog.hathitrust.org/Record/011984803. Accessed 2021–2022.

Donnell, E. J. *Chronological and Statistical History of Cotton.* New York: James Sutton, Printers, 1872.

Easterlin, Richard A. "Interregional Differences in per Capita Income, Population, and Total Income, 1840–1950." *Trends in the American Economy in the Nineteenth Century, Studies in Income and Wealth* 24 (1960): 73–100. Conference on Research in Income and Wealth, National Bureau of Economic Research. Princeton: Princeton University Press.

Eisterhold, John A. "Savannah: Lumber Center of the South Atlantic." *Georgia Historical Quarterly* 57/4 (1973): 526–43.

Estill, J. H. *Estill's Savannah Directory, for 1874–75.* Savannah: J. H. Estill, 1875.

FamilySearch. "John Stotesbury." https://ancestors.familysearch.org/KCVG-KRX/john-stotesbury-1790–1847. Accessed December 18, 2022.

Faragher, John, Mary Buhle, Daniel Czitrom, and Susan Armitage. *Out of Many.* Upper Saddle River, NJ: Prentice Hall, 2006.

Fields, Tara D., comp. "Camden County, Georgia, Slave Deed Abstracts." YUMPU.com. https://www.yumpu.com/en/document/view/15249775/camden-county-georgia-slave-deed-abstracts-afriquest. Accessed December 15, 2022.

———, comp. "The Crypt, History & Genealogy of Camden & Charlton Counties, Georgia." http://www.glynngen.com/thecrypt/. Accessed December 15, 2022.

Filby, P. W., ed. *Passenger and Immigration Lists Index, 1500s–1900s.* Farmington Hills, MI: Gale Research, 2003.

Fleetwood, William C., Jr. *Tidecraft: The Boats of South Carolina, Georgia and Northeastern Florida, 1550–1950.* Tybee Island, GA: WGB Marine Press, 1995.

Footner, Geoffrey M. *Tidewater Triumph: The Development and Worldwide Success of the Chesapeake Bay Pilot Schooner.* Centreville, MD: Tidewater Publishers, 1998.

Fraser, Walter J. *Savannah in the Old South.* Athens: University of Georgia Press, 2003.

Georgia Historical Society, comp. *The 1860 Census of Chatham County, Georgia.* Easley, SC: Southern Historical Press, 1979.

———. *Register of Deaths in Savannah, Georgia.* Vols. 2–6 (1807–June 1853). Savannah: Georgia Historical Society, 1984–1989.

Georgia Recorder Office. *Official Register of the Land Lottery of 1827.* Milledgeville: Grantland & Orme, 1827.

Gibson, Charles D., and E. Kay Gibson. *Dictionary of Transports and Combatant Vessels Steam and Sail Employed by the Union Army, 1861–1868.* Camden, ME: Ensign Press, 1995.

Goff, John H. "The Steamboat Period in Georgia." *Georgia Historical Quarterly* 12/3 (1928): 236–54.

Granger, Mary, ed. *Savannah River Plantations.* Savannah: Savannah Writers' Project, 1947. Reprint, Savannah: Oglethorpe Press, 1997.

Granger, Mary L. *Savannah Harbor: Its Origin and Development, 1733–1890.* Savannah: US Army Corps of Engineers, 1968.

Gray, Lewis C. *History of Agriculture in the Southern United States to 1860.* 2 vols. Washington: Carnegie Institute of Washington, 1933.

Grieve, Margaret. *A Modern Herbal.* Mineola, NY: Dover Publications, 1971.

Greenhill, Basil, and Sam Manning. *The Schooner Bertha L. Downs.* London: Conway Marine Press, 1995.

Grover, Kathryn. *The Fugitive's Gibraltar: Escaped Slaves and Abolitionism in New Bedford, Massachusetts.* Amherst: University of Massachusetts Press, 2001.

Hagy, James W. *People and Professions of Charleston, South Carolina, 1782–1802.* Baltimore: Genealogical Publishing, 1999.

Hahn, Harold M. *The Colonial Schooner, 1763–1775.* Annapolis: Naval Institute Press, 1981.

Hall, Henry. *Report on the Ship-Building Industry of the United States.* House Miscellaneous Documents, vol. 14, no. 42, pt. 8, Second Session of the Forty-Seventh Congress, 1882–1883. Washington: Government Printing Office, 1884.

Hardee, Charles T. H. *Reminiscences and Recollections of Old Savannah.* Savannah: privately printed, 1928.

Harden, William. *A History of Savannah and South Georgia.* Atlanta: Cherokee Publishing, 1969.

Harrison, Wayne, comp. "Noah B. Sisson, Harrison Family Tree2." https://www.ancestry.com/family-tree/person/tree/12829352/person/956818841/facts. Accessed December 16, 2022.

Haunton, Richard H. "Savannah in the 1850s." PhD diss., Emory University, 1968.

Heard, George A. "St. Simons Island during the War between the States." *Georgia Historical Quarterly* 22/3 (1938): 249–72.

Hedges, James B. *The Browns of Providence Plantations: Colonial Years.* Cambridge: Harvard University Press, 1952.

Helene, William, comp. "Helene Family Tree." https://www.ancestry.com/family-tree/person/tree/25310880/person/26141467876/facts. Accessed November 15, 2022.

Hemperley, Marion R., comp. "Federal Naturalization Oaths, Savannah, Georgia, 1790–1860." *Georgia Historical Quarterly* 51/4 (1967): 454–87.

Herndon, G. Melvin, "Naval Stores in Colonial Georgia." *Georgia Historical Quarterly* 52/4 (1968): 426–33.

Hibbard, Frederic A. *Genealogy of the Hibbard Family Who Are Descendants of Robert Hibbard of Salem, Massachusetts.* Salem: Higginson Book Company, 1992.

Hoffman, Charles, and Tess Hoffman. *North by South: The Two Lives of Richard James Arnold.* Athens: University of Georgia Press, 1988.

Hoffman, Paul E. *Florida's Frontiers.* Bloomington: Indiana University Press, 2002.

House, Albert Virgil, ed. *Planter Management and Capitalism in Ante-bellum Georgia: The Journal of Hugh Fraser Grant, Ricegrower.* Columbia University Studies in the History of American Agriculture 13. New York: Columbia University Press, 1954.

Hunt, Freeman, ed. "Commercial Cities and Towns of the United States—The City of Charleston." *Hunt's Merchants' Magazine and Commercial Review* 22 (1850): 499–515.

———, ed. "Commercial Statistics." *Hunt's Merchants' Magazine and Commercial Review* 27 (1852): 617–19.

Huston, James L. *The Panic of 1857 and the Coming of the Civil War.* Baton Rouge: Louisiana State University Press, 1999.

Huxford, Folks, comp. *Pioneers of Wiregrass Georgia.* Vol. 5. Homerville, GA: Huxford Genealogy Society, 1982.

Hvidt, Kristian, ed. *Von Reck's Voyage—Georgia in 1736.* Savannah: Beehive Press, 1980.

Jacobs, Major James Ripley, and Glenn Tucker. *The War of 1812: A Compact History.* New York: Hawthorn Books, 1969.

Jensen, Arthur L. *The Maritime Commerce of Colonial Philadelphia.* Madison: State Historical Society of Wisconsin for the Department of History, University of Wisconsin, 1963.

Johnson, Emory R. *History of Domestic and Foreign Commerce of the United States.* Washington: Carnegie Institute of Washington, 1915.

Jones, Charles Colcock, Jr. *The Dead Towns of Georgia.* 1878. Reprint, Savannah: Oglethorpe Press, 1997.

Kayser, John C. & Company, publishers. *Commercial Directory, Containing a Topographical Description, Extent and Production of Different Sections of the Union, Statistical Information Relative to Manufactures, Commercial and Port Regulations, A List of the Principal Commercial Houses, Tables of Imports and Exports, Foreign and Domestic, Tables of Foreign Coins, Weights and Measures, Tariff of Duties.* Philadelphia: John C. Kayser, J. Maxwell Printers, 1823.

Keber, Martha L. *Seas of Gold Seas of Cotton: Christophe Poulain DuBignon of Jekyll Island.* Athens: University of Georgia Press, 2002.

Kelly, D., comp. "Luther Cudworth, Stamper Family Tree." https://www.ancestry.com/family-tree/person/tree/8162494/person/-751098121/facts. Accessed November 18, 2022.

Kemble, Frances Anne. *Journal of a Residence on a Georgia Plantation in 1838–1839.* New York: Alfred A. Knopf, 1961.

Kilbourne, Elizabeth E. *Savannah, Georgia Newspaper Clippings (*Georgia Gazette*).* Vol. 3 (1786–1792). Savannah: privately printed, 2001.

King, Anna Matilda Page, and Melanie Pavich-Lindsay, ed. *Anna: The Letters of a St. Simons Island Plantation Mistress, 1817–1859.* Athens: University of Georgia Press, 2002.

Lambert, Frank. *James Habersham: Loyalty, Politics, and Commerce in Colonial Georgia.* Wormsloe Foundation Publications 24. Athens: University of Georgia Press, 2005.

Lee, F. D., and J. L. Agnew. *Historical Record of the City of Savannah.* Savannah: J. H. Estill, 1869.

Leech, Richard W., Jr., and Judy L. Wood. *Archival Research, Archaeological Survey and Site Monitoring: Back River, Chatham County, Georgia and Jasper County, South Carolina.* Savannah: US Army Corps of Engineers, 1994.

Leigh, Frances Butler. *Ten Years on a Georgia Plantation.* London: R. Bentley and Son, 1883.

Leonard, Mary H. *Mattapoisett and Old Rochester, Massachusetts: Being a History of These Towns and Also in Part of Marion and a Portion of Wareham.* 1907. Reprint, Salem, MA: Higginson Book, 1992.

Lieber, Francis, ed. *Encyclopaedia Americana: A Popular Dictionary of Arts, Sciences, Literature, History, Politics, and Biography.* Vol. 5. Boston: B. B. Mussey, 1851.

Lighthousefriends, comp. "Amelia Island." http://www.lighthousefriends.com/light.asp?ID=346. Accessed November 2022.

Lincoln, Charles Henry, comp. *Naval Records of the American Revolution, 1775–1788.* Washington: US Government Printing Office, 1906.

Lord, Walter. *The Dawn's Early Light.* New York: W. W. Norton, 1972.

Lovell, Caroline Couper. *Golden Isles of Georgia.* New York: Little, Brown, 1932.

Lyman, John. "Register Tonnage and Its Measurement." *American Neptune* 5 (1945): 223–34, 311–25.

MacGregor, David R. *Merchant Sailing Ships: 1775–1815, Sovereignty of Sail.* Annapolis: Naval Institute Press, 1985.

———. *The Schooner: Its Design and Development from 1600 to the Present.* London: Chatham Publishing, 1997.

Maganzin, Louis. "New York City's Coastwise Trade, 1713–1754." MA thesis, Georgetown University, 1961.

*Map of the Coast of Georgia, Bordering on Camden and Glynn Counties, Showing Also the Course and Soundings of the Altamaha, Turtle, Crooked, St. Mary's, Great*

*Satilla, and Little Satilla Rivers*. [map]. Scale not given. 178-?. Map G3922.C6 178-.M3. Geography and Map Division, Library of Congress, Washington. https://lccn.loc.gov/73691562.

Mathews, Marguerite M. "The Pacetti and Bonelly Families of New Smyrna, Florida." Unpublished and undated manuscript in the author's possession.

Martin, Sidney Walter. "A New Englander's Impression of Georgia in 1817–1818: Extracts from the Diary of Ebenezer Kellogg." *Journal of Southern History* 12 (1946): 247–62.

McKinnon, John. *Chart of Savannah River, 1825*. Map. Scale ca 1:33,350. Savannah: John McKinnon, 1825. Reproduced by the Georgia Historical Society, 1998.

Mid-Atlantic Technology. *Archaeological Data Recovery, Area 3, Fig Island Channel Site, Savannah Harbor, Georgia*. Prepared for Savannah District, US Army Corps of Engineers. Wilmington: Mid-Atlantic Technology, 1995.

Middleton, Arthur. *Tobacco Coast: A Maritime History of the Chesapeake Bay in the Colonial Era*. Newport News: The Mariners' Museum, 1953.

Mitchell, C. Bradford, ed. *Merchant Steam Vessels of the United States, 1790–1868: The Lytle-Holdcamper List*. Staten Island: Steamship Historical Society of America, 1975.

Mohl, Raymond A. "A Scotsman Visits Georgia in 1811." *Georgia Historical Quarterly* 55/2 (1971): 259–74.

Morrison, Carlton A. "Raftsmen of the Altamaha." MA thesis, University of Georgia, 1970.

Motes, Margaret P. *Migration to South Carolina: 1850 Census*. Baltimore: Genealogical Publishing Company, 2005.

Myers, Barton. "Sherman's Field Order No. 15." *New Georgia Encyclopedia*. https://www.georgiaencyclopedia.org/articles/history-archaeology/shermans-field-order-no-15/. Accessed December 16, 2022.

Myers, Robert M., ed. *Children of Pride: A True Story of Georgia and the Civil War*. New Haven: Yale University Press, 1972.

New England Historic Genealogical Society. *Rochester Births*. Vol. 2 of *Vital Records of Rochester, Massachusetts to the Year 1850*. Boston: New England Historic Genealogical Society, 1914.

Otto, Rhea C., comp. *1850 Census of Georgia, Bryan County*. Savannah: privately printed, 1975.

———, comp. *1850 Census of Georgia, Camden County*. Savannah: privately printed, 1974.

———, comp. *1850 Census of Georgia, Chatham County*. Savannah: privately printed, second printing, 1984.

———, comp. *1850 Census of Georgia, Glynn County*. Savannah: privately printed, 1973.

———, comp. *1850 Census of Georgia, Liberty County*. Savannah: privately printed, 1994.

———, comp. *1850 Census of Georgia, McIntosh County*. Savannah: privately printed, second printing, 1994.

———. "Naturalization Oaths of Allegiance Granted in Superior Court, Savannah, Georgia, 1790–1878 Inclusive." Unpublished manuscript on file at the Georgia Historical Society, Savannah, n.d.

Panamerican Consultants. *Archaeological Data Recovery, Area 2, Fig Island Channel Site, Savannah Harbor, Georgia.* Prepared for the Savannah District, US Army Corps of Engineers. Tuscaloosa: Panamerican Consultants, Inc., 1995.

Pares, Richard. *Yankees and Creoles: The Trade between North America and the West Indies before the American Revolution*. Cambridge: Harvard University Press, 1956.

Parscandola, John L. "Public Health Service." In *A Historical Guide to the U.S. Government*, edited by George T. Kurian, 487–93. New York: Oxford University Press, 1998.

Pearson, Charles E. "Captain Charles Stevens and the Antebellum Georgia Coasting Trade." *Georgia Historical Quarterly* 75/3 (1991): 485–506.

———. "Captain James Frewin of Frederica, St. Simons Island, Georgia." Appomattox: unpublished manuscript, 2001.

———. *Charles Stevens of St. Simons Island: Coasting Captain*. Appomattox: privately printed, 1998.

Pearson, Charles E., Stephen R. James, Jr., and J. Barto Arnold III. "The *Mary Summers*: An Early Iron Steamer in America." *American Neptune* 61/2 (2002):163–83.

Phillips, Ulrich B. *A History of Transportation in the Eastern Cotton Belt to 1860.* New York: Columbia University Press, 1908.

*Plan of the City of Savannah in Chatham County State of Georgia*. Map. Scale not given. Waring Maps, no. 1018, vol. 2, plate 14. Savannah: 1813. Georgia Historical Society, Savannah.

Purvis, George. An Accurate Survey of the Town and Commons of Brunswick in the County of Glynn and the State of Georgia. [map]. Scale not given. ca. 1797. Historic Maps, Surveyor General, RG 3-3-65, Georgia Archives, Atlanta.

Rahn, Ruby A. *River Highway for Trade*. Savannah: US Army Corps of Engineers, 1968.

Reddick, Marguerite G. *Camden's Challenge: A History of Camden County, Georgia.* St. Marys, GA: Camden County Historical Commission, 1976.

Reese, Trevor R. *Colonial Georgia: A Study in British Imperial Policy in the Eighteenth Century*. Athens: University of Georgia Press, 1963.

Rogers, George A., and R. Frank Saunders, Jr. *Swamp Water and Wiregrass: Historical Sketches of Coastal Georgia*. Atlanta: Mercer University Press, 1984.

Rogers, George C., Jr. *Charleston in the Age of the Pinckneys*. Norman: University of Oklahoma Press, 1969.

Rosbe, Judith W. *Maritime Marion, Massachusetts.* Mount Pleasant, SC: Arcadia Publishing, 2002.

Rosengarten, Theodore. *Tombee: Portrait of a Cotton Planter, with the Journal of Thomas B. Chaplin (1822–1890).* New York: William Morrow, 1986.

Rothbard, Murray N. *The Panic of 1819: Reactions and Policies.* New York: Columbia University Press, 1962.

Rowland, Lawrence S., Alexander Moore, and George C. Rogers, Jr. *The History of Beaufort County, South Carolina.* Vol. 1 (1514–1861). Columbia: University of South Carolina Press, 1996.

Rowlett, Russ. "Lighthouses of the United States: Georgia." https://www.ibiblio.org/lighthouse/ga.htm. Accessed December 15, 2022.

Ryder, Alice A. *Lands of Sippican on Buzzards Bay.* Marion, MA: Sippican Historical Society, 1975.

Sears, Louis M. *Jefferson and the Embargo.* Durham: Duke University Press, 1927.

Shelnutt, Cherri B. "Robert Habersham (1783–1870)." Savannah Biographies. Georgia Digital Commons. Special Collections at Lane Library, Georgia Southern University, 1986. https://digitalcommons.georgiasouthern.edu/cgi/viewcontent.cgi?article=1094&context=sav-bios-lane. Accessed December 12, 2022.

Sherwood, Adiel. *A Gazetteer of the State of Georgia*, 3rd ed. Washington: P. Force, 1837.

Sholes, A. E., pub. *Sholes' Directory of the City of Savannah.* Savannah: A. E. Sholes, 1879–1890.

Silva, James S. *Early Reminiscences of Camden County, Georgia.* Kingsland, GA: Southeast Georgian, 1954.

Smith, Alice Ravenel Huger. *A Carolina Rice Plantation of the Fifties.* New York: William Morrow, 1936.

———. "Loading the Rice Schooner." Charleston: Gibbs Art Gallery, ca. 1935.

Smith, Joseph W. *Visits to Brunswick, Georgia and Travels South.* Boston: Addison C. Getchell, 1907.

Smith, Julia Floyd. *Slavery and Rice Culture in Low Country Georgia, 1750–1860.* Knoxville: University of Tennessee Press, 1985.

SCIWAY (South Carolina Information Highway). "South Carolina Lighthouses-From North to South." https://www.sciway.net/tourism/lighthouse.html. Accessed December 16, 2022.

Spears, Robert, comp. "Claudius Leset." Spears Family Tree. https://www.ancestry.com/family-tree/person/tree/88716906/person/46598172531/facts. Accessed December 12, 2022.

Sprague, Waldo C., comp. *Genealogies of the Families of Braintree, Massachusetts, 1640–1850.* Boston: New England Historic and Genealogical Society, 2001.

Starbuck, Alexander. *History of the American Whale Fishery from its Earliest Inception to the Year 1876.* US Commission of Fish and Fisheries, part 4. Report of the Commissioner for 1875–1876. Appendix A. Sea Fisheries (The American

Whale Fisheries). US Senate, 44th Cong., 1st sess., Miscellaneous Document no. 107, Vol. 2. Washington: Government Printing Office, 1878.

Steel, David. *Steel's Elements of Mastmaking, Sailmaking and Rigging*. 1794. Reprint, New York: Edward W. Sweetman, 1932.

Steele, Edward M., Jr. *T. Butler King of Georgia*. Athens: University of Georgia Press, 1964.

Stein, Douglas L. *American Maritime Documents, 1776–1860*. Mystic Seaport, MA: Mystic Seaport Museum, 1992.

Stephens, S. G. "The Origin of Sea Island Cotton." *Agricultural History* 50 (1976): 391–99.

Stewart, Mart A. *"What Nature Suffers to Groe": Life, Labor, and Landscape on the Georgia Coast, 1680–1920*. Wormsloe Foundation Publication 19. Athens: University of Georgia Press, 2002.

Sullivan, Buddy, ed. *"All Under Bank," Roswell King, Jr. and Plantation Management in Tidewater Georgia*. Darien, GA: Liberty County Historical Society, 2003.

———. "Darien Becomes a Great Seaport." In *McIntosh County History Families Past and Present*, compiled by LaVerne A. Gardner, 5–7. Darien, GA: Curtis Media Group, 1992.

———. *Early Days on the Georgia Tidewater, The Story of McIntosh County and Sapelo*, 2nd ed. Darien, GA: McIntosh County Board of Commissioners, 1991.

———. *From Beautiful Zion to Red Bird Creek: A History of Bryan County, Georgia*. Pembroke, GA: Bryan County Board of Commissioners, 2000.

Taylor, George Rogers. *The Economic History of the United States*. Vol. 4 of *The Transportation Revolution: 1815–1860*. New York: Rinehart, 1951.

Thomas, David H. *St. Catherines: An Island in Time*. Georgia History and Culture Series. Atlanta: Georgia Endowment for the Humanities, 1988.

Thomas, Edward J. *Memories of a Southerner, 1840–1923*. Savannah: privately printed, 1923.

Thompson, Paul, comp. "Timothy Wightman III, Thompson-Kirslis Family." https://www.ancestry.com/family-tree/person/tree/111338365/person/372017144334/facts. Accessed December 12, 2022.

Thompson, Shirley Joiner. *The People of Camden County, Georgia: A Finding Index, Prior to 1850*. Kingsland, GA: self-published, 1982.

Thompson, Amanda R., and Rachel N. Horton. "A Short Summary of Health Data at the South End Plantation on Ossabaw Island, Georgia." *Early Georgia* 46/1 and 46/2 (2018): 177–99.

Tripp, Thomas, comp. *The Story of Fairhaven*. New Bedford, MA: Clement E. Daley, 1929.

Turnbull, Nicholas. "The Beginning of Cotton Cultivation in Georgia." *Georgia Historical Quarterly* 1/1 (1917): 39–45.

United Daughters of the Confederacy. *Patriot Ancestor Album*. Paducah, KY: Turner Publishing, 1999.

United States Census Bureau. "Life Expectancy by Age, 1850–2011." *Historical Statistics of the United States Online.* https://www.infoplease.com/us/health-statistics/life-expectancy-age-1850-2011. Accessed December 12, 2022.

———. "1790–1890 Federal Population Census—Part 2." Microfilm M637, NARA, Washington, DC. Database with images: https://www.familysearch.org/en/wiki/United_States_census. Accessed October 2024.

———. "U.S. Censuses, Slave Schedules, 1850 and 1860." Microfilm 432 and 653. NARA, Washington, DC. Database with images: https://www.familysearch.org/en/wiki/United_States_census_slave_schedules. Accessed October 2024.

———. "U.S. Censuses, Slave Schedules, 1850 and 1860." U.S. Federal Census Collection, https://www.ancestry.com/search/categories/usfedcen/. Accessed 2020–2022.

United States of America. *The Public Statutes at Large of the United States of America.* Vol. 1. Boston: Charles C. Little and James Brown, 1845.

United States Congress. *American State Papers.* 38 vols. Pdf. Buffalo: W. S. Hein, 1998. https://www.loc.gov/item/97080286/. Accessed March 10, 2022.

United States Department of State. *Compendium of the Enumeration of the Inhabitants and Statistics of the United States, as obtained at the Department of State, from the Return of the Sixth Census.* Enumeration of the Inhabitants of the United States by Counties and Principal Towns. Washington: Thomas Allen, 1841. https://www.census.gov/library/publications/1841/dec/1840c.html. Accessed October 2024.

United States. Naval War Records Office. *War of the Rebellion: Official Records of the Union and Confederate Navies.* Series 1 (27 vols.); Series 2 (3 vols.). Washington: Government Printing Office, 1894–1922. Cited as *ORN.*

United States. Secretary of the Interior. *Agriculture of the United States in 1860; Compiled from the Original Returns of the Eighth Census.* Washington: Government Printing Office, 1864. https://www2.census.gov/library/publications/decennial/1860/agriculture/1860b-01.pdf. Accessed October 2024.

United States War Department. *War of the Rebellion: A Compilation of the Official Records of the Union and Confederate Armies.* 70 volumes in 128 volumes. Washington: Government Printing Office, 1880–1901. Cited as *OR.*

Vanstory, Burnette. *Georgia's Land of the Golden Isles.* Athens: University of Georgia Press, 1970.

Vocelle, James T. *History of Camden County, Georgia.* 1914. Reprint, Kingsland, GA: Southeast Georgian, 1967.

Walker, William III. "Henry James Dickerson." Unpublished biographical manuscript. Georgia Historical Society, Savannah, 1980.

Wallis, John Joseph. "Depression of 1839 to 1843: States, Debts, and Banks." http://www.ltadvisors.net/Info/research/1839depression.pdf. Accessed November 20, 2022.

Wayne, Lucy B. "The Darien Waterfront: The Commercial Side of the Antebellum South." Paper presented at the Annual Conference on Historical and Underwater Archaeology, Savannah, GA, January 7–11, 1987.

Williamson, Samuel H. "Seven Ways to Compute the Relative Value of a U.S. Dollar Amount, 1790 to present." MeasuringWorth, 2024. https://www.measuringworth.com/calculators/uscompare/. Accessed August 15, 2024.

White, George. *Statistics of the State of Georgia: Including an Account of Its Natural, Civil, and Ecclesiastical History; Together with a Particular Description of Each County, Notices of the Manners and Customs of its Aboriginal Tribes, and a Correct Map of the State.* Savannah: W. Thorne Williams, 1849. Pdf. https://www.loc.gov/item/01007654/.

Wilkes, Robert L., J. H. Johnson, H. T. Stonner, and D. D. Bacon. *Soil Survey of Bryan and Chatham Counties, Georgia.* Washington: US Department of Agriculture, Soil Conservation Service, 1974.

Wise, Stephen R. *Lifeline of the Confederacy: Blockade Running During the Civil War.* Columbia: University of South Carolina Press, 1988.

Wood, Virginia Steele. *Live Oaking: Southern Timber for Tall Ships.* Annapolis: Naval Institute Press, 1995.

———, ed. *Robert Durfee's Journal and Recollections of Newport, Rhode Island; Freetown, Massachusetts; New York City & Long Island; Jamaica & Cuba, West Indies & Saint Simons Island, Georgia, ca. 1785–1810.* Marion, MA: Beldon Books, 1990.

Woodman, Harold D. *King Cotton & His Retainers: Financing & Marketing the Cotton Crop of the South; 1800–1925.* Lexington: University of Kentucky Press, 1968.

# INDEX